Forest Hydrology

Processes, Management and Assessment

Forest Hydrology

Processes, Management and Assessment

Edited by

Devendra M. Amatya

Center for Forested Wetlands Research,
USDA Forest Service,
Cordesville, South Carolina, USA

Thomas M. Williams

Baruch Institute of Coastal Ecology and Forest Science,
Clemson University,
Georgetown, South Carolina, USA

Leon Bren

Faculty of Science,
The University of Melbourne,
Creswick, Victoria, Australia

Carmen de Jong

Institute of Imagery, City and Environment,
Faculty of Geography and Spatial Planning,
University of Strasbourg,
Strasbourg, France

CABI

CABI is a trading name of CAB International

CABI
Nosworthy Way
Wallingford
Oxfordshire OX10 8DE
UK

Tel: +44 (0)1491 832111
Fax: +44 (0)1491 833508
E-mail: info@cabi.org
Website: www.cabi.org

CABI
745 Atlantic Avenue
8th Floor
Boston, MA 02111
USA

Tel: +1 (617)682-9015
E-mail: cabi-nao@cabi.org

A catalogue record for this book is available from the British Library, London, UK.

Library of Congress Cataloging-in-Publication Data

Names: Amatya, Devendra, editor.
Title: Forest hydrology : processes, management and assessment / [edited by] Devendra M Amatya, Thomas M Williams, Leon Bren, and Carmen de Jong.
Description: Boston, MA : CAB International, [2016] | Includes bibliographical references and index.
Identifiers: LCCN 2016013391 | ISBN 9781780646602 (alk. paper) | ISBN 9781780646626 (epub)
Subjects: LCSH: Forest hydrology.
Classification: LCC GB842 .F663 2016 | DDC 551.480915/2--dc23 LC record available at https://lccn.loc.gov/2016013391

ISBN-13: 978 1 78064 660 2

Commissioning editors: Nicki Dennis and Ward Cooper
Associate editor: Alexandra Lainsbury
Production editor: Tim Kapp

Typeset by SPi, Pondicherry, India
Printed and bound in the UK by CPI Group (UK) Ltd, Croydon, CR0 4YY

Contents

List of Contributors

Mary B. Adams, Fernow Experimental Forest, Northern Research Station, USDA Forest Service, Morgantown, West Virginia, USA. E-mail: mbadams@fs.fed.us

Devendra M. Amatya, Santee Experimental Forest/Center for Forested Wetlands Research, Southern Research Station, USDA Forest Service, Cordesville, South Carolina, USA. E-mail: damatya@fs.fed.us

Paul K. Barten, Department of Environmental Conservation, College of Natural Sciences, University of Massachusetts Amherst, Amherst, Massachusetts, USA. E-mail: pkbarten@eco.umass.edu

Leon Bren, Faculty of Science, The University of Melbourne, Creswick, Victoria, Australia. E-mail: l.bren@unimelb.edu.au

Erin Brooks, University of Idaho, Moscow, Idaho, USA. E-mail: ebrooks@uidaho.edu

Tamara Burenina, V.N. Sukachev Institute of Forest, Siberian Branch of the Russian Academy of Sciences, Krasnoyarsk, Russia. E-mail: burenina@ksc.krasn.ru

Peter Caldwell, Coweeta Hydrologic Laboratory, Southern Research Station, USDA Forest Service, Otto, North Carolina, USA. E-mail: pcaldwell02@fs.fed.us

John Campbell, Hubbard Brook Experimental Forest, Northern Research Station, USDA Forest Service, Durham, New Hampshire, USA. E-mail: jlcampbell@fs.fed.us

Nick A. Chappell, Lancaster Environment Centre, Lancaster University, Lancaster, UK. E-mail: n.chappell@lancaster.ac.uk

George M. Chescheir, Department of Biological and Agricultural Engineering, North Carolina State University, Raleigh, North Carolina, USA. E-mail: cheschei@unity.ncsu.edu

Carmen de Jong, Institute of Imagery, City and Environment, Faculty of Geography and Spatial Planning, University of Strasbourg, Strasbourg, France. E-mail: carmen.dejong@live-cnrs.unistra.fr

Mariana Dobre, University of Idaho, Moscow, Idaho, USA. E-mail: mdobre@uidaho.edu

Jean-Christophe Domec, Department of Forestry and Environmental Resources, North Carolina State University, Raleigh, North Carolina, USA; Bordeaux Sciences AGRO, UMR1391/ISPA/INRA, Gradignan, France. E-mail: jdomec@ncsu.edu

Kelly Elder, Fraser Experimental Forest, Rocky Mountain Research Station, USDA Forest Service, Fort Collins, Colorado, USA. E-mail: kelder@fs.fed.us

William Elliot, Forestry Sciences Laboratory, Rocky Mountain Research Station, USDA Forest Service, Moscow, Idaho, USA. E-mail: welliot@fs.fed.us

Grey R. Evenson, Oak Ridge Institute of Science and Education, c/o Office of Research and Development, National Exposure Research Laboratory, US Environmental Protection Agency, Cincinnati, Ohio, USA. E-mail: evenson.grey@epa.gov

Houbao Fan, Institute of Ecology & Environmental Science, Nanchang Institute of Technology, Nanchang, People's Republic of China. E-mail: hbfan@nit.edu.cn

Heather E. Golden, Office of Research and Development, National Exposure Research Laboratory, US Environmental Protection Agency, Cincinnati, Ohio, USA. E-mail: golden.heather@epa.gov

Georg Guggenberger, Leibniz University, Hannover, Germany. E-mail: guggenberger@ifbk.uni-hannover.de

Sherri Johnson, HJ Andrews Experimental Forest, Pacific Northwest Research Station, USDA Forest Service, Corvallis, Oregon, USA. E-mail: sherrijohnson@fs.fed.us

Hironari Kanamori, Institute for Space–Earth Environmental Research, Nagoya University, Nagoya, Japan. E-mail: kanamori@hyarc.nagoya-u.ac.jp

Elizabeth Keppeler, Caspar Creek Experimental Watershed, Pacific Southwest Research Station, USDA Forest Service, Fort Bragg, California, USA. E-mail: ekeppeler@fs.fed.us

Hyunwoo Kim, Department of Environmental and Energy Engineering, Anyang University, Anyang, Republic of Korea. E-mail: kim4ecohydro@anyang.ac.kr

Ken W. Krauss, Wetland and Aquatic Research Center, US Geological Survey, Lafayette, Louisiana, USA. E-mail: kraussk@usgs.gov

Tomo'omi Kumagai, Institute for Space–Earth Environmental Research, Nagoya University, Nagoya, Japan. E-mail: tomoomikumagai@gmail.com

Qiang Li, Department of Earth and Environmental Sciences, University of British Columbia (Okanagan Campus), Kelowna, British Columbia, Canada. E-mail: qiang.li@ubc.ca

Tim Link, University of Idaho, Moscow, Idaho, USA. E-mail: tlink@uidaho.edu

Wenfei Liu, Institute of Ecology & Environmental Science, Nanchang Institute of Technology, Nanchang, People's Republic of China. E-mail: liuwf729@126.com

Katherine L. Martin, Center for Integrated Forest Science, Southern Research Station, USDA Forest Service; Department of Forestry and Environmental Resources, College of Natural Resources, North Carolina State University, Raleigh, North Carolina, USA. E-mail: katie_martin@ncsu.edu

Eric Masson, Université des Sciences et Technologies de Lille, Lille, France. E-mail: eric.masson@univ-lille1.fr

Debasmita Misra, Caribou-Poker Experimental Forest, University of Alaska–Fairbanks, Fairbanks, Alaska, USA. E-mail: dmisra@alaska.edu

Anastasia Musokhranova, V.N. Sukachev Institute of Forest, Siberian Branch of the Russian Academy of Sciences, Krasnoyarsk, Russia. E-mail: nastya.krasn@mail.ru

Jami E. Nettles, Weyerhaeuser Company, Columbus, Mississippi, USA. E-mail: jami.nettles@weyerhaeuser.com

Tomasz Okruszko, Division of Hydrology and Water Resources, Warsaw University of Life Sciences, Gdańsk, Poland. E-mail: t.okruszko@levis.sggw.pl

Alexander Onuchin, V.N. Sukachev Institute of Forest, Siberian Branch of the Russian Academy of Sciences, Krasnoyarsk, Russia. E-mail: onuchin@ksc.krasn.ru

Sudhanshu S. Panda, Institute of Environmental Spatial Analysis, University of North Georgia, Gainesville, Georgia, USA. E-mail: sudhanshu.panda@ung.edu

Peter R. Robichaud, Forestry Sciences Laboratory, Rocky Mountain Research Station, USDA Forest Service, Moscow, Idaho, USA. E-mail: probichaud@fs.fed.us

Stephen Sebestyen, Marcell Experimental Forest, Northern Research Station, USDA Forest Service, Grand Rapids, Minnesota, USA. E-mail: ssebestyen@fs.fed.us

Sumit Sen, Indian Institute of Technology–Roorkee, Uttarakhand, India. E-mail: ssenhfhy@iitr.ac.in

Anatoly Shvidenko, V.N. Sukachev Institute of Forest, Siberian Branch of the Russian Academy of Sciences, Krasnoyarsk, Russia; International Institute for Applied Systems Analysis, Laxenburg, Austria. E-mail: shvidenk@iiasa.ac.at

R. Wayne Skaggs, Department of Biological and Agricultural Engineering, North Carolina State University, Raleigh, North Carolina, USA. E-mail: wayne_skaggs@ncsu.edu

Anurag Srivastava, University of Idaho, Moscow, Idaho, USA. E-mail: srivanu@uidaho.edu

John D. Stednick, Watershed Science Program, Colorado State University, Fort Collins, Colorado, USA. E-mail: John.Stednick@colostate.edu

Ge Sun, Eastern Forest Environmental Threat Assessment Center, Southern Research Station, USDA Forest Service, Raleigh, North Carolina, USA. E-mail: gesun@fs.fed.us

Shiying Tian, Department of Biological and Agricultural Engineering, North Carolina State University, Raleigh, North Carolina, USA. E-mail: stian@ncsu.edu

Charles A. Troendle (retired), Rocky Mountain Research Station, USDA Forest Service, Fort Collins, Colorado, USA. E-mail: corvtroendle@msn.com

James M. Vose, Center for Integrated Forest Science, Southern Research Station, USDA Forest Service Raleigh, North Carolina, USA; Department of Forestry and Environmental Resources, College of Natural Resources, North Carolina State University, Raleigh, North Carolina, USA. E-mail: jvose@fs.fed.us

Xiaohua Wei, Department of Earth and Environmental Sciences, University of British Columbia (Okanagan Campus), Kelowna, British Columbia, Canada. E-mail: adam.wei@ubc.ca

Thomas M. Williams, Baruch Institute of Coastal Ecology and Forest Science, Clemson University, Georgetown, South Carolina, USA. E-mail: tmwllms@clemson.edu

Peter Wohlgemuth, San Dimas Experimental Forest, Pacific Southwest Research Station, USDA Forest Service, Riverside, California, USA. E-mail: pwohlgemuth@fs.fed.us

Mohamed A. Youssef, Department of Biological and Agricultural Engineering, North Carolina State University, Raleigh, North Carolina, USA. E-mail: mohamed_youssef@ncsu.edu

Mingfang Zhang, School of Resources and Environment, University of Electronic Science and Technology of China, Chengdu, People's Republic of China. E-mail: mingfangzhang@uestc.edu.cn

Preface

The principles of forest hydrology were developed throughout the 20th century with the first book on forest hydrology published as *Principles of Forest Hydrology* by John D. Hewlett and Wade L. Nutter (University of Georgia, USA) in 1969, although a conference proceedings on Forest Hydrology was published by Pergomon Press in 1966. In Europe, this was followed in 1971 by the book *Wald, Wachstum und Umwelt – Waldklima und Wasserhaushalt* (*Forest, Growth and Environment – Forest Climatology and Forest Hydrology*) by G. Mitscherlich (J.D. Sauerländer Verlag, Germany). However, the context and concepts of forest landscape, land use and management, and human and natural disturbances have since changed and are continually changing. Accordingly, in recent years increasing attention has been paid towards advancing the science of forest hydrology to increase our understanding of forest hydrological processes, their interactions with other land uses and environments, their impacts on ecosystem functions and services in the face of changing climate, and their appropriate application at the watershed or basin scale. Advances in computing, sensors and information technology have accelerated this trend in the past few decades. To keep up with the growing knowledge of forests and water in a changing environment, the book *Watershed Hydrology* was published by Peter E. Black (Prentice Hall in 1991) followed by a summary of recent advances in Canadian forest hydrology by Buttle *et al.* (2000) in *Hydrological Processes* journal and a textbook for students, *Forest Hydrology: An Introduction to Water and Forests* by Mingteh Chang (CRC Press, USA), in 2003. An overview of a featured collection on forest hydrology in China was published by Sun *et al.* (2008) in *Journal of the American Water Resources Association*. In 2011 a new book on *Forest Hydrology and Biogeochemistry* edited by D.L. Levia, D. Carlyle-Moses and T. Tanaka (Springer) was published, linking hydrology to biogeochemistry. The newest one, *Forest Hydrology and Catchment Management: An Australian Perspective*, aimed primarily for students and land managers, was published by our own co-editor Leon Bren (Springer, December 2014).

In view of the large amount of new knowledge, data and information on forest hydrology being accumulated only in journals, proceedings papers, textbooks and reports around the world, Commissioning Editor Vicki Bonham at CABI in the UK recently saw the need for a new forest hydrology book. She asked Devendra Amatya at the US Department of Agriculture (USDA) Forest Service, USA to consider leading an effort to edit a new forest hydrology book focused on forest hydrology only. An editorial team led by Devendra Amatya, with Tom Williams at Clemson University, USA, Leon Bren at the University of Melbourne, Australia and Carmen de Jong at the University of Strasbourg, France, sincerely appreciated and formally accepted CABI's invitation in early 2015.

This new book with 17 chapters is unique and different from the previous forest hydrology books in that world-renowned international professors, scientists, engineers, managers and researchers with a long background and expertise in forest hydrology, management and applications have authored/contributed individual chapters focused on almost all aspects of forest hydrology. Chapters 2, 3, 4, 6, 7, 8, 12, 13, 14, 15 and 16 cover major advances in forest hydrology for areas ranging from tundra, taiga and mountains to tropics and from humid to dry climate forests, with new insights into landscape processes as affected by anthropogenic and natural disturbances such as extreme events (hurricanes, floods, droughts), wildfire, massive landslides and climate change. Chapter 12, with examples from Chapter 1, provides a review of past and current research on the hydrological effects of managing elements of the forest landscape. Chapter 11 discusses problems and statistical methods dealing with expanding knowledge gained from small watershed studies to much larger forest watersheds. Chapters 9 and 10 deal with numerical models and geospatial technology to address challenges of spatial scale, model uncertainties and assess impacts of disturbances and land-use change. Chapter 5 provides a European perspective on forest hydrology. The editors sincerely thank each author for accepting our invitation to lead the chapter of their expertise, and all other contributors for their time and dedication to accomplish this book. The editors believe, although this book in no way completely covers forest hydrological processes occurring in every single landscape situation or environment/biome around the world, it still has attempted to do so. Finally, the book ends with Chapter 17 highlighting the key points of forest hydrological processes in major biomes and providing recommendations for advancing forest hydrology in the remainder of the 21st century when humanity will be challenged by even more environmental complexity and in particular climate change. Throughout the book the terminology 'watershed' and 'catchment' with the same meaning are interchangeably used for the convenience of readers from around the world. The authors deeply acknowledge the external reviewers listed in the book for their time and effort reviewing chapters and providing valuable and constructive suggestions to improve quality while attempting to cover examples from around the world.

All four editors of the book have worked tirelessly on editing, proofreading and preparing this book throughout the process by communicating with all invited chapter contributors, reviewers, experts in the specific areas and the CABI commissioning editors to make this new book a reality. We therefore trust that the book will provide a good understanding of the basic principles of forest hydrology and hydrological processes to higher-level graduate students, professionals, land managers, practitioners and researchers for their application in contemporary issues of forest hydrology, watershed management and assessing potential global impacts.

We are thankful to Maureen Stuart and her editorial team at the USDA Forest Service Southern Research Station for providing assistance with editing of two chapters. We also thank Azal Amatya for help in preparing the Contents, Contributors and Reviewers lists for the book. We also sincerely acknowledge Jami Nettles at Weyerhaeuser Company for sponsoring the colour plate charges for some figures in this book. Last but not the least, we would like to greatly acknowledge CABI's former Commissioning Editor Vicki Bonham for inviting us to prepare this book; Alexandra Lainsbury, the current Associate Editor, for her great assistance and guidance in all steps of preparing this book; current Commissioning Editor Ward Cooper (and former Commissioning Editor Nicki Dennis); and all the CABI production staff for publishing the book.

Devendra M. Amatya, PhD, PE
Thomas M. Williams, PhD
Leon Bren, PhD
Carmen de Jong, Dr. rer. nat, habil.

Acknowledgements

Chapter Reviewers

Axel Anderson, University of Alberta, Alberta, Canada
Paul K. Barten, University of Massachusetts, Amherst, Massachusetts, USA
L.A. (Sampurno) Bruijnzeel, King's College, London, UK
Timothy J. Callahan, College of Charleston, Charleston, South Carolina, USA
Cole Green, US Bureau of Land Management, Salt Lake City, Utah, USA
Harald Grip, Swedish University of Agricultural Sciences, Umea, Sweden
Suat Irmak, University of Nebraska–Lincoln, Lincoln, Nebraska, USA
C. Rhett Jackson, University of Georgia, Athens, Georgia, USA
Megan Lang, Department of Geographical Sciences, University of Maryland, College Park, Maryland, USA
Dennis P. Lettenmaier, University of California, Los Angeles, California, USA
Dan Marion, USDA Forest Service, Hot Springs, Arkansas, USA
Patrick Meire, University of Antwerp, Antwerp, Belgium
Jami Nettles, Weyerhaeuser Company, Columbus, Mississippi, USA
Yue Qin, Tsinghua University, Beijing, People's Republic of China
Mark Robinson, Centre for Ecology and Hydrology, Wallingford, UK
Patrick Schleppi, Institute of Forest, Snow, and Landscape Research, Birmensdorf, Switzerland
James (Jamie) Shanley, US Geological Survey, Montpelier, Vermont, USA
Gary Sheridan, The University of Melbourne, Melbourne, Australia
Herbert Ssegane, Argonne National Laboratory, Lemont, Illinois, USA
Ralph Tiner, Institute for Wetland and Environmental Education and Research, Inc., Leverett, Massachusetts, USA
Yanhui Wang, Chinese Academy of Forestry, Beijing, People's Republic of China
Jimmy Williams, Texas A&M Agri-Life Research, College Station, Texas, USA
Dawen Yang, Tsinghua University, Beijing, People's Republic of China
Nicolas Zegre, West Virginia University, Charlottesville, West Virginia, USA
Lu Zhang, CSIRO Land and Water Flagship, Canberra ACT, Australia

1 An Introduction to Forest Hydrology

L. Bren*
The University of Melbourne, Creswick, Victoria, Australia

1.1 What is Forest Hydrology?

Forest hydrology is the study of the structure and function of watersheds and their influence on water movement and storage. In its purest form it is a quantitative discipline underpinned by conservation of mass and energy in connected, continuous media. However the application of such 'pure' theories is rendered difficult by the variations, both of inputs across space and time and in the properties of materials comprising the watersheds. Such difficulties are the stuff of forest hydrology.

In writing an overview of the discipline, one is struck by the vastness of the publications across what might be described as 'forest hydrology'. These encompass theory, observations, methodologies, processes, results and political advocacy. The scale of work ranges from molecular to effectively the size of the earth. Interests may be in the science, economics or politics of land-use management. Forest hydrology grades into the wider disciplines of meteorology, geology, hydrology, forestry, soil science and plant physiology. There is a diverse and voluminous worldwide literature in the discipline.

Interest in forest hydrology dates back to three sources; the first of these was intellectual curiosity about the way the world works. The second was the observation of landholders that actions such as clearing forests often generated consistent, observable and (in hindsight at least) predictable results in streamflow and sediment load. The third was an age-old concern about the 'sustainability' (as we would now define it) of land uses and, in particular, of rainfall. Underpinning this was and is, of course, the socio-economic importance of streamflow to the survival of communities and some harsh experiences when rainfall and consequent streamflow was either extremely low or extremely high.

1.2 Development of Forest Hydrology

1.2.1 Historical antecedents

In practical terms, the history of hydrology dates back to the earliest civilizations such as ancient Rome since they certainly had the ability to measure flows and to manage water with canals, drainage tunnels and dams. Scientific historians note the growth of hydrological science for many centuries but usually denote the starting point as the work of Frenchmen Pierre Perrault (1608–1680) and Edme Marriotte (1620–1684) in the

*E-mail: l.bren@unimelb.edu.au

period 1670–1680. This showed that the rainfall in the Seine Basin was entirely adequate to sustain the flow of the river (Biswas, 1970). Around 1700 the English astronomer Edmond Halley advanced the field further by providing the first quantitative estimates of what we would now call the hydrological cycle (Hubbart, 2011). Unfortunately there seems to be little information on who first formulated that key complementary idea to the rainfall – the watershed. McCulloch and Robinson (1993) suggest that the concept has been used for millennia. However the well-known scientist Cayley (1859) refined the concept of contours and slope lines and might well be viewed as an early scientific user. Examination of early dam-building projects in Australia, at least, suggest size was usually based on the size of the river feeding the dam, and that determination of the size and properties of the watershed usually came (much) later.

The emergence of forest hydrology as a subdiscipline of hydrology appears to owe much to the unfortunate victims of the guillotine in the French Revolution (Andreassian, 2004). This led to an unparalleled expansion of land clearing in France as 'the King's Forests' were cleared for settlement. Landholders then encountered many of the same problems – erosion, flooding, streams drying up, landslips or other forms of mass erosion, and sedimentation – now encountered in developing countries. At the time, France was probably the most technically advanced country in the world. The ills and possible remedies caused much discussion in intellectual circles of post-revolutionary France, although by modern standards the discussion was philosophical rather than scientific. Out of this came a view of the forested watershed as being something analogous to a 'sponge' (sometimes called the 'Law of Dausse' after Dausse, 1842) and this oversimplification still underpins the view of non-technical citizens.

Among other things, Dausse (1842) argued that 'Rain is formed when a warm and humid wind comes in contact with strata of cold air; and since the air of forests is colder and more humid than that of the open, rain must fall there in greater abundance'. The view was then expressed that the forests constitute 'a vast condensing apparatus'. This message became codified into 'trees bring rain', which was a worldwide catchcry of a century ago. Interestingly, satellite measurement of air temperatures in the last decade have at least confirmed that the air of forests is colder than surrounding agricultural land because of heat loss associated with greater transpiration (e.g. Mildrexler *et al.*, 2011), but the link to greater rainfall and condensation appears elusive and is a fertile field for future research using today's technology. This sort of approach can be viewed as a progenitor of more modern science applied to the same field. Subsequent chapters in this book will still explore some of the same ideas.

1.2.2 The era of hydro-mythology

In the latter part of the 19th century, views concerning the role of forests in hydrology began to become accepted and, indeed, were viewed as 'conventional wisdom'. These include 'trees bring rain', forests modify flooding, forests provide 'healthier water', forests provide increased dry-season flows and that forests reduce erosion. A century and a half later, such statements would be viewed as 'partly true', 'generalizations, 'sweeping statements' or 'unproven' but are still commonly cited by the media. In this period, data started to be collected to 'prove' such statements; the concepts of experimental design, rigorous measurement and hypothesis testing were yet to arrive in the world of forest hydrology.

By the start of the 20th century there was a body of advanced thought on the role of forests in protecting watersheds and some skilled observation, but little that we would now recognize as 'science'. Some authors (e.g. George Perkins Marsh, 1864; Raphael Zon, 1912) were far ahead of their time and contemporaries in examining the beneficial effects of the presence of large forests on streamflow. In retrospect, their work was a seminal contribution to the developing field of forest hydrology and watershed science. With the development of forestry science, stable forest management organizations and the advent of sophisticated and reliable instruments (e.g. water level, precipitation, air temperature and solar radiation recorders), the discipline was ripe for development.

1.2.3 The era of small watershed measurement

Around the middle of the 19th century the value of hydrological data was realized. In general this took the form of periodic readings of major river levels. Although informative, it was quickly realized that with this approach it was impossible to link rainfall and streamflow except in the crudest sense, and that large rivers were both difficult to measure flow on and too complex for simple water balance studies. This led to the first true 'watershed study' in the Bernese Emmental region of Switzerland in 1906. In this the hydrological responses of two watersheds of 0.6 km^2 were compared. These had different distributions of land use. Inferences on the hydrology of the slopes were drawn by comparison. In general, the results showed a moderating influence of the presence of forests on peak flows and a slower summertime recession from the forested watersheds (reflecting better slope storage). Measurement at Emmental still continues and the data set is a valuable asset for climate change researchers; Hegg *et al.* (2006) provide an overview of this project.

By contemporary standards, the early Emmental project was far from perfect. It relied on correlation between land use and outputs rather than experimental manipulation, data were sometimes discontinuous, and the project appears to have had a somewhat tenuous political existence. From this writer's distant viewpoint (in space and time) one has to admire the work and the people that made it happen – going out to the field on horseback or on foot, measuring in wet and cold conditions, countless hours of tedious calculations using hand calculators, logarithmic tables or slide rules, laborious hand-plotting of graphs, the constant struggle to maintain and upgrade equipment, and the ever-present demand from administrators of 'what is more data going to show you that you don't already know?' However the project did set the scene for the big advance in forest hydrology – paired watershed experiments.

1.2.4 That great leap forward; paired watershed experiments

The European experiences were not lost on a generation of US settlers, with massive efforts directed at controlling large rivers. The value of forests in protecting watersheds was explicitly recognized by the formation of the National Forest Service in 1891. However there was no clear basis of information beyond the earlier observations of George Perkins Marsh (1864) – a deficiency clearly evident to the early forestry scientists.

In 1910 the 'Wagon Wheel Gap' experiment was commenced in Colorado by the US Forest Service (Bates and Henry, 1921, 1928). This was the first formal examination of the effects of forest denudation on streamflow and sediment yield. This study ran until 1926 and was the prototype of hundreds of paired watershed experiments around the world; arguably this has been the most successful forest hydrology technique. In this, a 'to-be-treated' stream is 'calibrated' against a 'control' or reference stream. The forest on the first watershed is then altered and the effect on streamflow is determined by comparison with the flow in the 'control' stream. Their conclusions were based on mean values of study variables without the benefit of a statistical treatment of year-to-year variability.

Van Haveren (1988) revisited the data set produced by Bates and Henry (1928) to ascertain whether a more sophisticated 'modern' approach (including covariance and regression analysis) would give the same result as that of the older work. Table 1.1 summarizes his findings.

The results of this analysis showed that 'many of the original conclusions stated by Bates and Henry (1928) are statistically supportable. However a few of their conclusions could not be supported statistically'. The finding underlines the

Table 1.1. A comparison between the van Haveren (1988) examination of the Wagon Wheel Gap experimental logging and the conclusions reached by Bates and Henry (1928).

Hydrograph parameter	Original conclusion	Re-evaluation result
Average annual water yield	Increased 24 mm	Increased 25 mm
Annual maximum daily flow	Increased 50%	Increased an average of 50%
Date of the annual maximum flow	Advanced 3 days	Advanced 6 days (NS)
Starting date of snowmelt	Advanced 12 days	Advanced 5 days (NS)

NS, not significant.

discipline of the early researchers doing what is now viewed as 'computationally intensive' work in the pre-computer days. Study of the Bates and Henry (1928) work also shows tentative first steps in 'hydrograph analysis' – relating specific characteristics of the flow record to the land use or land-use change. This continues to be something of a specialty area in the discipline of forest hydrology today.

1.2.5 Proliferating paired watershed experiments

The success of Wagon Wheel Gap led to a large increase in paired watershed projects around the world; these can be generally classed as 'deforestation experiments' in which the effects of forest harvesting were studied or as 'afforestation experiments' in which the effects of plantation formation were measured. Figure 1.1 shows an example of such a project in which the native forest was cleared and replaced with radiata pine in Australia; this project is continuing. Brown *et al.* (2005) give a comprehensive list of projects around the world. In general, the data sets of matched streamflow and rainfall records have been invaluable in the development of modelling, testing of specific hypotheses and estimation of the effects of climate change.

A large body of experience has developed with this technique. Among other things it has shown that:

1. There is a rapid build-up of hydrological knowledge by the experimenters, with many gains peripheral to the main aims of the experiment (e.g. Hewlett *et al.*, 1969; for a quantitative example, see Bren and Lane, 2014).

Fig. 1.1. The experimental phase; a small watershed is converted from native eucalypt forest to a radiata pine plantation as part of a paired watershed project in north-eastern Victoria, Australia in 1980. The watershed is now on its second rotation of pine and measurement is continuing.

2. The projects serve as a great 'teaching tool' (usually self-education) for forest hydrologists (see Hewlett and Pienaar, 1973).
3. The results of the experiments are usually respected by courts and similar bodies as being 'trustworthy' and are not often attacked in courts. The author contends that this is partly due to the 'visual, tangible' nature of the experiment – people can see and visit the areas, and the concepts being explored are understandable.
4. The experiments involve a substantial capital cost and organizational commitment to get established. Once established, they are relatively inexpensive to maintain. This maintenance fits well with the routine of research organizations (Bren and McGuire, 2012).
5. Reflecting the nature of forests, the project may take many decades to bring to completion. In societies in which there is constant rearrangements of (or, worse, no) land-management agencies, the long-term management may prove difficult.

The major (technical) disadvantage of the technique is that the watersheds are small and that 'scaling up' of results to regional watersheds is difficult.

Paired watershed experimentation and experimental science?

Given that forest hydrology is a science, then concepts should be able to be quantified and hypotheses should be testable by experiment. Paired watershed projects are a specialized form of experiment in which a time-variant effect is measured relative to a reference state ('control'). From the scientific point of view there are some difficulties:

1. Replication of treatments is rarely feasible in economic or geographic terms. In general, there are many differences between watersheds and these can be viewed to some extent as uncontrolled differences. In any case, few organizations could afford the cost of replicating the treatment.
2. Concepts of 'blinding' in the experimental design (in which the analyst has no specific knowledge of treatments of individuals concerned) have been little investigated. This partly reflects that usually the experimenter ends up doing the necessary analysis.
3. Statistical analysis of data can be rendered difficult by non-normality of residuals from many models, thereby limiting appropriate testing. Hewlett and Pienaar (1973) noted that hydrologists were divided on the importance of this, and this schism still exists today.
4. Other than a null hypothesis (that there is no effect), development of a testable hypothesis can be difficult. To date, many of the paired catchment experiments have been 'exploratory' in the sense that the aim was to examine the deviations in stream properties from the 'normal' ones.

Although the design is not perfect, examination of the development of paired watershed projects over the years has usually indicated increasing rigour and sophistication, newer methods of dealing with non-normality of data residuals in testing, and use of hourly, daily or monthly data rather than annual data as a means of increasing degrees of freedom in the data (at the expense of autocorrelation); Watson *et al.* (2001) examine aspects of these very clearly.

1.2.6 'Closing' the water balance

Fundamental to paired watershed experiments are the measurement of rainfall and other precipitation entering the watershed and the measurement of water (or vapour) leaving the watershed. It is an axiom of forest hydrology that water entering and leaving a watershed can be viewed as forming a 'water balance' or water budget. Thus, over any period, the water entering the watershed = the volume of water leaving the watershed plus the change in water stored in the watershed. It follows that, by careful measurement of the processes and summation over a suitable period of time, one can compare inflows and outflows. Differences are a measure of error. This is called 'closing the water balance' and was an aim of forest hydrologists until relatively recently (e.g. Waichler and Wemple, 2005; Scott, 2010).

A number of difficulties are implicit in this scheme, leading to experimental uncertainty (e.g. Fisher *et al.*, 2005). All field-based measurement schemes have proved costly and laborious to maintain for years on end. Some variables such as rainfall and streamflow are relatively easy to measure. Others, such as evapotranspiration, have proven elusive, laborious and difficult to measure. Many parameters necessary to describe the hydrology have a wide stochastic variability. Formation of the water balance also depends on the key assumption that the actual watershed boundaries coincide with the surface boundaries; this is usually viewed as an axiom rather than a testable hypothesis since we have no way to test this premise. Similarly, leakage into or out of the watershed is assumed to be zero or, at most, a small, relatively constant value.

1.2.7 The search for experimental alternatives

A strength and weakness of the paired watershed approach is the sequential nature of the study. Thus if development of the forest takes a century then a paired watershed project following the full life of the forest will take at least this time. This is usually too long for most research organizations. In addition, the concept of a 'control watershed' remaining 'stationary' (i.e. unchanged for a century) is problematic and difficult.

One approach to speeding up the process has been the omission of a 'calibration period' before treatment. This has the disadvantage that it is difficult to set any statistical error limits (or, indeed, sometimes to form a view of just what is the treatment effect). An alternative approach is the use of plots to measure hydrological variables of interest. The experimenter is not bound by the sequential nature of measurement but, rather, can have many plots in different age classes. Sophisticated plot designs have the disadvantages that plots are difficult to sustain for long time periods and that the usual variable of interest – streamflow – is not commonly directly measureable. Thus streamflow effects must be inferred by water-balance differencing. It is arguable whether this is as satisfactory as a direct measurement of streamflow. However, with good statistical design, the errors involved can be quantified. An excellent example of plot use is the work of Benyon *et al.* (2006) in assessing the water use of pine and eucalypt plantations in sandy soil overlying groundwater in flat, karst country in southern Australia.

1.2.8 Coming to grips with the dynamics of watershed flows

As forest hydrology knowledge grew, there was a concomitant increase in knowledge in other fields of watershed hydrology. Some of this impetus came from the brilliant work in soil physics by Buckingham (1907); see Philip (1974) for a review of this and its later development. This provided the model of the slopes having a continuum of energy levels of water, manifesting themselves in saturated and unsaturated zones. Although the model was applicable to agricultural and forest soils, the inhomogeneity of the latter made it more difficult to apply.

Hydrology on non-forest land was substantially predicated on increases in flow during and after rain ('stormflow') being due to overland flow from an infiltrating surface. Groundwater was usually not considered as a contributing agent to stormflow (or even streamflow). The infiltrated water was considered as passing downwards though the pore structure of the soil with a portion reaching the aquifer to support 'baseflow'. A substantial base of theory linking these processes developed (e.g. Horton, 1945). Attempts to apply these formulations to forested watersheds were (and still are) unsatisfactory. Forest hydrologists found little evidence of overland flow; nor could correct infiltration values be developed from sieved soil samples due to rocks, root material and macropores found in forest soils.

The differences between agricultural and forested slopes were the subject of much research (and some acrimonious debates) in the 1950s to 1970s; since then the area has faded in its academic prominence, being viewed as 'difficult' and 'laborious'. In doing this research, the network of paired watershed projects provided both invaluable sites and data. The results of this research can be summarized as:

1. Usually forested slopes have a high infiltration capacity and true 'overland flow' is rarely generated. An account of where it did occur in massive rainfalls is given by Orr (1973), who noted large amounts of litter movement but little actual erosion.
2. Infiltrated water moves both through 'macropores' (holes) in the soil and through the soil matrix (see Aubertin, 1971). Because of the action of roots and forest biota the soil is constantly

'turning over'. It is difficult to characterize the hydraulic properties of such media using simple models. In particular, stochastic variation in soil texture, structure, and pore space geometry, anisotropy and air compression effects make water behaviour complex. Recent isotope work (e.g Brooks *et al.*, 2010; McDonnell, 2014) has shown that trees may preferentially remove water from the smaller pores only, leading to a 'two water worlds' model.

3. The watershed 'soil' is a complex medium composed of rocks, mineral soil and organic matter. Often the soil is better viewed as decomposing rock ('saprolites'). Thus, simple models based on agricultural soils are difficult to apply.

4. The hydrological response of the watershed may sometimes be generated many metres below the soil surface. Surface soil is generally a superconductive layer which may transmit water to substantial depths or to the stream. In general, relating surface soil properties in a forest to stream hydrographs is difficult.

5. Most slopes are characterized by a water-table aquifer at some depth below the surface. Elements of the behaviour can be approximated using groundwater theory (e.g. Troch *et al.*, 2003). However, as shown by workers such as Loague and Freeze (1985), this is never simple. Often reconciliation of forest watershed data and 'classic groundwater theory' hinges on subtle points of definition. Much of our understanding of the interaction of groundwater in forest hydrology has come from isotope signatures and 'end-member' mixing models of runoff chemistry (e.g. McDonnell, 2014). Other issues include stochastic variation, anisotropy, discontinuities, difficulty of specifying initial and boundary conditions, and how the presence of macropores and air flow may be incorporated (Morel-Seytoux, 1973). In general, thinking on these matters has not advanced much in the last few decades.

6. The role of small pores holding water at high tensions in watershed slopes is almost unknown. The recent finding of Brooks *et al.* (2010) (and others) using isotope ratios that forests in Mediterranean climates appear to obtain their water from these may lead to significant new insights into the nature of the watershed slope material and the forces acting on water in these.

Notwithstanding the difficulties of quantification, the research has given a clear picture of the watershed slopes recharging and discharging, with the presence of the trees providing a highly conductive surface layer and maintaining infiltration pathways to the subsurface water stores. The presence of the soil and the hydraulically rough forest floor allows impinging rainfall to infiltrate to below the surface. The effect of the forest and understorey vegetation is to maintain the favourable soil environment and, by transpiration, deplete the soil water content in the slope. This leads to lower storm responses at the next period of rainfall. The watershed behaviour at depth in soils and the interaction with tree roots is still a substantially unknown area.

1.2.9 Hydrograph analysis – 'the last refuge of the desperate hydrologist'

The hydrograph is the record of outflow of a watershed over time; ideally this is collected in conjunction with a 'hyetograph' – the record of rainfall intensity over time. The conventional (and still fundamental) approach to such records is to integrate over a year to obtain the volume or depth of both annual rainfall and annual streamflow. Integration smoothes errors and often makes long-term relationships apparent.

An alternative approach is to use the data directly or even to differentiate with respect to time; the latter process enhances both variability and errors in the data (Whittaker and Robinson, 1924). Generically, such operations come under the category of 'hydrograph analysis'. The doyen of forest hydrology, John Hewlett, is reputed to have quipped at a conference that 'hydrograph analysis was the last refuge of the desperate hydrologist'. The technique has provided much information on the dynamic behaviour of forest stream systems, but usually shows that streams emanating from forested watersheds have complex dynamic behaviour that cannot be encapsulated by simple equations or simple explanations. Thus, elegant formulations may explain some of the behaviour but cannot reproduce all facets of it.

The plethora of paired watershed experiments and associated data has made excellent sequences of rainfall and streamflow data available for testing and model calibration. Typical of such approaches was the 'quickflow separation' technique proposed by Hibbert and Cunningham (1967) and illustrated in Fig. 1.2. As initially envisaged, 'stormflow' (also known as 'quickflow') was delineated by an upward sloping line. Storm rainfall was the rainfall occurring between the initiation of the line and the intersection of the line with the receding hydrograph. The concept worked well for small storm hydrographs. However difficulties quickly manifested themselves:

1. By 1966 it was known that stormflow from forested watersheds was substantially a groundwater response. Thus the concept of 'quickflow' as delineating a particular and discrete process could not be sustained; rather it arbitrarily partitioned a part of a longer-lived slope response.
2. The process used the dependent variable (the storm hydrograph) to define the independent variable (storm rainfall). This is dubious in a statistical sense.
3. For large storms, the 'quickflow' separation line could take many days or weeks to intercept the receding hydrograph. Thus the concept of 'quickflow' becomes a confusing misnomer and is not really applicable to stormflows many days (or weeks) after the causal rainfall.

Alternative hydrograph separation procedures (sometimes generically known as 'baseflow–stormflow separation') suffer from similar issues. Thus it was (and is) unclear what was being delineated in hydrological terms. However the method did produce a national (Hewlett *et al.*, 1977) and an international data set (Hewlett *et al.*, 1984) that allowed a number of hypotheses arising from non-forest hydrology to be at least partly tested using data from forested watersheds. These showed that maximum short-term intensities had little impact on the depth or volume of stormflow, and that the depth of storm rainfall received was the best predictor of the stormflow arising from a forested watershed. Subsequent work (including Bren *et al.*, 1987) showed that rainfall intensity was, indeed, a factor in stormflow generation, but that simple measures such as maximum 15 min, 30 min or 1 h intensity were inadequate. Some 30 years after this was published, Howard *et al.* (2010) revisited this discussion using data from a watershed subject to very-high-intensity rainfall in a tropical zone, and suggested that the matter is still not resolved.

Hydrograph analysis and hydrograph separation still occupy an interesting place in forest hydrology, but are not subject to much active research at the moment. This partly reflects the difficulties of obtaining good matched data sets of rainfall and volumetric streamflow. The technique is very demanding in time.

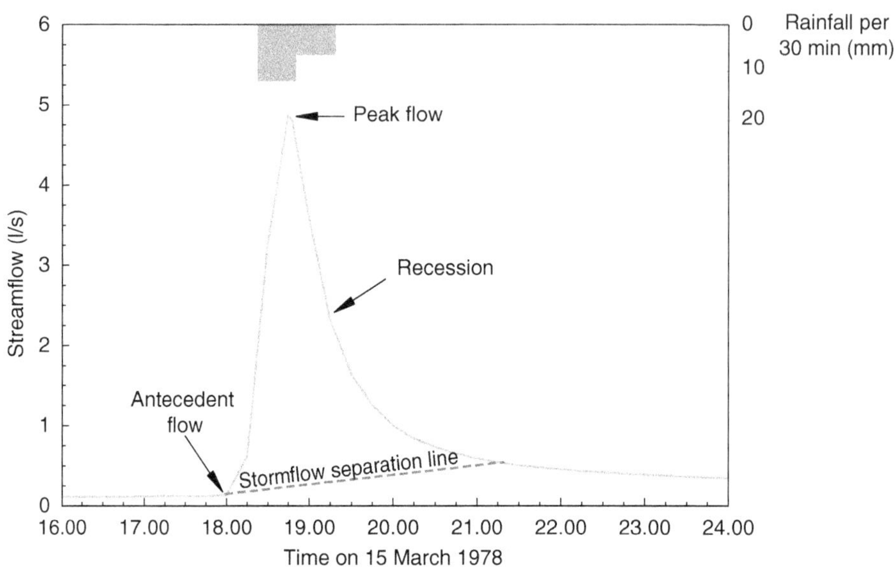

Fig. 1.2. Hydrograph terminology and the application of a stormflow separation procedure to a small hydrograph from a native forest watershed in north-eastern Victoria, Australia.

'Old' and 'new' water

During the 1970s, application of isotope-tagged water to forested slopes engendered a fast hydrological response, but the water entering the stream from the slopes was water that had been stored in the slopes for long periods and was commonly not the same water that 'rained' on the slopes. Thus the 'new' water was pushing out 'old' water. Effectively, this suggested an orderly process of replacement of slope water. Residence time could be weeks to months (e.g. Sklash and Farvolden, 1979; Pearce *et al.*, 1986).

At the time of this discovery, it was thought that stormflow separation based on hydrograph analysis would give new information on slope hydrology processes. However, as discussed by Burns (2002), this has not been the case; indeed the status of such studies was downgraded to 'just one more tool'. Difficulties relate to models of mixing, homogeneity of the slopes and, as always, the role of macropores in providing preferential flowpaths. The downgrading may have been premature; recent findings of Brooks *et al.* (2010) and McDonnell (2014) which used dual-isotope techniques to show that transpiration water came only from smaller pores are opening up a new area of research, but highlight many practical difficulties of sampling and techniques in what was already viewed as a 'difficult area'.

The 'variable source area' concept

An important – but somewhat ethereal and enigmatic – advance in forest hydrology was the concept of the 'variable source area' (VSA); Hibbert and Troendle (1988) present an account of this and some of the passions that went with it. The concept originated from work at the Coweeta Research Laboratory led by Hewlett and Nutter (1969). This had its origins in dissatisfaction with existing hydrological theory based on low rates of infiltration and predicted overland flow across the watershed surface. This theory stated that storm runoff was due to rainfall infiltrating into the watershed slopes near the stream (Fig. 1.3). This area would become saturated and contribute runoff to the stream fast. In heavy rainfall the source area would expand, and in drier periods it would contract – hence the 'variable source area'. In very large storms (e.g. Orr, 1973) the source area would expand to occupy the whole watershed.

The concept has been verified to some extent by studies in small watersheds. The 'variable source area' was and is a useful qualitative concept, but it is an abstraction of a more complex reality. McDonnell (2003) revisited this model some 40 years after Hewlett and colleagues articulated it. He noted that mathematical models of small watershed behaviour usually implicitly use a structure based on VSA concepts, but noted 'a disconnect' between modellers and field investigators which has slowed down attempts to link numerical modelling and VSA concepts. It is disappointing that, despite the tremendous growth in watershed hydrology knowledge since the first articulation of the theory, there has been no numerically based theory to develop this concept.

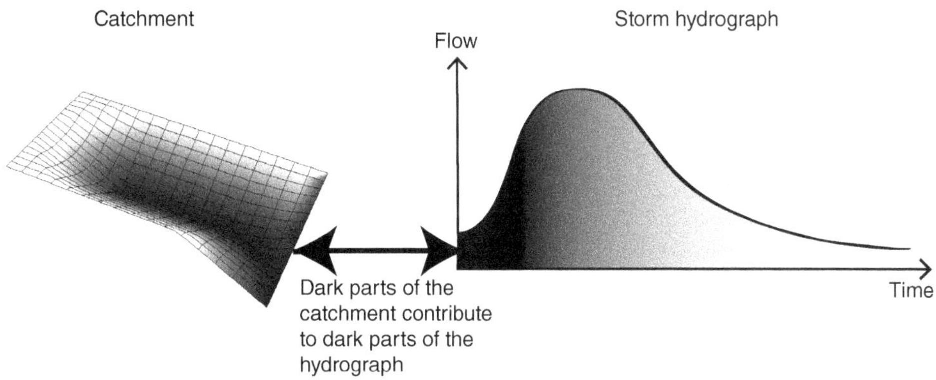

Fig. 1.3. The author's perception of the 'variable source area' (VSA) model; the interpretation is that the 'blacker' parts of the watershed have a higher probability of contributing to the 'blacker' parts of the hydrograph than the lighter colours. (From Bren, 2014.)

1.2.10 Forest fires and watershed hydrology

Forest fires have been around about as long as forests, but their effect on landscape formation has not been appreciated until relatively recently. In recent decades, Australia, the USA and Canada have experienced 'megafires' – large, destructive forest fires on a scale hitherto unknown (with area measured in hundreds or thousands of square kilometres). Some of these fires were fostered by the cumulative effect of fire suppression policies, unprecedented fuel loads, insect and disease infestations, and drier conditions linked to climate change. The impact of these high-intensity wildfires on the hydrology of forested landscapes has been of great importance. In general, the results can be summarized as:

1. Change of the forest age class and/or type, which may have long-term consequences on the hydrological regime. Thus, in Australia, the water use of the key commercial species mountain ash (*Eucalyptus regnans*) varies with forest age (see Bren, 2014 for a concatenation of results on this point). Mountain ash forests are killed by wildfire and a new, even-aged forest regenerates. Thus major fires in these forests introduce a long-term change in the water yield (relative to annual rainfall). Because of the economic importance of water from these forests, this is a major management concern.
2. After fires, 'spike hydrographs' in which very high rates of streamflow are generated for a short time are common (Brown, 1972). These have very high erosive power. Plate 1 shows an example of this after fire burnt the experimental watershed of Fig. 1.1.
3. The fire-induced erosion can have major consequences in degradation of important watersheds and appears to be altering the hydrology of large watersheds (Smith *et al.*, 2011). The relative importance of this in large watersheds which have many other agents of change is an important (and difficult) field.

A study of one such burnt watershed showed it took about 3 years to recover (Bren, 2012). The removal of undergrowth by burning showed many erosion features that appear to be associated with past burns dating back for unknown time periods. In this environment, on northern slopes at least, it is likely that the combination of fire and hydrology has accelerated the development of the Australian mountain landscapes (see, for instance, Nyman *et al.*, 2011). Definitive work on these processes is being done in many countries around the world at the time of writing.

1.2.11 The era of integration and the age of Budyko formulations

Integration of data has always been an effective way of subduing the influence of errors in data. Paired watershed projects produced annual volumes of input (rainfall) and output (streamflow), and tabulated versions of these were readily available. Additionally, the dynamic water yield behaviour of small watersheds has rarely been of interest to water supply managers compared with the annual outflow volumes of larger watersheds. Hence the use of integrated values over a year made sense. The 'year' was often a water year (from summer to summer) to avoid 'change of storage' effects. Thus in Australia this was from May to April because, at the end of April, the soil moisture and groundwater status of the watershed was predictably and consistently low.

Russian scientist Mihail Budyko developed an energy balance of the earth's climate (e.g. Budyko, 1982). This transformed climatology from a qualitative to a quantitative physical science. Aspects of the methodology have direct relevance to forest hydrology and have been widely applied in climate change modelling. In turn, the excellent small watershed data from paired watershed experiments have proved to be ideal for testing the theories of climate change (e.g. Donohue *et al.*, 2012). A widely used outcome of this approach in forest hydrology has been the work of Zhang *et al.* (2001) in producing generalized evaporation (or runoff) curves as a function of mean annual rainfall (Fig. 1.4). Such curves have been used to make coarse comparisons between the hydrology of mature forests and pasture.

More recent work has relied heavily on small watershed data to give estimates of E/P, where E is annual evapotranspiration and P is annual precipitation. These use aspects of the Budyko model to characterize regional hydrology (e.g. Donohue *et al.*, 2012). Of particular value has

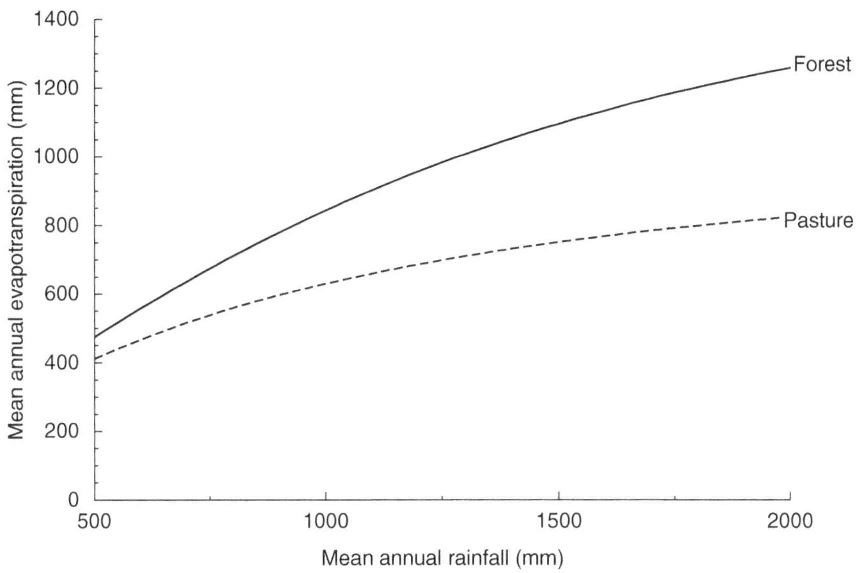

Fig. 1.4. The curves of Zhang *et al.* (2001) used to estimate evapotranspiration from pasture (grassland) and forest.

been the use of such models to allow characterization of regional and global trends in watershed hydrology. The work allows a link between the characteristics of the forests and their radiation environment. At the time of writing many publications on these are appearing. The growth of this theory holds promise for a direct linkage to climate change research.

1.3 Challenges for Forest Hydrology

As presented here, forest hydrology is an empirical discipline combining long-term field experimentation with some physical principles based substantially on the conservation of mass and the detailed accounting of volumes. Use of experiments for testing hypotheses allow it to meet a major criterion of science as expressed by Popper (2005), but the most important experimental design method is expensive to implement and does not easily meet widely accepted criteria of replication and reproducibility. In a statistical sense, paired watershed experiments are case studies; it is difficult to imagine study designs that involve 30 or more watersheds if the statistical criteria of other disciplines were to be met. Although, in an overall sense, the discipline has been successful in providing answers to questions posed by society, key challenges still exist. These are summarized below.

1.3.1 The curse of 0.8

Often, using hydrological data, it is relatively simple to derive a model with an R^2 (coefficient of determination) of about 0.8; thus 80% of the variation is explained. For example, in the hydrology of *E. regnans*, annual yield as a function of age and annual rainfall using some form of regression model gives about this value, with an error of about 80 mm (Bren *et al.*, 2010). Going beyond this (e.g. attaining an R^2 = 0.95) becomes difficult or impossible. Causes of this are usually viewed as being due to errors in the data and the spatial and temporal complexity of other factors that are not easily quantified. These might include the distribution of rainfall over a year, the soil properties, the composition of the forest – the list can be very long. The question of whether this is satisfactory and just what level of prediction or error should be attained is an area for future research. Various attempts to 'do better' using more complex models (e.g. 'Macaque' of Watson *et al.*, 1998) have not been markedly successful. First, the basic data used often have problems. Second, the parameterization of complex models poses formidable issues of measurement at (by human standards) substantial depths in the watershed slopes. The author believes that the time is now ripe for defining the ultimate prediction capability that is attainable in forest hydrology, and using this to assess past and future work.

1.3.2 Are we doomed to empiricism to generate predictive power?

Success in forest hydrology has usually been associated with predictions of the results of forest management based on past experiments. Thus a store of predictive power based on empiricism has been developed. It is relevant to consider whether this must remain the case in the future. The history of science has many cases of where a soundly developed body of observation has helped in the development of comprehensive theories of large predictive power, which then replaces the original empirical observations. Can and will this happen in forest hydrology?

The author's view is that this is both possible and desirable, but unlikely in the near future. First, errors in available data sets due to inadequate measurement of both rainfall and streamflow would need to be refined before there could be any reliability on the level of accuracy and precision. Resolution of such issues (e.g. what is the 'true' rainfall on a watershed?) are solvable but very expensive; most organizations do not have the resources to reduce errors to very low levels in data sets. Second, the most successful examples of overarching predictive theories involve a few variables; in contrast forest hydrology involves many variables. It can reasonably be argued that although not perfect, much forest hydrology data meets the needs at the usual level of observation of larger rivers. Until this also becomes more accurate, there would be little reward for increased accuracy and precision.

1.3.3 Climate change and forest hydrology

As presented to most forest hydrologists, climate change will result in a variation in the long-term mean quantity of water (or snow) falling on the watershed and perhaps a concomitant change in the amount of evapotranspiration. As pointed out by Stohlgren *et al.* (2007), climate change is not new to watersheds. Thus many watersheds have 'drainage lines' – dry stream beds formed in an era when the watershed had a rainfall that allowed such streams to be sustained. Most forests are resilient to both drought and excess rainfall. In the writer's home country of Australia, there is much discussion on the impact of climate change on ecosystems. If there is a climate change component, a major modifying factor is the impact of increased forest fire. The formulations of Zhang *et al.* (2001) allow some idea of the impact of lower rainfall on watershed outflows; these would lead to water supply constraints for many cities and towns. In general, a 10% change in annual rainfall would lead to about a 24% change in streamflow. The real challenge will be in the separation of 'climate change variability' from the large variability already inherent in hydrological records; see Mandelbrot and Wallis (1969) for an interesting view on the data needs to do this.

For many, climate change will translate to 'less water for use' or 'more floods'. De Jong (2015) has examined the impacts of such change on mountain hydrology and concluded that 'interaction between scientists, stakeholders, and decision-makers encompassing local stakeholder knowledge and historical evidence' will be required. The translation of the term 'climate change' into 'impacts on forests' is, and will be, challenging and difficult.

1.3.4 New technology of measurement

One does not need to be genius to note the explosion in field data-collecting capability associated with microprocessor devices; associated development in transducers is giving more and more data measurements to the scientist. LiDAR technology allows characterization of topography to an extent never before available. Associated with this are developments in remote sensing, allowing direct measurement of evapotranspiration, temperatures and other variables of interest. The combination of plot measurement and remote sensing will allow scientists to access the flows in forest hydrology as never before. The challenge will be to use such tools to form overviews of forest hydrology and to help hydrologists scale up from smaller-scale to larger-scale observations, then to convincingly validate the results.

Although the new technologies offer great potential, a common experience is that paired watershed experiments are used to 'test' the new methodologies. Thus it is unlikely that these technologies will, by themselves, make quality field experimentation and measurement obsolete.

1.3.5 More integrated modelling

The glittering prospect is that if one knew enough about plant–water–atmosphere relationships, the physics of water behaviour in watershed materials and the movement of water vapour in the atmosphere, and one had a big enough computer, then 'integrated modelling' would be entirely adequate to resolve any forest hydrology issue. This concept has been around for some decades but accomplishment still seems far away. An example of moving towards this is the inclusion of hydrology modules in forest growth/plant physiology models such as 3PG (e.g. Feikema *et al.*, 2010). Similarly GIS packages could include spatially distributed hydrology modules, allowing accurate prediction, and evapotranspiration from watersheds could be linked to climate models to allow answering of the long-standing question of whether transpiration is really water lost to the local forest.

To date, noticeable success in this direction has not been achieved. This reflects the complexity of parameterization of such large, integrated models. Often the models are 'calibrated' by setting most parameters to likely values and, if there are some data, optimizing the outputs using one or two values. This is a necessary procedure but is not really physically based 'deterministic modelling'. The concept of measuring parameters directly and inserting these into reliable 'integrated' models to estimate hydrological behaviour seems as distant as ever.

1.3.6 Rigorous hypothesis testing

Experimental science often uses hypothesis testing as a means of gaining knowledge. This has not always featured in forest hydrology because of the difficulty in framing statistically testable hypotheses. Growing knowledge of the field, attention to the design of measurement programmes and use of quantitative techniques such as isotope analysis do lend themselves to greater rigour in this respect. This could be interpreted as marking the transition of forest hydrology from a 'developing science' into a 'mature science'.

1.4 The Future of Forest Hydrology

Although forest hydrologists are enthusiastic about their discipline, there is surprisingly little prediction about the future; the one published commentary retrieved by search engines (McDonnell and Tanaka, 2001) is now 15 years old. The small amount of discussion included quantification and costing of ecosystem services, the need for 'over-arching theories' to get away from the 'idiosyncrasies of yet another catchment', and development in 'small scale understanding' and 'large scale modelling'.

To date, most forest hydrology has been concerned with small watershed behaviour. However the demand for water will make the behaviour of large watersheds much more important – thus forest managers will be concerned with the joint management of forests and flows of water. This will, in turn, lead to the application of efficiency 'benchmarks' for the performance of large water supply watersheds. Meeting these will be a requirement of forest management. This will place considerable stress on the levels of knowledge of watershed behaviour and raise difficult issues of forest policies necessary to meet such benchmarks. It is also possible that some water supply requirements will clash with biota conservation and other management requirements in watershed management, providing new challenges and defining needs for 'optimization' involving multiple resources. As shown by Barten *et al.* (2012), this can excite strong passions in communities.

As well as the forest hydrologist looking at the larger watersheds, the time is coming to look at regional, national or international hydrology. The implicit assumption of small-scale studies has always been that evapotranspiration is a 'watershed loss' – that the water vapour goes into a vast, global sink and that the contribution of a given watershed to precipitation elsewhere is impossible to detect. To date, there have been no 'tools' to allow testing of this. Integrated modelling, however difficult, does offer the prospect of addressing such questions in concert with field studies designed to fill key gaps in input data and model formulation. Thus it may well be that water evaporated from a particular watershed falls to earth at a predictable location downwind. It is possible that the age-old theories of 'rain following the plough' or 'trees bring rain' or the 'Law of Dausse' may be examined using this technology.

Finally, the question of the fate of the watershed forests in the new world of climate change is one that must be faced sooner or later (perhaps sooner than later). It is likely that the approach is one of modelling using the fine network of paired watershed experiments for calibration and estimates of error. Thus, although forest hydrology has come a long way towards meeting the needs of society, the pressure of society on earth's resources means that there is still a long way to go before we can ever say 'we know it all'.

References

Andreassian, V. (2004) Water and forests: from historical controversy to scientific debate. *Journal of Hydrology* 291, 1–27.

Aubertin, G.M. (1971) *Nature and Extent of Macropores in Forest Soils and Their Influence on Subsurface Water Movement*. USDA Forest Research Paper NE-192. USDA Forest Service, Northeastern Forest Experiment Station, Upper Darby, Pennsylvania.

Barten, P.K., Ashton, M.S., Boyce, J.K. and Brooks, R.T. (2012) *Review of the Massachusetts DWSP Watershed Forestry Program.* DWSP Science and Technical Advisory Committee, Massachusetts Division of Water Supply Protection, Department of Conservation and Recreation, Boston, Massachusetts.

Bates, C.G. and Henry, A.J. (1921) Streamflow at Wagon Wheel Gap, Colorado. *Monthly Weather Review* 49, 637–650.

Bates, C.G. and Henry, A.J. (1928) Second phase of streamflow experiment at Wagon Wheel Gap, Colorado. *Monthly Weather Review* 56, 79–85.

Benyon, R.G., Theiveyanathan, S. and Doody, T.M. (2006) Impacts of tree plantations on groundwater in south-eastern Australia. *Australian Journal of Botany* 54, 181–192.

Biswas, A.K. (1970) *History of Hydrology*. North-Holland Publishing Company, Amsterdam/London.

Bren, L. (2012) Hydrologic impact of fire on the Croppers Creek paired catchment experiment. In: Webb, A.A., Bonell, M., Bren, L., Lane, P.J.N., McGuire, D., Neary, D.J., Neary, D.G., Nettles, J., Scott, D.F., Stednick, J. and Wang, Y. (eds) *Revisiting Experimental Catchment Studies in Forest Hydrology (Proceedings of a Workshop held during the XXV IUGG General Assembly in Melbourne, June–July 2011).* IAHS Publication No. 353. International Association of Hydrological Sciences, Wallingford, UK, pp. 154–168.

Bren, L.J. (2014) *Forest Hydrology and Catchment Management: An Australian Perspective*. Springer, Dordrecht, the Netherlands.

Bren, L.J. and Lane, P.N.J. (2014) Optimal development of calibration equations for paired catchment projects. *Journal of Hydrology* 519, 720–731.

Bren, L.J. and McGuire, D. (2012) Paired catchment experiments and forestry politics in Australia. In: Webb, A.A., Bonell, M., Bren, L., Lane, P.J.N., McGuire, D., Neary, D.J., Nettles, J., Scott, D.F., Stednick, J. and Wang, Y. (eds) *Revisiting Experimental Catchment Studies in Forest Hydrology (Proceedings of a Workshop held during the XXV IUGG General Assembly in Melbourne, June–July 2011).* IAHS Publication No. 353. International Association of Hydrological Sciences, Wallingford, UK, pp. 106–116.

Bren, L.J., Farrell, P.W. and Leitch, C.J. (1987) Use of weighted integral variables to determine the relation between rainfall intensity and storm flow and peak flow generation. *Water Resources Research* 23, 1320–1326.

Bren, L.J., Lane, P.N.J. and Hepworth, G. (2010) Longer term water use of native eucalyptus forest after logging and regeneration; the Coranderrk Experiment. *Journal of Hydrology* 384, 52–64.

Brooks, J.R., Barnard, H.R., Coulombe, R. and McDonald, J.J. (2010) Ecohydrologic separation of water between trees and streams in a Mediterranean climate. *Nature Geoscience* 2009, 100–104, doi: 10.1038/NGEO0722 (accessed 18 March 2016).

Brown, A.E., Zhang, L., McMahon, T.A., Western, A.W. and Vertessy, R.A. (2005) A review of paired catchment studies for determining changes in water yield resulting from alterations in vegetation. *Journal of Hydrology* 310, 28–61.

Brown, J.A.H. (1972) Hydrologic effects of a bushfire in south eastern New South Wales. *Journal of Hydrology* 15, 72–96.

Buckingham, E. (1907) *Studies on the Movement of Soil Moisture*. USDA Bureau of Soils Bulletin 38. United States Department of Agriculture, Washington, DC.

Budyko, M.I. (1982) *The Earth's Climate: Past and Future*. Academic Press, New York.

Burns, D.A. (2002) Stormflow-hydrograph separation based on isotopes: the thrill is gone – what's next? *Hydrologic Processes* 16, 1515–1517.

Cayley, A. (1859) On contours and slope lines. *Philosophical Magazine Series 4* 18(120), 264–268.

Dausse, M. (1842) De la pluie et de l'influence des forets sur la course d'eau. *Annales des Ponts et Chaussees* Mars–Avril, 14–209.

De Jong, C. (2015) Challenges for mountain hydrology in the third millennium. *Frontiers in Environmental Science* 3, 38, doi: 10.3389/fenvs.2015.00038 (accessed 18 March 2016).

Donohue, R.J., Roderick M.L. and McVicar, T.R. (2012) Roots, storms and soil pores: incorporating key ecohydrological processes into Budyko's hydrological model. *Journal of Hydrology* 436, 35–50.

Feikema, P.M., Morris, J.D., Beverly, C.R., Collopy, J.J., Baker, T.G. and Lane, P.N.J. (2010) Validation of plantation transpiration in south-eastern Australia estimated using the 3PG+ forest growth model. *Forest Ecology and Management* 260, 663–678.

Fisher, J.B., de Biase, T.A., Qi, Y., Xu, M. and Goldstein, A.H. (2005) Evapotranspiration models compared on a Sierra Nevada forest ecosystem. *Environmental Modelling and Software* 20, 783–796.

Hegg, C., McArdell, B.W. and Badoux, A. (2006) One hundred years of mountain hydrology in Switzerland by the WSL. *Hydrological Processes* 20, 371–376.

Hewlett, J.D. and Nutter, W.L. (1969) *An Outline of Forest Hydrology*. University of Georgia, Athens, Georgia.

Hewlett, J.D. and Pienaar, L. (1973) Design and analysis of the catchment experiment. In: White, E.H. (ed.) *Proceedings of a Symposium on the Use of Small Watersheds in Determining Effects of Forest Land Use on Water Quality, May 22–23, 1973*. University of Kentucky, Lexington, Kentucky, pp. 88–106.

Hewlett, J.D., Lull, H.W. and Reinhart K.G. (1969) In defense of experimental watersheds. *Water Resources Research* 5, 306–315.

Hewlett, J.D., Fortson, J.C. and Cunningham, G.B. (1977) The effect of rainfall intensity on stormflow and peak discharge from forest land. *Water Resources Research* 13, 259–265.

Hewlett, J.D., Fortson, J.C. and Cunningham, G.B. (1984) Additional tests on the effect of rainfall intensity on storm flow and peak discharge from wild-land basins. *Water Resources Research* 20, 985–989.

Hibbert, A.R. and Cunningham, G.B. (1967) Streamflow data processing opportunities and application. In: Sopper, W.E. and Lull, H.W. (eds) *Forest Hydrology: Proceedings of a National Science Foundation Advanced Science Seminar held at the Pennsylvania State University, University Park, Pennsylvania, August 29–September 10, 1965*. Pergamon Press, Oxford, pp. 725–736.

Hibbert, A.R. and Troendle, C.A. (1988) Streamflow generation by variable source area. In: Swank, W. and Crossley, D.A. (eds) *Forest Hydrology and Ecology at Coweeta. Springer*, New York, pp. 111–127.

Howard, A.J., Bonell, M., Gilmour, D. and Cassells, D. (2010) Is rainfall intensity significant in the rainfall–runoff process within tropical rainforests of northeast Queensland? The Hewlett regression analyses revisited. *Hydrological Processes* 24, 2520–2537.

Horton, R. (1945) Erosional development of streams and their drainage basins; hydrophysical approach to quantitative morphology. *Geological Society of America Bulletin* 56, 275–380.

Hubbart, J.A. (2011) Origins of quantitative hydrology: Pierre Perrault, Edme Mariotte, and Edmund Halley. *Journal of the American Water Resources Association* 13(6), 15–17.

Loague, K.M. and Freeze, A.R. (1985) A comparison of rainfall–runoff modelling techniques on small upland catchments. *Water Resources Research* 21, 229–248.

Mandelbrot, B.B. and Wallis, J.R. (1969) Some long-run properties of geophysical records. *Water Resources Research* 5, 321–340.

Marsh, G.P. (1864) *Man and Nature: Or, Physical Geography as Modified by Human Action*. Belknap Press of Harvard University, Cambridge, Massachusetts, 1965 reprint with introduction by David Lowenthal.

McCulloch, J.S.G. and Robinson, M. (1993) History of forest hydrology. *Journal of Hydrology* 150, 189–216.

McDonnell, J.J. (2003) Where does water go when it rains? Moving beyond the variable source area concept of rainfall–runoff response. *Hydrologic Processes* 17, 1869–1875.

McDonnell, J.J. (2014) The two water worlds hypothesis; ecohydrological separation of water between streams and trees. *WIREs Water* 1, 323–329.

McDonnell, J.J. and Tanaka, T. (2001) On the future of forest hydrology and biogeochemistry. *Hydrologic Processes* 15, 2053–2055.

Mildrexler, D.J., Zhao, M. and Running, S.W. (2011) A global comparison between station air temperature and MODIS land temperature reveals the cooling role of forests. *Journal of Geophysical Research* 116(G3), G03025.

Morel-Seytoux, H.J. (1973) Two-phase flows in porous media. *Advances in Hydroscience* 9, 119–202.

Nyman, P., Sheridan, G.J., Smith, H.G. and Lane, P.N.J. (2011) Evidence of debris flow occurrence after wildfire in upland catchments of south-east Australia. *Geomorphology* 125, 383–401.

Orr, H.K. (1973) *The Black Hills (South Dakota) Flood of June 1972: Impacts and Implications*. General Technical Report RM-GTR-2. USDA Forest Service, Rocky Mountain Forest and Range Experiment Station, Fort Collins, Colorado.

Pearce, A.J., Stewart, M.K. and Sklash, M.G. (1986) Storm runoff generation in humid headwater catchments: 1. Where does the water come from? *Water Resources Research* 22, 1263–1272.

Philip, J.R. (1974) Fifty years progress in soil physics. *Geoderma* 12, 265–280.

Popper, K. (2005) *The Logic of Scientific Discovery*. Routledge, London/New York.

Scott, R.L. (2010) Using watershed water balance to evaluate the accuracy of eddy covariance evaporation measurements for three semiarid ecosystems. *Agricultural and Forest Meteorology* 150, 219–225.

Sklash, M.G. and Farvolden, R.N. (1979) The role of groundwater in storm runoff. *Developments in Water Science* 12, 45–65.

Smith, H.G., Sheridan, G.J., Lane, P.N., Nyman, P. and Haydon, S. (2011) Wildfire effects on water quality in forest catchments: a review with implications for water supply. *Journal of Hydrology* 396, 170–192.

Stohlgren, T., Jarnevich, C. and Kumar, S. (2007) Forest legacies, climate change, altered disturbance regimes, invasive species and water. *Unasylva* 58(229), 44–49.

Troch, P.A., Paniconi, C. and van Loon, E.E. (2003) Hillslope-storage Boussinesq model for subsurface flow and variable source areas along complex hillslopes: 1. Formulation and characteristic response. *Water Resources Research* 39, 1316, doi: 10.1029/2002WR001728 (accessed 18 March 2016).

Waichler, S.R. and Wemple, B.C. (2005) Simulation of water balance and forest treatment effects at the H.J. Andrews Experimental Forest. *Hydrological Processes* 19, 3177–3199.

Watson, F.G., Vertessy, R.A., Grayson, R.B. and Pierce, L.L. (1998) Towards parsimony in large scale hydrological modelling – Australian and Californian experience with the Macaque model. *EOS Transactions* 79(45), F260–F261.

Watson, F.G., Vertessy, R., McMahon, T., Rhodes, B. and Watson, I. (2001) Improved methods to assess water yield changes from paired-catchment studies; application to the Maroondah catchments. *Forest Ecology and Management* 143, 189–204.

Whittaker, E.T. and Robinson, G. (1924) *The Calculus of Observations*. Blackie, London.

Van Haveren, B. (1988) Notes: A re-evaluation of the Wagon Wheel Gap forest watershed experiment. *Forest Science* 34, 208–214.

Zhang, L., Dawes, W.R. and Walker, G.R. (2001) Repose of mean annual evapotranspiration to vegetation changes at catchment scale. *Water Resources Research* 37, 701–708.

Zon, R. (1912) *Forests and Water in the Light of Scientific Investigation*. US Government Printing Office, Washington, DC, Senate Document No. 469, 62nd Congress, reprinted in 1927 with a revised bibliography.

2 Forest Runoff Processes

T.M. Williams*
Clemson University, Georgetown, South Carolina, USA

2.1 Introduction

As illustrated in Chapter 1, the science of forest hydrology has been dominated by the quest to gain practical insight on how forest management activities alter the amount, timing and quality of water in streams emanating from managed forests. One difficulty with reviewing and interpreting this science has been the lack of precision in language and definitions applied to these investigations. There is no practical difference between discharge, runoff and streamflow as the word to designate water flowing from a forest. Throughout this chapter 'streamflow' is used to designate water rates or volumes as measured at a gauge defining the outlet of a watershed (or predicted at a point defining an ungauged watershed). 'Runoff' is used to designate water delivered to the stream throughout the watershed, including all surface or subsurface flows to the stream channel. Likewise, 'watershed' is used in the sense of all area draining to a chosen point. 'Hillslope' refers to the soils and underlying geological materials draining to a section of stream. A unit-width (x,z) hillslope cross-section (often used in illustration and model development) is called a 'hill section'. In this chapter I try to use somewhat more precise (although most certainly not universally accepted) definitions of runoff processes.

This chapter examines present understanding of two simple questions posed by observers of forests:

1. Where does the water go when it rains (McDonnell, 2003)?
2. Where does the water in the stream come from (Pearce *et al.*, 1986)?

The most basic goal of understanding runoff processes is to generate predictions of streamflow from the rate of rainfall and an index of wetness prior to the rain (Fig. 2.1). The simple linear model of Fig. 2.1 (flow = (rainfall – interception) × wetness index) is illustrative of the goal but treats the forest watershed as a uniform black box. The black box approach can work well with infiltration-excess overland flow (Horton, 1933) on small uniform watersheds. In this, functions of infiltration excess (Horton, 1940; Akan, 1992) are developed from soil physical parameters using infiltration equations of Green and Ampt (1911) or Richards (1931). These assumptions allow hydrograph separation (see Fig. 1.2 for an example) to separate baseflow, made up of groundwater, and stormflow originating as surface flows due to infiltration excess. Throughout the chapter, the term 'stormflow' is used to differentiate the upper portion of the hydrograph defined by the separation procedure of Fig. 1.2.

*E-mail: tmwllms@clemson.edu

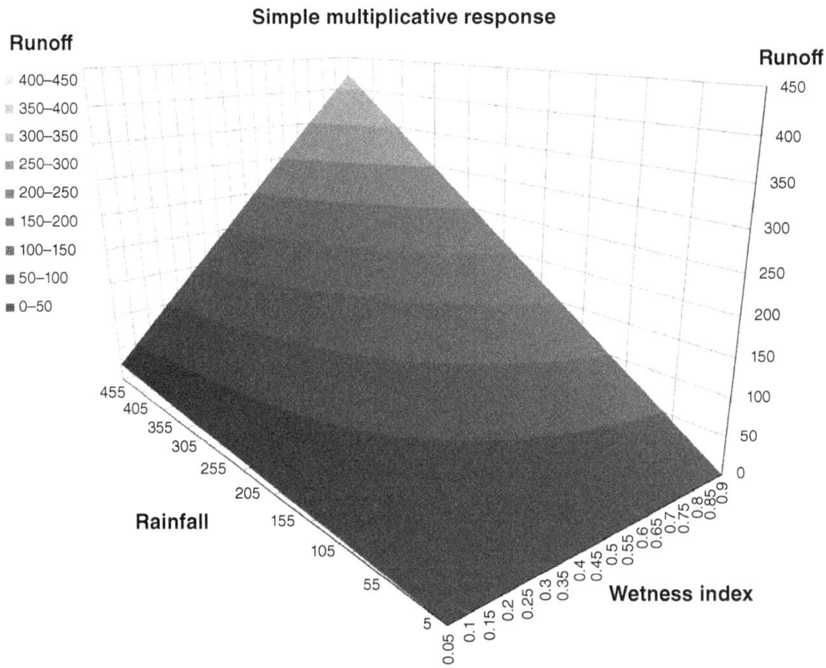

Fig. 2.1. Simplified flow response to rainfall and a wetness index. Wetness index varies from 0 at bone dry to 1 at saturated. Equation of the plane is: runoff = (0.95 rain – 5) × wetness), where runoff and rainfall are expressed in mm.

Infiltration excess rarely occurs in undisturbed temperate forests (Bonell, 1993) and efforts to model forested streamflows using infiltration excess assumptions performed poorly, requiring calibration of a 'partial' contributing area (Betson, 1964). In forested environments, infiltration-excess overland flow tends to occur only in 'special cases'. Often these are semi-arid watersheds with limited forested canopies, which develop water-repellent surfaces that greatly reduce infiltration (Puigdefabregas *et al.*, 1998). When the influence of the permeable forest floor is disturbed by logging or other management activities (Rab, 1994; Rivenbark and Jackson, 2004; Lang *et al.*, 2015) or fire (DeBano, 2000), infiltration-excess overland flow can also become important.

Hursh and Brater (1941) recognized that forested watersheds produced hydrographs with stormflow, but runoff in forests was primarily subsurface and the term 'subsurface stormflow' has become a staple of the forest hydrology vocabulary. Hewlett and Hibbert's (1963) early hill section examination at the Coweeta Hydrologic Laboratory, in the US southern Appalachians, stressed the importance of both saturated and unsaturated moisture movement of infiltrated water between rains. That research showed upslope areas did not directly contribute to stormflow. This work led them to produce the widely cited explanation of the 'variable source area' concept (Hewlett and Hibbert, 1967). Although infrequently cited, similar explanations were also developed in France (Cappus, 1960) and Japan (Tsukamoto, 1961).

Although widely viewed as a basis of modern forest hydrology, the variable source area concept did not necessarily define a particular runoff production mechanism. Hewlett (1982) emphasized the expansion of intermittent and ephemeral channels. Ambroise (2004) has translated quotes of Cappus (1960) that demonstrate he (Cappus) defined source areas in a similar manner to the Dunne *et al.* (1975) definition of source areas by the saturation-excess overland flow mechanism (Dunne and Black, 1970a,b). Mapping of watershed-scale areas of expanding saturation-excess-producing areas provided a concrete example of the variable source area. In many publications variable source area has come to mean the stream plus saturated areas surrounding the stream.

Preferential flowpaths are commonly found in forested soils (Bundt *et al.*, 2001). Voids caused by animal activity or root mortality are generally called 'macropores' and cause intact forest soils to display much higher vertical hydraulic conductivity than those obtained from sieved soil samples. Vertical macropore flow has been found to occur in most forest soils (Beven and Germann, 1982). The importance of lateral, or slope-parallel, macropore flow as a mechanism to produce stormflow was questioned in the USA by Hewlett (1982). Hewlett and Hibbert (1963) demonstrated matrix flow was sufficient to account for channel expansion and stormflow. Despite Beasley's (1976) clear demonstration that flow at the base of the hillslope in the Ouichta Mountains of Arkansas was derived primarily from outlets of larger soil voids, he (Hewlett, 1982) continued to argue Aubertin's (1971) assertion that water could not enter macroscopic soil pores until the soil was saturated.

A similar conflict was found at the Maimai research watersheds in New Zealand, where Mosley (1979) concluded soil voids (macropores) conveyed subsurface stormflow as the main runoff mechanism. Sklash and Farholven (1979) suggested that rain on riparian areas resulted in a rapid increase in the water table and enhanced groundwater runoff into streams, a process called 'groundwater ridging'. Using ^{18}O, Pearce *et al.* (1986) showed that water flowing from the macropores was 'old water' similar to that in the stream before rainfall started and water within the soil matrix. They argued that macropore flow could not be responsible for stormflow. McDonnell (1990) showed rapid exchange of 'new' and 'old' water as vertical flow quickly filled macropores near the soil–bedrock interface. McGlynn *et al.* (2002) discussed how concepts changed as more information was found to confirm a complex interaction of rainfall amount, macropore flow and bedrock surface topography. Graham *et al.* (2010) further elucidated flow on these watersheds, showing the role of bedrock topography in controlling the initiation of these soil voids. Downslope flow in soil macropores has been discovered in many other studies worldwide. Weiler and McDonnell's (2007) review of macropore flow studies led them to propose to restrict the term 'macropore flow' to vertical water movement in preferential channels. They propose to use the term 'soil pipe' to denote a macropore in which water moves in the downslope direction. This terminology is used throughout the chapter.

The prevalence of pipeflow in studies of forested hillslope hydrology led some authors to call this mechanism 'subsurface stormflow'. However, that ignores the contributions of groundwater flow. Sidle *et al.* (2001) found bedrock cracks to interact with soil pipes. Anderson *et al.* (2007) found flow in weathered cracked bedrock to be an important source of subsurface stormflow in the Oregon Coast range near Coos Bay. Gabrielli *et al.* (2012) also found subsurface stormflow occurred in bedrock cracks at Oregon's Andrews Experimental Forest watershed 10, despite its great similarity to the Maimai watershed in New Zealand where the primary source was soil pipes. Buttle and McDonald (2002) found that subsurface stormflows occurred primarily as saturated flow in a thin layer at the bedrock–soil interface on thin glaciated soils in Ontario.

Flow of water in carbonate aquifers is a major aspect of groundwater hydrology that has not been studied widely in forest hydrology. Most of the research on carbonate aquifers has been concentrated in geohydrology, geomorphology and contaminant flow (Kaçaroğlu, 1999; Ford and Williams, 2007; Williams, 2008). The most obvious geomorphical aspect of carbonate hydrology is karst topography with closed depressions, dry valleys, and losing or disappearing streams. In regions with shallow soils, epikarst (the upper highly weathered region of carbonate rock with numerous channels) forms the primary upper aquifer material and conduit to a deeper zone of fewer, larger conduits, supplying large springs. Hillslope or small watershed research is hampered by an inability to determine flowpaths or source areas without extensive drilling and water level monitoring (Jiang *et al.*, 2008). Epikarst provides rapid vertical transport-like macropores and may provide slope-parallel flow similar to soil pipes.

Bishop *et al.* (2011) describe a process they found in central Sweden somewhat similar to saturation-excess overland flow except that flow occurs entirely within the soil profile. On low-gradient watersheds with shallow soils over compacted till, groundwater flow is conveyed by the thin upper mineral soil and thick forest floor accumulations. They called the process a peculiar name, 'transmissivity feedback'. Transmissivity is the product of saturated hydraulic conductivity and aquifer thickness; this gives it the unusual dimension of length^2/time (e.g. m^2/s). The usefulness of this term is that flow to the stream is determined by the product of stream length and transmissivity. For most aquifers, transmissivity can be regarded as a constant, since neither the aquifer thickness nor the conductivity changes appreciably during precipitation. However, for the watershed they describe in Sweden, as the water table rises near the surface, the thickness of the aquifer increases substantially and the average hydraulic conductivity increases due to inclusion of the highly conductive surface organic layer. This transmissivity increase greatly increases subsurface stormflow to the stream (Seibert *et al.*, 2011).

2.2 Non-Linearity, Connectivity and Thresholds

As researchers observed the processes described above, they also found that both runoff and streamflow do not respond to rainfall in the smooth, linear manner envisioned in textbook hydrographs. Observation of infiltration excess on agricultural fields revealed spatial heterogeneity of even uniform agricultural fields, leading Betson (1964) to propose a partial source area of runoff. On agricultural fields, areas of infiltration excess are easily seen during a storm. These areas do not contribute to runoff until surface water has risen sufficiently to overtop microtopographic barriers and connect to the outlet. Similar surface connections were evident in watersheds with exposed rock, wetlands and lakes in the Canadian Northwest Territories (Phillips *et al.*, 2011). The idea that threshold runoff behaviour was caused by flowpath connections could be easily seen on watersheds with visible surface flows (Darboux *et al.*, 2002). Ambroise (2004) suggested runoff-producing areas are variable in both space and time. Using saturation excess as an example, he argued areas can be active (saturated) but not contributing (as in closed depressions) until some threshold value (rainfall intensity or depth, water table depth) allows connection to the stream. Spencer and Woo (2003), near Yellowknife, Northwest Territories, Canada (62°N), found intermittent saturation flow along a water path that became an ephemeral stream during wet conditions. Contrary to flows in more temperate regions the stream formed first in the headwaters and then formed progressively down the valley to the outlet. They termed this characteristic ‘fill and spill’, noting that upstream segments must saturate, ‘fill’, before overland flow can progress down the valley and ‘spill’.

While flowpath connections can be seen to be instrumental to threshold behaviour in surface flow (e.g. Jencso and McGlynn, 2011), they also have been found to be important in subsurface stormflow processes. Ali *et al.* (2013) review a large number of studies that describe threshold behaviour in runoff. Threshold behaviour has been found in conditions from Arctic permafrost to warm temperate areas and spanning annual rainfall rates of <350 to >2500 mm. Although threshold behaviour has been widely observed, it has generally not been experimentally examined. The group of studies outlined by Ali *et al.* (2013) generally found threshold behaviour in data collected in process studies intended for other purposes. Relatively few have combined examination of thresholds with studies of connectivity. Tromp-van Meerveld and McDonnell (2006a,b) examined a long record of hillslope flow at the Panola Mountain watershed in Georgia to examine both connectivity of soil pipes and the controls on runoff production. They found a 55 mm rainfall threshold was required to fill bedrock depressions controlling connection of soil pipes. Uchida *et al.* (2005) combined the Georgia data with watershed and hillslope studies in Japan, to find that a threshold value of precipitation was needed before pipeflow was initiated, and pipeflow was linearly related to total streamflow for storms larger than the threshold.

McGuire and McDonnell (2010) examined water age and flowpath at H.J. Andrews watershed in Oregon and found a 30 mm rainfall threshold before the hillslope delivered runoff. Detty and McGuire (2010a,b) examined flow at Hubbard Brook watershed in New Hampshire where the runoff mechanism was primarily saturation-excess overland flow. They found a strong threshold by combining stored soil moisture with event rainfall into an event index. Below an index of ~316 mm there was no relationship to runoff but above that value there was a strong linear association of index to streamflow (peak flow and stormflow volume as defined by the Hewlett separation procedure of Fig. 1.2). They also found a similar relationship of runoff ratio to an averaged normalized water table with an index of 0.48 (0 being the deepest water table at a point and 1 being the shallowest).

McDonnell (2013) expressed the idea that all runoff processes may be explained by a few universal controlling concepts that are expressed in different relationships depending on climate, vegetation and geology. He built on the 'fill and spill' hypothesis suggesting all runoff process occurs by overtopping some type of storage reservoir in the watershed. He tacitly accepted Aubertin's (1971) contention that all water flowing in macropores, pipes, cracks, etc. must have a moisture potential (hydrostatic head) greater than atmospheric. He postulates that processes are driven and controlled by establishment of a, at least momentary, perched water table. Connectivity within the watershed is spatially arranged by a controlling surface such as the soil surface, the top of a slowly permeable layer or the water table. With this view, one could state that infiltration excess is merely the extreme where a water table perches at the soil surface, and baseflow is the other extreme where the stream intersects a zone of permanent saturation.

2.3 Distribution of Processes

Figure 2.2 is a schematic used to summarize the runoff process described here and also to add personal ideas of how these processes can contribute to a theory of connection-determined threshold hydrology of forested watersheds. The figure is arranged as a series of reservoirs

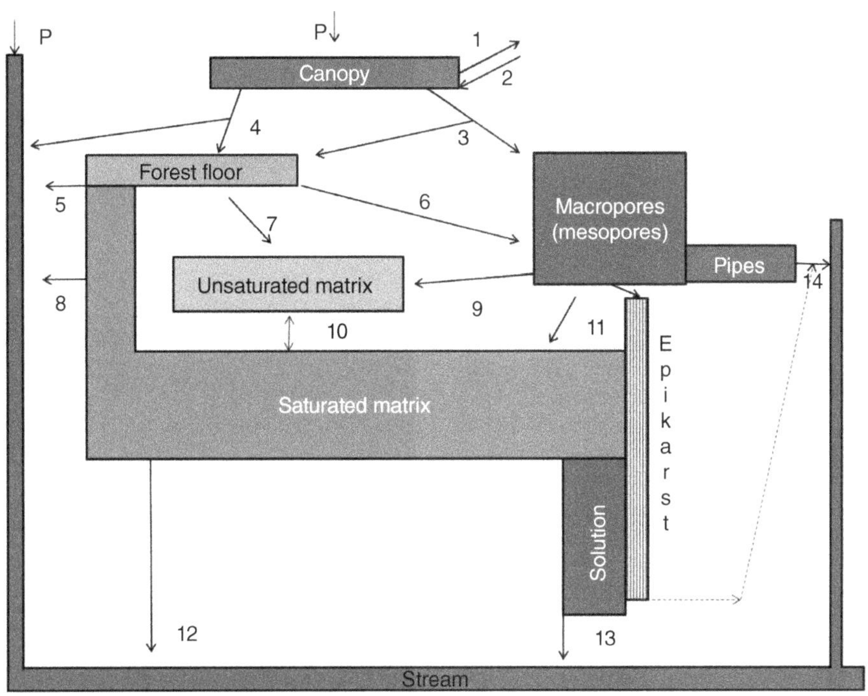

Fig. 2.2. Schematic representation of possible flowpaths within a forested watershed. Incoming rain P is distributed through the watershed reservoirs by various flows represented as numbered arrows: 1 = evaporation of intercepted rain; 2 = condensation on the forest canopy; 3 = stemflow; 4 = throughfall; 5 = saturated forest floor to stream; 6 = forest floor to macropores; 7 = forest floor directly to unsaturated soil; 8 = groundwater ridging; 9 = transfer from macropores to unsaturated soil; 10 = flow from unsaturated soil to saturated matrix; 11 = transfer from macropores to saturated matrix or epikarst; 12 = baseflow; 13 = flow from carbonate rock springs; 14 = flow in soil pipes to streams.

represented as separate boxes and flows as numbered arrows (defined below). For this discussion an undisturbed forested watershed with full canopy cover is assumed, and all flow is assumed to be liquid water. Interactions with snow are covered in chapters later in this volume.

2.3.1 Interception

Canopy interception is the first interaction of the forest with precipitation. Flows from the canopy can occur in four directions: evaporation (arrow 1, Fig. 2.2), condensation (arrow 2), stemflow (arrow 3) and throughfall (arrow 4). The relative sizes of these flows may result in a moderate threshold in the production of runoff. Evaporation and condensation are exchanges of intercepted water with the atmosphere and primarily controlled by humidity levels during and after rainfall. Evaporated interception can be modelled by the popular Gash (Gash, 1979; Gash *et al.*, 1995) or Rutter (Rutter *et al.*, 1971) model. Maximum evaporation occurs when the canopy dries completely between rains (a condition assumed in Gash models), enhanced by well-separated storms, rough canopies and high turbulence. In Western Europe the maximum evaporative loss from interception may approach 30% of open ground precipitation. Maximum condensation occurs with saturated atmosphere, low wind velocity and maximum exposed leaf area. Such conditions may occur in warm temperate, subtropical and tropical, maritime climates. The most extreme conditions occur in tropical montane cloud forests where interception of wind-driven fog input may augment rain to exceed open ground rainfall by 20% (Bruijnzeel *et al.*, 2011).

Throughfall and stemflow are residuals of the balance of flows 1 and 2. A number of factors contributed to high variability in estimates of each (Crockford and Richardson, 2000). Tree species, size, density and canopy roughness all interacted. Integrated measures such as the leaf area index (LAI) and canopy storage (crown surface area) were good indicators while stem basal area was not. The most important climatic variables were rain (quantity, intensity and duration), wind speed and direction during rain, and air temperature and humidity. Stemflow was also influenced by leaf angle, branch angle and wind exposure of exposed stems. In general, conifers had higher interception (18–25% of gross rainfall) than hardwoods (10–15%). Tropical and subtropical forests showed a wide variation of interception due to wide differences in climatic influence. The lowest value (9%) was determined for an Amazon rainforest (Lloyd and Marques, 1988) and highest (39%) for Puerto Rico mountains with mostly low-intensity rain (Scatena, 1990). In most studies mentioned by Crockford and Richardson (2000) stemflow was from 1 to 4%, but Crockford and Richardson (1990) measured a value of 8.9% for *Pinus radiata*.

Stemflow measures show values are generally a small percentage of total rainfall (1–4%) but several studies found values up to 20% for certain forest types. Even higher values were obtained for arid-region shrubs (27–45%) (Levia and Frost, 2003). Stemflow quantity increased as storm volume increased and wind increased, but decreased as intensity increased. High intensity increased branch drip as intercepted water exceeded the rate it could flow down branches and bole. The angle branches joined the bole was important to tree species differences. *Acacia* species with steep branch angles have particularly high values of stemflow: *Acacia holoserica* with 16% (Langkamp *et al.*, 1982), *Acacia auriculiformis* with 6.2–7.9% (Bruijnzeel and Wiersum, 1987) and *Acacia aneura* with 16% (Pressland, 1973).

Throughfall deposits from 60 to 90% of incoming rain with patterns associated with canopy gaps and areas of branch drip. Throughfall on the stream surface, along with direct precipitation (which may become significant on glaciated landscapes with numerous lakes) form the most constant part of stormflow, dependent only on the small threshold of canopy interception. Most throughfall will be transferred to the forest floor as it is unlikely water will fall directly into a macropore. Stemflow is more likely to flow directly into micropores near stumps. High stemflow and direct macropore recharge may be important to regions with distinct wet and dry seasons. There, preferential flows may produce hillslope runoff early in the wetting phase by entirely bypassing the dry soil matrix. Canopy interception is usually a small threshold (perhaps 0 to 10 mm) that depends on the prevalence of evaporative losses.

2.3.2 Forest floor flows

The forest floor is the focus of litter accumulation, fine root activity, soil faunal activity and microbial activity. The litter layer and upper few centimetres of mineral soil can be thought of as a biological mat that covers the hillslope or watershed. The most important aspect of the forest floor is its high porosity and hydraulic conductivity. Torres *et al.* (1998) needed 'unreasonable' rainfall simulator intensities (>5700 mm/h) to cause ponding on a forested soil in Oregon. That value is nearly three times the claimed maximum measured rainfall intensity (1.23 inches per minute or 1847 mm/h; Engelbrecht and Brancato, 1959). Transfer of water into the unsaturated soil (arrow 7, Fig. 2.2) follows the Richards equation for unsaturated flow and is quite well understood.

Transfer from the forest floor to soil macropores (arrow 6, Fig. 2.2) is not well understood. Rapid downslope transport within the forest floor has been observed in temperate rain-dominated forests of New Zealand (McDonnell *et al.*, 1991) and Japan (Terajima and Moriizumi, 2013), tropical forests of Ecuador (Goller *et al.*, 2005; Crespo *et al.*, 2012) and summer high flows in New York (Brown *et al.*, 1999). Such rapid transport may allow short-distance flow within the forest floor that would be free to move into macropores. Luxmoore (1981) coined a term 'mesopore' for drainable pores <0.1 mm (Luxmoore *et al.*, 1990) that have been found to produce vertical flow of 3.4×10^{-4} m/s (1.2 m/h) on forested watersheds in eastern Tennessee (Wilson and Luxmoore, 1988). Such small pores may be continuous into the forest floor, allowing transfer into larger pores. Sidle *et al.* (2001) suggested pores of all sizes self-organized under increasing wetness to produce pipeflow at the bottom of the hillslope. Whatever the mechanism, rapid vertical preferential flow is common to most forest ecosystems.

Saturation-excess overland flow is one form of transfer directly through the forest floor to the stream (arrow 5, Fig. 2.2). Brown *et al.* (1999) found that much of the event water in a New York stream had dissolved organic carbon concentrations, suggesting flow within the forest floor rather than above it. Only when the entire soil profile is saturated will water be transported a significant distance laterally. This process is likely to occur everywhere the stream is separated from the hillslope by a relatively flat riparian zone. The extent of such a zone depends on the steepness of the slope, upslope area, slope conductivity, overall water balance, and rates of evaporation and transpiration from vegetation within the riparian zone (Burt *et al.*, 2002). Since topography is important to this mechanism it has been modelled using variants of TOPMODEL (Beven and Kirkby, 1979) that incorporate terms for soil depth and hydraulic conductivity into the basic topographic index calculation (Frankenberger *et al.*, 1999; Walter *et al.*, 2002; Lyon *et al.*, 2004).

Bishop *et al.* (2011) called flow within a thick organic layer over less-permeable till 'transmissivity-feedback' as a form of groundwater flow. However, if one considers the mineral soil of the surface, such flow would be called saturation-excess overland flow. Similar flows have been found on watersheds with organics over till in Vermont (Kendall *et al.*, 1999) and Ontario (Montieth *et al.*, 2006). Skaggs *et al.* (2011) found the upper 90 cm of soil with a forest litter layer resulted in an observed hydraulic conductivity three orders of magnitude greater on drained forested watersheds than on comparable drained agricultural fields; on these the observed hydraulic conductivity was similar to published data for that soil series. Skaggs *et al.* (2006) also found that logging did not change conductivity, but bedding for a new plantation reduced observed hydraulic conductivity to published values and resulted in significant overland flow. Detty and McGuire (2010a,b) found a clear 'hockey stick' (see Ali *et al.*, 2013 for threshold response types) threshold response with a high threshold of 316 mm. Epps *et al.* (2013) found a similar 'hockey stick' response explained stormflows on a low gradient (slope 0.001) watershed in the south-eastern US lower coastal plain, with soils similar to those studied in Skaggs *et al.* (2006). On such watersheds, the blade of the hockey stick (constant rainfall–runoff ratio) reflects stormflow generated within the riparian zone, while the handle (steep increase in rainfall–runoff ratio) reflects growing connectivity across the watershed with increasing rain. The riparian zone response may be due to throughfall on the stream and near-stream saturated area, and occurs with a relatively small threshold as long there is baseflow in the stream. The watershed response requires filling the unsaturated matrix on a substantial portion of the watershed. This needs 100–300 mm of excess in either event rain or previous soil water content. The slope of the hockey stick handle will depend on the spatial distribution of saturated, or near-saturated, area on the watershed. Since shallow water tables are relatively easily measured, these watersheds may present an opportunity to easily examine the partitioning surface of 'fill and spill' depicted in McDonnell (2013, Figure 5).

2.3.3 Unsaturated and saturated matrix

Flow to the saturated matrix from the unsaturated matrix (arrow 10, Fig. 2.2) has been long studied and well modelled by the Richards equation. Likewise, the flow from the saturated aquifer to the stream (arrow 12) is matrix groundwater flow and can be modelled by the Darcy equation. This is the primary path supplying flow between storms. It occurs as long as the stream channel bottom is below the aquifer water table or piezometric potential if the aquifer is semi-confined. Segregating this flow in a storm hydrograph can be done computationally, with isotopes, and with dissolved minerals. Often the three techniques do not agree as each technique has limitations (Klaus and McDonnell, 2013).

In addition to baseflow, under geological conditions discussed in the introduction, groundwater can be a large component of stormflow. On forested areas with underlying carbonate rocks large channels form by solution of the aquifer (arrow 13, Fig. 2.2). Larger springs from carbonate rocks have subdued stormflow where a water table formed in the epikarst provides storage between rains. Smaller streams from carbonate aquifers may show stormflow response through connections to other processes. Streams that originate from the epikarst zone may show behaviour similar to soil pipes or deeper carbonate systems. However, difficulty in determining source areas and flowpaths makes research in this area difficult.

McDonnell and Buttle's (1998) strong response to Jayatilaka and Gillham's (1996) claim that groundwater input to stormflow by groundwater ridging (arrow 8, Fig. 2.2) was a widespread mechanism, indicated that groundwater ridging is limited to specific geological conditions. Drainable porosity and hydraulic conductivity limit soil texture to fine sand or coarser to convey large quantities of stormflow. In fine sand the capillary fringe extends only about 1 m above the water table. Therefore, the increase in head caused by the ridging can be no more than 1 m. Such a head change can result in no more than doubling of baseflow unless the baseflow stream is less than 1 m deep. It seems that to deliver large stormflows, the process requires a connection from the stream bottom to a semi-confined aquifer. That aquifer must also have a relatively good connection to a point on an upper slope where a perched water table forms. In that case, hydrostatic head on the semi-confined aquifer could increase several metres as the perched water table forms. Such a mechanism may have occurred in the study described by Katsura *et al.* (2014). Near-stream ridging will produce small stormflow rates and the threshold will be similar to near-stream saturation excess flow, probably differing only in isotope or solute signature. If the semi-confined aquifer case exists, it can produce larger stormflows and may have thresholds similarly to pipeflow.

2.3.4 Macropores and soil pipes

McDonnell (1990) showed that exchange of macropore and unsaturated matrix water (arrow 9, Fig. 2.2) could resolve the 'old water problem'. He found event water flowing from soil pipes quickly mixed with matrix water near the boundary of soil and bedrock. Water within the macropore had isotopic and chemical signatures matching matrix water rather than rain or throughfall. Rapid transfer of water in macropores to the saturated matrix (arrow 11) is common on many forested landscapes (Beven and Germann, 1982; Weiler and McDonnell, 2007). Rapid vertical flow during larger rains may result in saturation of the soil from the bottom up, by rising water tables, rather than from the top down by wetting fronts. This connection may explain why threshold behaviour can often be explained equally well with a soil moisture index or water table (Detty and McGuire, 2010b). Although matrix permeability may suggest flow (arrow 9) could be limited, transfer from macropores to the unsaturated matrix may be quite large by way of the saturated matrix.

Unless groundwater can move in highly conductive layers such as fractured rocks, epikarst or highly weathered zone at the bedrock–soil interface, vertical macropore flow (arrow 11, Fig. 2.2) will result in rise of a perched water table. Horizontal (downslope) flow in macropores (called 'soil pipes' by Weiler and McDonnell, 2007) can continue down the slope to the stream (arrow 14), to the riparian zone, or emerge at the surface somewhere along the slope. Water returning to the surface due to reduced conductivity either in matrix flow or from soil pipes has been called 'return flow'. In the riparian zone, return flow may increase the size of the saturated riparian zone or form an ephemeral channel to the stream. Likewise, on the slope, return flow will either infiltrate back into the soil or form an ephemeral stream. Return flow which infiltrates is of little consequence other than by impeding extrapolation from small plots to hillslopes. In the riparian zone it may increase the likelihood of soil saturation at the edge of the hillslope. Hewlett (1982) suggested increase in stream length by addition of ephemeral channels was a significant part of the variable source area.

Soil pipes also play a significant role in the geomorphical development of landscapes. Jones (2010) was adamant that the term 'soil pipe' be only used with an older geomorphical interpretation as a soil pore that has been altered by water flow. This is similar to Beven and Germann (1982), who defined pipes as being more than 4 cm in diameter. Uchida *et al.* (2001) explored pipeflow in relationship to both stormflow and generation of shallow landslides, stressing primarily length rather than diameter for calling a horizontal macropore a pipe. That review found pipeflow greater in large-diameter pipes, on steep slopes, with greater wetness and larger storms. They found pipes would generally decrease slope water content and lessen landslide potential. However, larger sediment-carrying pipes could enhance landslide potential if they collapsed or clogged. Fujimoto *et al.* (2008) examined slope convergence and determined pipes were more concentrated in streamhead hollows, followed by convergent slopes and least on planar slopes. Concentration of flow in hollows increased the likelihood of high flow and erosion of soil pipes in these positions.

Sidle *et al.* (2001) summarized various studies of soil macropore, frequency, diameter, length, tortuosity and orientation on the Hitachi Ohta Experimental Watershed in Japan. In this, relatively short (10–50 cm) macropores organized into pipe systems with increasing wetness. The organization was facilitated by inclusions of thick organic matter (deep litter, overturned litter in windfalls, decaying logs and roots) and bedrock cracks that connected pores. This organization appeared to create two threshold responses after 40 mm and 110 mm of rain. They concluded that the behaviour of individual basins was determined by the macropore/pipe connections with varying soil wetness. In Malaysia, Negishi *et al.* (2007) found similar threshold behaviour in larger (2.5 to 7.5 cm diameter) pipes. Although deep pipes flowed more often, when the shallowest pipes were active they produced an order of magnitude greater flow. The control of runoff seemed similar to that described by McGlynn *et al.* (2002) in New Zealand, where rapid vertical flow produced a saturated layer at the soil regolith surface. As the saturated layer thickened more soil pipes became active.

There seems to be no agreement on the definition of 'soil pipe' although many researchers have described large-diameter pipes that are important in the generation of stormflow. However, at other locations, smaller-pore flowpaths were similarly important in stormflow production. Since biological activity results in a range of sizes of soil pores in all forested regions, it seems that for forest hydrology there is no need to restrict a definition of soil pipe to only those sizes that are altered by water flow. Conversely, in most forested watersheds the primary movement of water, dissolved elements and sediments is in soil macropores and these are likely to be the main agents of landscape erosion and geomorphical structuring. Soil pipes that are large diameter and water sculpted are also an agent of landscape erosion and geomorphical structuring in many non-forested environments (Jones, 1994). 'Soil pipe' can then mean a series of macropores of biological and physical origin that can interact with each other, bedrock cracks or porous inclusions to form slope-parallel flowpaths that can rapidly transport runoff to a stream to be included in stormflow. 'Soil pipe' may also mean larger (>4 cm) water-sculpted openings that may extend upslope tens to hundreds of metres and are sites of physical or chemical erosion that form landscapes.

2.4 Summary

'Are all runoff processes the same?' McDonnell (2013) went a long way to resolve many processes researchers have found into a set of principles that are applicable to the many forested watersheds found on the earth. The following is my attempt to expand on those thoughts.

The 'fill and spill' explanations (Spencer and Woo, 2003; Tromp-van Meerveld and McDonnell, 2006b) are a three-dimensional explanation of the simple bucket storage model shown by Hewlett (1982). The bucket (Fig. 2.3) represents all moisture storage on the watershed, with the full bucket being a completely saturated watershed. Each pipe extending from the bucket represents a possible runoff process, with the size roughly proportional to the size of stormflow hydrograph that process may produce. The position of the pipe represents roughly the amount of storage threshold needed to be filled for that process to activate.

Any perennial stream has a channel that intersects a permanently saturated aquifer, producing baseflow. Baseflow is dependent on elevation of the stream channel in relation to the water table elevation in that aquifer. The balance of aquifer inflow and outflow, in relation to total aquifer storage, determines the water table elevation or piezometric head of a leaky confined aquifer where that provides flow to the stream. For shallow alluvial aquifers that could vary with rainfall on a weekly or monthly basis; or for larger deep alluvium, porous rock, or for large carbonate aquifers, variations may be on an annual or decade-long time frame.

Throughfall on the stream will occur after any rainfall that exceeds the threshold of interception storage. It is variable in that the stream will expand into intermittent and ephemeral channels as the watershed storage fills. In three dimensions the extent of channel expansion is relatively well modelled by a topographic index such as that employed in TOPMODEL.

The threshold for flow from saturated riparian zones is probably close to that of throughfall on the stream. The groundwater ridging process will add water in proportion to the rate of water table rise. As long as the soil surface is within the capillary fringe this process will add to stormflow before the riparian soil fully saturates. Saturation overland flow, which, I believe, moves primarily in the forest floor, begins as soon as the soil near the stream saturates. Runoff from this

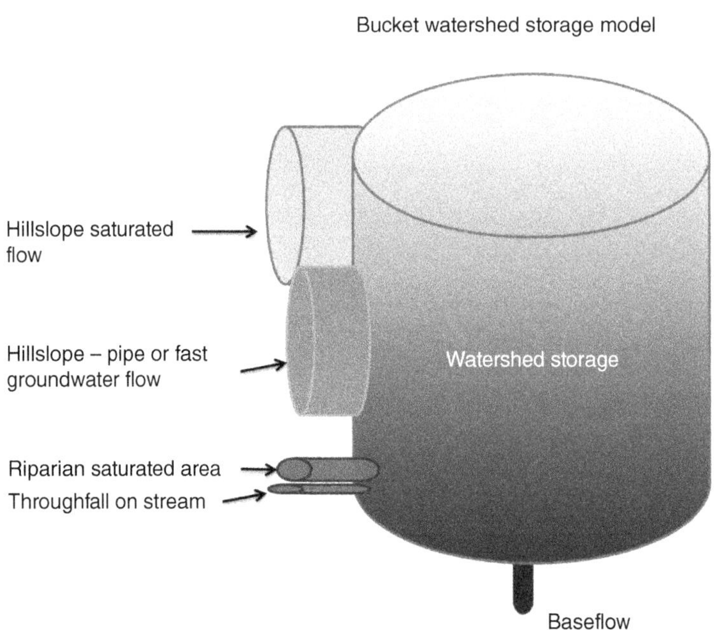

Fig. 2.3. The watershed as a storage bucket derived from Hewlett (1982). This can also be thought of as a non-dimensional representation of fill and spill threshold behaviour.

source will be roughly proportional to rainfall. Wickel *et al.* (2008) found that to be true even in a tropical Amazon watershed receiving over 2000 mm of rain.

Soil pipes or, on those watersheds with a highly conductive layer at the soil–bedrock interface fast groundwater flow, is the primary subsurface stormflow mechanism of most studied watersheds. Most studied watersheds were fairly small and steep, with soils on relatively impermeable bedrock. Soil pipes produce hillslope runoff after rainfall exceeds a threshold determined by the surface of the bedrock–soil interface. Macropore flow produces perched water tables in depressions of the bedrock and pipeflow begins when these perched water tables overtop barriers in the interface contours. The 'fill and spill' theory holds that the thresholds will be lower and spill will occur more quickly on steeper slopes. Thresholds of 20–30 mm have been found on steep watersheds while thresholds of 50–60 mm were found on more gentle slopes.

The highest threshold of process initiation occurs for saturation-excess overland flow outside the riparian zone. This mechanism is common in formerly glaciated regions where forest soils are thin over compacted till or fresh bedrock. On moderate slopes visible surface flows (called return flows) occur at points where flow in the forest floor is greater than transmissivity of that layer. Such flows may re-infiltrate or form ephemeral streams. High levels of slope water storage are needed for this process to occur. Accumulations of over 300 mm were needed to activate this process in New Hampshire. On gentler slopes with thicker forest floor/organic soil horizon such flow can remain within the surface organic materials and may not emerge as visible surface flow. A specific threshold has not been reported for this process but was associated with snow melt or heavy rain. On low-gradient watersheds (<10 m total relief), saturated soil can be found throughout the watershed when trees are dormant (see Chapter 7 and Santee Watershed in Chapter 14, this volume, for example). On such watersheds, stormflows are produced with storms as small as 20 mm during the dormant wet period or may be absent with storms of 100 mm late in dry growing seasons.

In addition to evaluating forest management activities, forest hydrology research has found a number of different ways rainfall (or snow; see Chapters 4 and 9, this volume) becomes streamflow. As with all natural processes, each answer comes with two new questions. In the 50 years since we first understood that rain on forested watersheds does not act the same everywhere, we have begun to understand the process and principles that may lead to real understanding of forest runoff processes.

References

Akan, A.O. (1992) Horton infiltration equation revisited. *Journal of Irrigation and Drainage Engineering* 18, 828–830.

Ali, G., Oswald, C.J., Spence, C., Cammeraat, E.L.K., McGuire, K.J., Meixner, T. and Reaney, S.M. (2013) Towards a unified threshold-based hydrological theory: necessary components and recurring challenges. *Hydrological Processes* 27, 313–318.

Ambroise, B. (2004) Variable 'active' versus 'contributing' areas or periods: a necessary distinction. *Hydrological Processes* 18, 1149–1155.

Anderson, S.P., Dietrich, W.E., Torres, R., Montgomery, D.R. and Loague, K. (1997) Subsurface flow paths in a steep, unchanneled catchment. *Water Resources Research* 33, 2637–2653.

Aubertin, G.M. (1971) *Nature and Extent of Macropores in Forest Soils and Their Influence on Subsurface Water Movement*. USDA Forest Research Paper NE-192. USDA Forest Service, Northeastern Forest Experiment Station, Upper Darby, Pennsylvania.

Beasley, R.S. (1976) Contribution of subsurface flow from the upper slopes of forested watersheds to channel flow. *Soil Science Society of America Journal* 40, 955–957.

Betson, R.P. (1964) What is watershed runoff? *Journal of Geophysical Research* 69, 1541–1552.

Beven, K. and Germann, P. (1982) Macropores and water flow in soils. *Water Resources Research* 18, 1311–1325.

Beven, K.J. and Kirkby, M.J. (1979) A physically-based, variable contributing area model of basin hydrology. *Hydrological Sciences Bulletin* 24, 43–69.

Bishop, K., Seibert, J., Nyberg, L. and Rodhe, A. (2011) Water storage in a till catchment II: implications of transmissivity feedback for flow paths and turnover times. *Hydrological Processes* 2, 3950–3959.
Bonell, M. (1993) Progress in the understanding of runoff generation in forests. *Journal of Hydrology* 150, 217–275.
Brown, V.A., McDonnell, J.J., Burns, D.A. and Kendall, C. (1999) The role of event water, a rapid shallow flow path component, and catchment size in summer stormflow. *Journal of Hydrology* 217, 179–190.
Bruijnzeel, L.A. and Wiersum, K.F. (1987) Rainfall interception by a young *Acacia auriculiformis* (A. Cunn.) plantation forest in West Java, Indonesia: application of Gash's analytical model. *Hydrological Processes* 1, 309–319.
Bruijnzeel, L.A., Mulligan, M. and Scatena, F.N. (2011) Hydrometeorolgy of tropical montane cloud forests: emerging patterns. *Hydrological Processes* 25, 465–498.
Bundt, M., Widner, F., Pesaro, W., Zeyer, J. and Blaser, P. (2001) Preferential flow path: biological hot spots in soils. *Soil Biology and Biochemistry* 33, 729–738.
Buttle, J.M. and McDonald, D.J. (2002) Coupled vertical and lateral preferential flow on a forested slope. *Water Resources Research* 38, 1060, doi: 10.1029/2001wr000773 (accessed 19 March 2016).
Burt, T.P., Pinay, G., Matheson, F.E., Haycock, N.E., Butturini, A., Clement, J.C., Danielescu, S., Dowrick, D.J., Hefting, M.M., Hillbricht-Ilkowska, A. and Maitre, V. (2002) Water table fluctuations in the riparian zone: comparative results from a pan-European experiment. *Journal of Hydrology* 265, 129–148.
Cappus, P. (1960) Bassin experimental d'Alrance – Étude des lois de l'écoulement –Application au calcul et à la prévision des débits. *La Houille Blanche A*, 493–520.
Crespo, P., Bücker, A., Feyen, J., Vaché, K.B., Frede, H.G. and Beuer, L. (2012) Preliminary evaluation of the runoff processes in a remote cloud forest basin using mixing model analysis and mean transit time. *Hydrological Processes* 26, 3896–3910.
Crockford, R.H. and Richardson, D.P. (1990) Partitioning of rainfall in a eucalypt forest and pine plantation in southeastern Australia: II. Stemflow and factors affecting stemflow in a dry sclerophyll eucalypt forest and a *Pinus radiata* plantation. *Hydrological Processes* 4, 145–155.
Crockford, R.H. and Richardson, D.P. (2000) Partitioning of rainfall into throughfall, stemflow, and interception: effect of forest type, ground cover, and climate. *Hydrological Processes* 14, 2903–2920.
Darboux, F., Davy, P.H., Gascuel-Odoux, C. and Haung C. (2002) Evolution of soil surface roughness and flowpath connectivity in overland flow experiments. *Catena* 46, 135–139.
DeBano, L.F. (2000) The role of fire and soil heating on water repellency in wildland environments: a review. *Journal of Hydrology* 231–232, 195–206.
Detty, J.M. and McGuire, K.J. (2010a) Threshold changes in storm runoff generation at a till-mantled headwater catchment. *Water Resources Research* 46, W07525, doi: 10.1029/2009WR008102 (accessed 19 March 2016).
Detty, J.M. and McGuire, K.J. (2010b) Topographic controls on shallow groundwater dynamics: implications of hydrologic connectivity between hillslopes and riparian zones in a till-mantled catchment. *Hydrological Processes* 24, 2222–2236.
Dunne, T. and Black, R.D. (1970a) An experimental investigation of runoff production in permeable soils. *Water Resources Research* 6, 478–490.
Dunne, T. and Black, R.D. (1970b) Partial area contributions to storm runoff in a small New-England watershed. *Water Resources Research* 6, 1296–1308.
Dunne, T., Moore, T.R. and Taylor, C.H. (1975) Recognition and prediction of runoff-producing zones in humid regions. *Hydrological Sciences Bulletin* 20, 305–327.
Engelbrecht, H.H. and Brancato, G.N. (1959) World record one-minute rainfall at Unionville Maryland. *Monthly Weather Review* 87, 303–306.
Epps, T.H., Hitchcock, D.R., Jayakaran, A.D., Loftis, D.R., Williams, T.M. and Amatya, D.M. (2013) Characterization of storm flow dynamics of headwater streams in the South Carolina lower coastal plain. *Journal of the American Water Resources Association* 49, 76–89.
Ford, D.C. and Williams, P.W. (2007) *Karst Hydrogeology and Geomorphology*. Wiley, Chichester, UK.
Frankenberger, J.R., Brooks, E.S., Walter, M.T., Walter, M.F. and Steenhuis, T.S. (1999) A GIS-based variable source area model. *Hydrological Processes* 13, 804–822.
Fujimoto, M., Ohte, N. and Tani, M. (2008) Effects of hillslope topography on hydrological responses in a weathered granite mountain, Japan: comparison of the runoff response between the valley-head and the side slope. *Hydrological Processes* 22, 2581–2594.
Gabrielli, C.P., McDonnell, J.J. and Jarvis, W.T. (2012) The role of bedrock groundwater in rainfall–runoff response at hillside and catchment scales. *Journal of Hydrology* 450–451, 117–133.

Gash, J.H.C. (1979) An analytical model of rainfall interception by forests. *Quarterly Journal of the Royal Meteorological Society* 105, 43–55.
Gash, J.H.C., Lloyd, C.R. and Lachaud, G. (1995) Estimating sparse forest rainfall interception with an analytical model. *Journal of Hydrology* 48, 89–105.
Goller, R., Wilcke, W., Leng, M.J., Tobscall, H.J., Wagner, K., Valarezo, C. and Zech, W. (2005) Tracing water paths through small catchments under a tropical montane rain forest in South Ecuador by an oxygen isotope approach. *Journal of Hydrology* 308, 67–80.
Graham, C.B., Woods, R.A. and McDonnell, J.J. (2010) Threshold response to rainfall: a field-based forensic approach. *Journal of Hydrology* 393, 65–76.
Green, W.H. and Ampt, C.A. (1911) Studies on soil physics, 1. Flow of air and water through soils. *Journal of Agricultural Science* 4, 1–21.
Hewlett, J.D. (1982) *Principles of Forest Hydrology*. University of Georgia Press, Athens, GA.
Hewlett, J.D. and Hibbert, A.R. (1963) Moisture and energy conditions within a sloping soil mass during drainage. *Journal of Geophysical Research* 68, 1081–1087.
Hewlett, J.D. and Hibbert, A.R. (1967) Factors affecting the response of small watersheds to precipitation in humid areas. In: Sopper, W.E. and Lull, H.W. (eds) *Forest Hydrology*. Pergamon Press, New York, pp. 275–291.
Horton, R.E. (1933) The role of infiltration in the hydrologic cycle. *Transactions of the American Geophysical Union* 14, 446–460.
Horton, R.E. (1940) An approach toward a physical interpretation of infiltration capacity. *Soil Science Society of America Journal* 5, 399–417.
Hursh, C.R. and Brater, E.F. (1941) Separating stream hydrographs from small drainage areas into surface and subsurface flow. *Transactions of the American Geophysical Union* 22, 863–871.
Jayatilaka, C.J. and Gillham, R.W. (1996) A deterministic-empirical model of the effect of the capillary fringe on near-stream area runoff 1. Description of the model. *Journal of Hydrology* 184, 299–315.
Jencso, K.G. and McGlynn, B.L. (2011) Hierarchical controls on runoff generation: topographically driven hydrologic connectivity, geology, and vegetation. *Water Resources Research* 47, W11527, doi: 10.1029/2011WR010666 (accessed 19 March 2016).
Jiang, G., Guo, F., Wu, J., Li, H. and Sun, H. (2008) The threshold value of epikarst runoff in forest karst mountain area. *Environmental Geology* 55, 87–93.
Jones, J.A.A. (1994) Soil piping and its hydrogeomorphic function. *Cuaternario y Geomorfologıa* 8(3–4), 77–102.
Jones, J.A.A. (2010) Soil piping and catchment response. *Hydrological Processes* 24, 1548–1566.
Kaçaroğlu, F. (1999) Review of groundwater pollution and protection in karst areas. *Water, Air, and Soil Pollution* 113, 337–356.
Katsura, S., Kosugi, K., Yamakawa, Y. and Mizuyama, T. (2014) Field evidence of groundwater ridging in a slope of a granite watershed without the capillary fringe effect. *Journal of Hydrology* 511, 703–718.
Kendall, K.A., Shanley, J.B. and McDonnell, J.J. (1999) A hydrometric and geochemical approach to test the transmissivity feedback hypothesis during snowmelt. *Journal of Hydrology* 219, 188–205.
Klaus, J. and McDonnell, J.J. (2013) Hydrograph separation using stable isotopes: review and evaluation. *Journal of Hydrology* 505, 47–64.
Lang, A.J., Aust, W.M., Bolding, M.C, Barrett, S.M., McGuire, K.J. and Lakel III, W.A. (2015) Streamside management zones compromised by stream crossings, legacy gullies, and over-harvest in the Piedmont. *Journal of the American Water Resources Association* 51, 1153–1164.
Langkamp, P.J., Farnell, G.K. and Dalling, M.J. (1982) Nutrient cycling in a stand of *Acacia holoserica* (A. Cunn. ex G. Don). I. Measurement of precipitation, interception, seasonal acetylene reduction, plant growth and nitrogen requirement. *Australian Journal of Botany* 30, 87–106.
Levia, D.F. and Frost, E.E. (2003) A review and evaluation of stemflow literature in the hydrologic and biogeochemical cycles of forested and agricultural ecosystems. *Journal of Hydrology* 274, 1–29.
Lloyd, C.R. and Marques, A.d.O. (1988) Spatial variability of throughfall and stemflow measurements in Amazonian rainforest. *Agricultural and Forest Meteorology* 42, 63–73.
Luxmoore, R.J. (1981) Micro-, meso-, and macro-porosity of soil. *Soil Science Society of America Journal* 45, 671–672.
Luxmoore, R.J., Jardine, P.M., Wilson, G.V., Jones, J.R. and Zelazny, L.W. (1990) Physical and chemical controls of preferred path flow through a forested hillslope. *Geoderma* 46, 139–154.
Lyon, S.W., Walter, M.T., Gerard-Marchant, P. and Steenhuis, T.S. (2004) Using a topographic index to distribute variable source area runoff predicted with the SCS curve-number equation. *Hydrological Processes* 18, 2757–2771.

McDonnell, J.J. (1990) A rationale for old water discharge through macropores in a steep, humid catchment. *Water Resources Research* 26, 2821–2832.
McDonnell, J.J. (2003) Where does water go when it rains? Moving beyond the variable source area concept of rainfall–runoff response. *Hydrological Processes* 17, 1869–1875.
McDonnell, J.J. (2013) Are all runoff processes the same? *Hydrological Processes* 27, 4103–4111.
McDonnell, J.J. and Buttle, J.M. (1998) Comment on 'A deterministic-empirical model of the effect of the capillary-fringe on near-stream area runoff. 1. Description of the model' in Journal of Hydrology, 184, 299–315. *Journal of Hydrology* 207, 280–285.
McDonnell, J.J., Owens, I.F. and Stewart, M.K. (1991) A case study of shallow flow paths in a steep zero-order basin. *Journal of the American Water Resources Association* 27, 679–685.
McGlynn, B.L., McDonnell, J.J. and Brammer, D.D. (2002) A review of the evolving perceptual model of hillslope flowpaths at the Maimai catchments, New Zealand. *Journal of Hydrology* 257, 1–26.
McGuire, K.J. and McDonnell, J.J. (2010) Hydrological connectivity of hillslopes and streams: characteristic time scales and nonlinearities. *Water Resources Research* 46, W10543, doi: 10.1029/2010WR009341 (accessed 19 March 2016).
Montieth, S.S., Buttle, J.M., Hazlett, P.W., Beall, F.D., Senkin, R.G. and Jefferies, D.S. (2006) Paired basin hydrologic response in harvested and undisturbed hardwood forests during snowmelt in central Ontario: streamflow, groundwater and flowpath behavior. *Hydrological Processes* 20, 1095–1116.
Mosley, M.P. (1979) Streamflow generation in a forested watershed, New Zealand. *Water Resources Research* 15, 795–806.
Negishi, J.N., Nogushi, S., Sidle, R.C., Ziegler, A.D. and Nik, A R. (2007) Stormflow generation involving pipe flow in a zero-order basin of Peninsular Malaysia. *Hydrological Processes* 21, 789–806.
Pearce, A.J., Stewart, M.K. and Sklash, M.G. (1986) Storm runoff generation in humid headwater catchments. I. Where does the water come from? *Water Resources Research* 22, 1263–1272.
Phillips, R.W., Spence, C. and Pomeroy, J.W. (2011) Connectivity and runoff dynamics in heterogeneous basins. *Hydrological Processes* 25, 3061–3975.
Pressland, A.J. (1973) Rainfall partitioning by an arid woodland (*Acacia anuera* F. Muell.) in south-western Queensland. *Australian Journal of Botany* 21, 235–245.
Puigdefabregas, J., del Barrio, G., Boer, M.M., Gutiérrez, L. and Solé, A. (1998) Differential response of hillslope and channel elements to rainfall events in a semi-arid area. *Geomorphology* 23, 337–351.
Rab, M.A. (1994) Changes in physical properties of soil with logging of *Eucalyptus regnans* forest in south-eastern Australia. *Forest Ecology and Management* 70, 215–229.
Richards, L.A. (1931) Capillary conduction of liquids in porous mediums. *Physics* 1, 318–333.
Rivenbark, B.L. and Jackson, C.R. (2004) Concentrated flow breakthroughs moving through silvicultural streamside management zones: southeastern Piedmont, USA. *Journal of the American Water Resources Association* 40, 1043–1052.
Rutter, A.J., Kershaw, K.A., Robins, P.C. and Morton, A.J. (1971) A predictive model of rainfall interception in forests. I. Derivation of the model from observations in a plantation of Corsican pine. *Agricultural Meteorology* 9, 367–384.
Scatena, F.N. (1990) Watershed scale rainfall interception on two forested watersheds in the Luquillo Mountains of Puerto Rico. *Journal of Hydrology* 113, 89–102.
Seibert, J., Bishop, K., Nyberg, L. and Rodhe, A. (2011) Water storage in a till catchment. I: Distributed modeling and relationship to runoff. *Hydrological Processes* 25, 3937–3949.
Sidle, R.C., Noguchi, S., Tsuboyama, Y. and Laursen, K. (2001) A conceptual model of preferential flow systems in forested hillslopes: evidence of self-organization. *Hydrological Processes* 15, 1675–1692.
Skaggs, R.W., Amatya, D.M., Chescheir, G.M. and Blanton, C.D. (2006) *Soil Property Changes during Loblolly Pine Production*. ASABE Paper No. 068206. American Society of Agricultural and Biological Engineers, St Joseph, Michigan.
Skaggs, R.W., Chescheir, G.M., Fernandez, G.P., Amatya, D.M. and Diggs, J. (2011) Effects of land use on soil properties and hydrology of drained coastal plain watersheds. *Transactions of the ASABE* 54, 1357–1365.
Sklash, M.G. and Farvolden, R.N. (1979) The role of groundwater in storm runoff. *Journal of Hydrology* 43, 45–65.
Spencer, C. and Woo, M. (2003) Hydrology of the sub-arctic Canadian shield: soil filled valleys. *Journal of Hydrology* 279, 151–166.

Terajima, T. and Moriizumi, M. (2013) Temporal and spatial changes in dissolved organic carbon concentration and fluorescence intensity of fulvic acid like materials in mountainous headwater catchments. *Journal of Hydrology* 479, 1–12.
Torres, R., Dietrich, W.E., Montgomery, D.R., Anderson, S.P. and Loague, K. (1998) Unsaturated zone response of a steep, unchanneled catchment. *Water Resources Research* 34, 1865–1878.
Tromp-van Meerveld, H.J. and McDonnell, J.J. (2006a) Threshold relations in subsurface stormflow: 1. A 147-storm analysis of the Panola hillslope. *Water Resources Research* 42, W02410, doi: 10.1029/2004wr003778 (accessed 19 March 2016).
Tromp-van Meerveld, H.J. and McDonnell, J.J. (2006b) Threshold relations in subsurface stormflow: 2. The fill and spill hypothesis. *Water Resources Research* 42, W02411, doi: 10.1029/2004WR003800 (accessed 19 March 2016).
Tsukamoto, Y. (1961) An experiment on sub-surface flow. *Journal of the Japanese Forestry Society* 43, 62–67 (in Japanese with an English abstract).
Uchida, T., Kosugi, K. and Mizuyana, T. (2001) Effects of pipeflow on hydrological process and its relation to landslide: a review of pipeflow studies in forested headwater catchments. *Hydrological Processes* 15, 2151–2174.
Uchida, T., Tromp-van Meerveld, I. and McDonnell, J.J. (2005) The role of lateral pipe flow in hillslope runoff response: an intercomparison of nonlinear hillslope response. *Journal of Hydrology* 311, 117–133.
Walter, M.T., Steenhuis, T.S., Mehta, V.K., Thongs, D., Zion, M. and Schneiderman, E. (2002) Refined conceptualization of TOPMODEL for shallow subsurface flows. *Hydrological Processes* 16, 2041–2046.
Weiler, M. and McDonnell, J.J. (2007) Conceptualizing lateral preferential flow and flow networks and simulating the effects on gauged and ungauged hillslopes. *Water Resources Research* 43, W03403, doi: 10.1029/2006WR004867 (accessed 19 March 2016).
Wickel, A.J., van der Giesen, N.C. and Sá, T.D.d.A. (2008) Stormflow generation in two headwater catchments in eastern Amazonia, Brazil. *Hydrological Processes* 22, 3285–3293.
Williams, P.W. (2008) The role of the epikarst in karst and cave hydrogeology: a review. *International Journal of Speleology* 37, 1–10.
Wilson, G.V. and Luxmoore, R.J. (1988) Infiltration, macroporosity, and mesoporosity distributions on two forested watersheds. *Soil Science Society of America Journal* 52, 329–335.

3 Forest Evapotranspiration: Measurement and Modelling at Multiple Scales

G. Sun[1]*, J.-C. Domec[2,3] and D.M. Amatya[4]

[1]USDA Forest Service, Raleigh, North Carolina, USA; [2]North Carolina State University, Raleigh, North Carolina, USA; [3]Bordeaux Sciences AGRO, UMR1391/ISPA/INRA, Gradignan, France; [4]USDA Forest Service, Cordesville, South Carolina, USA

3.1 Introduction

Compared with traditional engineering hydrology, forest hydrology has a relatively long history of studying the effects of vegetation in regulating streamflow through evapotranspiration (Hewlett, 1982; Swank and Crossley, 1988; Andreassian, 2004; Brown *et al.*, 2005; Amatya *et al.*, 2011, 2015, 2016; Sun *et al.*, 2011b; Vose *et al.*, 2011). It is estimated that more than half of the solar energy absorbed by land surfaces is used to evaporate water (Trenberth *et al.*, 2009). Evapotranspiration (ET), the sum of evaporation from soil (E), canopy and litter interception (I), and plant surface and plant transpiration (T), is critical to understanding the energy, water and biogeochemical cycles in forests (Baldocchi *et al.*, 2001; Levia *et al.*, 2011).

The linkage among energy, water and carbon balances at a forest-stand level over a long time period (Fig. 3.1), in which ET plays a key role, can be described conceptually in the following interlinked formulae (Sun *et al.*, 2010, 2011a).

Water balance:

$$P = ET + Q \tag{3.1}$$

Energy balance:

$$R_n = L_E + H = ET \times L + H \tag{3.2}$$

Carbon balance:

$$\text{NEP} = \text{GPP} - R_e - L_C = \text{ET} \times \text{WUE} - R_e - L_C \tag{3.3}$$

In the above, P is precipitation (mm), Q is runoff (mm), R_n is net radiation (W/m^2), L_E is latent heat (W/m^2) that represents the energy used to evaporate the amount of water by ET assuming a constant conversion factor called the latent heat of vaporization of water (L = 539 cal/g H_2O = 2256 kJ/kg H_2O), H is sensible heat that is consumed to heat the air near the forest canopy. The net ecosystem productivity (NEP; g C/m^2) is the carbon balance between carbon gain by gross ecosystem productivity (i.e. plant photosynthesis) and carbon loss by ecosystem respiration (R_e; g C/m^2) and lateral export in stream runoff (L_C; g C/m^2). The magnitude of both gross primer productivity (GPP; g C/m^2) and R_e is much larger than that of NEP and L_C, and all four variables are influenced by soil moisture and the hydrology. In many cases, ET explains the majority of the seasonal variability of GPP for all ecosystems

*Corresponding author; e-mail: gesun@fs.fed.us

 Forest Hydrology: Processes, Management and Assessment (eds D.M. Amatya, T.M. Williams, L. Bren and C. de Jong)

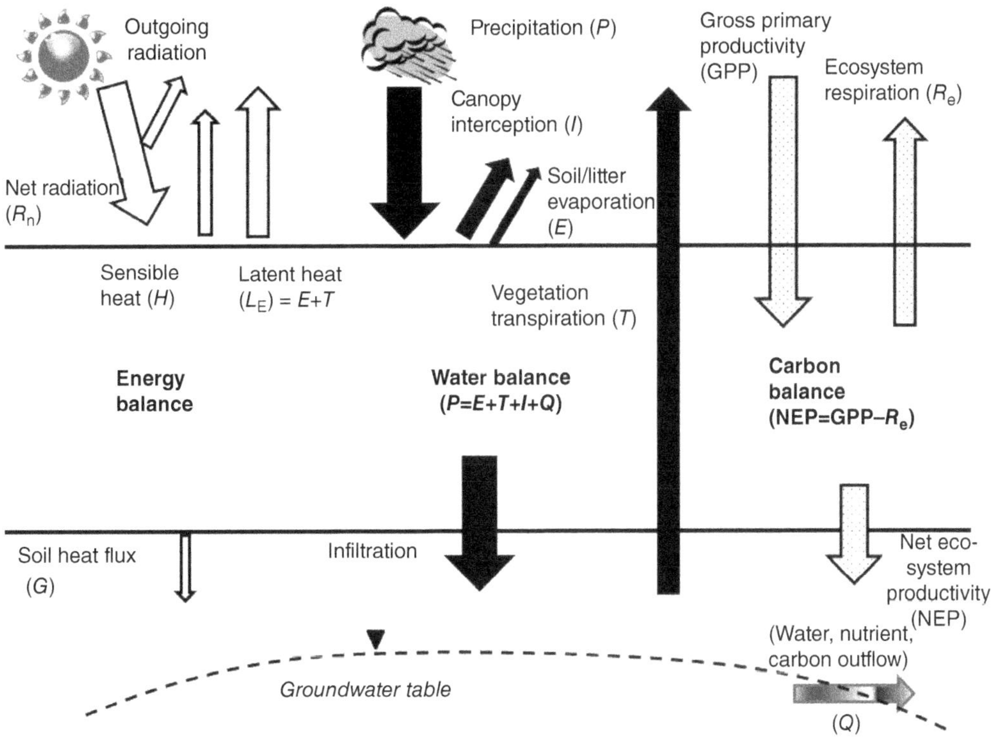

Fig. 3.1. Linkages among energy, water and carbon cycles in a forest ecosystem on the lower coastal plain of North Carolina in the USA. Note that net radiation (R_n) is a result of total incoming minus reflected shortwave radiation, along with the absorbed minus emitted longwave radiation.

(Law *et al.*, 2002; Sun *et al.*, 2011b). The ratio GPP/ET is termed water-use efficiency (WUE) and has been used as an important variable to understand the linkages of water–carbon coupling (Law *et al.*, 2002; Gao *et al.*, 2014; Frank *et al.*, 2015).

3.1.1 Understanding ecosystem processes

ET is a key variable linking meteorology, hydrology and ecosystem sciences (Baldocchi *et al.*, 2000; Oishi *et al.*, 2010; Sun *et al.*, 2011b). Plant transpiration T is a key variable directly coupled with ecosystem productivity (Rosenzweig, 1968) and carbon sequestration (Aber and Federer, 1992). This is easy to understand by the simple fact that CO_2 intake during plant photosynthesis uses the same pores, stomata, as the water loss, transpiration, uses (Canny, 1998). However, although E and T are both driven by atmospheric demand, T is actively controlled by stomatal regulation. ET is the only variable that links hydrology and biological processes in many ecosystem models (Aber and Federer, 1992). ET is also highly linked to ecosystem productivity and net ecosystem exchange of CO_2 because both photosynthesis and ecosystem respiration are controlled by soil water availability (Law *et al.*, 2002; Jackson *et al.*, 2005; Huang *et al.*, 2015).

3.1.2 Constructing water balances

ET is a large component of the water budget. Worldwide, mean annual ET rates are estimated to be about 600 mm (Jung *et al.*, 2010; Zeng *et al.*, 2014), or 60–70% of precipitation (Oki and Kanae, 2006; Teuling *et al.*, 2009). In the USA, more than 70% of the annual precipitation returns to the atmosphere as ET (Sanford and Selnick, 2013). Annual forest ET can exceed precipitation in the humid southern USA (Sun *et al.*, 2002, 2010) in dry years and it is not uncommon that

ET exceeds precipitation during the growing season in forests. Vegetation affects watershed hydrology and water balances through ET (Zhang *et al.*, 2001; Oudin *et al.*, 2008; Ukkola and Prentice, 2013; Jayakaran *et al.*, 2014). Land-use conversion (i.e. bioenergy crop expansion) can dramatically change plant cover and biomass, affecting transpiration and evaporation rates, and therefore site water balances (King *et al.*, 2013; Albaugh *et al.*, 2014; Amatya *et al.*, 2015; Christopher *et al.*, 2015), including streamflow quantity (Ford *et al.*, 2007; Palmroth *et al.*, 2010; Amatya *et al.*, 2015) and quality such as total sediment loading (Boggs *et al.*, 2015).

3.1.3 Understanding climate change, variability and feedbacks

The ET processes are closely linked to energy partitioning, water balances and climate systems (Betts, 2000; Bonan, 2008). ET is tightly coupled to land-surface energy balance and thus influences vegetation–climate feedbacks (Bonan, 2008; Cheng *et al.*, 2011). Changes in ET directly affect runoff, soil water storage, and local precipitation and temperature at the regional scale (Liu, 2011). The cooling or warming effects of reforestation are due to the increase in ET by planted trees or altered surface albedo (Peng *et al.*, 2014). ET may be considered an 'air conditioner'.

Global climate change, in turn, directly affects the local water resources through ET (Sun *et al.*, 2000, 2008). An increase in air temperature generally means an increase in vapour pressure deficit and evaporative demand or potential ET, resulting in an increase in water loss by ET, and thus a decrease in groundwater recharge and soil water availability to ecosystems and human water supply. Regions that are experiencing more warming would see more severe hydrological droughts regardless of changes in precipitation (Mann and Gleick, 2015).

3.1.4 Modelling regional ecosystem biodiversity

ET has long been regarded as an index to represent the available environmental energies and ecosystem productivity by bioclimatologists. Thus, ET has been used to explain the large regional variations in plant and animal species' richness and biodiversity. For example, the variability in species richness in vertebrate classes could be statistically explained by a monotonically increasing function of a single variable, potential evapotranspiration (PET) (Currie, 1991). In contrast, regional tree richness was more closely related to actual ET (Currie, 1991; Hawkins *et al.*, 2003).

3.2 Evapotranspiration Processes

Forest ET processes are inherently complex due to the many ecohydrological interactions within a forest ecosystem that often consists of multiple plant species with heterogeneous spatial distribution and variable microclimate over space and time (Canny, 1998). Both the physiological (e.g. stomata control) and physical processes (e.g. water potential control) influence the water vapour movements from plant organs of roots, xylem and leaf, to stands and landscapes (i.e. watersheds). Since soil evaporation can be minor in closed-canopy forests (McCarthy *et al.*, 1992; Domec *et al.*, 2012b), this chapter focuses on the processes that control canopy and litter interception (I) and transpiration (T), and methods to quantify these two major components of ET.

3.2.1 Canopy and litter interception

The quantity of canopy and litter interception (I) in forests can be a large component of the ET and water balances, depending on forest structure characteristics such as leaf area index (LAI) and canopy holding capacity, and the amount of litter and litter water-holding capacity, respectively (Gash, 1979; Deguchi *et al.*, 2006). In addition, the frequency of storms and the drying and wetting cycles affect total canopy and litter interception. Although interception can be 20–50% of the precipitation, most hydrological models do not simulate this process explicitly (Gerrits *et al.*, 2007).

The earliest studies by Horton (1919) showed highly variable interception rates between and across species, with the spruce–fir–hemlock forest type the highest, followed by pines and then hardwoods. Helvey (1974) reported annual canopy interception as

17% for red pine (*Pinus resinosa* Ait.), 16% for ponderosa pine (*Pinus ponderosa* Dougl. ex. Laws.), 19% for eastern white pine and 28% for the spruce–fir–hemlock forest type. The difference in canopy interception rates between hardwood and conifer forests partially explained the observed difference in streamflow (Swank and Miner, 1968). Summer interception rates of deciduous forests in the south-eastern USA ranged from 8 to 33%, with a mean of 17%, and winter rates ranged from 5 to 22%, with a mean of 12% (Helvey and Patric, 1965). Annual canopy interception rate was 18% for wetland sites, 20% for hardwood sites and a longleaf pine (*Pinus palustus* Mill.) plantation and 23% for pine-dominated forests in the south-eastern USA (Bryant *et al.*, 2005). Thinning of a loblolly pine (*Pinus taeda* L.) plantation forest reduces basal area and subsequent leaf area, resulting in a decrease in canopy interception (McCarthy *et al.*, 1992). Interception rates vary between 10–35% and 5–25% for un-thinned versus thinned loblolly pine stands, respectively (Gavazzi *et al.*, 2015). Forests in tropical and subtropical regions could intercept 6 to 42% of precipitation (Bryant *et al.*, 2005). In the USA, reported annual values of litter precipitation interception rate for eastern forests vary by about 2–5%, generally less than 50 mm per year (Helvey and Patric, 1965). However, litter interception may be higher than canopy interception in other forest ecosystems (Gerrits *et al.*, 2007).

3.2.2 Transpiration

The transpiration process (T) represents water loss through leaf stomata, the tiny openings found on one side or both sides of the tree leaves (Canny, 1998). Because T is an inevitable consequence of CO_2 assimilation by plants through photosynthesis, maintaining of leaf tissue turgidity and plant nutrient uptake, together with soil evaporation, T represents an ecosystem water loss and thus is a 'necessary evil' for net ecosystem productivity.

A global synthesis study indicates that T accounts for 61 ± 15% of total ET and returns approximately 39 ± 10% of incident precipitation to the atmosphere, playing a great role in the global water cycle (Schlesinger and Jasechko, 2014). The T/ET ratios are highest in tropical rainforests (70 ± 14%) and lowest in steppes, shrublands and deserts (51 ± 15%). Transpiration is the major component of the total evapotranspiration in global hydrological cycles and ET is highly dependent upon biophysical parameters like stomatal conductance (Jasechko *et al.*, 2013). Therefore, changes in transpiration due to increasing CO_2 concentrations, land-use changes, shifting ecozones, air pollution and climate warming may have significant impacts on water resources (Schlesinger and Jasechko, 2014). An increase in CO_2 concentrations may reduce plant leaf stomata conductance and increase WUE, but T can arise from increased leaf area in addition to lengthened growing seasons and enhanced evaporative demand in a warming climate with increased CO_2 concentration (Frank *et al.*, 2015).

Carbon and water fluxes are coupled through the stomata activities: water vapour exits the stomata along with oxygen; carbon dioxide flows into the stomata and is absorbed by the photosynthesis process to produce carbohydrate (Crétaz and Barten, 2007). Transpiration is an active water translocation process that occurs only when water exists continuously along the soil–root–stem–branch–leaf–stomata flow pathway (Kumagai, 2011). However, transpiration rates differ tremendously among different tree species and ages (Plate 2). For example, a *Quercus rubra* tree with a 50 cm trunk diameter transpires an average of 30 kg H_2O/day, but *Betula lenta* can transpire as high as 110 kg H_2O/day under the same climate in the southern Appalachians in the south-eastern USA (Vose *et al.*, 2011). A review of 52 whole-tree water use studies for 67 tree species worldwide using different techniques concluded that maximum daily water use rates for trees averaging 21 m in height were within 10–200 kg/day (Wullschleger *et al.*, 1998).

The transpiration rates are controlled by numerous biophysical factors such as microclimatic characteristics, atmospheric CO_2 concentration, soil water potential, stand characteristics (e.g. leaf area, species compositions, tree density) and hydraulic transport properties of plant tissues (Domec *et al.*, 2009, 2010, 2012a). The species compositions of forests change over space and time due to natural regeneration or in response to climatic change and/or human

Table 3.1. A comparison of major methods for estimating evapotranspiration (ET) at multiple scales.

	Method	Strength	Weakness	Source
Direct field-based	Porometer and cuvette	Leaf-level physiological process	Difficult to scale up due to uncertainty on the influence of boundary layers and variability of leaf age, radiation and humidity	Olbrich (1991)
	Weighing lysimeter	Single whole-tree water use	High cost	
	Heat balance/heat dissipation sapflow	Allows routine unsupervised measurement accurately at single plant scale	Large-scale measurement errors are determined by sample size and the variability of samples	Granier (1987)
	Eddy covariance	Measuring fluxes continuously, offering data with high temporal resolution	High cost in instrumentation, gap filling required, energy imbalance problems	Baldocchi *et al.* (2001)
	Bowen ratio	Works for both crops and natural vegetation	Relies on several assumptions, errors associated with low gradients	Irmak *et al.* (2014)
	Catchment water balance	Easy to measure	Only long-term average is reliable	Ukkola and Prentice (2013)
Remote sensing	MODIS	Provides high-resolution spatial, continuous and temporal data	Uncertainties due to errors generated by measurement of sparse canopies, data mostly from clear-sky conditions	Kalma *et al.* (2008)
Mathematical modelling for ET alone or the full hydrological cycle	Theoretical models (e.g. Penman–Monteith equation)	Widely tested, including all conditions, low cost	Requires site-specific parameters, not easy to apply to data-poor regions	McMahon *et al.* (2013)
Isotopes	Stable isotope H and O	Process-based understanding of water source of ET; partitioning of evaporation and transpiration	Cost and scaling up to stand level	Good *et al.* (2015)

activities such as silviculture (i.e. reforestation, afforestation). In addition, forest ecosystem structure changes in both above-ground characteristics, including leaf (i.e. leaf biomass) and stem (i.e. sapwood area) (Domec *et al.*, 2012a; Komatsu and Kume, 2015), and below ground (i.e. root biomass) over time. Little is known about water pathways between soil water and roots and the water uptake mechanism of deep roots in response to drought (Meinzer *et al.*, 2004; Warren *et al.*, 2007).

Different from croplands, forests have multiple canopies and the understorey vegetation is an important component of a forest stand by intercepting and transpiring a significant amount of water. For example, over 20% of the total ET for a 17-year-old pine plantation was from understories (Domec *et al.*, 2012b). Emergent understorey vegetation soon after harvest in the humid coastal plain was shown to have a substantial LAI, potentially affecting water balance for 4–5 years until the planted pine seedlings dominated the understorey (Sampson *et al.*, 2011).

3.2.3 Hydraulic redistribution by roots: exchange of water at the soil–root interface

Plants can reduce water stress by extracting water from deeper and moist soil layers through plant roots and storing it in the upper, drier soil layers for use by shallow roots. The bidirectional (upward and downward) processes are termed 'hydraulic redistribution' (HR) (Burgess *et al.*, 1998). The HR process occurs widely in all water-limited vegetated environments (Meinzer *et al.*, 2004; Neumann and Cardon, 2012). HR is a passive process that depends on the soil suction head (soil water potential) and the root distribution within the soil column. HR by roots acts as a large water capacitor, increasing the efficiency of whole-plant water transport, buffering the seasonality of ET against water stress during seasonal water deficits, and representing 20–40% of whole-stand water use (Domec *et al.*, 2010). Even when HR represents only a relatively small amount of ecosystem water use (e.g. <0.5 mm/day) and just a fraction (e.g. 5–10%) of total ET during the dry period, the partial recharge of upper soil moisture by HR is important to slow down the decline of soil water content and thus maintain water availability in topsoil layers (Warren *et al.*, 2007). The influx of soil water maintains root water-uptake capacity and extends root functioning later into the drought period (Domec *et al.*, 2004), influencing forest productivity (Domec *et al.*, 2010).

3.2.4 Total evapotranspiration

The total ET rates at the ecosystem or watershed landscape level are controlled mainly by regional energy and water availability (Douglass, 1983; Zhang *et al.*, 2001), but also are influenced by other anthropogenic management factors such as site fertilization (CO_2 effects and N deposition) (Tian, H.Q., *et al.*, 2012; Frank *et al.*, 2015), tree genetic improvement, species conversion (Swank and Douglass, 1974), artificial drainage (Amatya *et al.*, 2000) and irrigation (Amatya *et al.*, 2011). During the course of the forest stand development, site-level energy and water availability also vary, resulting in dramatic seasonal changes in total ET and its partitioning into sensible heat and other energy balance variables (Sun *et al.*, 2010).

Forested watershed ET generally decreases soon after removal of the canopy by either harvesting or other natural disturbances (hurricanes, invasive species, fires, wind and snow storms, etc.) as a result of reduced canopy interception and transpiration (Sun *et al.*, 2010; Tian, S.Y., *et al.*, 2012; Jayakaran *et al.*, 2014; Boggs *et al.*, 2015). However, ET generally tends to increase soon after plantation (afforestation/reforestation) and after natural regeneration (Sun *et al.*, 2010; Jayakaran *et al.*, 2014). Figures 3.2 and 3.3 present an example of increase in annual ET after planting a harvested watershed (Amatya *et al.*, 2000; Amatya and Skaggs, 2001, 2011; Tian, S.Y., *et al.*, 2012) and after natural regeneration of a watershed (Jayakaran *et al.*, 2014) substantially impacted by hurricane force winds. The inter-annual variability of ET was a result of precipitation variability at both the sites, consistent with other studies (Sun *et al.*, 2002, 2010; Ukkola and Prentice, 2013).

Forest ET rates also vary dramatically across space and time on a heterogeneous terrain.

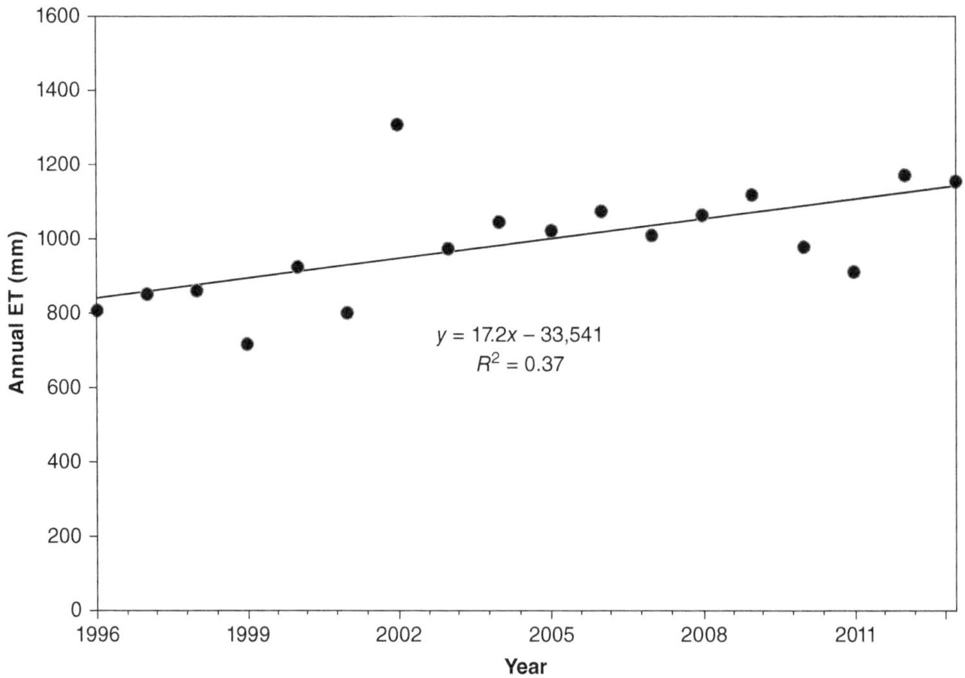

Fig. 3.2. Annual forest ecosystem evapotranspiration (ET) calculated as the differences between measured precipitation and measured streamflow for an experimental watershed. The ET rate increases gradually following tree/forest harvest in 1995 and replanting with loblolly pine in 1997 in Carteret County, coastal North Carolina, USA.

For example, ET rates of a forest stand are higher in the sunny side or/and near the ridges in a mountain watershed due to more solar radiation available (Douglass, 1983; Emanuel *et al.*, 2010). Forest thinning practices reduce forest biomass, thus canopy interception and transpiration from remaining trees (Boggs *et al.*, 2015), but do not necessarily reduce total ET (Sun *et al.*, 2015).

3.3 Direct Measurement of Evapotranspiration

Forest ET processes have been quantified at multiple temporal and spatial scales from leaf to watershed, and even to global scale, using various methods from the hand-held cuvette method to the remote sensing approach (Table 3.1). The porometer method has been used to understand the environmental control on gas (CO_2 and H_2O) exchange at the leaf level (Olbrich, 1991). Other methods to measure T include ventilated chambers (Denmead *et al.*, 1993), complex models parameterized by leaf-scale physiological traits and three-dimensional tree architecture (Kumagai *et al.*, 2014), or sap flux density based on thermal dissipation and heat transport theories (Granier *et al.*, 1996; Granier, 1987).

The sapflow technique has the advantage of not being limited by landform heterogeneity (Granier, 1987). The sapflow method measures water use by a single plant or tree, and thus answers questions on water use at the species and whole-stand levels. Components of forest water loss may be determined by measuring differences between total ET and tree sapflow, providing insights in terms of the response of water use by plants to climatic variability and stand development (Domec *et al.*, 2012a). Sapflow measurements provide a powerful tool for quantifying plant water use and physiological responses of plants to environmental conditions (Domec *et al.*, 2009).

In contrast, the eddy covariance technique measures forest ET by calculating the covariance

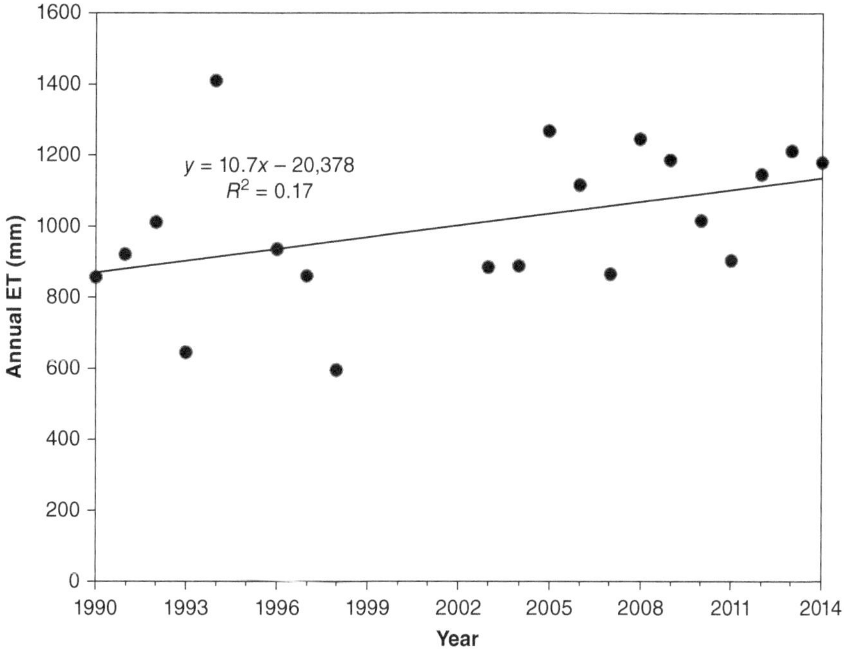

Fig. 3.3. Recovery of annual evapotranspiration (ET) calculated as the differences between measured precipitation and streamflow for a forested watershed that was naturally regenerated after the impact of Hurricane Hugo in 1989 at Santee Experimental Forest in coastal South Carolina, USA.

between fluctuations in vertical eddy velocity and the specific water vapour content above forest canopies (Baldocchi and Ryu, 2011). The method is designed to understand the gas exchange at the boundary layer between vegetation and the atmosphere, and answers questions at the landscape scale (the footprint of the flux tower) (Baldocchi *et al.*, 1988). The method relies on several assumptions such as an extensive fetch over a homogeneous surface.

Global participation in flux measurements through the FLUXNET (over 500 sites) (http://www.fluxnet.ornl.gov) since the 1990s has been a major driving force for advancing ET science (Baldocchi *et al.*, 2001).

The Bowen ratio methods have been used in quantifying ET in croplands under various soil (tillage), crop and irrigation management (sprinklers, subsurface drip, gravity irrigation, etc.) practices through the NEBFLUX project (Irmak, 2010) and have similar accuracy to the eddy flux methods (Irmak *et al.*, 2014). The method estimates ET from the ratio of sensible heat to latent heat, using air temperature and humidity gradients measured above the canopy, net radiation and soil heat flux. The fetch requirements for the Bowen ratio method are less than those for the eddy covariance method.

In addition to micrometeorological methods, stable isotopes have been used as tracers for identifying the sources of water uptake in ecosystems and evaluating quantitatively the relationships among water, energy and isotopic budgets. For example, tree-ring ^{13}C is used to identify changes in WUE and soil water stress (McNulty and Swank, 1995), and ^{18}O assists in determining whether those changes in WUE are due to changes in photosynthetic rate or stomatal conductance. Vegetation affects water/energy balance and isotopic budget through transpiration. Recently, using the D/H isotope ratios of continental runoff and evapotranspiration, independent of terrestrial hydrological partitioning, Good *et al.* (2015) demonstrated that globally the transpired fraction of evapotranspiration is estimated to be 56 to 74% (25th to 75th percentile), with a median of 65% and mean of 64%. Furthermore, studies across an ecosystem gradient in the USA and Mexico provided evidence of ecohydrological separation, whereby different subsurface compartmentalized pools of water supply either plant transpiration fluxes or the combined fluxes of groundwater and streamflow (Evaristo *et al.*, 2015).

Estimating regional ET using satellite remote sensing data has emerged since the 1980s when there was an increasing interest in spatial dynamics in water use at the landscape scale (Kalma *et al.*, 2008). Remote sensing ET products such as MODIS (Moderate Resolution Image Spectroradiometer) (Mu *et al.*, 2011) have provided spatially and temporally continuous ET estimates at a 1 km resolution for understanding regional hydrology and environmental controls. However, uncertainties in modelling effective surface emissivity and effective aerodynamic exchange resistance, and sparse canopies and cloud conditions may make the remote sensing methods less reliable (Shuttleworth, 2012). Coupling energy balance models with remotely sensed land-surface temperature retrieved from thermal infrared imagery provides proxy information regarding the surface moisture and vegetation growth status (Anderson *et al.*, 2012). Models such as the regional Atmosphere–Land Exchange Inverse (ALEXI) and the associated flux disaggregation model (DisALEXI) are based on the Two Source Energy Balance (TSEB) land-surface representations (Kustas and Norman, 1996). These modelling systems have recently been applied in a lower coastal plain in North Carolina and show promise to map high-resolution ET (e.g. daily, 30 m) for a landscape with mixed land uses with natural wetland forests, drained pine forest with multiple stand ages, and croplands (see also Chapter 9, this volume).

Long-term and annual watershed water balance ET are generally estimated using a simple water balance as the difference between measured precipitation and streamflow, assuming a negligible change in storage (Wilson *et al.*, 2001; Sun *et al.*, 2005; Amatya and Skaggs, 2011; Ukkola and Prentice, 2013). Watershed-scale ET is also dependent upon its land use or the area covered by vegetation (Amatya *et al.*, 2015) in addition to the broader controls of precipitation and potential ET. Using observed data from 109 river basins during 1961–1999, Ukkola and Prentice (2013) showed strong control by precipitation followed by vegetation processes on ET trends and variability.

A few studies comparing multiple ET methods found that each method has its own limitations (Wilson *et al.*, 2001; Ford *et al.*, 2007; Domec *et al.*, 2012b). The eddy covariance method measures fluxes continuously, offering time series data with high temporal resolution, but data availability is limited by costly site instrumentation, gap filling issues and extensive data corrections issues. In addition, the eddy covariance method may underestimate ET by as much as 30% due to a lack of energy balance closure (Wilson *et al.*, 2002). The eddy covariance technique has also been shown to be problematic to underestimate ET on wet days because the sonic anemometer and infrared gas analyser must be dry to function properly (Wilson *et al.*, 2001).

3.4 Indirect Estimates of Evapotranspiration

3.4.1 Methods based on potential evapotranspiration

Due to the high cost of trained personnel requirements for measuring ET directly at field and larger scales, mathematical modelling has been widely used to estimate ET (McMahon *et al.*, 2013). ET models can be roughly divided into two groups: biophysical (theoretical) and empirical models. The former type of models refers to those developed based on physical and physiological principles describing energy and water transport in the soil–plant–atmosphere continuum (SPAC). Many theoretical models have evolved from the famous Penman (1948) and later from the Penman–Monteith model (Monteith, 1965) that represents the most advanced process-based ET model. The Penman–Monteith model estimates ET as a function of available energy, vapour pressure deficit, air temperature and pressure, and aerodynamic and canopy resistance. In contrast, empirical ET models are models developed using empirical observed ET data, land cover type, biophysical variables of plant characteristics such as LAI, soil moisture and atmospheric conditions. Empirical ET models do not intend to describe the processes of vaporization, but can give a reasonable estimate with limited environmental information.

In practice, it is often rather difficult to parameterize the process-based ET models to estimate actual ET. To simplify calculations, the concept of potential ET (PET) was introduced in the 1940s. For any ecosystem, PET represents the potential maximum water loss when soil water is not limiting. Actual ET then can be scaled down from the hypothetical PET by limiting canopy conductance and soil moisture, and correlates to pan evaporation (Grismer *et al.*, 2002). Such PET models are often embedded in hydrological models that can simulate the dynamics of soil moisture, a major control on soil evaporation and transpiration (Sun *et al.*, 1998; Tian, S.Y., *et al.*, 2012). McMahon *et al.* (2013) provide a comprehensive review on conceptual PET models and the techniques to estimate actual ET from open-surface waters, landscapes, catchments, deep lakes, shallow lakes, farm dams, lakes covered with vegetation, irrigation areas and bare soils.

Existing PET models can be classified into five groups (Lu *et al.*, 2005): (i) water budget; (ii) mass transfer; (iii) combination; (iv) radiation; and (v) temperature-based. There are approximately 50 models available to estimate PET that are developed considering input data availability and regional climate characteristics. The models give inconsistent values due to their different assumptions and input data requirements, or because they were often developed for specific climatic regions.

Numerous studies have suggested that different PET methods may give significantly different results (Amatya *et al.*, 1995; Lu *et al.*, 2005; McMahon *et al.*, 2013), so the standardized grass-reference PET method (Allen *et al.*, 2005), ET_o, is recommended to achieve comparable results across sites. Details of the computation procedures for ET_o are found in Allen *et al.* (1994). A computer program is available for public use (http://www.agr.kuleuven.ac.be/lbh/lsw/iupware/downloads/elearning/software/EtoCalculator.pdf). Once ET_o is calculated, actual ET for a particular ecosystem type can be estimated by simply multiplying by a 'crop coefficient, K_c' developed for that crop using ET measured by lysimeter or some other method (Allen *et al.*, 2005; Irmak, 2010). The K_c method works well in irrigation agriculture for various croplands that have uniform phenology. However, for forests, this method can be problematic given the large variability of species composition of a forest, leaf biomass dynamics throughout the season, and the age and density effects on tree biomass and water transport properties (canopy conductance, sapwood area). In addition, the reference ET concept may be misleading, because actual forest ET rates in humid climates often exceed the ET_o (Sun *et al.*, 2010). A casual use of ET_o as the maximum ET in a hydrological model may result in underestimation of actual ET (Amatya and Harrison, 2016). A recent study suggests that K_c for any forest type may vary tremendously and latitude, precipitation and LAI are the best predictors of K_c (Liu *et al.*, 2015). Forests generally have higher K_c values than other ecosystem types (Fig. 3.4).

3.4.2 Empirical evapotranspiration models

Empirical ET models are derived from direct ET measurements at the ecosystem scale. Empirical models may be best used as a first-order approximation of mean climatic conditions. The following model was derived from field data collected at 13 sites using a variety of methods (Sun *et al.*, 2011a). The model estimates monthly ET as a function of LAI, ET_o (mm/month) and precipitation P (mm/month) (see equation 3.4 at the bottom of the page), where ET_o is the FAO (Food and Agriculture Organization) reference ET as discussed above.

Other forms of the ET model use Hamon's potential ET (*PET*) instead of the more data-demanding FAO reference ET method (Sun *et al.*, 2011b) (see equation 3.5 at the bottom of the page).

$$ET = 11.94 + 4.76 \times \text{LAI} + ET_o(0.032 \times \text{LAI} + 0.0026 \times P + 0.15) \quad (3.4)$$

$$ET = 0.174 \times P + 0.502 \times PET + 5.31 \times \text{LAI} + 0.0222 \times PET \times \text{LAI} \quad (3.5)$$

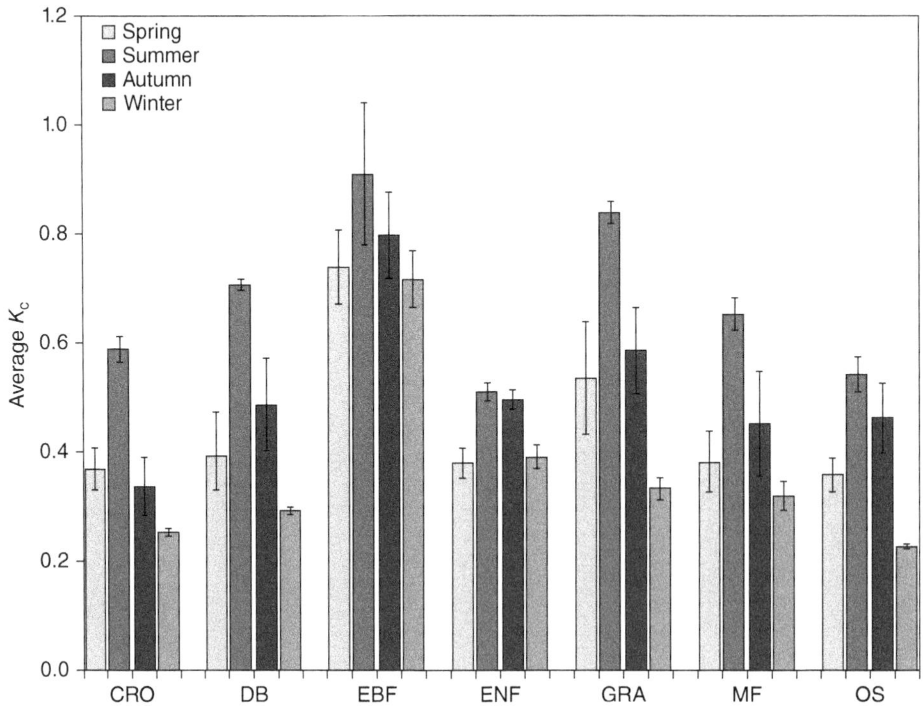

Fig. 3.4. A comparison of seasonal mean crop coefficient (K_c) calculated from global eddy flux measurements for cropland (CRO), deciduous broadleaf forest (DB), evergreen broadleaf forest (EBF), evergreen needle-leaf forest (ENF), grassland (GRA), mixed forest (MF) and open shrub land (OS). K_c is estimated as measured ET divided by the grass reference ET (ET_o) calculated by the standardized FAO-56 method; error bars represent standard deviation.

Using a similar concept and a 250 FLUXNET synthesis data set, Fang *et al.* (2015) developed the two monthly ET models (Eqns 3.6 and 3.7) that require different input variables (see equation 3.6 at the bottom of the page), where *PET* is monthly potential ET (mm) calculated by Hamon's method, VPD is vapour pressure deficit (hundreds of Pascals) that can be estimated from relative air humidity, R^2 is the coefficient of determination and RMSE is root-mean-squared error. Since R_n is rarely available at the regional scale, another model that uses more commonly available data was developed (see equation 3.7 at the bottom of the page).

A series of ecosystem-specific monthly scale ET models was also developed using the global eddy flux data (Fang *et al.*, 2015) (Table 3.2). An empirical annual ET model was developed by combining a water balance method with a climate and land cover regression equation to estimate mean annual ET across the conterminous USA (Sanford and Selnick, 2013). The climate variables included mean annual daily maximum and daily minimum air temperature and mean annual precipitation. The land cover types included developed, forest, shrubland, grassland, agriculture and marsh.

$$ET = 0.42 + 0.74 \times PET - 2.73 \times \mathrm{VPD} + 0.10 \times R_n \quad \left(R^2 = 0.73, \mathrm{RMSE} = 17.0\ \mathrm{mm/month}\right) \tag{3.6}$$

$$ET = -4.79 + 0.75 \times PET + 3.92 \times \mathrm{LAI} + 0.04 \times P \quad \left(R^2 = 0.68, \mathrm{RMSE} = 18.1\ \mathrm{mm/month}\right) \tag{3.7}$$

Table 3.2. Empirical models by land cover type developed using three commonly measured biophysical variables.

Land cover type	Model	RMSE	R^2	*n*
Shrubland	$ET = -3.11 + 0.39 \times PET + 0.09 \times P + 11.127 \times \text{LAI}$	12.5	0.80	193
Cropland	$ET = -8.15 + 0.86 \times PET + 0.01 \times P + 9.54 \times \text{LAI}$	20.9	0.70	653
Grassland	$ET = -1.36 + 0.70 \times PET + 0.04 \times P + 6.56 \times \text{LAI}$	16.8	0.66	803
Deciduous forest	$ET = -14.82 + 0.98 \times PET + 2.72 \times \text{LAI}$	23.7	0.74	754
Evergreen needle-leaf forest	$ET = 0.10 + 0.64 \times PET + 0.04 \times P + 3.53 \times \text{LAI}$	17.8	0.68	1382
Evergreen broadleaf forest	$ET = 7.71 + 0.74 \times PET + 1.85 \times \text{LAI}$	16.8	0.76	233
Mixed forest	$ET = -8.763 + 0.95 \times PET$	13.1	0.79	259
Savannah	$ET = -5.66 + 0.18 \times PET + 0.10 \times P + 44.63 \times \text{LAI}$	11.1	0.68	36

ET = evapotranspiration (mm/month); *P* = precipitation (mm/month); *PET* = potential ET estimated by Hamon's method (mm/month); LAI = leaf area index; RMSE = root-mean-squared error; R^2 = coefficient of determination; *n* = sample size.

The long-term mean ET in a region is controlled mainly by water availability (precipitation) and atmosphere demand (potential ET), and this relationship is well described in the Budyko framework (Budyko *et al.*, 1962; Zhang *et al.*, 2001; Zhou *et al.*, 2015). Using the same concept, Zhang *et al.* (2001) analysed watershed balances data for over 250 catchments worldwide and developed a simple two-parameter ET model. The model offers a practical tool that can be readily used for assessing the long-term average effect of vegetation changes on catchment evapotranspiration:

$$ET = P \times \frac{1 + w(PET/P)}{1 + w(PET/P) + (P/PET)} \quad (3.8)$$

where *w* is the plant-available water coefficient which represents the relative difference in plant water use for transpiration. *PET* can be estimated by the Priestley and Taylor (1972) model. *P* is annual precipitation. The best fitted value of *w* for forest and grassland is 2.0 and 0.5, respectively, when PET is estimated using the Priestley and Taylor (1972) model (Zhang *et al.*, 2001). Sun *et al.* (2005) suggested that *w* can be as high as 2.8 when the Hamon PET method is used in applying the model for the humid south-eastern USA, consistent with a study for a managed pine forest in the Atlantic coastal plain (Amatya *et al.*, 2002). Kumagai *et al.* (Chapter 6, this volume) modified the above equation to obtain ET for tropical forests.

By combining remote sensing and climate data for 299 large river basins, Zeng *et al.* (2014) developed an annual ET model that has been used to estimate global ET (see equation 3.9 at the bottom of the page), where *ET* is basin-averaged annual evapotranspiration (mm/year), *P*, *T* and NDVI are annual precipitation (mm/year), mean annual temperature (°C) and annual normalized difference vegetation index, respectively. Similarly, an empirical model was developed using only mean annual temperature from 43 catchment water balance data sets in Japan (Komatsu *et al.*, 2008).

3.5 Future Directions

3.5.1 Response to climate change

Climate change is the largest environmental threat to forest ecosystems in the 21st century (Vose *et al.*, 2012). Climate warming and the

$$ET = 0.4(\pm 0.02) \times P + 10.62(\pm 0.39) \times T + 9.63(\pm 2.27) \times \text{NDVI} + 31.58(\pm 7.89) \qquad (R^2 = 0.85) \quad (3.9)$$

increased variability of precipitation form, amount and timing are expected to have rippling effects on forest ecosystem structure and functions through directly or indirectly altering ET processes. However, because precipitation, a key environmental control of tree transpiration and soil evaporation, is uncertain and difficult to predict, we have little capacity to project ET changes at the local scale.

3.5.2 Managing evapotranspiration in a water-shortage world

Accurate quantification of watershed water budgets including water use by trees and shrubs is becoming increasingly important given the growing competition for water resources among all users, from agricultural irrigation and bioenergy development to domestic water withdrawals by cities, in the Anthropocene (Sun *et al.*, 2008). We need better simulation models to reliably account for the role of forest ET in regulating streamflow and other ecosystem services (carbon fluxes) in large basins. Land managers have long asked the question: is it practical to manage upland headwater forests to meet future water supply demand in an urbanizing world (Douglass, 1983)? We know a lot of the basic relationships among forest cover, ET and water yield, but applying the knowledge to management remains a challenge (Vose and Klepzig, 2014). The services provided by forests in regulating local and regional climate (e.g. urban heat island, or cooling effects) through influencing the local energy balances, ET and precipitation patterns have been studied using computer simulation models (Liu, 2011), but these regional climate models need further parameterization, validation and refinement to enhance their prediction accuracy.

3.5.3 Measuring evapotranspiration everywhere all the time

Although large progress has been made in the past two decades towards measuring ET 'everywhere all the time' (Baldocchi *et al.*, 2001; Baldocchi and Ryu, 2011), the study of ET is still regarded as an imprecise science (Shuttleworth, 2012). Research is needed to scale up or scale down among plot, watershed, regional and global scales to integrate methods and data (Amatya *et al.*, 2014). In recent years remote sensing and radar technologies have advanced rapidly and enhanced our capability to accurately quantify water use and irrigation scheduling for croplands. However, the remote sensing applications in forest water management and water supply monitoring are rare. In fact, few studies have examined the accuracy of remote sensing-based ET products for forested areas. Forest ET measurements on the ground for calibrating remote sensing models are costly and the remote sensing techniques are often hampered by cloud cover and the complexity of multilayered tree canopies that vary spatially and temporally. For example, leaf clustering and light saturation problems are often problematic in estimating LAI for forests. Although images with high spatial and temporal resolution obtained from unmanned aerial vehicles may potentially play a role for precision agriculture and irrigation scheduling in the future, the validity of this method in estimating forest ET requires a significant amount of research (Amatya *et al.*, 2014). The best approach to estimate ET for large watersheds is achieved by combining field hydrological measurements with high-resolution remote sensing and energy balance-based land-surface modelling (Wang *et al.*, 2015).

3.5.4 New generation of ecohydrological models

Field measurements of ET at the leaf, tree, stand and landscape scale are essential to parameterize process-based hydrological models that have often not been validated with spatial and temporal distribution of various ET components (Sun *et al.*, 2011b). The so-called 'equifinality' in hydrological models is common, partially due to the lack of understanding of ET processes or the lack of ET data for model verification. To develop reliable predictive models, there is a great need for better understanding of the interactions and feedback mechanisms of ET and other ecohydrological processes (Evaristo *et al.*, 2015), including the canopy resistance factor used in the Penman–Monteith based ET models. More information is needed about how forest ET may be affected by species, density, stand age and management (managed versus natural forests, fertilization, thinning) in various ecoregions. Budyko's framework has been widely used to explain the mean spatial patterns of ET under land cover change (Zhou *et al.*, 2015) and climate change (Creed *et al.*, 2014). However, the model needs to be extended to finer temporal scale such as daily or seasonal to fully capture the dynamics of ET over time (Zhang *et al.*, 2008; Wang *et al.*, 2011). A new generation of ecohydrological models that combine the effects of CO_2 on ET processes and couple the physical and biological processes such as soil moisture redistribution, hydraulic distribution, photosynthesis, canopy conductance and tree growth is needed to fully understand the atmosphere–vegetation–soil processes mechanistically (Cheng *et al.*, 2014). Such models can provide better information to regional land-surface and climate models for quantifying the feedbacks of forest cover change to regional and global climate systems. Oversimplified model designs in the ET processes likely contain errors in the computation of dry-season water balances and the associated heat fluxes, and thus in the possible feedbacks between soil moisture and climate (Bonetti *et al.*, 2015).

References

Aber, J.D. and Federer, C.A. (1992) A generalized, lumped-parameter model of photosynthesis, evapotranspiration and net primary production in temperate and boreal forest ecosystems. *Oecologia* 92, 463–474.

Albaugh, J.M., Domec, J.C., Maier, C.A., Sucre, E.B., Leggett, Z.H. and King, J.S. (2014) Gas exchange and stand-level estimates of water use and gross primary productivity in an experimental pine and switchgrass intercrop forestry system on the Lower Coastal Plain of North Carolina, USA. *Agriculture and Forest Meteorology* 192, 27–40.

Allen, R.G., Smith, M., Perrier, A. and Pereira, L.S. (1994) An update for the definition of reference evapotranspiration. *ICID Bulletin* 43, 1–34.

Allen, R.G., Pereira, L.S., Smith, M., Raes, D. and Wright, J.L. (2005) FAO-56 dual crop coefficient method for estimating evaporation from soil and application extensions. *Journal of Irrigation and Drainage Engineering-ASCE* 131, 2–13.

Amatya, D.M. and Harrison, C.A. (2016) Grass and forest potential evapotranspiration comparison using five methods in the Atlantic coastal plain. *ASCE Journal of Hydraulic Engineering* (in press).

Amatya, D.M. and Skaggs, R.W. (2001) Hydrologic modeling of a drained pine plantation on poorly drained soils. *Forest Science* 47, 103–114.

Amatya, D.M. and Skaggs, R.W. (2011) Long-term hydrology and water quality of a drained pine plantation in North Carolina. *Transactions of the ASABE* 54, 2087–2098.

Amatya, D.M., Skaggs, R.W. and Gregory, J.D. (1995) Comparison of methods for estimating Reference-ET. *Journal of Irrigation and Drainage Engineering-ASCE* 121, 427–435.

Amatya, D.M., Gregory, J.D. and Skaggs, R.W. (2000) Effects of controlled drainage on storm event hydrology in a loblolly pine plantation. *Journal of the American Water Resources Association* 36, 175–190.

Amatya, D.M., Chescheir, G.M., Skaggs, R.W., Fernandez, G.P. and Gilliam, J.W. (2002) A watershed analysis & treatment evaluation routine spreadsheet (WATERS). In: *Total Maximum Daily Load (TMDL) Environmental Regulations: Proceedings of the March 11–13, 2002 Conference (Fort Worth, Texas, USA)*. American Society of Agricultural and Biological Engineers, St Joseph, Michigan, pp. 490–495.

Amatya, D.M., Douglas-Mankin, K.R., Williams, T.M., Skaggs, R.W. and Nettles, J.E. (2011) Advances in forest hydrology: challenges and opportunities. *Transactions of the ASABE* 54, 2049–2056.

Amatya, D., Sun, G. and Gowda, P. (2014) Evapotranspiration: challenges in measurement and modeling. *Eos, Transactions of the AGU* 95(28), 256.

Amatya, D.M., Sun, G., Rossi, C.G., Ssegane, H.S., Nettles, J.E. and Panda, S. (2015) Forests, land use change, and water. In: Zolin, C.A. and Rodrigues, R.d.A.R. (eds) *Impact of Climate Change on Water Resources in Agriculture*. CRC Press/Taylor & Francis, Boca Raton, Florida, pp. 116–153.

Amatya, D.M., Tian, S., Dai, Z., Sun, G. and Trettin, C. (2016) Long-term potential evapotranspiration and actual evapotranspiration of two different forests on the Atlantic coastal plain. *Transactions of the ASABE* (in press).

Anderson, M.C., Allen, R.G., Morse, A. and Kustas, W.P. (2012) Use of Landsat thermal imagery in monitoring evapotranspiration and managing water resources. *Remote Sensing of Environment* 122, 50–65.

Andreassian, V. (2004) Waters and forests: from historical controversy to scientific debate. *Journal of Hydrology* 291, 1–27.

Baldocchi, D.D. and Ryu, Y. (2011) A synthesis of forest evaporation fluxes – from days to years as measured with eddy covariance. *Ecological Studies: Analysis and Synthesis* 216, 101–116.

Baldocchi, D.D., Hincks, B.B. and Meyers, T.P. (1988) Measuring biosphere–atmosphere exchanges of biologically related gases with micrometeorological methods. *Ecology* 69, 1331–1340.

Baldocchi, D., Kelliher, F.M., Black, T.A. and Jarvis, P. (2000) Climate and vegetation controls on boreal zone energy exchange. *Global Change Biology* 6, 69–83.

Baldocchi, D., Falge, E., Gu, L.H., Olson, R., Hollinger, D., Running, S., Anthoni, P., Bernhofer, C., Davis, K., Evans, R., *et al.* (2001) FLUXNET: a new tool to study the temporal and spatial variability of ecosystem-scale carbon dioxide, water vapor, and energy flux densities. *Bulletin of the American Meteorological Society* 82, 2415–2434.

Betts, R.A. (2000) Offset of the potential carbon sink from boreal forestation by decreases in surface albedo. *Nature* 408, 187–190.

Boggs, J., Sun, G. and McNulty, S.G. (2015) Effects of timber harvest on water quantity and quality in small watersheds in the piedmont of North Carolina. *Journal of Forestry* 114, 27–40.

Bonan, G.B. (2008) Forests and climate change: forcings, feedbacks, and the climate benefits of forests. *Science* 320, 1444–1449.

Bonetti, S., Manoli, G., Domec, J.C., Putti, M., Marani, M. and Katul, G.G. (2015) The influence of water table depth and the free atmospheric state on convective rainfall predisposition. *Water Resources Research* 51, 2283–2297.

Brown, A.E., Zhang, L., McMahon, T.A., Western, A.W. and Vertessy, R.A. (2005) A review of paired catchment studies for determining changes in water yield resulting from alterations in vegetation. *Journal of Hydrology* 310, 28–61.

Bryant, M.L., Bhat, S. and Jacobs, J.M. (2005) Measurements and modeling of throughfall variability for five forest communities in the southeastern US. *Journal of Hydrology* 312, 95–108.

Budyko, M.I., Yefimova, N.A., Aubenok, L.I. and Strokina, L.A. (1962) The heat-balance of the surface of the Earth. *Soviet Geography: Review and Translation* 3, 3–16.

Burgess, S.S.O., Adams, M.A., Turner, N.C. and Ong, C.K. (1998) The redistribution of soil water by tree root systems. *Oecologia* 115, 306–311.

Canny, M.J. (1998) Transporting water in plants. *American Scientist* 86, 152–159.

Cheng, L., Xu, Z.X., Wang, D.B. and Cai, X.M. (2011) Assessing interannual variability of evapotranspiration at the catchment scale using satellite-based evapotranspiration data sets. *Water Resources Research* 47, W09509, doi: 10.1029/2011WR010636 (accessed 20 March 2016).

Cheng, L., Zhang, L., Wang, Y.P., Yu, Q., Eamus, D. and O'Grady, A. (2014) Impacts of elevated CO_2, climate change and their interactions on water budgets in four different catchments in Australia. *Journal of Hydrology* 519, 1350–1361.

Christopher, S.F., Schoenholtz, S.H. and Nettles, J.E. (2015) Water quantity implications of regional-scale switchgrass production in the southeastern US. *Biomass and Bioenergy* 83, 50–59.

Creed, I.F., Spargo, A.T., Jones, J.A., Buttle, J.M., Adams, M.B., Beall, F.D., Booth, E.G., Campbell, J.L., Clow, D., Elder, K., *et al.* (2014) Changing forest water yields in response to climate warming: results from long-term experimental watershed sites across North America. *Global Change Biology* 20, 3191–3208.

Crétaz, A.L.d.l. and Barten, P.K. (2007) *Land Use Effects on Streamflow and Water Quality in the Northeastern United States*. CRC Press/Taylor & Francis, Boca Raton, Florida.

Currie, D.J. (1991) Energy and large-scale patterns of animal-species and plant-species richness. *American Naturalist* 137, 27–49.

Deguchi, A., Hattori, S., Park, H.T. (2006) The influence of seasonal changes in canopy structure on interception loss: application of the revised Gash model. *Journal of Hydrology* 318, 80–102.

Denmead, O.T., Dunin, F.X., Wong, S.C. and Greenwood, E.A.N. (1993) Measuring water use efficiency of Eucalypt trees with chambers and micrometeorological techniques. *Journal of Hydrology* 150, 649–664.

Domec, J.C., Warren, J.M., Meinzer, F.C., Brooks, J.R. and Coulombe, R. (2004) Native root xylem embolism and stomatal closure in stands of Douglas-fir and ponderosa pine: mitigation by hydraulic redistribution. *Oecologia* 141, 7–16.

Domec, J.C., Noormets, A., King, J.S., Sun, G., McNulty, S.G., Gavazzi, M.J., Boggs, J.L. and Treasure, E.A. (2009) Decoupling the influence of leaf and root hydraulic conductances on stomatal conductance and its sensitivity to vapour pressure deficit as soil dries in a drained loblolly pine plantation. *Plant and Cell Environment* 32, 980–991.

Domec, J.C., King, J.S., Noormets, A., Treasure, E., Gavazzi, M.J., Sun, G. and McNulty, S.G. (2010) Hydraulic redistribution of soil water by roots affects whole-stand evapotranspiration and net ecosystem carbon exchange. *New Phytologist* 187, 171–183.

Domec, J.C., Lachenbruch, B., Pruyn, M.L. and Spicer, R. (2012a) Effects of age-related increases in sapwood area, leaf area, and xylem conductivity on height-related hydraulic costs in two contrasting coniferous species. *Annals of Forest Science* 69, 17–27.

Domec, J.C., Sun, G., Noormets, A., Gavazzi, M.J., Treasure, E.A., Cohen, E., Swenson, J.J., McNulty, S.G. and King, J.S. (2012b) A comparison of three methods to estimate evapotranspiration in two contrasting lob lolly pine plantations: age-related changes in water use and drought sensitivity of evapotranspiration components. *Forest Science* 58, 497–512.

Douglass, J.E. (1983) The potential for water yield augmentation from forest management in the eastern-United-States. *Water Resources Bulletin* 19, 351–358.

Emanuel, R.E., Epstein, H.E., McGlynn, B.L., Welsch, D.L., Muth, D.J. and D'Odorico, P. (2010) Spatial and temporal controls on watershed ecohydrology in the northern Rocky Mountains. *Water Resources Research* 46, W11553, doi: 10.1029/2009WR008890 (accessed 20 March 2016).

Evaristo, J., Jasechko, S. and McDonnell, J.J. (2015) Global separation of plant transpiration from groundwater and streamflow. *Nature* 525, 91–94.

Fang, Y., Sun, G., Caldwell, P., McNulty, S.G., Noormets, A., Domec, J.-C., King, J., Zhang, Z., Zhang, X., Lin, G., *et al.* (2015) Monthly land cover-specific evapotranspiration models derived from global eddy flux measurements and remote sensing data. *Ecohydrology*, doi: 10.1002/eco.1629 (accessed 20 March 2016).

Ford, C.R., Hubbard, R.M., Kloeppel, B.D. and Vose, J.M. (2007) A comparison of sap flux-based evapotranspiration estimates with catchment-scale water balance. *Agricultural and Forest Meteorology* 145, 176–185.

Frank, D.C., Poulter, B., Saurer, M., Esper, J., Huntingford, C., Helle, G., Treydte, K., Zimmermann, N.E., Schleser, G.H., Ahlstrom, A., *et al.* (2015) Water-use efficiency and transpiration across European forests during the Anthropocene. *Nature Climate Change* 5, 579.

Gao, Y., Zhu, X.J., Yu, G.R., He, N.P., Wang, Q.F. and Tian, J. (2014) Water use efficiency threshold for terrestrial ecosystem carbon sequestration in China under afforestation. *Agricultural and Forest Meteorology* 195, 32–37.

Gash, J.H.C. (1979) Analytical model of rainfall interception by forests. *Quarterly Journal of the Royal Meteorological Society* 105, 43–55.

Gavazzi, M.G., Sun, G., McNulty, S.G. and Treasure, E.A. (2015) Canopy rainfall interception measured over 10 years in a coastal plain loblolly pine (*Pinus taeda* L.) plantation. *Transactions of the ASABE* (in press).

Gerrits, A.M.J., Savenije, H.H.G., Hoffmann, L. and Pfister, L. (2007) New technique to measure forest floor interception – an application in a beech forest in Luxembourg. *Hydrology and Earth System Sciences* 11, 695–701.

Good, S.P., Noone, D., Kurita, N., Benetti, M. and Bowen, G.J. (2015) D/H isotope ratios in the global hydrologic cycle. *Geophysical Research Letters* 42, 5042–5050.

Granier, A. (1987) Evaluation of transpiration in a Douglas-fir stand by means of sap flow measurements. *Tree Physiology* 3, 309–320.

Granier, A., Huc, R. and Barigah, S.T. (1996) Transpiration of natural rain forest and its dependence on climatic factors. *Agricultural and Forest Meteorology* 78, 19–29.

Grismer, M.E., Orang, M., Snyder, R. and Matyac, R. (2002) Pan evaporation to reference evapotranspiration conversion methods. *Journal of Irrigation and Drainage Engineering–ASCE* 128, 180–184.

Hawkins, B.A., Field, R., Cornell, H.V., Currie, D.J., Guegan, J.F., Kaufman, D.M., Kerr, J.T., Mittelbach, G.G., Oberdorff, T., O'Brien, E.M., *et al.* (2003) Energy, water, and broad-scale geographic patterns of species richness. *Ecology* 84, 3105–3117.

Helvey, J.D. (1974) Summary of rainfall interception by certain conifers on North America. In: *Proceedings of the 3rd International Seminar for Hydrology Professors*. NSFASS, West Lafayette, Indiana, pp. 103–113.

Helvey, J.D. and Patric, J.H. (1965) Canopy and litter interception of rainfall by hardwoods of eastern united states. *Water Resources Research* 1, 193–206.

Hewlett, J.D. (1982) *Principles of Forest Hydrology*. University of Georgia Press, Athens, Georgia.

Horton, R.E. (1919) Rainfall interception. *Monthly Weather Review* 47, 16.

Huang, M.T., Piao, S.L., Sun, Y., Ciais, P., Cheng, L., Mao, J.F., Poulter, B., Shi, X.Y., Zeng, Z.Z. and Wang, Y.P. (2015) Change in terrestrial ecosystem water-use efficiency over the last three decades. *Global Change Biology* 21, 2366–2378.

Irmak, S. (2010) Nebraska Water and Energy Flux Measurement, Modeling, and Research Network (NEBFLUX). *Transactions of the ASABE* 53, 1097–1115.

Irmak, S., Skaggs, K.E. and Chatterjee, S. (2014) A review of the Bowen ratio surface energy balance method for quantifying evapotranspiration and other energy fluxes. *Transactions of the ASABE* 57, 1657–1674.

Jackson, R.B., Jobbagy, E.G., Avissar, R., Roy, S.B., Barrett, D.J., Cook, C.W., Farley, K.A., le Maitre, D.C., McCarl, B.A. and Murray, B.C. (2005) Trading water for carbon with biological sequestration. *Science* 310, 1944–1947.

Jasechko, S., Sharp, Z.D., Gibson, J.J., Birks, S.J., Yi, Y. and Fawcett, P.J. (2013) Terrestrial water fluxes dominated by transpiration. *Nature* 496, 347–350.

Jayakaran, A.D., Williams, T.M., Ssegane, H., Amatya, D.M., Song, B. and Trettin, C.C. (2014) Hurricane impacts on a pair of coastal forested watersheds: implications of selective hurricane damage to forest structure and streamflow dynamics. *Hydrology and Earth System Sciences* 18, 1151–1164.

Jung, M., Reichstein, M., Ciais, P., Seneviratne, S.I., Sheffield, J., Goulden, M.L., Bonan, G., Cescatti, A., Chen, J.Q., de Jeu, R., *et al.* (2010) Recent decline in the global land evapotranspiration trend due to limited moisture supply. *Nature* 467, 951–954.

Kalma, J.D., McVicar, T.R. and McCabe, M.F. (2008) Estimating land surface evaporation: a review of methods using remotely sensed surface temperature data. *Surveys in Geophysics* 29, 421–469.

King, J.S., Ceulemans, R., Albaugh, J.M., Dillen, S.Y., Domec, J.C., Fichot, R., Fischer, M., Leggett, Z., Sucre, E., Trnka, M. and Zenone, T. (2013) The challenge of lignocellulosic bioenergy in a water-limited world. *Bioscience* 63, 102–117.

Komatsu, H. and Kume, T. (2015) Changes in the sapwood area of Japanese cedar and cypress plantations after thinning. *Journal of Forest Research* 20, 43–51.

Komatsu, H., Maita, E. and Otsuki, K. (2008) A model to estimate annual forest evapotranspiration in Japan from mean annual temperature. *Journal of Hydrology* 348, 330–340.

Kumagai, T. (2011) Transpiration in forest ecosystems. In: Levia, D.F., Carlyle-Moses, D. and Tanaka, T. (eds) *Forest Hydrology and Biogeochemistry: Synthesis of Past Research and Future Directions*. Springer, New York, pp. 389–406.

Kumagai, T., Tateishi, M., Miyazawa, Y., Kobayashi, M., Yoshifuji, N., Komatsu, H. and Shimizu, T. (2014) Estimation of annual forest evapotranspiration from a coniferous plantation watershed in Japan (1): Water use components in Japanese cedar stands. *Journal of Hydrology* 508, 66–76.

Kustas, W.P. and Norman, J.M. (1996) Use of remote sensing for evapotranspiration monitoring over land surfaces. *Hydrological Sciences Journal* 41, 495–516.

Law, B.E., Falge, E., Gu, L., Baldocchi, D.D., Bakwin, P., Berbigier, P., Davis, K., Dolman, A.J., Falk, M., Fuentes, J.D., *et al.* (2002) Environmental controls over carbon dioxide and water vapor exchange of terrestrial vegetation. *Agricultural and Forest Meteorology* 113, 97–120.

Levia, D.F., Carlyle-Moses, D. and Tanaka, T. (2011) *Forest Hydrology and Biogeochemistry: Synthesis of Past Research and Future Directions*. Springer, New York.

Liu, C., Sun, G., McNulty, S.G. and Kang, S. (2015) An improved evapotranspiration model for an apple orchard in northwestern China. *Transactions of the ASABE* 58, 1253–1264.

Liu, Y.Q. (2011) A numerical study on hydrological impacts of forest restoration in the southern United States. *Ecohydrology* 4, 299–314.

Lu, J.B., Sun, G., McNulty, S.G. and Amatya, D.M. (2005) A comparison of six potential evapotranspiration methods for regional use in the southeastern United States. *Journal of the American Water Resources Association* 41, 621–633.

Mann, M.E. and Gleick, P.H. (2015) Climate change and California drought in the 21st century. *Proceedings of the National Academy of Sciences USA* 112, 3858–3859.

McCarthy, E.J., Flewelling, J.W. and Skaggs, R.W. (1992) Hydrologic model for drained forest watershed. *Journal of Irrigation and Drainage Engineering–ASCE* 118, 242–255.

McMahon, T.A., Peel, M.C., Lowe, L., Srikanthan, R. and McVicar, T.R. (2013) Estimating actual, potential, reference crop and pan evaporation using standard meteorological data: a pragmatic synthesis. *Hydrology and Earth System Sciences* 17, 1331–1363.

McNulty, S.G. and Swank, W.T. (1995) Wood $\delta^{13}C$ as a measure of annual basal area growth and soil-water stress in a *Pinus strobus* forest. *Ecology* 76, 1581–1586.

Meinzer, F.C., James, S.A. and Goldstein, G. (2004) Dynamics of transpiration, sap flow and use of stored water in tropical forest canopy trees. *Tree Physiology* 24, 901–909.

Monteith, J.L. (1965) Evaporation and the environment. *Symposium of the Society of Experimental Biology* 19, 205–234.

Mu, Q.Z., Zhao, M.S. and Running, S.W. (2011) Improvements to a MODIS global terrestrial evapotranspiration algorithm. *Remote Sensing of Environment* 115, 1781–1800.

Neumann, R.B. and Cardon, Z.G. (2012) The magnitude of hydraulic redistribution by plant roots: a review and synthesis of empirical and modeling studies. *New Phytologist* 194, 337–352.

Oishi, A.C., Oren, R., Novick, K.A., Palmroth, S. and Katul, G.G. (2010) Interannual invariability of forest evapotranspiration and its consequence to water flow downstream. *Ecosystems* 13, 421–436.

Oki, T. and Kanae, S. (2006) Global hydrological cycles and world water resources. *Science* 313, 1068–1072.

Olbrich, B.W. (1991) The verification of the heat pulse velocity technique for estimating sap flow in *Eucalyptus grandis*. *Canadian Journal of Forest Research* 21, 836–841.

Oudin, L., Andreassian, V., Lerat, J. and Michel, C. (2008) Has land cover a significant impact on mean annual streamflow? An international assessment using 1508 catchments. *Journal of Hydrology* 357, 303–316.

Palmroth, S., Katul, G.G., Hui, D.F., McCarthy, H.R., Jackson, R.B. and Oren, R. (2010) Estimation of long-term basin scale evapotranspiration from streamflow time series. *Water Resources Research* 46, W10512, doi: 10.1029/2009WR008838 (accessed 20 March 2016).

Peng, S.S., Piao, S.L., Zeng, Z.Z., Ciais, P., Zhou, L.M., Li, L.Z.X., Myneni, R.B., Yin, Y. and Zeng, H. (2014) Afforestation in China cools local land surface temperature. *Proceedings of the National Academy of Sciences USA* 111, 2915–2919.

Penman, H.L. (1948) Natural evaporation from open water, bare soil and grass. *Proceedings of the Royal Society of London Series–A, Mathematical and Physical Sciences* 193, 120–145.

Priestley, C.H.B. and Taylor, R.J. (1972) On the assessment of surface heat-flux and evaporation using large-scale parameters. *Monthly Weather Review* 100, 81–92.

Rosenzweig, M.L. (1968) Net primary productivity of terrestrial communities – prediction from climatological data. *American Naturalist* 102, 67–74.

Sampson, D.A., Amatya, D.M., Lawson, C.D.B. and Skaggs, R.W. (2011) Leaf area index (LAI) of loblolly pine and emergent vegetation following a harvest. *Transactions of the ASABE* 54, 2057–2066.

Sanford, W.E. and Selnick, D.L. (2013) Estimation of evapotranspiration across the conterminous united states using a regression with climate and land-cover data. *Journal of the American Water Resources Association* 49, 217–230.

Schlesinger, W.H. and Jasechko, S. (2014) Transpiration in the global water cycle. *Agricultural and Forest Meteorology* 189, 115–117.

Shuttleworth, W.J. (2012) *Terrestrial Hydrometeorology*. Wiley, Hoboken, New Jersey.

Sun, G., Riekerk, H. and Comerford, N.B. (1998) Modeling the forest hydrology of wetland-upland ecosystems in Florida. *Journal of the American Water Resources Association* 34, 827–841.

Sun, G., Amatya, D.M., McNulty, S.G., Skaggs, R.W. and Hughes, J.H. (2000) Climate change impacts on the hydrology and productivity of a pine plantation. *Journal of the American Water Resources Association* 36, 367–374.

Sun, G., McNulty, S.G., Amatya, D.M., Skaggs, R.W., Swift, L.W., Shepard, J.P. and Riekerk, H. (2002) A comparison of the watershed hydrology of coastal forested wetlands and the mountainous uplands in the southern US. *Journal of Hydrology* 263, 92–104.

Sun, G., McNulty, S.G., Lu, J., Amatya, D.M., Liang, Y. and Kolka, R.K. (2005) Regional annual water yield from forest lands and its response to potential deforestation across the southeastern United States. *Journal of Hydrology* 308, 258–268.

Sun, G., McNulty, S.G., Moore-Myers, J.A. and Cohen, E.C. (2008) Impacts of multiple stresses on water demand and supply across the southeastern United States. *Journal of the American Water Resources Association* 44, 1441–1457.

Sun, G., Noormets, A., Gavazzi, M.J., McNulty, S.G., Chen, J., Domec, J.C., King, J.S., Amatya, D.M. and Skaggs, R.W. (2010) Energy and water balance of two contrasting loblolly pine plantations on the lower coastal plain of North Carolina, USA. *Forest Ecology and Management* 259, 1299–1310.

Sun, G., Alstad, K., Chen, J.Q., Chen, S.P., Ford, C.R., Lin, G.H., Liu, C.F., Lu, N., McNulty, S.G., Miao, H.X., *et al.* (2011a) A general predictive model for estimating monthly ecosystem evapotranspiration. *Ecohydrology* 4, 245–255.

Sun, G., Caldwell, P., Noormets, A., McNulty, S.G., Cohen, E., Moore Myers, J., Domec, J.-C., Treasure, E., Mu, Q., Xiao, J., *et al.* (2011b) Upscaling key ecosystem functions across the conterminous United States by a water-centric ecosystem model. *Journal of Geophysical Research-Biogeosciences* 116, G00J05.

Sun, X.C., Onda, Y., Kato, H., Gomi, T. and Komatsu, H. (2015) Effect of strip thinning on rainfall interception in a Japanese cypress plantation. *Journal of Hydrology* 525, 607–618.

Swank, W.T. and Crossley, D.A. (eds) (1988) *Forest Hydrology and Ecology at Coweeta*. Ecological Studies Vol. 66. Springer, New York.
Swank, W.T. and Douglass, J.E. (1974) Streamflow greatly reduced by converting deciduous hardwood stands to pine. *Science* 185, 857–859.
Swank, W.T. and Miner, N.H. (1968) Conversion of hardwood-covered watersheds to white pine reduces water yield. *Water Resources Research* 4, 947–954.
Teuling, A.J., Hirschi, M., Ohmura, A., Wild, M., Reichstein, M., Ciais, P., Buchmann, N., Ammann, C., Montagnani, L., Richardson, A.D., *et al.* (2009) A regional perspective on trends in continental evaporation. *Geophysical Research Letters* 36, L02404, doi: 10.1029/2008GL036584 (accessed 20 March 2016).
Tian, H.Q., Chen, G.S., Zhang, C., Liu, M.L., Sun, G., Chappelka, A., Ren, W., Xu, X.F., Lu, C.Q., Pan, S.F., *et al.* (2012) Century-scale responses of ecosystem carbon storage and flux to multiple environmental changes in the southern United States. *Ecosystems* 15, 674–694.
Tian, S.Y., Youssef, M.A., Skaggs, R.W., Amatya, D.M. and Chescheir, G.M. (2012) Modeling water, carbon and nitrogen dynamics for two drained pine plantations under intensive management practices. *Forest Ecology and Management* 264, 20–36.
Trenberth, K.E., Fasullo, J.T. and Kiehl, J. (2009) Earth's global energy budget. *Bulletin of the American Meteorological Society* 90, 311–323, doi: 10.1175/2008BAMS2634.1 (accessed 20 March 2016).
Ukkola, A.M. and Prentice, I.C. (2013) A worldwide analysis of trends in water-balance evapotranspiration. *Hydrology and Earth System Sciences* 17, 4177–4187.
Vose, J.M. and Klepzig, K.D. (eds) (2014) *Climate Change Adaptation and Mitigation Management Options: A Guide for Natural Resource Managers in Southern Forest Ecosystems*. CRC Press, Boca Raton, Florida.
Vose, J.M., Sun, G., Ford, C.R., Bredemeier, M., Otsuki, K., Wei, X.H., Zhang, Z.Q. and Zhang, L. (2011) Forest ecohydrological research in the 21st century: what are the critical needs? *Ecohydrology* 4, 146–158.
Vose, J.M., Peterson, D. and Patel-Weynand, T. (eds) (2012) *Effects of Climatic Variability and Change on Forest Ecosystems: A Comprehensive Science Synthesis for the US Forest Sector*. US Forest Service, Pacific Northwest Research Station, Portland, Oregon.
Wang, Q.J., Pagano, T.C., Zhou, S.L., Hapuarachchi, H.A.P., Zhang, L. and Robertson, D.E. (2011) Monthly versus daily water balance models in simulating monthly runoff. *Journal of Hydrology* 404, 166–175.
Wang, S., Pan, M., Mu, Q., Shi, X., Mao, J., Brümmer, C., Jassal, R.S., Krishnan, P., Li, J. and Black, T.A. (2015) Comparing evapotranspiration from eddy covariance measurements, water budgets, remote sensing, and land surface models over Canada. *Journal of Hydrometeorolgy* 16, 1540–1560.
Warren, J.M., Meinzer, F.C., Brooks, J.R., Domec, J.C. and Coulombe, R. (2007) Hydraulic redistribution of soil water in two old-growth coniferous forests: quantifying patterns and controls. *New Phytologist* 173, 753–765.
Wilson, K.B., Hanson, P.J., Mulholland, P.J., Baldocchi, D.D. and Wullschleger, S.D. (2001) A comparison of methods for determining forest evapotranspiration and its components: sap-flow, soil water budget, eddy covariance and catchment water balance. *Agricultural and Forest Meteorology* 106, 153–168.
Wilson, K., Goldstein, A., Falge, E., Aubinet, M., Baldocchi, D., Berbigier, P., Bernhofer, C., Ceulemans, R., Dolman, H., Field, C., *et al.* (2002) Energy balance closure at FLUXNET sites. *Agricultural and Forest Meteorology* 113, 223–243.
Wullschleger, S.D., Meinzer, F.C. and Vertessy, R.A. (1998) A review of whole-plant water use studies in trees. *Tree Physiology* 18, 499–512.
Zeng, Z.Z., Wang, T., Zhou, F., Ciais, P., Mao, J.F., Shi, X.Y. and Piao, S.L. (2014) A worldwide analysis of spatiotemporal changes in water balance-based evapotranspiration from 1982 to 2009. *Journal of Geophysical Research–Atmosphere* 119, 1186–1202.
Zhang, L., Dawes, W.R. and Walker, G.R. (2001) Response of mean annual evapotranspiration to vegetation changes at catchment scale. *Water Resources Research* 37, 701–708.
Zhang, L., Potter, N., Hickel, K., Zhang, Y.Q. and Shao, Q.X. (2008) Water balance modeling over variable time scales based on the Budyko framework – model development and testing. *Journal of Hydrology* 360, 117–131.
Zhou, G., Wei, X., Chen, X., Zhou, P., Liu, X., Xiao, Y., Sun, G., Scott, D.F., Zhou, S., Han, L. and Su, Y. (2015) Global pattern for the effect of climate and land cover on water yield. *Nature Communications* 6, 5918.

4 Forest Hydrology of Mountainous and Snow-Dominated Watersheds

W.J. Elliot[1]*, M. Dobre[2], A. Srivastava[2], K.J. Elder[3], T.E. Link[2] and E.S. Brooks[2]
[1]USDA Forest Service, Moscow, Idaho, USA; [2]University of Idaho, Moscow, Idaho, USA; [3]USDA Forest Service, Fort Collins, Colorado, USA

4.1 Introduction

This chapter addresses snow and hydrological processes of steep forested watersheds. Many of the hydrological principles described in other chapters are applicable to steep, high elevations, but the relationships among surface runoff, shallow lateral flow and baseflow warrant a special discussion in the context of snow processes and steep watersheds. The principles presented in this chapter are generally applicable to forested ecosystems overlapping with chaparral or agriculture at lower elevations, and subalpine ecosystems at higher elevations.

High-elevation forests are often considered the 'water towers' for much of the world. From the Americas, throughout Europe, Africa, Asia and Australia, large urban areas and irrigation projects look to runoff from higher-elevation forests to meet their water supply needs (Viviroli and Weingartner, 2004). These areas not only provide surface water to keep streams flowing and lakes and reservoirs filled, but also are often important areas for recharging groundwater reservoirs that are later tapped by wells or bore holes for their precious water. Worldwide, it is estimated that more than a billion people rely on snow-covered areas and glaciers for their water supply (Bales *et al.*, 2006).

The accumulation of snow within high-elevation forests is of greater importance in areas with low dry-season precipitation (Viviroli and Weingartner, 2004). In these watersheds, the high-elevation mountain snowpack becomes a major component of the hydrological cycle recharging groundwater and providing low flows throughout the dry season. In continental climates, convective storms can also contribute to snow processes in the winter, but the high-elevation areas generally receive greater amounts of precipitation in the summer due to orographic effects. This chapter describes how precipitation is slowly routed through forested watersheds as lateral flow and baseflow, rather than surface runoff, as is more common in non-forested watersheds or watersheds following a wildfire.

The dominant hydrological processes in most high-elevation forests are snow accumulation and melt in the winter, increased amounts of high-elevation precipitation in the summer compared with lower elevations, shallow subsurface flow from slowly melting snow in the spring, and a continuous baseflow from deep seepage that lasts through the dry season.

Some of the above processes were introduced in Chapters 1 and 2. This chapter focuses mainly on the hydrological processes observed in the mountainous and subalpine temperate coniferous

*Corresponding author; e-mail: welliot@fs.fed.us

 Forest Hydrology: Processes, Management and Assessment (eds D.M. Amatya, T.M. Williams, L. Bren and C. de Jong)

forests in the Western part of North America, in which we broadly include the states of Idaho, Oregon, Washington, northern California, Montana – west of the Continental Divide, in the USA, and the southern British Columbia province in Canada. This area is characterized mainly by maritime climate on the west coast and continental climate inland. There is a transition zone between the maritime and continental climates that may experience a maritime climate during some storms or seasons, and a continental climate during others, depending on the air masses that are dominant at the time (Hubbart *et al.*, 2007). The major distinction between watersheds within the two types of climate is in the amount and form of winter precipitation.

The maritime climate is distributed along the Pacific coast and is dominated by relatively warm moist maritime air masses from the Pacific Ocean. This type of climate can also be found in countries bordering the Mediterranean Sea and, to a lesser extent, in other regions strongly influenced by maritime air masses, or other waterbodies, in the winter (Peel *et al.*, 2007). Further east, continental convective storms throughout the year tend to dominate the hydrology. The continental climate is not moderated by seas or oceans, and is characterized by significant differences in temperature, with colder winters and hotter summers. Although we refer mainly to the hydrological processes in North America, the principles described in this chapter are applicable to other steep, snow-impacted forested areas throughout the world with similar climates.

4.2 Snow Processes

The dominant snow processes in forested watersheds occur in the atmosphere, in the canopy or on the ground. In the atmosphere, snow crystals form in the clouds and, as they fall, may melt and become rainfall, remain crystallized as snow, or become a mix of the two, before reaching the forest canopy or the ground. Temperatures of the upper and lower atmosphere influence this process. The canopy can intercept snowfall, which then falls off, melts or sublimates. On the ground, the snowpack can accumulate or melt. Melt rates are dependent on energy input from shortwave radiation from the sun, longwave radiation from clouds or the forest canopy, convective energy from warm air, or latent heat energy from humid air.

Snow accumulation depends greatly on the amount of fallen snow, wind speed and direction, and snow interception by the canopy, while snow melt is driven mostly by solar radiation (Gelfan *et al.*, 2004). Forest canopy structure (tree height, canopy density and tree spacing) affects the degree of shading and thus the amount of incoming shortwave radiation that reaches the forest floor. In dense forests, the decreases in the amount of shortwave radiation may be offset by increases in the longwave radiation from the canopy (Pomeroy *et al.*, 2009; Lundquist *et al.*, 2013). In less dense forests, the incoming all-wave solar radiation can be affected by both the size of the gap and solar angle, resulting in either 'hotspots' or 'cold holes' (Lawler and Link, 2011). In these situations, the snow within the gaps can melt faster or slower than both open sites and areas under dense canopy, depending on vegetation, topographic and microclimatic conditions (Berry and Rothwell, 1992).

Alteration of forest canopy can be used to obtain an increase in streamflow discharge; however, the response of snow accumulation and melt to management activities is not the same for all high-elevation forests. Interactions among the canopy, snow interception, snowmelt rates and runoff vary depending on the elevation and the dominant air mass (maritime or continental). This categorization is complicated in that a given location may demonstrate the characteristics of one of these categories one year, and a different category in another, or a combination of responses within a single year.

4.2.1 Maritime climates

In maritime climates, the snow accumulation patterns depend on forest clearing size and forest type (Lundquist *et al.*, 2013). Coniferous trees tend to intercept more snow than broadleaf trees with few significant differences among coniferous species. Storck *et al.* (2002) demonstrated that the canopy cover of a Douglas fir (*Pseudotsuga menziesii*) can intercept as much as 60% of the snowfall, with most of it being removed quickly by meltwater drip and mass release. Sublimation can occur in maritime climates, but the amount of snow lost to sublimation is minimal (Storck *et al.*, 2002; DeWalle and Rango, 2008).

In maritime climates, the snowpack is dynamic and snow accumulation and melt can occur in frequent episodes throughout the winter season. Rain-on-snow (ROS) events are typical in these areas and they often occur during midwinter, when warm rains from the oceans fall directly on shallow snowpacks. Depending on the conditions of the snowpack and the amount of rain over a long period of time, these ROS events can generate large runoff events with high-intensity peak flow rates, driven mainly by rainfall, but with additional runoff from melting snow, and often coupled with saturated soils (see Section 4.4).

Under typical winter conditions in high-elevation maritime zones, snow accumulates throughout the cold winter months and melts slowly at the beginning of spring – driven by net radiation and slowly increasing air temperature and humidity – to generate a prolonged period of low streamflow. Maritime air currents can sometimes travel over snow-covered areas bringing warmer rains and higher humidity. These air masses contain energy from being warm (sensible heat) and being moist (latent heat). When this occurs, the convective transfer of sensible (warm air warms the snowpack) and latent (condensation of water from the air on to the snow warms the snowpack) heat energy from the atmosphere can quickly melt the snowpack. If the soil is saturated, this can lead to high-intensity runoff peaks (Harr, 1986; Marks *et al.*, 1998). Canopy cover removal in these areas causes wind speed to increase within the forest gaps. Higher wind speeds increase turbulent energy exchanges at the snow surface, resulting in faster snowmelt rates from both sensible and latent heat transfers. Some research has shown that there are situations when the snowmelt resulting from ROS events is still mainly driven by the net radiation; however, the sensible and latent heat fluxes play the major role in the rapid snowmelt during ROS events (Marks *et al.*, 1998).

In ROS events, the amount of rainfall is still the principal cause for the quick increase in hydrograph peaks; however, in a few events, water from melting of the snowpack can saturate the soil in days preceding the storm and add up to 30% extra water to the runoff beyond the base rainfall (Harr, 1986). The largest events, however, are generally dominated by rainfall (Marchi *et al.*, 2010) because snowmelt rates seldom exceed a few millimetres per hour, whereas a large rainfall event can deliver in excess of 25 mm of precipitation in an hour.

The snowpack can have a natural water-holding capacity, depending on the initial conditions (e.g. prior melting and consolidation, depth, prior rains), and can store some or most of smaller rainfall events. In such situations, it is common to observe several days' delay in runoff following a winter rainfall event. Most large ROS events are the result of a prolonged period of rainfall when several days of rain warm the snowpack and saturate both the snowpack and the soil. When an additional day of rain occurs, the snow melts rapidly and runs off quickly (see Section 4.4), resulting in a major runoff event (Marks *et al.*, 1998). The majority of ROS events take place in the transient snow zone (where snow accumulates and melts more than once each winter) when temperatures are often just above freezing (Berris and Harr, 1987; Jefferson, 2011).

4.2.2 Continental climates

Snow–vegetation interaction in the forests from continental climates can be very different from those found in maritime climates. For example, continental climates generally receive more precipitation in the form of snow than maritime climates and have longer winter seasons with temperatures below 0°C for many consecutive days. Studies in the Canadian boreal forests demonstrate that the canopy cover can intercept up to 60% of the snowfall for up to 1 month (Pomeroy and Schmidt, 1993; Hedstrom and Pomeroy, 1998). Snow interception by the canopy also depends on temperature. At temperatures near 0°C, the snow is more cohesive and can easily attach to needles and branches but the interception efficiency declines with increasing amounts of snowfall (Hedstrom and Pomeroy, 1998). Sublimation losses for complete coniferous canopies are high, with sublimation reaching 30–50% of the annual snowfall (Lundberg and Halldin, 2001).

In addition, wind plays a major role in these cold forests in removing snow from the canopy.

While topography and the size and amount of forest canopy gaps are important for snow accumulation, the snowmelt is driven primarily by the net available radiant energy, which accounts for a large proportion of the total snowmelt energy. The radiative fluxes for snowmelt are incoming shortwave radiation (direct and diffuse radiation) and longwave radiation (diffuse). In addition, longwave radiation from forest canopy and trunks is also significant (Lundquist *et al.*, 2013). The proportion of each of these components that reaches the snow surface is dependent on many factors such as elevation, aspect, latitude, day length, cloudiness and solar angle.

During clear-sky conditions, incoming shortwave radiation is intercepted by the forest canopy and further transmitted as longwave radiation below the canopy. Fresh snow has a high shortwave radiation reflectance or albedo (0.8–0.9) and therefore reflects a greater proportion of the incoming shortwave radiation than a forest with an albedo of about 0.15 (Manninen and Stenberg, 2009). The intercepted snow within the forest canopy has little effect on the canopy albedo in a boreal forest under winter clear-sky conditions (Pomeroy and Dion, 1996). These results were contradicted in a subalpine forest stand in Switzerland where the authors showed an increase in canopy albedo with snow interception (Stähli *et al.*, 2009); however, this increase had no effect on the melting of the snow below canopy. Particles in the atmosphere can accumulate on snow, reducing the albedo and increasing melt rates (Warren and Wiscombe, 1980).

4.2.3 Forest management to improve water yield

In higher-elevation forests in temperate climates, water in the form of snow is stored in the forests throughout the winter and released slowly in the spring and summer when water demands for agriculture and human consumption are higher (Mote *et al.*, 2005). In the continental Rocky Mountains, 70 to 80% of summer flow comes from snowmelt from the high-elevation alpine and subalpine forest zones (Troendle, 1983). In the last few centuries, the demand for water has increased greatly with the number of people, which requires water managers to find ways to augment the downstream water available for municipalities and farmers. Alteration of forest canopy to increase stream runoff is possible, and some recent analyses have demonstrated that the economic benefits from increased water usage are sufficient to offset the cost of thinning in mountainous watersheds (Podolak *et al.*, 2015).

Alteration of forest canopy cover has been widely researched as a method to increase water yield (Troendle, 1983; Troendle *et al.*, 2001). The effect of partial or total removal of the vegetation on streamflow is twofold. First, vegetation removal reduces the tree evapotranspiration losses. Second, in the absence of forest cover, all fallen snow reaches the ground, which is ideal from a water management perspective. However, without the shading provided from a forest canopy, snow melts earlier in the season, before the vegetation is transpiring, and is transported as runoff downstream filling reservoirs, but without as much water stored in the snowpack. Conversely, in a dense forest with a closed canopy, snow interception by the canopy is high, which decreases the depth of the snowpack on the ground and consequently the amount of water that could potentially enter the system. However, the trees in these dense forests provide snowpack shading, thus slowing down the melt process. Therefore, land and water managers have considered altering the forest canopies through clearcutting and thinning to accumulate sufficient snow in the watersheds during the winter that will melt slowly when temperatures increase in order to generate a long-duration hydrograph with low peak flows.

Results from various studies generally recommend a minimum of 20% reduction in forest cover in order to obtain an increase in the water yield (Stednick, 1996). This also means that forest activities that result in less than 20% canopy reduction are unlikely to result in any significant offsite impacts on runoff from forest management (Podolak *et al.*, 2015). Many studies have focused on identifying the optimum size of forest thinning gaps in order to allow an increase in snow accumulation, but also to provide sufficient shading to prevent early melting. Results from these studies are mixed, but there is a consensus that the maximum accumulation of snow – in the absence of strong winds – occurs in clearings of 2 to 5*H*, where *H* is the height of the surrounding trees (Golding and Swanson, 1978; Varhola *et al.*, 2010). Similarly, both field observations and theoretical work demonstrate that snowmelt rates are lower in clearings of 1*H* and 2*H* than in open and fully covered forests (Lawler and Link, 2011). However, natural factors such as topography (i.e. elevation and aspect) and year-to-year weather variability (amount of precipitation in the form of snow, temperature and solar radiation) tend to overshadow effects of forest management on snowpack accumulation and melt rates.

The processes that control the snow accumulation and melt in forests are well understood, but the variability in year-to-year weather makes predictions challenging. Considerable advancements have been made in translating this knowledge to computer models in order to better understand forest hydrology in snow-dominated areas and to improve runoff predictions (Chapter 9, this volume; Flerchinger *et al.*, 1994; Flerchinger, 2000; Gelfan *et al.*, 2004; Elliot *et al.*, 2010; Srivastava *et al.*, 2015).

4.3 Forests and Avalanches

One of the more highly publicized concerns related to snow on steep slopes is snow avalanches. Atmospheric forms of snow crystals are extremely unstable in most ambient conditions on the surface of the earth. As soon as snow is deposited on the ground, it undergoes changes where the rates and processes are driven by temperature, vapour pressure and other complex factors. Some processes promote cohesion and bonding of the developing snowpack, while others result in relatively weak, cohesionless layers. In most mountain environments, a complex snowpack develops during the accumulation season that may contain multiple strong and weak layers in a single profile. Snowpack properties vary greatly in both space and time. This natural spatial and temporal variability gives rise to the possibility of snow avalanches and also explains the difficulty in predicting them.

Landscapes in snowy regions share complex relationships with avalanches that are driven by interconnections of terrain, weather, climate, ecology and land use. Forests may ameliorate or exacerbate avalanche issues, and avalanches may influence forest dynamics. These complex interactions were first identified and documented in Europe nearly five centuries ago when people observed increased avalanche devastation on towns following removal of local forests. The same physiographical factors that promote avalanche activity are also often associated with rich mineral deposits in North America and many of the greatest avalanche tragedies have been related to mining operations in the western USA and Canada (Armstrong, 1977). Many of these catastrophic events were in forested basins where timber practices perpetrated by the miners themselves were largely responsible for the related fatalities.

4.3.1 Snow avalanche anatomy and characteristics

Snow avalanches occur where there is a significant snowpack capable of sliding, terrain capable of producing an avalanche, a trigger causing failure, and gravity to drive the movement. The primary terrain factor is slope, although aspect and roughness also play important roles. A snowpack capable of avalanching usually requires a cohesive layer overlying a weak layer (slab avalanche) or a weak matrix capable of downslope disaggregation (loose snow avalanche). Avalanches occur when stress due to gravitational forces exceeds the shear strength within the snowpack (Perla, 1971). A trigger may be an increase in stress or a decrease in strength. Increased stress may consist of any loading event (precipitation, wind loading, skiers, explosives, etc.). A decrease in strength may result from relatively subtle snow metamorphism processes (e.g. faceting) or a more dramatic weather forcing (temperature change or liquid water movement). In either case, failure occurs when stress exceeds the snowpack's ability to resist shear. Local failure in the snowpack increases the stress on surrounding grains and layers. If the failure propagates over a significant area, an avalanche results.

Avalanche paths have a starting zone, track and runout zone (Fig. 4.1). The starting zone is

Fig. 4.1. Typical avalanche paths near Loveland Pass, Colorado, USA. Path A is located entirely above the treeline. Path B intersects the treeline and deposits debris in the valley bottom. Both paths have starting zones (1), tracks (2) and runout zones (3), which are delineated approximately by the black lines in this figure. The transition between the zones and the margin and extent of the paths themselves vary with the magnitude and season of individual events. In extreme events all of path A may be part of the starting zone of path B. In minor events the starting zone alone of path B may contain a starting zone, track and runout zone as shown in path A. Note the vegetation growth and recovery in the heal zone of the upper right of the runout zone in path B (B3). A high-frequency zone remains free of larger woody vegetation just to the left of this heal zone. A high-magnitude, low-frequency event may remove this heal zone.

the area of snow accumulation, as well as the location where failure usually occurs (Fig. 4.1, zone 1). The track is a portion of the path where transport of snow from the starting zone primarily takes place, although additional mass is also often entrained (Fig. 4.1, zone 2). The runout zone is where the avalanche decelerates and stops, depositing the avalanche debris (Fig. 4.1, zone 3). In large, well-defined paths, the different zones are obvious, particularly if they lie in the forested zone (Fig. 4.1, path B). In small paths or paths located above the treeline, the delineation of the different zones may be subtle (Fig. 4.1, path A). Starting zones are often located above the treeline in alpine zones where wind loading on lee slopes enhances snow deposition. Starting zones above the treeline feed tracks that intersect the treeline and flow through a well-defined path and runout zone without significant tree cover. Paths that lie wholly below the treeline often have obvious starting zones, tracks and runout zones defined by forest margins or avalanche gullies that have developed over time. Snowpack character, terrain and vegetation may all affect path morphology and extent, and many variations in avalanche paths are manifest in a variety of snow climates from the maritime to continental and high arctic. Avalanche effects on trees and vegetation are discussed in greater detail below. Greater detail regarding avalanche anatomy and phenomena can be found in McClung and Schaerer (2006).

In maritime climates, avalanches tend to be a two-step process. The first step is the accumulation of a snowpack of sufficient depth to initiate a landslide. This is generally followed by a rainfall event, which could be a continuation of the snowmelt event, until the snowpack is sufficiently saturated that, as with the slab avalanche, the internal stresses from the heavy wet snow exceed the resistance to stress from the rainfall-weakened snowpack (Conway and Raymond, 1989). Such avalanches more typically are confined to historical chutes, maintained by frequent landslides.

4.3.2 Avalanche effects on forests

Avalanche path ecology has been studied across the globe in avalanche-prone areas (Walsh *et al.*, 1994). From a forestry perspective, avalanches have profound effects. Avalanches prevent forests from establishing mature trees in paths, remove mature trees and stands in extreme events, and control species composition within paths. Indeed, avalanche paths below treeline can be characterized in a variety of ways by the trees located in, along the margins, and surrounding the path. Event frequency, magnitude and impact can be reconstructed using a variety of methods including dendrochronology (Burrows and Burrows, 1976; Elder *et al.*, 2014).

The high-frequency zone is defined as the part of a path that is subject to recurring events capable of maintaining a treeless, or largely treeless, cover. High-frequency events make tree establishment difficult and these portions of the path are often covered by annual herbaceous vegetation, small woody vegetation or no vegetation at all. The margin between the high-frequency zone and the path boundary is defined by the trim line, an obvious change from no trees to a mature forest. In large paths, paths with less frequent large events, or in paths with complicated terrain features that control flow some of the time, there is often a heal zone. The heal zone is readily identified by regeneration of tree species well suited to disturbance or young trees typical of the surrounding forests.

Extreme events are capable of leaving well-defined paths in the track or runout zone and, on rare occasions, may even cause significant damage to the forest in the starting zone or track. These extreme events remove mature timber and redefine the boundaries of the path. Because they are extreme events, they are infrequent by nature, and the extended boundaries of the path are often quickly occupied by regenerating forest and become a heal zone. This recovery may be a slow process because once the mature forest is removed, smaller-magnitude events may more easily pass the former path margins. While forests may suffer extensive damage from avalanches, they also offer the most effective and widespread protection from avalanches. Thus, avalanches may have profound effects on forests, but forests also affect avalanches.

4.3.3 Forest effects on avalanches

Forests affect avalanches in both passive and active processes. Passive processes include forest influences on snowpack accumulation and distribution (Section 4.2), which may define a starting zone, as well as controls of snowpack energy balance. Trees may act as anchors for snow on slopes, but they may also cause localized weak points where avalanches can start. Forest canopies may intercept up to 60% of the annual snowfall (Storck *et al.*, 2002), but the fate of intercepted snow is species- and weather-dependent and may follow many paths including sublimation, melt and drip, and unloading. Losses due to sublimation coupled with the anchor effect explain why avalanches seldom release from dense forest: the necessary mass does not accumulate and what does accumulate is pinned in place by tree boles.

As described in Section 4.2, forests also have a profound effect on the local energy balance of the snowpack. Incoming shortwave radiation is diminished at the snowpack surface under a canopy, longwave radiation is increased, and turbulent energy exchanges are typically reduced as wind is suppressed. The snow surface energy balance as an upper boundary layer, coupled with the ground surface below the snowpack as a lower boundary layer, drives snow metamorphism within the snowpack. Snow metamorphism ultimately controls the snowpack structure creating strong and weak layers, a stable snowpack or a snowpack capable of avalanching. The difference in snowpack metamorphism and resultant stratigraphy is often great between forested and unforested slopes.

Active processes affected by trees include avalanche track delineation, roughness or friction on snow in motion, entrainment of woody material, mass in the avalanche flow, and runout zone processes that resist or arrest flow. Trees represent the greatest source of friction or flow impediment for avalanches, followed by significant terrain features. Flow through trees dissipates kinetic energy and alters velocity. Small trees may bend, break or be buried by flow, with reduced effect on the moving mass. Avalanche flow may be stopped completely by downslope stands of large trees. Large trees on the margins of runout zones may experience frequent events that are incapable of causing significant damage, but longer-return events may periodically devastate the established boundaries (Schlappy *et al.*, 2014). Entrainment of woody materials from anywhere in the path, including the runout zone, may significantly increase the destructive force of avalanches as they continue their journey to rest.

4.3.4 Avalanches and hydrology

Snow avalanches are an effective mechanism for moving large volumes and masses of snow from one location to another. Two notable differences of hydrological significance occur when snow is moved by avalanches. The first is that snow is moved from one energy balance regime to another. Snow deposited in high-alpine starting zones is subject to high values of incoming solar radiation, but also to large losses of outgoing longwave radiation. Temperatures are typically cold and winds are relatively high. When snow is moved to a valley bottom, it encounters reduced incoming solar radiation and reduced longwave losses from reduced atmospheric transmittance and because of increased shading from valley side walls. Temperatures are warmer and winds are usually lower than in alpine areas. Overall, one can expect snow to melt faster in the valley bottom or the runout zone environment.

The second hydrological effect is that the structure and distribution of the snowpack are altered. Snowpacks in the starting zone are often relatively shallow and snow is distributed over a large area. Densities are low, except in wind deposits. Avalanching generally transforms this low-density, shallow, extensive snowpack into a high-density, deep snowpack with a smaller surface area.

The hydrological effect of avalanching depends on the relative change in the snowpack properties between the starting zone and the deposition area, and the relative difference in the energy balance between the two locations. Snow typically melts more rapidly and earlier in the season at lower elevations, but often deep avalanche deposits in mountain valley bottoms last well into the melt season. Indeed, these deposits often outlast the surrounding snowpack by weeks or even months. Martinec and de Quervain (1971) found an avalanche in the Swiss Alps that increased early-season streamflow and decreased late-season streamflow compared with modelled non-avalanche results. Other researchers have observed different results (e.g. Sosedov and Seversky, 1965). Effects of avalanching on runoff regimes vary and require local, path- or event-specific investigations.

The movement of snow by avalanches also changes the hydrological pathways of snow that would have otherwise melted in the starting zone. Snowmelt in the upper reaches of a basin undergoes a number of hillslope processes before reaching the stream channel, including overland flow, infiltration, subsurface flow and maybe loss to evapotranspiration (Section 4.4). Snowmelt from avalanche deposits in the valley bottom has a short or direct path to the stream channel. Snowmelt from the two different locations may have very different biogeochemical signatures given the difference in pathways, residence time and exposure to soils, vegetation and geology (Sánchez-Murillo *et al.*, 2015).

Snow avalanches may alter runoff through damming of streams. Flow may be impounded and released catastrophically, leading to downstream flooding after initially decreasing runoff. Flood response may alter downstream vegetation and channel morphology, and have significant impact on life and infrastructure. Finally, avalanches may impact ice-covered lake surfaces causing a plunger effect, rapid expulsion of stored water and downstream flooding (Williams *et al.*, 1992).

4.4 Hydrology in Steep Watersheds

High-elevation steep watersheds include montane and subalpine regions with or without forests that have similar soil water and baseflow processes, but differ in vegetation impacts on hydrology. Chapter 2 provided a good introduction to forest hydrological processes. This chapter expands on those fundamental hydrological processes by applying those principles to watersheds where snow accumulation and melt usually dominate the hydrological response. The focus in this chapter is on hydrological processes common in small steep watersheds or hillslopes. Figure 4.2 is a diagram of the dominant flow processes that are described in this section.

Upland forested watersheds are characterized by steep slopes, shallow soils, absence of flood plains, high precipitation and low evapotranspiration, in contrast to lower-elevation watersheds. Consequently, streams respond quickly to storms. Water from rainfall or snowmelt generally moves through hillslopes into streams following the pathways of overland flow, subsurface flow and baseflow (Fig. 4.2). For all of these pathways, as slope steepness increases, so does the velocity of the flow. Each of these pathways responds differently to snowmelt than to rainfall in generating runoff amount, peak flows and timing of runoff contributions to streamflow. Each runoff pathway is influenced by the complex interactions among climate, vegetation, topography, soil characteristics and geology. In forest soils, the presence of macropores, soil pipes from decayed tree roots, lateral and vertical tree root systems, animal burrows, large cracks, etc. provides highly permeable conduits for rapid movement of water in the direction of the macropores, further contributing to subsurface flow processes (Aubertin, 1971).

4.4.1 Rain- versus snow-dominated hydrological systems

In steep mountainous regions, generally, precipitation increases and temperatures decreases with elevation. The air and surface temperatures directly affect the phases of precipitation (rain, sleet or snow). Higher-elevation and/or high-latitude watershed processes are driven by snowfall and snowmelt events while at moderate elevations and latitudes the more dominant watershed processes are rain or ROS events. Lower-elevation runoff response in many regions is constrained mostly by low precipitation. Both snow- and rain-dominated processes are influenced by large seasonal variability and both processes are usually highly interactive with vegetation. In the more snowmelt-dominated regions, the source of runoff for streamflow usually occurs in the late spring or early summer. Generally, under snowmelt conditions, a significant amount of water percolates

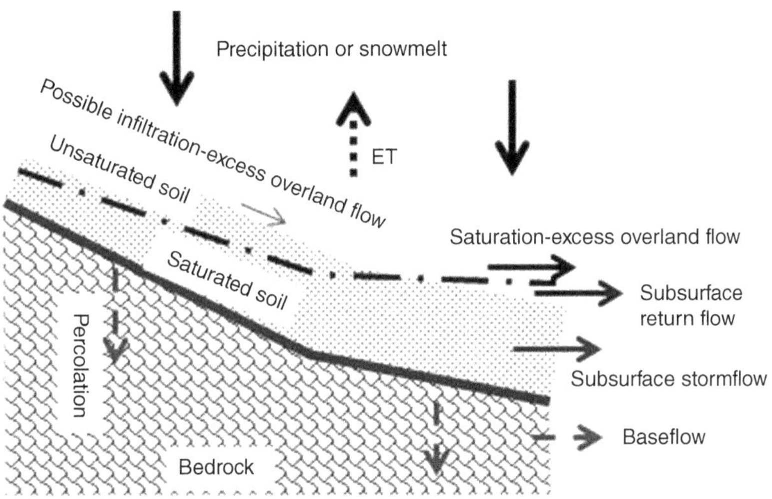

Fig. 4.2. Hydrological pathways on a steep, concave forested hillslope (ET = evapotranspiration).

into the bedrock underlying the soils (Wilson and Guan, 2004). In rainfall- or ROS-dominated watersheds, peak flows are linked to the seasonal weather patterns. In these watersheds peak flows can occur at any time, including summer periods from intense rainfall events.

4.4.2 Infiltration and runoff processes

Understanding hydrological response from snow accumulation and melt is critical for the assessment of water resources and to identify sources of sediment and nutrients. In steep undisturbed forested watersheds, infiltration-excess overland flow is rare (Bachmair and Weiler, 2011). The presence of an organic litter layer covering the ground surface, together with highly permeable litter and soils, result in a high infiltration capacity that generally exceeds the rainfall intensity (Elliot, 2013). Water preferentially moves vertically down into subsurface soil layers instead of over the surface. However, in some cases rainfall or snowmelt rates can exceed infiltration capacity, resulting in 'infiltration-excess' overland flow (Fig. 4.2; Luce, 1995). This is more likely to occur on forest roads and logging trails due to soil compaction (Elliot, 2013), following wildfire where surface soil infiltration capacity is reduced due to litter loss and water repellency effects (Chapter 13, this volume; Elliot *et al.*, 2010), and on frozen soils. Soils that have a high clay fraction are uncommon in steep watersheds but, as long as the ground cover is not disturbed, will seldom have surface runoff (Conroy *et al.*, 2006). Soils that are high in rock content or rock outcrops are also more likely to generate surface runoff (Arnau-Rosalén *et al.*, 2008; Brooks *et al.*, 2016; Robichaud *et al.*, 2016). Infiltration-excess runoff is also common in areas dominated by continental climates with high-intensity thunderstorm or monsoonal rainfall events.

In addition to infiltration excess, the other process that can lead to surface runoff is 'saturation-excess' (Luce, 1995). A soil can become saturated because of a combination of a prolonged wet spell, high infiltration on a shallow soil or location on the landscape, usually near the bottom of a hill. At the bottom of steep hills, the slope begins to flatten and subsurface flow can be directed towards the surface (Fig. 4.2). Under these saturated conditions, the soil may be unable to absorb as much rainfall or snowmelt as is available, resulting in surface runoff even though the melt rate of snow or rainfall intensity is below the saturated hydraulic conductivity of the soil (Luce, 1995). In some cases, the soil can become sufficiently saturated that soil water seeps to the surface and becomes 'subsurface return flow' (Fig. 4.2). Such seepage locations are sometimes referred to as spring lines on the landscape.

Baseflow is generally the return of water that has percolated into fissured bedrock beneath forest soils and/or water accumulated on top of the soil–bedrock interface forming a perched water table that drains slowly to the nearest channel or seeps into the groundwater. This water can take from days to years to find its way to forest streams. The rate of return depends on the amount of water that has percolated and the geology (Sánchez-Murillo *et al.*, 2014; Brooks *et al.*, 2016).

Because there is seldom infiltration-excess overland flow in steep forests, subsurface stormflow, saturation-excess runoff, subsurface return flow and baseflow are the dominant runoff generation mechanisms (Brooks *et al.*, 2004; Srivastava *et al.*, 2013). The relative role of each pathway is dependent on site conditions, including soil properties, bedrock permeability, topography and soil water content.

4.4.3 Soil water-holding capacity

In steep forest hydrology, important soil properties include soil thickness, bedrock permeability, drainable porosity (difference between soil water content at saturation and field capacity), topography, soil water conditions at the time of the rainfall or snowmelt event (antecedent conditions) and plant rooting depth. Spatial variability of these properties within the hillslope directly affects soil storage capacity, water availability for evapotranspiration, and vertical and lateral movement of water within and below the soils, defining the runoff generation processes.

Soil texture interacts with soil thickness to influence hydrological response on steep forested hillslopes. Coarse-textured soils have higher infiltration rates and higher hydraulic conductivities, but may not be able to retain as much water as finer-textured soils (Teepe *et al.*, 2003). In many steep watersheds, soils higher in rock content will not be able to retain as much water as soils with fewer rocks (van Wesemael *et al.*, 2000).

Soil thickness and other properties that affect availability of soil water are considered one of the most important influences on the relative proportion of subsurface stormflow, the availability of soil water for percolation into the bedrock, and for the uptake of water by vegetation. The topography plays an important role in determining the variability of soil thickness and texture (van Wesemael *et al.*, 2000). Both at the hillslope and watershed scale, generally, shallower soils are more common on the upland steep slopes and deeper soils on gently sloping lowlands. Using a water balance approach, the major hydrological processes in shallower and deeper soils on the sloping terrain can be contrasted. To understand the effect of soil depth, assume that both shallow and deep soils are well-drained, are distributed above low-permeability bedrock and receive the same precipitation inputs. In soils with less water-holding capacity, soil water content will approach saturation faster. As a result, there will be more subsurface stormflow and bedrock infiltration, and low seasonal evapotranspiration because less water is available water (difference between soil water content at field capacity and wilting point) for shallow-rooted vegetation to extract (van Wesemael *et al.*, 2000). Depending on the weather, the shallow soil is more likely to become saturated during a wet spell, leading to saturation-excess runoff. In contrast, in soils with high water-storage capacity, the soil is less likely to become fully saturated and more likely to hold more of the infiltrated water, resulting in less subsurface stormflow and bedrock percolation. Evapotranspiration will be greater on deeper soils because more water is available for deeper-rooted vegetation to extract the additional available soil water. Figure 4.3 shows the comparison of cumulative observed and simulated hillslope hydrological responses from two watersheds (with different soil thickness) located in the Lake Tahoe Basin, Nevada, USA (Brooks *et al.*, 2016). The Blackwood Creek watershed with shallow soils showed the major contribution to streamflow from subsurface stormflow in May and June (Fig. 4.3a) in addition to baseflow. The General Creek watershed with deep soils showed little subsurface stormflow following snowmelt, with the major contribution to streamflow coming from baseflow (Fig. 4.3b).

4.4.4 Bedrock permeability

The importance of bedrock permeability on subsurface stormflow and bedrock percolation can be understood by comparing watershed response from highly permeable bedrock to that from bedrock with low permeability. Soils occurring above permeable bedrock will generate less subsurface stormflow and more bedrock percolation as gravity will more quickly route water from the soil layer into the bedrock (Fig. 4.2). Seasonal evapotranspiration will be less because the soil will retain less water for plants to access later in the growing season. In comparison, soils overlying less permeable bedrock will have less deep percolation into the bedrock, leaving more water in the soil column. This can lead to higher seasonal evapotranspiration because deep seepage rates from the soil are lower, increasing the water available for plants longer into the growing season.

4.4.5 Topography

The hillslope topography plays a major role in mountainous hydrological processes. Based on slope length and the variability in steepness, most of the hillslopes can be characterized as convex, concave or planar (Fig. 4.4). Different sections within the hillslope profile can significantly affect soil saturation and flow behaviour along the hillslope. Figure 4.2 shows an idealized concave hillslope configuration with a steep slope at the top and a gentler slope at the bottom. Figure 4.2 assumes the hillslope has well-drained soils of uniform thickness with low-permeability bedrock and it shows the pathways of subsurface stormflow and saturation-excess overland flow or subsurface return flow to the stream. Before a storm event, baseflow from an unconfined aquifer will define a water table near the stream. During or after the storm event, the combination of direct precipitation falling on the valley bottom

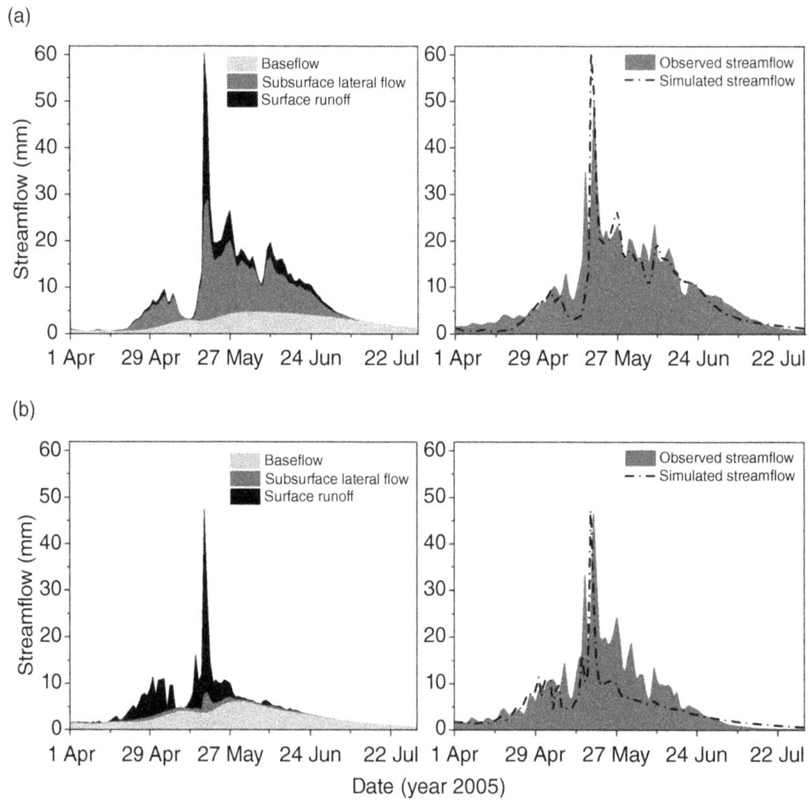

Fig. 4.3. Simulated and observed hydrological responses from two watersheds located in the Lake Tahoe Basin: (a) Blackwood Creek watershed (shallow soils); (b) General Creek watershed (deep soils). (Courtesy of Brooks *et al.*, 2016.)

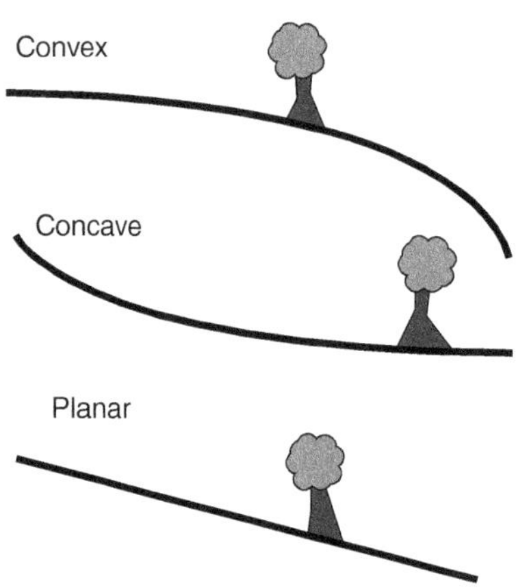

Fig. 4.4. Convex, concave and planar slope shapes.

and water moving downhill as subsurface stormflow towards the valley bottom will elevate the water table and increase the soil water content. For small or moderate rain or snowmelt events, the soil is unlikely to become saturated, even at the bottom of the hill. Therefore, saturation-excess overland flow will be rare and subsurface stormflow will dominate the volume of runoff to streamflow. However, during large rain or snowmelt events, more of the soil profile will become saturated. In this case, saturation-excess runoff and subsurface return flow will likely become the dominate pathways for runoff to stream. The soil thickness will influence the degree of soil saturation near the valley bottom and the relative contributions of each of the runoff pathways. Convex and planar slopes are less likely to experience saturation-excess or subsurface return flows unless the underlying geology diverts subsurface flow to the surface.

At a landscape scale, soil depth is not uniform, but rather tends to be deeper in upland swales and less deep on upland ridges. Soils tend to be deeper on the lower parts of concave slopes and less deep on shoulder areas of convex slopes. Underlying geology can further affect soil depth, leading to a diversity of potential seepage points as lateral flows are forced to the surface or as surface runoff seeps back into the soil profile (McDonnell *et al.*, 2007).

4.4.6 Antecedent soil water conditions

Most of the runoff generation mechanisms are dependent on soil water content at the time of the rainfall or snowmelt event. This is referred to as 'antecedent soil water'. In general, the higher the soil water content, the faster and more intense will be the runoff response to a precipitation or snowmelt event. Continuous processes such as soil evaporation, plant transpiration, subsurface stormflow and bedrock percolation gradually reduce the soil water content in the days following an event until the next event occurs. Therefore, runoff generation mechanisms are dependent not only on soil water conditions during a precipitation event, but also on the hydrological history of the site.

Soil water conditions are strongly associated with climatic conditions (precipitation, temperature, solar radiation, etc.) and seasonal variability in weather. In high-elevation snow-dominated watersheds, low temperatures in the winter generally result in snow accumulation. Soil water content during this period is frequently low because the evapotranspiration from the previous growing season had used up available soil water, and temperatures dropped below freezing before the onset of winter precipitation. In the following spring, a gradual increase in temperature typically results in snowmelt that increases soil water content and subsequently activates the runoff pathways in Fig. 4.2, often helped by early-spring rain on snow, or later-spring rainfall events on wet soil. In the summer or dry season, evapotranspiration reduces soil water content, especially in climates with dry summers as typical of the western USA and southern Europe. Soil water content is a highly dynamic process that is controlled not only by climatic conditions but also is influenced by soil properties (drainable porosity, thickness, texture), vegetation (species, age, rooting depth), underlying geology (bedrock permeability, soil depth) and topography (van Wesemael *et al.*, 2000).

The complex interactions just described make it challenging to fully understand high-elevation steep hydrological processes and to explain different hydrological responses observed on similar watersheds. In recent decades, researchers have focused on studying detailed hydrological processes operating on small watersheds. This has led to the development of new theories on the reasons for variability in hydrological responses from these watersheds, such as the importance of spatial variability of soil and vegetative properties, microclimates, snowmelt dynamics, and surface and groundwater interactions (Bachmair and Weiler, 2011). Current research on mountainous hydrological processes relies heavily on the development and application of complex computer models that can aid in understanding the hydrological interactions described in this section (Chapter 9, this volume; Kirchner, 2006; Elliot *et al.*, 2010).

4.5 Likely Effects of a Changing Climate on Watershed Processes

Snow processes, avalanche occurrences and steep slope hydrology are all dependent on weather patterns. The climates that were the basis of the knowledge presented in this chapter are not constant, however (Milly *et al.*, 2008). Even though the principles presented are valid, the watershed responses to climate change are already occurring, particularly the snow processes. During the 20th century, the air temperatures have increased (Folland *et al.*, 2001), decreasing the mountain snowpack, especially at elevations below 1800 m, and threatening water resources (Mote, 2003; Mote *et al.*, 2005; Milly *et al.*, 2008). The increase in temperature will not only result in a decrease in the snow accumulation, but will also affect the timing and volume of spring snowmelt.

Predictions of future climates around the world suggest a continued increase in temperatures and an increase or decrease in precipitation (Stocker *et al.*, 2013). A less understood change, but likely to have a significant impact on steep forest hydrological processes, is an increase in precipitation intensity, sometimes described as 'intensification'. Intensification is generally associated with increased within-storm peak rainfall intensity, fewer days with precipitation, but greater amounts of precipitation on days when precipitation does occur, and longer periods (dry spells) between precipitation events (Bayley *et al.*, 2010). Increased intensification will likely lead to increased surface runoff and lateral flow, and decreased deep seepage and baseflow.

If the increase in temperature predicted by all future climate scenarios is considered, then major changes can be anticipated in the amount and the distribution of snow-covered areas (Beniston *et al.*, 2003; Zierl and Bugmann, 2005; Manninen and Stenberg, 2009). Although high-elevation mountains will experience changes associated with temperature increase, the snow accumulation areas most sensitive to climate change will be the snow-covered mid- or low-elevation mountains that lie within the extent of the rain–snow transition zone. In the western USA, it is estimated that by the middle of the 21st century, a change in precipitation phase from snow to rain will cause a 30% decrease in the winter snow-dominated areas (Klos *et al.*, 2014). These predicted conditions can have a dramatic effect on the overall hydrology in maritime climates, with likely increases in the number and magnitude of high peak flow events. Streams will have greater early-spring flows and summer flows are likely to decline. In some areas, there will likely be more streams that dry up completely before the end of the dry season (Rauscher *et al.*, 2008). Another concern about the decrease in snowpack is a decrease in albedo, which is complicating energy balance analyses in climate change models (Manninen and Stenberg, 2009). As landscape-scale albedo decreases with decreasing snowpack, so the energy within the earth's atmosphere will increase, likely leading to greater increases in temperature.

The effect of a reduced snowpack on peak flows is less clear, and will likely be climate-dependent. In some areas, where peak flows are frequently associated with rain falling on a melting snowpack, the reduced snowpack could result in a reduced peak flow rate. In other areas, heavy early- and late-winter rain events on saturated soils rather than snow could lead to an increase in flooding.

Intensification of rainfall events with more rainfall on wet days is likely to lead to an increase in days with high streamflow rates, followed by periods of very low streamflow. There will likely be fewer days when the soils are saturated, reducing the amount of groundwater recharge and baseflow rates later in the summer.

The increasing temperatures will also encourage vegetation to begin transpiring earlier in the growing season and will likely lead to soil water depletion earlier in the growing season. In drier forests this could alter the plant communities, with trees and shrubs that are more tolerant of dry soils gradually replacing plants that thrived under wetter conditions (Rehfeldt *et al.*, 2006). The earlier loss of soil water will also lead to an increased risk of wildfire (Schumacher and Bugmann, 2006; Westerling *et al.*, 2006). This increase is already apparent with increased fire frequency and intensity in forests in both the northern and southern hemispheres (Jolly *et al.*, 2015). Wildfire often leads to increased peak flows and sediment delivery from burned forest watersheds (Chapter 13, this volume; Elliot *et al.*, 2010).

4.6 Summary

Mountainous and snow-dominated forests are major sources of water on every continent. Snow accumulation and melt rates are big drivers of the hydrological processes. Temperature, radiation, and the interaction between rain and snow are major factors in generating runoff. Surface and subsurface flow processes route water from saturated hillsides to deeper groundwater storage or directly to streams, resulting in hydrographs with long duration and low peak flow. Flooding is associated with high rainfall and snowmelt rates, and ROS events occurring when soils are saturated. Future climates will likely generate a greater impact on floods from rainfall events and less snow accumulation. The impact that future climate scenarios have on streamflows cannot be generalized, but will depend on elevation, source of air masses, and complex interactions among the climate, topography, soil and vegetation. Hydrological models will continue to play an important role in better understanding these complex interactions.

References

Armstrong, B. (1977) *Avalanche Hazard in Ouray County, Colorado, 1877–1976*. Institute of Arctic and Alpine Research Occasional Paper No. 24. University of Colorado, Boulder, Colorado.
Arnau-Rosalén, E., Calvo-Cases, A., Boix-Fayos, C., Lavee, H. and Sarah, P. (2008) Analysis of soil surface component patterns affecting runoff generation. An example of method applied to Mediterranean hillslopes in Alicante (Spain). *Geomorphology* 101, 595–606.
Aubertin, G.M. (1971) *Nature and Extent of Macropores in Forest Soils and Their Influence on Subsurface Water Movement*. USDA Forest Research Paper NE-192. USDA Forest Service, Northeastern Forest Experiment Station, Upper Darby, Pennsylvania.
Bachmair, S. and Weiler, M. (2011) New dimensions of hillslope hydrology. In: Levia, D.F., Carlyle-Moses, D. and Tanaka, T. (eds) *Forest Hydrology and Biogeochemistry: Synthesis of Past Research and Future Directions*. Ecological Studies Vol. 216. Springer, Heidelberg, Germany, pp. 455–481.
Bales, R.C., Molotch, N.P., Painter, T.H., Dettinger, M.D., Rice, R. and Dozier, J. (2006) Mountain hydrology of the western United States. *Water Resources Research* 42(8), 1–13.
Bayley, T., Elliot, W., Nearing, M.A., Guertin, D.P., Johnson, T., Goodrich, D. and Flanagan, D. (2010) Modeling erosion under future climates with the WEPP model. In: Hydrology and Sedimentation for a Changing Future: Existing and Emerging Issues; Proceedings of the 2nd Joint Federal Interagency Conference, 27 June–1 July, Las Vegas, Nevada. Available at: http://acwi.gov/sos/pubs/2ndJFIC/Contents/2B_Elliot_3_25_10_paper.pdf (accessed 1 April 2016).
Beniston, M., Keller, F., Koffi, B. and Goyette, S. (2003) Estimates of snow accumulation and volume in the Swiss Alps under changing climate conditions. *Theoretical and Applied Climatology* 76, 125–140.
Berris, S.N. and Harr, R.D. (1987) Comparative snow accumulation and melt during rainfall in forested and clear-cut plots in the Western Cascades of Oregon. *Water Resources Research* 23, 135–142.
Berry, G.J. and Rothwell, R.L. (1992) Snow ablation in small forest openings in southwest Alberta. *Canadian Journal of Forest Research* 22, 1326–1331.
Brooks, E.S., Boll, J. and McDaniel, P.A. (2004) A hillslope-scale experiment to measure lateral saturated hydraulic conductivity. *Water Resources Research* 40, W04208, doi: 10.1029/2003WR002858 (accessed 20 March 2016).
Brooks, E.S., Dobre, M., Elliot, W.J., Boll, J. and Wu, J.Q. (2016) Assessing the sources and transport of fine sediment in response to management practices in the Tahoe Basin using the WEPP Model. *Journal of Hydrology* 533, 389–402.
Burrows, C. and Burrows, V. (1976) *Procedures for the Study of Snow Avalanche Chronology using Growth Layers of Woody Plants*. Institute of Arctic and Alpine Research Occasional Paper No. 54. University of Colorado, Boulder, Colorado.
Conroy, W.J., Hotchkiss, R.H. and Elliot, W.J. (2006) A coupled upland-erosion and instream hydrodynamic-sediment transport model for evaluating sediment transport in forested watersheds. *Transactions of the ASABE* 49, 1713–1722.
Conway, H. and Raymond, C.F. (1989) *Prediction of Snow Avalanches in Maritime Climates*. WA-RD 203.2 Technical Report. Washington Department of Transport, Olympia, Washington.
DeWalle, D.R. and Rango, A. (2008) *Principles of Snow Hydrology*. Cambridge University Press, New York.
Elder, L., Elder, D., McCorkel, N., Markey, E., Starr, B., McClain, D. and Elder, K. (2014) Field observations and mapping of an avalanche event in Second Creek, Berthoud Pass, Colorado. In: ISSW 2014: Proceedings of the 19th International Snow Science Workshop, Banff, Alberta, Canada, pp. 232–238. Available at: http://arc.lib.montana.edu/snow-science/item.php?id=2169 (accessed 1 April 2016).
Elliot, W.J. (2013) Erosion processes and prediction with WEPP technology in forests in the Northwestern US. *Transactions of the ASABE* 52, 563–579.
Elliot, W.J., Miller, I.S. and Audin, L. (eds) (2010) *Cumulative Watershed Effects of Fuel Management in the Western United States*. General Technical Report RMRS-GTR-231. USDA Forest Service, Rocky Mountain Research Station, Fort Collins, Colorado.
Flerchinger, G.N. (2000) *The Simultaneous Heat and Water (SHAW) Model: User's Manual*. Technical Report NWRC 2000-10. USDA Agricultural Research Service, Northwest Watershed Research Center, Boise, Idaho. Available at: http://www.ars.usda.gov/SP2UserFiles/Place/20520000/ShawUsersManual.pdf (accessed 20 March 2016).
Flerchinger, G.N., Cooley, K.R. and Deng, Y. (1994) Impacts of spatially and temporally varying snowmelt on subsurface flow in a mountainous watershed: 1. Snowmelt simulation. *Hydrological Sciences Journal* 39, 507–520.

Folland, C.K., Karl, T.R., Christy, J.R., Clarke, R.A., Gruza, G.V., Jouzel, J., Mann, M.E., Oerlemans, J., Salinger, M.J. and Wang, S.-W. (2001) Observed climate variability and change. In: Houghton, J.T., Ding, Y., Griggs, D.J., Noguer, M., van der Linden, P.J., Dai, X., Maskell, K. and Johnson, C.A. (eds) *Climate Change 2001: The Scientific Basis. Contribution of Working Group I to the Third Assessment Report of the Intergovernmental Panel on Climate Change*. Cambridge University Press, Cambridge, pp. 99–181.

Gelfan, A.N., Pomeroy, J.W. and Kuchment, L.S. (2004) Modeling forest cover influences on snow accumulation, sublimation and melt. *Journal of Hydrometeorology* 5, 785–803.

Golding, D.L. and Swanson, R.H. (1978) Snow accumulation and melt in small forest openings in Alberta. *Canadian Journal of Forest Research* 8, 380–388.

Harr, R.D. (1986) Effects of clearcutting on rain-on-snow runoff in western Oregon: a new look at old studies. *Water Resources Research* 22, 1095–1100.

Hedstrom, N.R. and Pomeroy, J.W. (1998) Measurements and modelling of snow interception in the boreal forest. *Hydrological Processes* 12, 1611–1625.

Hubbart, J.A., Link, T.E., Gravelle, J.A. and Elliot, W.J. (2007) Timber harvest impacts on water yield in the continental/maritime hydroclimate region of the US. *Journal of Forestry* 53, 169–180.

Jefferson, A.J. (2011) Seasonal versus transient snow and the elevation dependence of climate sensitivity in maritime mountainous regions. *Geophysical Research Letters* 38, 16402, doi: 10.1029/2011GL048346 (accessed 20 March 2016).

Jolly, M.W., Cochrane, M.A., Freeborn, P.H., Holden, Z.A., Brown, T.J., Williamson, G.J. and Bowman, D.M.J.S. (2015) Climate-induced variations in global wildfire danger from 1979 to 2013. *Nature Communications* 6, 7537.

Kirchner, J.W. (2006) Getting the right answers for the right reasons: linking measurements, analyses, and models to advance the science of hydrology. *Water Resources Research* 42, W03S04, doi: 10.1029/2005WR004362 (accessed 20 March 2016).

Klos, P.Z., Link, T.E. and Abatzoglou, T.J. (2014) Extent of the rain-on-snow transition zone in the western US under historic and projected climate. *Geophysical Research Letters* 41, 4560–4568.

Lawler, R.R. and Link, T.E. (2011) Quantification of incoming all-wave radiation in discontinuous forest canopies with application to snowmelt prediction. *Hydrological Processes* 25, 3322–3331.

Luce, C.H. (1995) Forests and wetlands. In: Ward, A.D. and Elliot, W.J. (eds) *Environmental Hydrology*. CRC/Lewis Publishers, Boca Raton, Florida, pp. 253–284.

Lundberg, A. and Halldin, S. (2001) Snow interception evaporation. Review of measurement techniques, processes, and models. *Theoretical and Applied Climatology* 70, 117–133.

Lundquist, J.D., Dickerson-Lange, S.E., Lutz, J.A. and Cristea, N.C. (2013) Lower forest density enhances snow retention in regions with warmer winters: a global framework developed from plot-scale observations and modeling. *Water Resources Research* 49, 6356–6370.

Manninen, T. and Stenberg, P. (2009) Simulation of the effect of snow covered forest floor on the total forest albedo. *Agricultural and Forest Meteorology* 149, 303–319.

Marchi, L. Borga, M., Preciso, E. and Gaume, E. (2010) Characterisation of selected extreme flash floods in Europe and implications for flood risk management. *Journal of Hydrology* 394, 118–133.

Marks, D., Kimball, J., Tingey, D. and Link, T. (1998) The sensitivity of snowmelt processes to climate conditions and forest cover during rain-on-snow: a case study of the 1996 Pacific Northwest flood. *Hydrological Processes* 12, 1569–1587.

Martinec, J. and de Quervain, M. (1971) The effect of snow displacement by avalanches on snow melt and runoff. In: *Proceedings of the Snow and Ice Symposium, Moscow, Russia, August 1971*. IAHS–AISH Publication No. 104. International Association of Hydrological Science, Wallingford, UK, pp. 364–377.

McClung, D. and Schaerer, P.A. (2006) *The Avalanche Handbook*, 3rd edn. The Mountaineers Books, Seattle, Washington.

McDonnell, J.J., Sivapalan, M., Vaché, K., Dunn, S., Grant, G., Haggerty, R., Hinz, C., Hooper, R., Kirchner, J., Roderick, M.L., *et al.* (2007) Moving beyond heterogeneity and process complexity: a new vision for watershed hydrology. *Water Resources Research* 43, W07301, doi: 10.1029/2006WR005467 (accessed 20 March 2016).

Milly, P.C.D., Betancourt, J., Falkenmark, M., Hirsch, R.M., Kundzewicz, Z.W., Lettenmaier, D.P. and Stouffer, R.J. (2008) Stationarity is dead: whither water management? *Science* 319, 573–574.

Mote, P.W. (2003) Trends in snow water equivalent in the Pacific Northwest and their climatic causes. *Geophysical Research Letters* 30, 1601.

Mote, P.W., Hamlet, A.F., Clark, M.P. and Lettenmaier, D.P. (2005) Declining mountain snow pack in western North America. *Bulletin of the American Meteorological Society* 86, 39–49.

Peel, M.C., Finlayson, B.L. and McMahon, T.A. (2007) Updated world map of the Köppen–Geiger climate classification. *Hydrology and Earth System Sciences* 11, 1633–1644.
Perla, R.I. (1971) The slab avalanche. PhD dissertation, University of Utah, Salt Lake City, Utah.
Podolak, K., Edelson, D., Kruse, S., Aylward, B., Zimring, M. and Wobbrock, N. (2015) Estimating the water supply benefits from forest restoration in the Northern Sierra Nevada. A report by The Nature Conservancy and Ecosystem Economics, San Francisco, California. Available at: http://www.nature.org/ourinitiatives/regions/northamerica/unitedstates/california/forest-restoration-northern-sierras.pdf (accessed 1 September 2015).
Pomeroy, J.W. and Dion, K. (1996) Winter radiation extinction and reflection in a boreal pine canopy: measurements and modelling. *Hydrological Processes* 10, 1591–1608.
Pomeroy, J.W. and Schmidt, R.A. (1993) The use of fractal geometry in modeling intercepted snow accumulation and sublimation. In: Proceedings of the 50th Eastern Snow Conference, Quebec City, 1993. Available at: http://www.usask.ca/hydrology/papers/Pomeroy_Schmidt_1993.pdf (accessed 1 April 2016).
Pomeroy, J.W., Marks, D., Link, T., Ellis, C., Hardy, J., Rowlands, A. and Granger, R. (2009) The impact of coniferous forest temperatures on incoming longwave radiation to melting snow. *Hydrological Processes* 23, 2513–2525.
Rauscher, S.A., Pal, J.S., Diffenbaugh, N.S. and Benedetti, M.M. (2008) Future changes in snowmelt-driven runoff timing over the western US. *Geophysical Research Letters* 35, L16703, doi: 10.1029/2008GL034424 (accessed 20 March 2016).
Rehfeldt, G.E., Crookston, N.L., Warwell, M.V. and Evans, J.S. (2006) Empirical analyses of plant–climate relationships for the Western United States. *International Journal of Plant Sciences* 167, 1123–1150.
Robichaud, P.R., Elliot, W.J., Lewis, S.A. and Miller, M.E. (2016) Validation of a probabilistic post-fire erosion model. *International Journal of Wildland Fire* 25, 337–350.
Sánchez-Murillo, R., Brooks, E.S., Elliot, W.J., Gazel, E. and Boll, J. (2014) Baseflow recession analysis in the inland Pacific Northwest of the United States. *Hydrogeology Journal* 23, 287–303.
Sánchez-Murillo, R., Brooks, E.S., Elliot, W.J. and Boll, J. (2015) Isotope hydrology and baseflow geochemistry in natural and human-altered watersheds in the Inland Pacific Northwest, USA. *Isotopes in Environmental and Health Studies* 51, 231–254.
Schlappy, R., Eckert, N., Jomelli, V., Stoffel, M., Grancher, D., Brunstein, D., Naaim, M. and Deschatres, M. (2014) Validation of extreme snow avalanches and related return periods derived from a statistical-dynamical model using tree-ring techniques. *Cold Regions Science and Technology* 99, 12–26.
Schumacher, S. and Bugmann, H. (2006) The relative importance of climatic effects, wildfires and management for future forest landscape dynamics in the Swiss Alps. *Global Change Biology* 12, 1435–1450.
Sosedov, I. and Seversky, I. (1965) On the hydrological role of snow avalanches in the northern slope of the Zailiysky Alatua. In: *Proceedings of the International Symposium on the Scientific Aspects of Snow and Ice Avalanches, 5–10 April 1965, Davos, Switzerland*. IAHS–AISH Publication No. 69. International Association of Hydrological Science, Wallingford, UK, pp. 78–85.
Srivastava, A., Dobre, M., Wu, J.Q., Elliot, W.J., Bruner, E.A. Dun, S., Brooks, E.S. and Miller, I.S. (2013) Modifying WEPP to improve streamflow simulation in a Pacific Northwest watershed. *Transactions of the ASABE* 56, 603–611.
Srivastava, A., Wu, J.Q., Elliot, W.J. and Brooks, E.S. (2015) Enhancements to the water erosion prediction project (WEPP) for modeling large snow-dominated mountainous forest watersheds. In: Proceedings of the 2015 EWRI Watershed Management Conference, 4–7 August 2015, Reston, Virginia. Available at: http://ascelibrary.org/doi/abs/10.1061/9780784479322.019 (accessed 1 April 2016).
Stähli, M., Jonas, T. and Gustafson, D (2009) The role of snow interception in winter-time radiation processes of a coniferous sub-alpine forest. *Hydrological Processes* 23, 2498–2512.
Stednick, J. D. (1996) Monitoring the effects of timber harvest on annual water yield. *Journal of Hydrology* 176, 79–95.
Stocker, T.F., Qin, D., Plattner, G.-K., Tignor, M.M.B., Allen, S.K., Boschung, J., Nauels, A., Xia, Y., Bex, V. and Midgley, P.M. (eds) (2013) *Climate Change 2013: The Physical Science Basis. Contribution of Working Group I to the Fifth Assessment Report of the Intergovernmental Panel on Climate Change*. Cambridge University Press, Cambridge.
Storck, P., Lettenmaier, D. P. and Bolton, S. M. (2002) Measurement of snow interception and canopy effects on snow accumulation and melt in a mountainous maritime climate, Oregon, United States. *Water Resources Research* 38, 1223.
Teepe, R., Dilling, H. and Beese, F. (2003) Estimating water retention curves of forest soils from soil texture and bulk density. *Journal of Plant Nutrition and Soil Science* 166, 111–119.

Troendle, C.A. (1983) The potential for water yield augmentation from forest management in the Rocky Mountain Region. *Water Resources Bulletin* 19, 359–373.

Troendle, C.A., Wilcox, M.S., Bevenger, G.S. and Porth, L.S. (2001) The Coon Creek water yield augmentation project: implementation of timber harvesting technology to increase streamflow. *Forest Ecology and Management* 143, 179–187.

van Wesemael, B., Mulligan, M. and Poesen, J. (2000) Spatial patterns of soil water balance on intensively cultivated hillslopes in a semi-arid environment: the impact of rock fragments and soil thickness. *Hydrological Processes* 14, 1811–1828.

Varhola, A., Coops, N.C., Weiler, M. and Moore, R.D. (2010) Forest canopy effects on snow accumulation and ablation: an integrative review of empirical results. *Journal of Hydrology* 392, 219–233.

Viviroli, D. and Weingartner, R. (2004) The hydrological significance of mountains: from regional to global scale. *Hydrology and Earth System Sciences Discussions, European Geosciences Union* 8, 1017–1030.

Walsh, S., Butler, D., Allen, T. and Malanson, G. (1994) Influence of snow patterns and snow avalanches on the alpine treeline ecotone. *Journal of Vegetation Science* 5, 657–672.

Warren, S.G. and Wiscombe, W.J. (1980) A model for the spectral albedo of snow. II: Snow containing atmospheric aerosols. *Journal of Atmospheric Science* 37, 2734–2745.

Westerling, A.L., Hidalgo, H.G., Cayan, D.R. and Swetnam, T.W. (2006) Warming and earlier spring increase western US forest wildfire activity. *Science* 313, 940–943.

Williams, M., Elder, K., Soiseth, C. and Kattelmann, R. (1992) Aquatic ecology as a function of avalanche runout into an alpine lake. In: *Proceedings of the 1992 International Snow Science Workshop, Breckenridge, Colorado, USA,* pp. 47–56. Available at: http://arc.lib.montana.edu/snow-science/item.php?id=1234 (accessed 1 April 2016).

Wilson, J.L. and Guan, H. (2004) Mountain-block hydrology and mountain-front recharge. In: Hogan, J.F., Phillips, F.M. and Scanlon, B.R. (eds) *Groundwater Recharge in a Desert Environment: The Southwestern United States.* Water Science and Applications Series. Vol. 9. American Geophysical Union, Washington, DC, pp. 113–137.

Zierl, B. and Bugmann, H. (2005) Global change impacts on hydrological processes in Alpine catchments. *Water Resources Research* 41, W02028, doi: 10.1029/2004WR003447 (accessed 20 March 2016).

5 European Perspectives on Forest Hydrology

C. de Jong*
University of Strasbourg, Strasbourg, France

5.1 Introduction

Information on European forest hydrology is highly dispersed and sectoral. This is partially due to the fact that Europe consists of 50 different countries, 28 of which form part of the EU with 24 official languages to overcome. Europe stretches from the Mediterranean including Spain, Italy and Greece in the south to Scandinavia including Iceland in the north, and from Ireland in the west to Ukraine and European Russia in the east. Modern Europe is centred on the EU, which evolved from the European Community in 1993 as a political, economic and peace-making entity in reaction to World War II. The EU lies half way between a federation and confederation, and has its own Parliament, court and central bank. EU policies are mandatory and are reinforced at the European, national, regional and local levels. Even though European Russia covers roughly one-third of Europe's surface, it does not belong to the EU and therefore abides by its own laws and regulations. The same is true for Norway, Iceland, Liechtenstein and (not least) Switzerland, an ideal island of comparison within Europe. Taking into account Europe's wide range of climatic and hydrological regimes, long-term studies are clustered in a surprisingly very narrow temperate/alpine belt. This chapter deals with case studies from ten different European countries.

With the advent of EU-financed projects accompanied by individual project reports, literature on forest hydrology has become increasingly segmented. Despite the fact that different EU regulations have been enforced on water quality, water quantity and adaptation to and mitigation of extreme hydrological events in forest hydrology, there is neither a single encompassing book nor a report or a book chapter summarizing its recent evolution and perspectives. This chapter can, therefore, only be seen as an attempt to summarize European perspectives on forest hydrology based on EU and non-EU examples.

European literature on quantitative forest hydrology seems limited compared with studies from the USA, Australia and Japan (Schleppi, 2011). This may be due to the small-scale ownership structures in Europe (P. Schleppi, Birmensdorf, Switzerland, personal communication, 2015) and the lack of large tracts of government land (L. Bren, Melbourne, Australia, personal communication, 2015) that make basin-wide experiments feasible. Andréassian (2004a) corroborates this hypothesis with his observation that the unprecedented development of experimental basins in the 20th century

*E-mail: carmen.dejong@live-cnrs.unistra.fr

occurred mainly in the USA. Europe has produced some notable forest hydrologists and this is still an active field. However the number of forest hydrologists as compared with forest biologists is restricted, in particular those working in alpine, arid and arctic regions. The paired catchment approach was initiated in Europe in the Swiss Sperbelgraben and Rappengraben area and then spread to the USA (see Chapter 1, this volume). Later, the concept was taken up in the Plynlimon catchment in Wales, UK (Blackie and Robinson, 2007). Whereas earlier work focused on floods, there has been a recent shift towards droughts and low flows as well as water quality and protection of drinking-water.

One of the oldest European examples of experimental forest hydrology comes from the Swiss Sperbelgraben and Rappengraben. These were instrumented as paired watersheds (with nearly 100% and 66% forest cover, respectively) in 1902 as a consequence of large floods in the 1860s and 1870s that were thought to be the result of large-scale deforestation and infrequent high rainfall events (Keller, 1988). As early as 1907, the first forest lysimeter station worldwide began investigations into the water budget of young trees on the Drachenkopf Mountain, Eberswalde, Germany (Müller and Bolte, 2009). Somewhat later, in 1948, instrumentation of the Harz Mountains in Germany began after heavy deforestation (Hermann and Schumann, 2009). Early Swiss research showed that forests are effective in strongly reducing peak discharge (by up to 50%) for short and intensive rainfall events but that the difference between forested and non-forested catchments diminishes totally for increasingly strong rainfall events (Fig. 5.1; Hegg, 2006; Schleppi, 2011). Comparison is limited by parameters such as catchment size, morphology, exposition, soil and topography. Schleppi (2011) found that the catchment size and vegetation type affected the frequency of low flow more than peak flow in the two catchments and that, in general, forests deliver less water than other vegetation during droughts. Similarly, results from the now completely afforested Lange Bramke catchment (Harz) in Germany show that forests do not protect against catastrophic floods and can only partially decrease the impacts of smaller floods.

Across Europe, the planning and evolution of flood protection in forested catchments is proceeding at different paces, depending on the political history and status of the countries concerned within the EU. It is well known that flood risk results from the natural characteristics of a catchment, anthropogenically changed characteristics through urbanization and infrastructural development (Ristić *et al.*, 2011) and climate change. In the future we will be facing much stronger effects of anthropogenic changes,

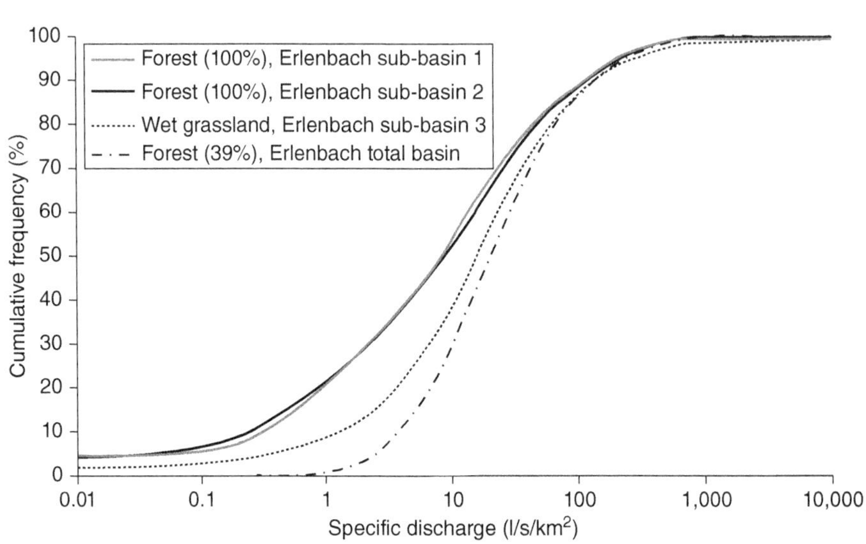

Fig. 5.1. Flow duration curves for forested and non-forested catchments in the Erlenbach and surrounding three small experimental catchments. (Data from Patrick Schleppi.)

such as accelerated or altered flowpaths, in addition to stronger hydrological variability due to climate change (de Jong, 2015). Examples from Eastern Europe for flood protection include the Jelanisca catchment, Serbia where plans are underway to restore forest and introduce protective land uses. Broadleaved forested surfaces are to be increased by 2.4% to above 40%, forest protective belts and silt-filtering strips created, and non-irrigated land reclaimed. The main aim is to reduce flood peaks and excessive sediment transport.

Increasing water yield as a result of deforestation and decreasing water yield by afforestation, despite its variability, is the main hydrological dilemma in European forest management today (Flörke *et al.*, 2011). It is well established that partial or complete removal of the tree cover accelerates water discharge, increasing the risk of flood during the rainy season and drought in the dry season (Flörke *et al.*, 2011). The larger the area of deforestation or reforestation, the stronger the effect on the annual water balance, with up to 700 mm in total difference. However, compared with deforestation, afforestation experiments are limited, probably due to the long timescales in observation involved (Schleppi, 2011). In addition, tree cutting experiments are limited in that they do not allow comparison of forestry with other types of land use.

In his publication on European experiences in long-term hydrology research, Keller (1988) indicates that not all scientific problems have been solved even after nearly 90 years of investigations – in particular, the question of extreme floods in forested areas. Studies in Europe generally show that there is an increase in discharge and peak flow in deforested catchments, after forest fires or insect outbreaks (Schleppi, 2011). However, other experimental studies in Switzerland and Sweden demonstrate that the role of forests in reducing peak flow is proportionally less than for low flows. The water-retention capacity of a forest soil is exhausted rapidly during an extreme event and therefore pre-event soil moisture plays a more important role than vegetation type, especially in alpine catchments (Hegg, 2006; Schleppi, 2011). In fact, runoff coefficients of rainstorms increase with pre-event groundwater levels. Thus a comparison of three small catchments in Switzerland showed that there was no direct relationship between the proportion of forest cover and peak flow (Burch *et al.*, 1996 in Schleppi, 2011).

In Europe, as for the USA, there was a shift from quantitative to qualitative forest hydrology in the 1980s (Schleppi, 2011). European literature on forest-related water quality is more abundant for obvious reasons. The main problems include suspended sediment concentration from erosion, floods, pollutants and acid deposition. Forests may concentrate pollutants and decrease water quality by retaining atmospheric pollution both via the canopy structure and through evaporation (Schleppi, 2011). The deep roots associated with forests may be helpful in counteracting this. In Central Europe, negative impacts of forest monoculture, such as spruce, on water quality have been recognized. Measures improving water body biocenosis are under way in accordance with the WFD (Meesenburg et al., 2005).

In recent decades forest hydrology in Europe has become much more interdisciplinary, with interdisciplinary forest faculties (e.g. Forest, Geological and Hydrological Sciences, Forest and Environmental Sciences, Forest Ecology and Hydrology). Finland hosts a European Forest Institute (EFI) (http://www.efi.int/portal/contact_us/, accessed 23 April 2016) with Mediterranean, Central European, Atlantic European, Central-East, South-East European and North European Regional Offices. However, experimental forest hydrology research in Europe still stands in the shadow of the USA. This may be due to greater financial resources devoted to environmental research in the USA, bringing forward 'some of the most noteworthy contributions to catchment area research' worldwide according to McCulloch and Robinson (1993). It may also be due to a higher standing of experimental forest hydrology in the USA or Australia. As such, recent decades have witnessed more emphasis on theoretical and modelling work in Europe with a more segmented and specialized, rather than basin-wide, approach towards forest hydrology.

Despite the importance of forests, there is no common forest policy in Europe. Forest cover in Europe has increased by 17 million hectares since 1990 through a combination of afforestation and land abandonment. Land is being abandoned in rural and mountain regions (in particular in the Alps and Pyrenees) although the population is growing because urban areas are acting as magnets of ever-increasing population density (Plate 3). Land abandonment has major hydrological impacts (García-Ruiz and

Lana-Renault, 2011). This land usually remains private property in EU countries and therefore complicates forest and water management. However, at the same time, forests face growing pressure from fragmentation, expanding urban areas, climate change and loss of biodiversity (SOER, 2015). Despite the efforts to halt loss of biodiversity, 80% of forest habitat assessments still have unfavourable conservation status, with the worst situation existing in the boreal zone. In Europe, forest areas designated for the protection of soil, water and other ecosystem services cover 12% in North, 18% in Central-West, 25% in Central-East, 42% in South-West, 10% in South-East and 20% in the rest of the EU, respectively.

An increase in water scarcity has led to a focus on the provision of drinking-water from forests. Following efforts in recent years, more than 20% of European forests are dedicated to protect water and soils, mainly in mountainous areas. The EU's Forest Strategy highlights the importance of European forests as key providers of ecosystem services such as soil and water protection. A coherent policy approach to European governance of forest resources is needed to protect and maintain forests and their functions within sustainable limits. Monitoring at the European level is essential to build a knowledge base on forests. Forest data and information are collected at national levels, but this information is not easily available and seldom comparable from country to country. The EU's Forest Strategy calls for such coordination of forest information and suggests using national forest inventories and monitoring systems (SOER, 2015), as summarized appropriately in the French–Swiss Interreg IV 2008 'Bois du Jura' ('Wood from the Jura'), 'La forêt ignore la frontière' ('forest ignores frontiers').

This chapter summarizes different European forest hydrology hypotheses concerning floods and droughts, impacts of land-use change, effects of important European policies and national regulations, water-sensitive forest management geared towards improving water quantity and quality, as well as future challenges linked to climate and anthropogenic change.

5.2 Floods and the Protective Role of Forests

The protective role of forests in Europe in preventing large floods has recently been put into question. Although forest restructuring and forestation have been recognized as providing an important contribution towards mitigating small floods, no protection can be provided against damage caused by catastrophic flood events (Calder *et al.*, 2007; Kubatzsch, 2007; Hall *et al.*, 2014). Thus the importance of forest cover in regulating hydrological flows has often been overestimated and the impacts of forest cover removal are evident only at the micro level and in association with short-duration and low-intensity rainfall events (Flörke *et al.*, 2011).

As rainfall duration or intensity increases and the distance down the watershed and river basin becomes greater, other factors start to override or dwarf the effects experienced close to the deforested area (Hamilton, 2008). Accordingly, for meso-scale catchments in the Swiss Alps (10–500 km^2, $n = 37$), no clear correlation between percentage of forest cover and specific mean annual flood discharge was found (Aschwanden and Spreafico, 1995) because other factors like slope, soil characteristics, altitude, precipitation and snow dynamics play into this relationship (Allewell and Bebi, 2011). This is corroborated by results from the Global Forest Resources Assessment (Hamilton, 2008), where natural processes – rather than land management in the upper watershed – are held responsible for flooding at the macro scale. It concludes that 'although there are many good reasons for reforesting watersheds (e.g. reducing soil loss, keeping sediments out of streams, maintaining agricultural production, wildlife habitat), reducing flood risk ... is certainly not one of them' and further that 'reforestation to prevent or reduce floods is effective at only a local scale of a few hundred hectares'.

The role of forest cover over temporal timescales is also variable and depends on forest structure. As for Switzerland, an intense programme of reforestation believed to reduce floods was implemented in France beginning in the mid-19th century (Mather *et al.*, 1993). However, there is not much evidence ascertaining its effectiveness (Humbert and Najar, 1992). Long-term simulations over 45 years in the now forested Coalburn catchment of northern England show that following afforestation, the development of mature forests has produced a decrease of about 250–300 mm in the annual streamflow compared with the original upland grassland vegetation (Birkinshaw *et al.*, 2014). A decrease of about 350 mm in the annual streamflow was observed compared with when the site was ploughed and the trees planted.

Long-term results show that peak flood discharges for medium-sized events are higher for watersheds with smaller trees compared with those with taller trees (Birkinshaw *et al.*, 2014). However, the results suggest that the bigger the event the smaller the difference, i.e. there is absolute convergence for the two different scenarios at higher flood discharges. Simulation results also show that for large discharge events there is an approximately 50% increase in the frequency of a given discharge for a cover of smaller trees compared with taller trees. The future challenges identified include considering the effects of parameter uncertainty on the simulated results from ensembles of feasible parameters and creating more long-term analyses.

In the frame of the WaReLa (Water Retention by Land-use, 2003–2006) project, water retention measures were investigated as to their effectiveness in reducing or temporally delaying floods within small forested watersheds in Germany (Schüler, 2006). A positive effect can be achieved as long as the water-storage capacity of forest sites is not exceeded. Discharge-accelerating linear structures such as roads and ditches have to be identified. A significant increase in runoff was established with an increasing road density from 20 to 50 m/ha. Despite this, recommendations by the Forest Administration Rheinland-Pfalz remain very variable, ranging between a road density of 16.7 and 62.5 m/ha.

WaReLa issued a series of recommendations. These include:

1. Since road orientation is hydrologically significant, forest roads should be aligned slope-parallel (Schüler, 2006).
2. Logging trails that reduce soil permeability should be kept as short as possible.
3. To delay discharge as long as possible, water retention management must focus on retaining water in sufficiently large retention areas such as streams and river valleys.
4. Water pathways, riverbed and bank structures as well as vegetation in the valleys should be kept as natural as possible.

The project suggests that if all small catchments in a larger watershed are managed with a view to water retention, the occurrence of damaging floods may be reduced. Finally, a DSS (Decision Support System) was developed that included an evaluation tool for the economic consequences and the eco-efficiency of flood-precaution measures.

Andréassian (2004b) identifies that watershed-scale research is still required to advance our understanding of forest impact on hydrology. He suggests seven future research issues including: (i) varying watershed size; (ii) improving models; (iii) establishing forest descriptors; (iv) taking gradual changes into account; (v) evaluating long-term impacts; (vi) distinguishing forest stands from forest soil impacts; and (vii) varying the number of watersheds.

In forest hydrology, one of the last remaining challenges is defining a threshold above which forest cover is no longer effective in reducing a flood in terms of precipitation intensity and discharge return intervals. In the European literature, with few exceptions, the quantitative definition of such threshold remains vague. Bathurst (2014) states that forests do not prevent floods and that they do not appear to affect the magnitude of larger floods. He observes that above a certain magnitude (or frequency) of rainfall event, there is little difference in the peak discharges of forested versus non-forested catchments for those with a surface area larger than 1500 km^2.

Both field data and model studies support the general trend toward either absolute or relative convergence (depending on antecedent soil moisture conditions) for large events. The level of event at which the two responses converge appears to be a rainfall return period of about 10 years (Fig. 5.2). According to the frequency-pairing approach, forests can reduce the frequency with which a given flood peak occurs and this effect may be greater for larger floods than for smaller floods. Vice versa, the frequency of a given flood magnitude does increase following removal of forest. In this discussion it is important to take into account the differences generated by rain- and snow-dominated regimes. For Swiss test sites, the duration of snowmelt discharge is reduced in forests, with less water loss from forested areas than from grasslands.

Although forested slopes often have high infiltration rates and peaks are insensitive to short-term rainfall intensities (Hewlett *et al.*, 1984), particularly heavy precipitation events could cause Hortonian flow to manifest itself where infiltration is poor (e.g. shallow soils, fine-textured soils, saturated soils, rock outcrops and compacted road surfaces).

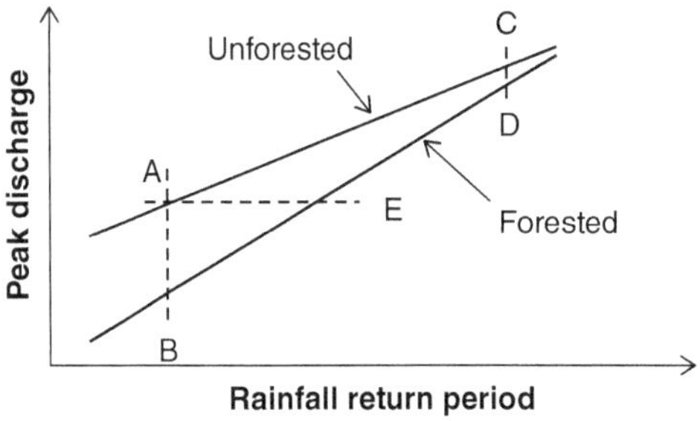

Fig. 5.2. Convergence of peak discharge for forested and non-forested catchments with increasing flood return period. A represents the discharge–rainfall relationship in the unforested and B in the forested catchment. For small but frequent rainfall events, the unforested catchment has a higher peak discharge (line AB). For the forested catchment to achieve the same peak discharge, it would have to be subjected to a rainfall event with a larger return period (line AE). For large but infrequent rainfall events, the two responses converge (line CD) so that the same peak discharge is produced in each catchment by the same rainfall event. (From Bathurst, 2014.)

Schüler (2006) recommends that flood precautions should not only be restricted to forestry management concepts but should be integrated in land and infrastructural planning. Cooperation with the water, agriculture and viticulture sectors should be a requirement. Management should thus be combined with domestic policy.

5.3 Drought and Forest Interactions

Drought periods severely affect forest productivity, decrease tree vigour and reduce tree growth, and are an important trigger for forest decline and mortality as well as for decline-induced vegetation shifts worldwide and in forest ecosystems (WSL Irrigation Experiment Pfynwald, 2015) (http://www.wsl.ch/fe/walddynamik/projekte/irrigationpfynwald/index_EN, accessed 23 April 2016). Furthermore, several studies have shown that forested areas may produce a lower runoff coefficient and that this may cause water supply deficiencies in times of increased drought stress under future climate change (Allewell and Bebi, 2011).

Ongoing EU projects under the 7th Framework Programme such as 'Drought' (Fostering European Drought Research and Science–Policy Interfacing) address the impacts of droughts on forest ecosystems primarily in the Mediterranean region (Andreu *et al.*, 2015). Often the effects of drought on forests have received less attention than in agriculture as they are less well understood (Domingo *et al.*, 2015). Analysis of spatio-temporal drought patterns should be seen as a key input to forest-related management. Most importantly, meteorological droughts have a statistically significant influence on wildfires in forests (Stagge *et al.*, 2015). Estimates of wildfire severity based on monthly area burned are documented in the European Forest Fire Information System (EFFIS). In the Mediterranean, wildfires dominate as a single, large, peak fire-danger period in late summer whereas the temperate regions of Central Europe produce two distinct fire peaks occurring in the spring and again in late summer. In the far northerly regions there is no distinct peak, but rather a consistent likelihood for the period when land is snow-free.

Nowadays alpine forests are not exempt from drought. Simulations show that even relatively small climatic shifts could result in large negative drought-related impacts on forest ecosystem services (Beniston and Stoffel, 2013). A long-term irrigation experiment (from 2003 to 2022) on drought and drought release effects on alpine forest is being carried out in the Pfynwald, situated in the inner-alpine dry valley of Wallis in Switzerland (WSL Irrigation Experiment Pfynwald, 2015). This large-scale drought field experiment investigates plant water stress in young and mature forests. First results show a significant shortening of the growth period by 2–5 weeks in the non-irrigated trees in comparison to the irrigated trees. The irrigation treatment was stopped on some selected sub-plots within each of the irrigated plots to simulate ecosystem response and resilience to drier conditions. In correspondence with the climate warming-induced increase in evaporation, a change in water supply is expected with increased frequency of summer heatwaves, but also an increase in frequency and intensity of precipitation events with strong surface runoff, probably further enhancing drought stress for plants.

Some alpine forests in the Wallis were even irrigated during summer droughts as a preventive measure following wildfires in adjacent forests in 2011.

5.4 The Bureaucracy of Forest-Relevant EU Policies

5.4.1 Evolution of European forest policies

EU policies concerned with forests include: the Water Framework Directive; the Habitat Directive; the EU Biodiversity Action Plan; the Rural Development Regulation; the Common Agricultural Policy (CAP); the EU Forest Action Plan (FAP); Natura 2000; and the Biomass Action Plan (BAP). Common policies affecting forests include the CAP and environmental, energy, industry, trade, research and cohesion policies including regional policy. These often exhibit a lack of coherence with regard to forest protection (European Parliament, 2011). The 1998 EU Forest Strategy led to the non-binding 2006 EU Forest Action Plan. Apart from the EU, organizations such as the International Union for Conservation of Nature (IUCN) have direct access to policy making through project funding and the provision of expertise via their own water projects (Flörke *et al.*, 2011). The EU Water Scarcity and Drought Policy (European Commission, 2012) should have substantial impacts on forest hydrology but as yet there is little information available (European Commission, 2013a).

The European Agricultural Fund for Rural Development (EAFRD) regulation (http://eur-lex.europa.eu/legal-content/EN/TXT/?uri=URISERV:l60032, accessed 23 April 2016) is the principal instrument for the implementation of the EU Forest Strategy and the EU Forest Action Plan (2007–2011). Eight of its 40 measures are forest-specific. All of these contribute to the EU-level priority objectives of biodiversity, water and climate change. Nevertheless, in 2011 there was significant under-spending (with less than 15% of an already reduced budget spent), particularly in terms of the allocation to the forest environment and Natura 2000 measures. This could indicate a lack of awareness of the importance of forest management related to the hydrological cycle and climate change. Indeed there are only few countries in Europe and the Middle East that have specific Ministries of Water and Forest. These include Romania, Bosnia-Herzegovina and Turkey.

Many measures supported by the Rural Development Programme of CAP (such as 'Axis 2, improving the environment and the countryside') are directly linked to forestry protection and rehabilitation measures, including forest environment payments introduced for voluntary commitments to maintenance of water resources and water quality. These are mostly short- to medium-term measures (for the next 25 years) (Flörke *et al.*, 2011).

In 2010, a *Green Paper on Forest Protection and Information in the EU: Preparing Forests for Climate Change* was elaborated (European Commission, 2010). It recognized that forests 'regulate freshwater supplies and that forests play a major role in the storage, purification and release of water to surface water bodies and subsurface aquifers'. More generally soils are said to 'buffer large quantities of water, reducing flooding'. For example, in Belgium, water from the Ardennes forest area is the principal supply source for Brussels and Flanders. In Germany, two-thirds of the 'Wasserschutzgebiete' (drinking-water perimeters) for abstraction of high-quality drinking-water are under forest cover. In Spain, forests in upper river catchments have been given special conservation status because of their capacity to improve water quality.

In 2011 the European Parliament issued a report on the Commission's Green Paper on forest protection (European Parliament, 2011), where it urges the Commission to compile and monitor indicators relating to the protective functions of forests such as soil retention and water capacity. In order to achieve the objectives of the EU 2020 strategy (http://ec.europa.eu/europe2020/index_en.htm, accessed 23 April 2016) with regard to national forest action plans, it requests that each Member State develop a forest strategy including reforestation of river banks, capture of rainwater and production of research results for selection of traditional plant and tree varieties and species best adapted to drought. Payments for ecosystem services (PES) should be formalized, building on the success of forest and water projects.

In 2013 a new EU Forest Strategy for forests and the forest-based sector was published (European Commission, 2013b) as a response to growing demands on and threats to forests over the past 15 years. Its aims are to protect forests and biodiversity from the transnational effects of storms and fires and increasingly sparse water resources. The Commission recognizes that the increasing number of forest-related policies creates a complex and fragmented forest policy environment (European Commission, 2013b). Forests are the focus of a range of different targets with increasingly competing claims on forests. As noted by Pülzl *et al.* (2014):

> When there is an emphasis on forest ecosystem services beyond biomass production, such as water provision, protection, and recreation, trade-offs at the regional level occur. As a result, fostering strategies to simultaneously intensify resource use while also wanting to reduce it inevitably leads to constraints and challenges.

Forest management plans (FMPs) based on the principles of sustainable forest management are key instruments in delivering multiple goods and services according to the European Commission (2013b). FMPs are at the core of both the EU 2020 Biodiversity Strategy and EU Rural Development funding. Member States should maintain and enhance forest cover to ensure soil protection, water quality and quantity regulation by integrating sustainable forestry practices in the Programme of Measures of River Basin Management Plans under the EU Water Framework Directive and in the Rural Development Programmes.

In 2015 the European Parliament produced a new resolution for the 2013 EU Forest Strategy (European Commission, 2013b) issued by the European Commission (European Parliament, 2015). Among new challenges, it proposes that the:

> EU needs a new comprehensive strategy to tackle cross-border challenges such as forest fires, climate change, natural disasters or invasive alien species, but also to strengthen forest-based industries and improve efficient use of raw materials such as timber, cork or textile fibers.

It is recognized that sustainable forest management has positive impacts on combating climate change, maintaining biodiversity and contributing to the objectives of the Europe 2020 strategy.

The new EU Forest Strategy is seen as 'a much-needed response to growing demands on forests and significant societal and political changes that have affected them over the last 15 years by politicians' (http://www.europarl.europa.eu/pdfs/news/expert/infopress/20150424IPR45802/20150424IPR45802_en.pdf, accessed 23 April 2016). Following the slogan 'Member states manage, EU coordinates', Members of the European Parliament back the European Commission's plan to develop, in close cooperation with EU Member States, local authorities and forest owners, an ambitious and objective set of criteria for managing forests sustainably. The resolution states that the EU must strive to coordinate its forestry-related policies better, but should not make forestry a matter of EU policy. However, other highly relevant policies such as the EU Water Scarcity and Drought Policy (European Commission, 2012) does not yet tackle forests explicitly. In the meantime the European Climate Change Adaptation Platform has produced a Water Sensitive Forest Management adaptation option for the next 25 years (Climate-Adapt, 2015).

The most recently established forest-related political platform in Europe is the Resolution on Forests and Water, adopted in 2007 by the Ministerial Conference for the Protection of Forests in Europe (Flörke *et al.*, 2011). This resolution consists of four parts: (i) sustainable management of forests with relation to water; (ii) coordinating policies on forests and water; (iii) forests, water and climate change; and (iv) economic valuation of water-related forest services.

5.4.2 Effects of policies on forest hydrology

As yet there is surprisingly little information available on the effects of policies on forest hydrology in Europe (Meesenburg *et al.*, 2005). Müller (2012) describes the positive hydroecological impacts of new guidelines for the creation of stable mixed pine–beech stands and a nature-oriented approach to forest structures by German Forestry in the north-eastern lowlands. Stemflow on beech trees in the winter half-year has the advantage of generating additional deep soil seepage and an increased amount of stemflow in the summer half-year linked to more trees with larger diameters increasing topsoil moisture. With an annual precipitation of below 600 mm and sandy soils with a low water-storage capacity, water availability is limited and at risk from more frequent future droughts. For some decades groundwater levels have been falling. Counteracting this trend through targeted forest conversion is a challenge for forestry in north-east Germany (Friedmann and Müller, 2010). Furthermore, ameliorating the regional water balance could also have positive spin-off effects such as the protection of 'forest mires' (wetlands). Similarly, afforestation contributes to improving drinking-water and groundwater quality in the vicinity of urban agglomerations.

European policies have different hydrological objectives that may not all be compatible with each other to the extent of generating different forest hydrological impacts. For example, whereas the EU Water Framework Directive (WFD) aims to bring waterbodies into a good ecological status and seeks mainly to combat water pollution, the European Flood Directive focuses specifically on flood prevention through forest management at the catchment scale. Thus Pülzl *et al.* (2014) note:

> The likely provision of water-related ecosystem services by forests is not clearly recognized in the WFD, and the complex interplay between water protection management and forestry is neglected. While timber-production-oriented forestry is considered a risk in reaching a good ecological water status, especially of local water bodies, the potential benefits of forests and forest management in achieving a good ecological status for waters and their catchments are not recognized.

The European Forest Institute (2009) even suggests establishing an EU Forest Framework Directive to strengthen coordination of forest-related aspects by EU regulation, keeping in mind implementation problems of other European directives. MOUNTFOR (Preserving and Enhancing the Multifunctionality of Mountain Forests), initiated in 2013 in cooperation with EFI (http://www.efi.int/portal/about_efi/structure/project_centres/mountfor/, accessed 23 April 2016), is one of the few alpine projects that assesses the potential impacts of forest management and land-use changes on mountain hydrology and the availability/quality of water resources. In the meantime, the landscape approach is becoming more important in forest management. This is the case in the Netherlands, where water management plans are integrated into spatial planning. Landscape areas are redesigned in participatory workshops, for example to integrate water safety (Pülzl *et al.*, 2014).

A case study from the north-eastern Ore Mountains in Saxony, Germany carried out in the frame of the 6th EU-FP project FLOODsite evaluates the impacts of the European Flood Directive on forest land management and, in turn, water dynamics (Wahren and Feger, 2011). Since the disastrous floods of the River Elbe in 2002, a new water law exists for Saxony with regulations concerning flood-originating areas. Natural water retention is to be conserved and improved, and soils are to remain unsealed and afforested, if possible, with compensation measures in case of reduction or loss. Due to its mining history the area has only a 20% forest cover. 'The catchments traditionally provide drinking water from reservoirs and are known for their lower specific runoff under forest. Competitive goals of future land-use planning like flood protection, profitable food/wood production, water supply and water protection' create conflicts in decision making facing an uncertain future (Wahren and Feger, 2011).

The case-study authors investigated impacts of land use on runoff generation at different scales and under different scenarios of reforestation (Wahren and Feger, 2011). Although afforestation and 'near-natural' silviculture will increase water retention for smaller flood events (i.e. those with a recurrence interval (RI) of 25 years) by up to 20%, the effects are negligible for extreme events (RI > 100 years). Thus the impacts of the land use on flood formation decrease with increasing rainfall intensity and the benefits of land use for optimized flood protection are (mostly) not directly noticeable. The quantitative role of non-structural flood risk management measures with respect to event size remains a controversial topic. The authors challenge present models which suggest that socio-economic methods have to be combined with state-of-the-art hydrological modelling and that integrated modelling approaches should deal with all competitive requirements of future land use, demographic and climate change. They draw attention to the fact that EU subsidies for land-use change are primarily issued by the CAP, followed only in second place by structural and cohesion policies. As long as the EU Flood Directive does not have clear subsidy policies, flood protection remains at best an additional benefit but not a target, they argue (Wahren and Feger, 2011).

The European Commission recognizes that forests are particularly important in Mediterranean countries because of their ability to balance the water cycle and, therefore, considers that reforestation should be preceded by scientific studies to identify the most suitable rainwater catchments. The Commission acknowledges that mountain forests accounting for one-third of the total forest area in the EU are essential for soil protection and regulating water supply. However, with respect to forest fragmentation and resulting forest dieback, it fears that reduced ability to dampen runoff peaks generated in mountain catchments can impact floods and water quality.

Manser (2013) challenges the role of scientists in forest policy in a changing climate. He postulates that climate change will proceed at a rate faster than the natural adaptation capacity of forests and that this poses a serious challenge for forest policy. Among future challenges he identifies: (i) understanding the reaction of forest stands to climate change; (ii) its consequence for forest goods and services; (iii) the benefits and risks involved in adaptation strategies and ways to increase the adaptive capacity of forests; (iv) overcoming long research timescales when climate change strategies require short-term decision making; and (v) how models and tools can be optimized for practitioners. Concerning Central Europe, the greatest challenge identified is the response of trees to prolonged and repeated drought (Psidova *et al.*, 2013). Their study identifies effects of drought stress on European beech and concludes that the higher-altitude trees are less resistant to water deficit than lower-altitude stands already growing in a drier climate.

5.5 Emerging Issues

5.5.1 Drinking-water and groundwater protection

At the EU level, a new focus has been put on the provision of drinking-water from forests (SOER, 2015), which primarily addresses water quality. Groundwater pollution is identified as a new challenge due to land-use change. In the Harz, Germany, new voluntary drinking-water protection associations have been created at the initiative of forestry (Rüping *et al.*, 2012). Foresters and farmers ideally cooperate closely with water managers to engender stronger responsibility of local stakeholders more familiar with local conditions and best solutions. The main aim is suspended sediment concentration in the drinking-water reservoirs of the West Harz mountains that act as water towers for the surroundings countries (Länder) (www.umwelt.niedersachsen.de/aktuelles/pressemitteilungen/knapp-14-millionen-euro-fuer-neu-gegruendete-trinkwasserschutzkooperation-westharz-111447.html). Catchments as large as 29,000 ha are mainly forest-covered. In future, forest and agriculture stakeholders will be advised by experts from the Harz Water Works and accompanied by the provincial office for water management, coastal and nature protection.

In Austria, there is a new trend towards economic incentives for water management such as transfer payments for water protection by forests around population agglomerations (e.g. Vienna). Proposed action plans for South-Eastern Europe include avoiding clearcutting within drinking-water protected areas, ensuring a continuous forest cover and stable forest ecosystem, and limiting silviculture-related road construction (CC-Ware, 2014). The main challenges are that in most countries within South-Eastern Europe drinking-water protection legislation is not homogeneous, often not implemented properly and that practical implementation might be confusing. In future, an improved applicable legal framework integrating specific subjects of relevance for drinking-water supply (DWS), ecosystem services (ESS), land use (LU) and climate change (CC) is required.

Schüler *et al.* (2011) identify the prediction of forest management and environmental impacts on groundwater quality as one of our strongest present-day challenges: 'Forest management in a changing environment subject to global warming or air pollution may diminish the protective functions of forests with regard to groundwater quality'. Moreover, interactions between forests and water and impacts of forestry on groundwater quality and runoff remain a scientific grey zone.

It is to be expected that in future the frequency and timing of forest wildfires will alter with drought patterns and significantly change the temporal and spatial patterns of forest hydrology.

5.5.2 Water scarcity

Since the Mediterranean region is particularly vulnerable to water scarcity, it may require adaptive forest management (AFM) in future to adapt forests to water availability by means of artificial regulation of the forest structure and density (González-Sanchis *et al.*, 2015). To date there is no clear linkage between forest management and droughts. Such an approach would, for example, enable optimization of the hydrological cycle of an Aleppo pine forest under normal and future global change conditions. The aims would be to reduce water interception and plant transpiration (green water) and increase water runoff and/or percolation (blue water).

5.5.3 Runoff from ski runs and mountain resorts

Concerning floods, preferential runoff paths that accelerate and increase flood peaks in forested catchments have so far been attributed only to forest roads, timber harvesting and log trails. Ski runs, in addition to new roads constructed to access water reservoirs for snowmaking and constantly expanding ski resorts with their highly reduced infiltration capacity, are rarely mentioned and not sufficiently integrated into modelling approaches (Plate 4).

For the USA, Wemple *et al.* (2007) distinguish four factors associated with ski area development that may affect watershed processes and that are distinct from those associated with traditional forest management practices:

> First, forest clearings created for ski trails are oriented along gravitational flow paths, enhancing the potential for efficient down slope routing of water, solutes and particulates. Second, forest clearings for ski trails are intended to persist over time and represent a relatively permanent alteration of the forest landscape. Third, certain activities associated with ski area development, particularly artificial snowmaking, are not present in traditional forest management operations. Finally, other practices, including creation of impervious surfaces and development of drainage infrastructure are more extensive than those associated with traditional forest management practices.

They remark that there is little scientific literature specifically addressing the effects of resort development in mountain settings. This is the case as much in the USA as in Europe.

Catastrophic flooding and debris flows in 2005 and 2015 generated in and above the treeline in catchments with ski runs in the Paznaun valley in Austria are thought to be linked to loss in soil permeability and vegetation cover and preferential flood routing. Of the few studies available on this topic in Europe, most are clustered in Austria. Pötzelsberger and Hasenauer (2015) found that ski runs increased runoff considerably in the 10 km^2 alpine Schmittental catchment near Salzburg covered principally by Norway spruce. Land use evolved from sparse to dense forest from 1890 to 1965, but has increasingly been dissected by ski runs since the 1970s. At present the catchment is covered by 71% forest (520 ha) and 28% grassland (200 ha), of which about 14% (100 ha or 77 km length) is used as ski slopes. The high average runoff coefficient of 0.74 is attributed primarily to dense soils but also to ski slopes that increase surface runoff due to reduced infiltration, in particular when ski runs are groomed and when topsoil has been removed for ski run remodelling (Hagen, 2003).

Runoff produced on ski runs amounts to as much as 18.4% of precipitation. Furthermore, about 500,000 m^3 of water is introduced into the catchment every year to produce artificial snow for the ski industry, mainly from the downstream Zeller Lake. This surplus water input probably increases peak snowmelt runoff from ski runs in spring by 20–25% (Pötzelsberger and Hasenauer, 2015). In eastern Serbia, ski resorts have triggered more frequent extreme events in the Zubska River headwater due to deforestation and enhanced surface runoff on ski runs and access roads (Ristić *et al.*, 2012). In the European Alps, the extent of ski runs in forested areas has increased substantially and is still increasing. Therefore, winter resorts in forested regions should be identified as an emerging hydrological challenge.

5.5.4 Climate change

One of the most serious challenges of forest hydrology is coping with climate change, including in particular the effects of increased temperature and CO_2 concentration and decreased snow duration where relevant (Schleppi, 2011), but also concerning windfalls and pests (Andreu *et al.*, 2015). Recently, Frank *et al.* (2015) have shown that despite decreased stomatal opening, there has been a 5% increase in European forest transpiration calculated over the 20th century. Consequently, both catchment- and plot-scale experiments should be carried out, although costs may limit larger-scale experiments. Ideally the global scale should be recognized. According to Schleppi (2011), a big challenge is combing plot-scale and long-term experiments as well as unravelling the causal factors. He proposes flow-proportional sampling schemes for reducing errors in flux measurements and above all maintaining long-term monitoring experiments at the catchment scale. The analyses show that in future new questions and adapted methods have to be tackled.

Zimmermann *et al.* (2006) conclude that open questions concerning tree species composition, forest development and range in distribution pose a major challenge both for forest practice and research. Changes are proceeding at such a fast rate that researchers can only partially resort to existing knowledge. Since it is neither possible to investigate entire ecosystems experimentally nor is sufficient time available, a good cooperation between practitioners and researchers is inevitable. Strong changes are to be expected but uncertainty has to be confronted in future climate change and the way in which endemic forests will react. At the moment no single solution is available but the preservation of higher diversity of tree species is important.

5.6 Adaptation and Water-Sensitive Forest Management

In the frame of the European project ClimWatAdapt (Climate Adaptation – modelling water scenarios and sectoral impacts, 2010–2011) adaptation measures such as water-sensitive forest management, a technical measure related to green infrastructure, were assessed (Flörke *et al.*, 2011). The main climate threats identified for forests include: (i) not enough water (water scarcity and droughts); (ii) too much water (flooding, sea-level rise and coastal erosion); and (iii) deteriorating water quality and biodiversity (Flörke *et al.*, 2011).

The project recognized that forest management measures can increase water yield, regulate water flow and reduce drought stress for a forest under both present as well as future low-flow conditions. Some measures that have been put into practice to support forests' water regulation role include: (i) reduced density of stand stocking; (ii) shorter length of the cutting cycles; (iii) planting hardwood species; and (iv) regeneration from seedlings rather than sprouts. It was found that afforestation, in particular near watercourses, brings benefits for the regulation of water flow, the maintenance of water quality and the severity of droughts. Concerning floods, forest buffers have generally not provided substantial flood reduction and, if so, only at a very local scale. Flörke *et al.* (2011) point out that digital classification of forest sites is useful for analysis, consultation and developing adaptation recommendations. However, among the strategies aimed to achieve a water-sensitive forest management, stakeholders highlighted the limitations posed by the digital classification of the forest sites and expressed a need to improve this measure.

Adaptation of management rules in silviculture in order to improve tree water balance was difficult to put into practice. Stakeholders pointed out its potential limitations including undesired side-effects and the costs of the strategy. Afforestation was considered highly valuable with a high benefit even in the case of less pronounced climate change impacts (Flörke *et al.*, 2011). According to the literature review carried out on forest hydrology in Europe within ClimWatAdapt, an increasing number of studies have challenged the popular idea that more forests imply more and better water. Identification and correct application of forest management to reduce water use is therefore seen as a crucial aspect regarding water scarcity.

The Silvistrat project (Response Strategies to Climatic Change in Management of European Forests) of the EFI developed adaptive management strategies between 2000 and 2003 for sustainable forest management in European forests under global climate change. It analysed AFM strategies aimed at reducing the impacts of drought and other adverse effects of climate change, and suggested the substitution of species sensitive to drought and to late spring frosts with more drought-tolerant and frost-resistant tree species.

Within Central Europe, forests in Slovakia and Hungary were identified as most prone to droughts (Hlásny *et al.*, 2014). Adaptive measures to increased drought risk include artificial regeneration to enrich local gene pools and increase the drought tolerance of stands. A stronger focus is put on risk management and disturbance monitoring systems.

A shortcoming in adaptation strategies is available subsidies. Funding for water retention in drought-endangered agriculture and forest landscapes was included in Germany's 2008 *Report on Active Climate Protection in the Agriculture, Forestry and Food Industries and on Adaptation of Agriculture and Forestry to Climate Change* (Federal Ministry of Food, Agriculture and Consumer Protection, 2008). It was suggested that the federal government must offer such incentives (Flörke *et al.*, 2011).

The aim of German forestry is the creation of stable mixed stands and a nature-oriented approach to forest structures. In this context the hydrological functions of forest conversion play an important role in the fields of regional water budget, water supply and water distribution (Müller, 2012).

In Sweden, it has been recognized that trees and forests will play an increasingly important role in regulating the hydrological cycle in different landscapes and climates, but as yet 'integration of water management in the day-to-day management of forests is a fairly new practice in Sweden' (Samuelson *et al.*, 2015). Sweden's role is to develop societal strategies to restore and/or maintain forests and trees for the benefit of strategies such as water regulation and management. One of the primary aims is to 'initiate bilateral and multilateral activities to build resilient landscapes'. Among others, a water management toolbox for forest planners has been developed. Apart from water resources management, forest policies and management strategies have started to integrate climate change mitigation and adaptation.

5.7 Challenges and Future Research Needs

It is difficult to entangle future research needs for Europe from those of other continents worldwide from general texts on forest hydrology in the literature.

Concerning boreal forests, Lindroth and Crill (2011) notice that comparatively little research has been carried out on both hydrology and biogeochemistry of the different component ecosystems but do not state whether this is specific for Europe. Given that strongest climate change is to be expected in the boreal forest latitudes, they suggest that future research should consider the links between hydrological, energy and biochemical processes. Furthermore, they expect that the most significant effects will occur at the seasonal transition periods from winter to spring and autumn to winter. Growing season will be affected by thawing permafrost, with shorter or longer duration of snow cover resulting in longer or shorter growing season, respectively.

Lindroth and Crill (2011) predict that shifting patterns of temperature and precipitation will lead to changes in fire frequency and intensity and will have consequences for drought frequency/waterlogging. Boreal forests are considerably vulnerable to a warming of climate mainly due to low surface albedo during the snow season, which offsets the negative climate forcing due to carbon sequestration (Bonan, 2008 in Allewell and Bebi, 2011). The latter might be extrapolated to winter-time conditions of alpine forests even though scientific evidence is missing so far (Allewell and Bebi, 2011).

Regarding temperate forests, Ohte and Tokuchi (2011) are concerned that there are few studies of in-stream biogeochemical processes and that in future an assessment of the contributions of scale effects to the hillslope and in-stream biogeochemical processes in regions under various climatic conditions is required. They suggest collecting more data from sites with high summer precipitation and discharge. In addition, more extensive survey of previously published literature as well as conventional and project-based databases is proposed.

Several European Framework projects have dealt with or are dealing with forest hydrology aspects such as ecosystem experimentation and adaptation and vulnerability to climate change, in particular droughts and impacts of hydropeaking from dams on riparian forests.

The 'forest hydrological hypothesis' states that forests increase baseflow (Fig. 5.3). However, Allewell and Bebi (2011) conclude in two studies for alpine catchments that surface runoff can

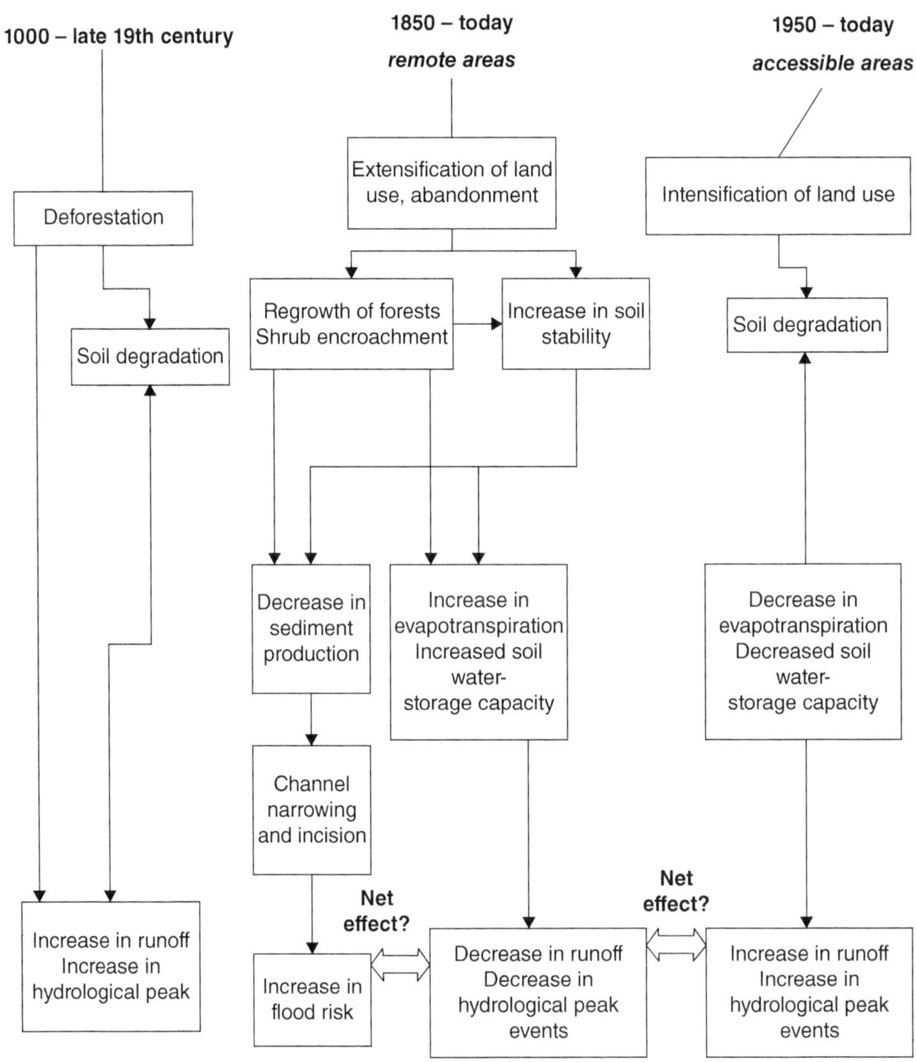

Fig. 5.3. Hypothesis on forest–water interactions in the European Alps, applicable to other European regions. (From Allewell and Bebi, 2011.)

be considered to be generally lower in forests due to high infiltration rates of humus layers and higher tree evapotranspiration compared with grasslands. Thus, forests provide protection against peak runoff events especially during the critical time period of snowmelt. Greatest relative differences between evapotranspiration rates of mountain forests in the Alps and grasslands were simulated in early spring when spruce forests are already transpiring while grasslands are still under closed snow cover and in winter when precipitation interception is effective in forests and/or forests are transpiring during warmer winter days.

The alpine forest water balance dilemma still needs to be solved. For the Swiss Alps, Allewell and Bebi (2011) conclude that a complete regrowth of forests in the Urseren Valley within forests' climatic boundaries would only insignificantly influence hydrograph dynamics because of the relatively small area of regrowth compared with the forest-free area above the treeline. Long-term measurements in the Reuss catchment in the Urseren Valley (runoff data since 1904) indicate a reduction in runoff by about 30% during the last 10 years, particularly during the summer months, compared with long-term means over the last 60 years (BAFU, 2007). However, it is unclear to what extent these runoff changes are caused by shrub encroachment (30% increase in shrub area between 1959 and 2004), soil degradation, climate change or changes in the waterways and management.

According to Allewell and Bebi (2011), natural forest growth and regrowth in the Alps is highly heterogeneous with some areas likely to experience almost complete forest regrowth in future while others will be limited by natural and socio-economic factors. In the Swiss Alps the typical pattern is land abandonment and forest regrowth in remote areas as the population moves into the cities. Abandonment of pasture in the Alps has reduced discharge from springs, torrents and rivers, and increased evapotranspiration (Dumas, 2011; Van den Bergh *et al.*, 2014). It is postulated that, in many European mountain ecosystems, completely new ecosystem dynamics will form and rural landscapes will be lost (Allewell and Bebi, 2011). For complex mountain environments it is assumed that forest regrowth will reduce runoff and the magnitude of hydrological peak events. However, the authors warn that the influence of forest cover on runoff remains highly variable according to site conditions and forest type. Furthermore, the future regrowth of forested areas might in some areas be impeded by the combined effects of increased insect disturbance and increased risk of droughts and wildfires under future climate change.

While factors relevant for hydrological processes in forests are increasingly respected in the management of existing protection of forests, land-use planning and policy measures at the interface between agriculture, land and forests are hardly adapted to these spatial variations (Allewell and Bebi, 2011). Allewell and Bebi (2011) suggest that contemporary alpine forest governance should encompass regionally adapted measures to avoid land abandonment or to manage abandoned land. Forest management should become decentralized to meet the growing demands on forest ecosystem services such as food, biofuel, timber and disaster protection, but avoid private ownership. A shift in competence from the federal to regional or possibility local level is advisable.

According to Bathurst (2014) it is the responsibility of the forest hydrologist 'to provide a quantitative context for water resources, river engineering and, more generally, riparian civil engineering projects in catchments subjected to large-scale changes in forest cover.'

When floods are considered a major risk, afforestation is considered as a solution; yet, as known, this causes a significant reduction of annual streamflow and exacerbates droughts (Birkinshaw *et al.*, 2014). A difficult future challenge will be to merge contradictory approaches of forest hydrology management with respect to droughts and floods.

5.8 Acknowledgements

My acknowledgements go to James Bathurst, Newcastle; Jürgen Müller, Eberswalde; Friedhard Knolle, Goslar; and Patrick Schleppi, Birmensdorf.

References

Allewell, C. and Bebi, P. (2011) Forest development in the European alps and potential consequences on hydrological regime. In: Bredemeier, M., Cohen, S., Godbold, D.L., Lode, E., Pichler, V. and Schleppi, P. (eds) *Forest Management and the Water Cycle: An Ecosystem-Based Approach*. Ecological Studies Vol. 212. Springer, New York, pp. 111–126.

Andréassian, V. (2004a) Couvert forestier et comportement hydrologique des bassins versants. *La Huille Blanche* 2, 31–35.

Andréassian, V. (2004b) Waters and forests: from historical controversy to scientific debate. *Journal of Hydrology* 291, 1–27.

Andreu, J., Solera, A., Paredes-Arquiola, J., Haro-Monteagudo, D. and van Lanen, H. (eds) (2015) *Drought: Research and Science–Policy Interfacing. Proceedings of the International Conference on Drought: Research and Science–Policy Interfacing, Valencia, Spain, 10–13 March, 2015*. Taylor & Francis, London.

Aschwanden, H. and Spreafico, M. (1995) Hochwasserabflüsse – Analyse langer Messreihen. In: *Hydrologischer Atlas der Schweiz*. Bundesland für Umwelt, Bern, Table 5.6.
BAFU (2007) *Hydrologisches Jahrbuch der Schweiz 2005*. Umwelt-Wissen, Vol. 0713. Bundesamt für Umwelt, Bern.
Bathurst, J.C. (2014) Comment on 'A paradigm shift in understanding and quantifying the effects of forest harvesting on floods in snow environments' by K.C. Green and Y. Alila. *Water Resources Research* 50, 2756–2758.
Beniston, M. and Stoffel, M. (2013) Assessing the impacts of climatic change on mountain water resources. *Science of the Total Environment* 493, 1129–1137.
Birkinshaw, S.J., Bathurst, J.C. and Robinson, M. (2014) 45 years of non-stationary hydrology over a forest plantation growth cycle, Coalburn catchment, Northern England. *Journal of Hydrology* 519, 559–573.
Blackie, J.R. and Robinson, M. (2007) Development of catchment research, with particular attention to Plynlimon and its forerunner, the East African catchments. *Hydrology and Earth System Sciences* 11, 26–42.
Bonan, G.B. (2008) Forests and climate change: forcings, feedbacks, and the climate benefits of forests. *Science* 320, 1444–1449.
Burch, H., Forester, F. and Schleppi, P. (1996) Zum Einfluss des waldes auf die Hydrologie der Flysch-Einzugsgebiete des Alptals (On the influence of forest on hydrology in the Flysch catchments of the Alptal). *Schweizer Zeitschrift für Forstwesen* 147, 925–938.
Calder, I., Hofer, T., Vermont, S. and Warren, P. (2007) Towards a new understanding of forests and water. *Unasylva* 229, 58.
CC-Ware (2014) *Towards an Action Plan for the SEE-Region. Guidance for Water Resources under Climate Change*. ETC/SEE Project, Federal Ministry of Agriculture, Forestry, Environment and Water Management (BMLFUW), Forestry Department, Vienna.
Climate-Adapt (2015) Water Sensitive Forest Management. European Climate Adaptation Platform. Available at: http://climate-adapt.eea.europa.eu/viewmeasure?ace_measure_id=626 (accessed 28 March 2016).
De Jong, C. (2015) Challenges for mountain hydrology in the third millennium. *Frontiers in Environmental Sciences* 3, 38, doi: 10.3389/fenvs.2015.00038 (accessed 28 March 2016).
Domingo, C., Pons, X., Cristóbal, J., Ninyerola, M. and Wardlow, B. (2015) Integration of climate time series and MODIS data as an analysis tool for forest drought detection. In: Andreu, J., Solera, A., Paredes-Arquiola, J., Haro-Monteagudo, D. and van Lanen, H. (eds) *Drought: Research and Science–Policy Interfacing. Proceedings of the International Conference on Drought: Research and Science–Policy Interfacing, Valencia, Spain, 10–13 March, 2015*. Taylor & Francis, London, pp. 97–104.
Dumas, D. (2011) The impact of forests on the evolution of water resources in the mid-altitude Alps from the middle of the 19th century (Chartreuse Massif, France). *Journal of Alpine Research* 99(4), 2–14.
European Commission (2010) Green Paper on Forest Protection and Information in the EU: Preparing Forests for Climate Change. SEC(2010)163 final. Available at: http://eur-lex.europa.eu/LexUriServ/LexUriServ.do?uri=COM:2010:0066:FIN:EN:PDF (accessed 28 March 2016).
European Commission (2012) Water Scarcity & Droughts in the European Union, EU Action on Water Scarcity and Drought, Water Scarcity and Droughts – Policy Review. Available at: http://ec.europa.eu/environment/water/quantity/eu_action.htm (accessed 28 March 2016).
European Commission (2013a) Report on the Review of the European Water Scarcity and Droughts Policy. Available at: http://ec.europa.eu/environment/water/quantity/pdf/COM-2012-672final-EN.pdf (accessed 28 March 2016).
European Commission (2013b) A New EU Forest Strategy: For Forests and the Forest-Based Sector. COM(2013)659 final. Available at: http://eur-lex.europa.eu/legal-content/EN/TXT/?uri=CELEX:52013DC0659 (accessed 28 March 2016).
European Forest Institute (2009) Interaction between Water and Forest – Challenge to Water Policies and Forest Management. Available at: http://www.agriculture-xprt.com/news/interaction-between-water-and-forest-challenge-to-water-policies-and-forest-management-43088 (accessed 28 March 2016).
European Parliament (2011) Report on the Commission Green Paper on Forest Protection and Information in the EU: Preparing Forests for Climate Change (2010/2106(INI)). Committee on the Environment, Public Health and Food Safety, Rapporteur: Kriton Arsenis. Available at: http://www.europarl.europa.eu/sides/getDoc.do?pubRef=-//EP//TEXT+REPORT+A7-2011-0113+0+DOC+XML+V0//EN (accessed 28 March 2016).
European Parliament (2015) A new EU Forest Strategy: for forests and the forest-based sector. Available at: http://www.europarl.europa.eu/sides/getDoc.do?type=TA&reference=P8-TA-2015-0109&language=EN&ring=A8-2015-0126,http://www.europarl.europa.eu/oeil/popups/summary.do?id=1387355&t=d&l=en (accessed 23 April 2016).

Federal Ministry of Food, Agriculture and Consumer Protection (2008) Report on Active Climate Protection in the Agriculture, Forestry and Food Industries and on Adaptation of Agriculture and Forestry to Climate Change. Germany, September 2008. Available at: http://www.bmel.de/SharedDocs/Downloads/EN/Agriculture/ClimateReport2008.pdf?__blob=publicationFile (accessed 23 April 2016).
Flörke, M., Wimmer, F., Laaser, C., Vidaurre, R., Tröltzsch, J., Dworak, T., Stein, U., Marinova, N., Jaspers, F., Ludwig, F., *et al.* (2011) Assessment of adaptation measures: Factsheets. Annex 11 to the Final Report for the project Climate Adaptation – modelling water scenarios and sectoral impacts (ClimWatAdapt), CESR, 173. Available at: http://climwatadapt.eu/sites/default/files/Annex11_ClimWatAdapt_Factsheets.pdf (accessed 29 March 2016).
Frank, D.C., Poulter, B., Saurer, M., Esper, J., Huntingford, C., Helle, G., Treydte, K., Zimmermann, N.E., Schleser, G.H., Ahlström, A., *et al.* (2015) Water-use efficiency and transpiration across European forests during the Anthropocene. *Nature Climate Change* 5, 579–583.
Friedmann, G. and Müller, J. (2010) Auswirkungen des Waldumbaus im Waldgebiet der Schorfheide auf die Entwicklung der Grundwasserhöhen und den Zustand der Waldmoore. *Naturschutz und Landschaftspflege in Brandenburg* 19, 158–166.
García-Ruiz, J.M. and Lana-Renault, N. (2011) Hydrological and erosive consequences of farmland abandonment in Europe, with special reference to the Mediterranean region – a review. *Agriculture, Ecosystems & Environment* 140, 317–338.
González-Sanchis, M., Del Campo, A.D., Molina, A.J. and Fernandes, T.J.G. (2015) Modeling adaptive forest management of a semi-arid Mediterranean Aleppo pine plantation. *Ecological Modelling* 308, 34–44.
Hagen, K. (2003) *Wildbacheinzugsgebiet Schmittenbach (Salzburg), Analyse des Niederschlags- und Abflussgeschehens 1977–1998*. BFW Report No. 129. Bundesforschungszentrum für Wald, Vienna, pp. 5–106.
Hall, J., Arheimer, B., Borga, M., Brázdil, R., Claps, P., Kiss, A., Kjeldsen, T.R., Kriaučiūnienė, Z.W., Kundzewicz, M., Lang, M.C., *et al.* (2014) Understanding flood regime changes in Europe: a state-of-the-art assessment. *Hydrology and Earth System Sciences* 18, 2735–2772.
Hamilton, F.C. (2008) *Forests and Water. A Thematic Study Prepared in the Framework of the Global Forest Resources Assessment 2005*. FAO Forestry Paper No. 155. Food and Agriculture Organization of the United Nations, Rome.
Hegg, C. (2006) Waldwirkung auf Hochwasser (Forest impacts on floods). *LWF Wissen (Bayerisches Landesamt für Wald und Forstwirtschaft)* 55, 29–33.
Hermann, A. and Schumann, S. (2009) Investigations of the runoff formation process as a mechanism for monitoring the basin turnover in the Lange Bramke catchment, Upper Harz Mountains. *Hochwasser* 53, 64–79.
Hewlett, J.D., Fortson, J.C. and Cunningham, G.B. (1984) Additional tests on the effect of rainfall intensity on storm flow and peak flow from wild-land basins. *Water Resources Research* 20, 985–989.
Hlásny, T., Mátyás, C., Seidl, R., Kulla, L., Merganičová K., Trombik, J., Dobor, L., Barcza, Z. and Konôpka, B. (2014) Climate change increases the drought risk in Central European forests: what are the options for adaptation? *Lesnicky Casopis Forestry Journal* 60, 5–18.
Humbert, N. and Naijar, G. (1992) *Influence de la forêt sur le cycle de l'eau en domaine tempéré: une analyse de la littérature francophone*. Université Louis Pasteur, Centre d'Études et de Recherches Éco-Géographiques, Strasbourg, France.
Keller, H.M. (1988) European experiences in long-term forest hydrology research. In: Swank, W.T. and Crossley, D.A. (eds) *Forest Hydrology and Ecology at Coweeta*. Ecological Studies Vol. 66. Springer, New York, pp. 407–414.
Kubatzsch, A. (2007) *Waldwirkung und Hochwasser. Ein Leitfaden für Landnutzer und Entscheidungsträger*. Arbeitsgruppe Wald und Hochwasser, Initiative Weißeritz-Regio, Bärenfels, Germany.
Lindroth, A. and Crill, P. (2011) Hydrology and biogeochemistry of boreal forests. In: Levia, D.F., Carlyle-Moses, D. and Tanaka, T. (eds) *Forest Hydrology and Biogeochemistry: Synthesis of Past Research and Future Directions*. Ecological Studies Vol. 216. Springer, New York, pp. 321–339.
Manser, R. (2013) Forest policy in a changing climate – what are we expecting from scientists? In: Wohlgemuth, T. and Priewasser, K. (eds) *ClimTree 2013, Proceedings of the International Conference on Climate Change and Tree Responses in Central European Forests, Zurich, Swizterland, 1–5 September 2013*. Swiss Federal Research Institute WSL, Birmensdorf, Switzerland, p. 18 (abstract).
Mather, A.S., Fairbairn, J. and Needle, C.L. (1998) The course and drivers of the forest transition: the case of France. *Journal of Rural Studies* 15, 65–90.
McCulloch, J.S.G. and Robinson, M. (1993) History of forest hydrology. *Journal of Hydrology* 150, 189–216.

Meesenburg, H., Jansen, M., Döring, C., Besse, F., Rüping, U., Möhring, B., Hentschel, S., Meiwes, K.J. and Spellmann, H. (2005) Konzept zur Beurteilung der Auswirkungen forstlicher Maßnahmen auf den Gewässerzustand nach den Anforderungen der EG-Wasserrahmenrichtlinie. In: *Wasservorsorge in bewaldeten Einzugsgebieten Heft 62*. Berichte Freiburger Forstliche Forschung, Freiburg, Germany, pp. 171–180.

Müller, J. (2012) Auswirkungen von waldstrukturellen Veränderungen auf die hydroökologischen Bedingungen in den Beständen im Zuge des Waldumbaus. In: Grünewald, U., Bens, O., Fischer, H., Hüttl, R.F., Kaiser, K. and Knierim, A. (eds) *Anpassungsmaßnahmen an den Landschafts- und Klimawandel.* Schweizerbart, Stuttgart, Germany, pp. 280–291.

Müller, J. and Bolte, A. (2009) The use of lysimeters in forest hydrology research in north-east Germany. *Landbauforschung Volkenrode* 59, 1–10.

Ohte, N. and Tokuchi, N. (2011) Hydrology and biogeochemistry of temperate forests. In: Levia, D.F., Carlyle-Moses, D. and Tanaka, T. (eds) *Forest Hydrology and Biogeochemistry: Synthesis of Past Research and Future Directions*. Ecological Studies Vol. 216. Springer, New York, pp. 261–283.

Pötzelsberger, E. and Hasenauer, H. (2015) Forest–water dynamics within a mountainous catchment in Austria. *Natural Hazards* 77, 625–644.

Psidova, E., Ditmarova, L., Jamnicka, G., Kurjak, D., Majerova, J., Czajkowski, T. and Bolte, A. (2013) Photosynthetic response of beech seedlings of different origin to water deficit. *Photosynthetica* 53, 187–194.

Pülzl, H., Hogl, K., Kleinschmit, D., Wydra, D., Arts, B., Mayer, P., Palahí, M., Winkel, G. and Wolfslehner, B. (eds) (2014) *European Forest Governance: Issues at Stake and the Way Forward. What Science Can Tell Us*. European Forest Institute, Joensuu, Finland.

Ristić, R., Radić, B., Vasiljević, N. and Nikić, Z. (2011) Land use change for flood protection – a prospective study for the restoration of the river Jelašnica watershed. *Bulletin of the Faculty of Forestry* 103, 115–130.

Ristić, R., Kasanin-Grubin, M., Radić, B., Nikić, Z. and Vasiljević, N. (2012) Land degradation at the Stara Planina Ski Resort. *Environmental Management* 49, 580–592.

Rüping, U., Gutsche, C. and Möhring, B. (2012) *SILVAQUA – Auswirkungen forstlicher Bewirtschaftungsmaßnahmen auf den Zustand von Gewässern in bewaldeten Einzugsgebieten am Beispiel der Oker im Nordharz*. Beiträge aus der Nordwestdeutschen Forstlichen Versuchsanstalt Band 9. Nordwestdeutsche Forstliche Versuchsanstalt, Universitätsverlag Göttingen, Göttingen, Germany, pp. 189–204.

Samuelson, L., Bengtsson, K., Celander, T., Johansson, O., Jägrud, L., Malmer, A., Mattsson, E., Schaaf, N., Svending, O. and Tengberg, A. (2015) *Water, Forests, People – Building Resilient Landscapes.* SIWI Report No. 36. Stockholm International Water Institute, Stockholm.

Schleppi, P. (2011) Forested water catchments in a changing environment. In: Bredemeier, M., Cohen, S., Godbold, D.L., Lode, E., Pichler, V. and Schleppi, P. (eds) *Forest Management and the Water Cycle: An Ecosystem-Based Approach*. Ecological Studies Vol. 212. Springer, New York, pp. 89–110.

SOER (2015) The European environment – state and outlook 2015/European briefings/Forests. Available at: http://www.eea.europa.eu/soer-2015/europe/forests (accessed 29 March 2016).

Schüler, G. (2006) Identification of flood-generating forest areas and forestry measures for water retention. *Forest, Snow and Landscape Research* 80, 99–114.

Schüler, G., Pfister, L., Vohland, M., Seeling, S. and Hil, J. (2011) Large scale approaches to forest and water interactions. In: Bredemeier, M., Cohen, S., Godbold, D., Lode, E., Pichler, V. and Schleppi, P. (eds) *Forest Management and the Water Cycle: An Ecosystem-Based Approach*. Ecological Studies Vol. 212. Springer, New York, pp. 435–452.

Stagge, J.H., Dias, S., Rego, F.C. and Tallaksen, L.M. (2015) Integration of climate time series and MODIS data as an analysis tool for forest drought detection. In: Haro-Monteagudo, D., Momblanch-Benavent, M., Andreu-Alvarez, J., Solera-Solera, A., Parades-Arquiola, J., and Van Lanen, H. (eds) *International Conference on Drought: Research and Science–Policy Interfacing, 10–13 March 2015, Valencia (Spain)*. Technical Report No. 23. Universidad Politecnica de Valencia (UPVLC), Valencia, Spain and Wageningen University (WU), Wageningen, the Netherlands, pp. 51–53.

Van den Bergh, T., Hiltbrunner, E. and Körner, C. (2014) Land abandonment increases evapotranspiration in mountain grassland. In: Prasanna, H.G., Nettles, J. and Devendra, A.M. (eds) *Evapotranspiration: Challenges in Measurement and Modeling from Leaf to the Landscape Scale and Beyond, Conference Proceedings, 7–10 April 2014, Raleigh, North Carolina*. American Society of Agricultural and Biological Engineers, St Joseph, Michigan, p. 84 (abstract 1887374).

Wahren, A. and Feger, K.H. (2011) Model-based assessment of forest land management on water dynamics at various hydrological scales – a case study. In: Bredemeier, M., Cohen, S., Godbold, D.L., Lode, E., Pichler, V. and Schleppi, P. (eds) *Forest Management and the Water Cycle: An Ecosystem-Based Approach*. Ecological Studies Vol. 212. Springer, New York, pp. 453–468.

Wemple, B., Shanley, J., Denner, J., Ross, D. and Mills, K. (2007) Hydrology and water quality in two mountain basins of the northeastern US: assessing baseline conditions and effects of ski area development. *Hydrological Processes* 21, 1639–1650.

Zimmermann, N.E., Bollinger, J., Gehrig-Fasel, J., Guisan, A., Kienast, F., Lischke, H., Rickebusch, S. and Wohlgemuth, T. (2006) Wo wachsen die Bäume in 100 Jahren? *Forum für Wissen* 2006, 63–73.

6 Tropical Forest Hydrology

T. Kumagai[1]*, H. Kanamori[1] and N.A. Chappell[2]
[1]*Nagoya University, Nagoya, Japan;* [2]*Lancaster University, Lancaster, UK*

6.1 Introduction

One may imagine that tropical forest regions are in general characterized by higher incident radiative energy, constantly higher temperature and a large amount of ecosystem water resources, enough for supporting their higher primary productivity and active water cycling. In reality, tropical forests play a significant role in the global carbon budget (e.g. Beer *et al.*, 2010; Pan *et al.*, 2011) and are a major source of global hydrological fluxes, profoundly influencing both global and regional climates (e.g. Avissar and Werth, 2005; Spracklen *et al.*, 2012; Poveda *et al.*, 2014).

Tropical forest regions may be roughly classified into two types based upon seasonal variations in precipitation, although each region can be classified into more climate and forest types (e.g. Walsh, 1996; Tanaka *et al.*, 2008): (i) tropical evergreen rainforest with a 'rainforest climate'; and (ii) tropical seasonal forest with a 'monsoon climate' (Kumagai *et al.*, 2009). Tropical evergreen rainforests exist only in regions of ample water resources (Kumagai *et al.*, 2005; Kumagai and Porporato, 2012b), while tropical seasonal forests have to cope with seasonal droughts (e.g. Vourlitis *et al.*, 2008; Miyazawa *et al.*, 2014; Kumagai *et al.*, 2015). Thus, since the existence of these tropical forests achieves a delicate balance in terms of their ecosystem water resources, hydrological change in either region could result in significant impacts on ecological patterns and processes (Malhi *et al.*, 2009; Phillips *et al.*, 2009; Kumagai and Porporato, 2012a), in turn affecting feedbacks to the atmosphere (Meir *et al.*, 2006; Bonan, 2008; Kumagai *et al.*, 2013; van der Ent *et al.*, 2014). Here, we should note that recent evidence shows marked change in regional climate will occur first in the tropics, making tropical forest ecosystems particularly vulnerable in the future (Mora *et al.*, 2013).

Humans have been modifying the tropical forest land cover for food and energy production and for the development of the tropical countries. Consequently, such modifications (i.e. land-use changes) are being combined with climate change and should be anticipated to impact the regional hydro-climate as well as the local freshwater resources (Bruijnzeel, 1990; Giambelluca, 2002; van der Ent *et al.*, 2010; Wohl *et al.*, 2012; Kumagai *et al.*, 2013). What distinguishes the current modifications are the intensity and global reach, where the entire hydro-climate is now subject to these modifications. The consequences of land-use and climate change in the tropical forest regions are considered among the

*Corresponding author; e-mail: tomoomikumagai@gmail.com

greatest environmental concerns for the survival of the human population today, and include changes in streamflow and flood frequency, loss of productive soils, changes in nutrient fluxes, a warmer and drier climate, and concomitant changes in hydrological ecosystem services at the local scale (e.g. Bruijnzeel, 2004; Chappell *et al.*, 2004; Krusche *et al.*, 2011; Lawrence and Vandecar, 2015). To tackle all aspects of these hydrological problems, we have broad recognition and tacit acceptance that progress on these complex issues benefits from fundamental knowledge on all hydrological components in various types of tropical forest and synthesizing them in the global implications (see Bruijnzeel, 2004).

The main objective of this chapter, therefore, is to provide the state of knowledge on the characteristics of the hydrological components of precipitation, evapotranspiration and streamflow generation in tropical forest regions by a review of cornerstone tropical literature and a pan-tropical perspective partly via the use of pan-tropical maps of the hydrological components. Then, we suggest research needs and strategies for the study of tropical forest hydrology in the context of a dramatically changing world.

6.2 Pan-Tropical Climatic Regime and Forest Type

Pan-tropical maps of precipitation characteristics are derived from the PREC/L long-term gridded precipitation data set (Chen *et al.*, 2002) (Plate 5). High radiative energy received around the equator generates the updraft of hot and humid air. The convergence of the hot and humid air (i.e. the low atmospheric pressure system) results in the belt of precipitation (Plate 5a) associated with the Inter-Tropical Convergence Zone (ITCZ) around the equator. Updraft in the ITCZ moves both northward and southward in the tropopause, and descends around the Tropic of Cancer and the Tropic of Capricorn. Around both tropics (the so-called 'horse latitudes') the divergence of hot and dry air typically creates arid areas and deserts (Plate 5a). This circulation of air masses ascending around the equator and descending around tropics is called the Hadley circulation.

According to the Köppen climate classification, tropical climates are defined as the condition in which air temperatures do not fall below 18°C in the coolest month where the annual mean precipitation is more than the Köppen arid boundary ($P_K = 20 \times (T_Y + x)$ in mm, where T_Y is annual mean air temperature (°C) and x is an index representing the precipitation pattern such as 0, 7 and 14 for dry summer and wet winter, wet throughout the year, and dry winter and wet summer, respectively). These climates are further subdivided into rainforest (Af), monsoon (Am) and savannah (Aw) classes. The rainforest climate occurs in the zone where all 12 months have mean precipitation of at least 60 mm. The monsoon climate and the savannah climate have mean precipitation less than 60 mm in the least precipitation month, but they have the least monthly precipitation more and less than 100–0.04 × (annual mean precipitation in mm), respectively.

As Feng *et al.* (2013) suggested, the areas with the largest amount of precipitation always exist in the regions with the most aseasonal precipitation patterns, but not vice versa (Plate 5). The rainforest climate (Af) is characterized by both a large annual amount of precipitation (Plate 5a) and little or negligible seasonal variation in precipitation (Plate 5b), caused by a stationary low atmospheric pressure system around the equator. Tropical evergreen forests can exist in this region because of their constant water requirements throughout the year. High rainfall intensity of short duration is distinctive in the rainforest climate, and in both diurnal and seasonal variation senses, there is strongly spatial heterogeneous rainfall distribution over tropical rainforests at the scale of single to thousands of square kilometres (Krusche *et al.*, 2011; Kumagai and Kume, 2012; Kanamori *et al.*, 2013).

The monsoon (Am) and the savanna (Aw) climates are, by contrast, characterized by their degree of seasonality in precipitation, caused primarily by seasonal fluctuations of the ITCZ. It should be noted that annual precipitation decreases with the increased seasonality of precipitation, but not vice versa (see Feng *et al.*, 2013; Plate 5). There is a less intense dry season and a larger amount of precipitation in the monsoon climate than in the savannah climate. Tropical seasonal forests can exist in areas of the monsoon with climate adaptations that cope with water stress in the dry season, e.g. full- or semi-deciduous traits, seasonal variations in stomatal behaviour, etc. This, in turn, means that tropical seasonal forest can vary from evergreen to deciduous forests corresponding to the local precipitation regimes such as influenced by the altitude and topography (e.g. Tanaka *et al.*, 2008).

The tropics are alternatively divided into three regions based on biotic and geological history (US DOE, 2012): (i) the neo-tropical ecozone (NEO) of South America, Central America and the Caribbean; (ii) the Afro-tropical ecozone (AFR) of sub-Saharan Africa; and (iii) the Indo-Malay-Australasia tropical ecozone (IMA) including regions of India, South-East Asia, southern China, New Guinea and northern Australia. The forests of the Amazon Basin in the neo-tropics represent the single largest block of intact tropical forests, occupying ~40% of forest biomass in the tropics (Saatchi *et al.*, 2011). South and Central American and Caribbean forests often experience hurricane-induced disturbance (e.g. Boose *et al.*, 1994, 2004; Negrón-Juárez *et al.*, 2010). Among the Afro-tropical forests, those of the Congo Basin represent the second-largest intact block of tropical forests (Pan *et al.*, 2011). Central African forests do not experience tropical cyclones but are still subject to large storm events, and suffer from severe drought and special wet periods particularly during the El Niño and the La Niña phenomena (US DOE, 2012). While the dominant plant family in neo-tropical and Afro-tropical forests is *Leguminosae*, within South-East Asian forests *Dipterocarpaceae* dominate. Dipterocarps are among the most valuable timber species globally, leading to intense logging pressure within the Indo-Malay-Australasia tropical forests. Indeed over the past 50 years, more timber was exported from Borneo Island in this region than from neo-tropical and Afro-tropical regions combined (Curran *et al.*, 2004). Although South-East Asian tropical forests represent only about 11% of the world's tropical forests in terms of area, they have the highest relative deforestation rate in the tropics (e.g. Canadell *et al.*, 2007; Pan *et al.*, 2011). An analysis of climatic trends in global tropical rainforest regions over the period 1960–1998 showed that precipitation has declined more significantly in South-East Asia than in Amazonia, and that the El Niño–Southern Oscillation (ENSO) is a particularly important driver of drought in South-East Asia (Malhi and Wright, 2004).

6.3 Pan-Tropical Evapotranspiration

Evapotranspiration, i.e. water vapour evolution from the forest canopy, can be simply partitioned into wet canopy evaporation or canopy interception and dry canopy evaporation or transpiration. Kume *et al.* (2011) reviewed tropical forest canopy interception ratios, i.e. the wet canopy evaporation divided by gross precipitation, from the data of 40 tropical forests including both rainforests and seasonal forests, and reported that most ratios measured in tropical forests were in the range of 10–20% with a mean of 17%. On the other hand, the reported transpiration rates in tropical forests have a large temporal and spatial variation in value as 2.3–4.6 mm/day (e.g. Shuttleworth *et al.*, 1984; Roberts *et al.*, 1993; Cienciala *et al.*, 2000; Kumagai *et al.*, 2004). Bruijnzeel (1990) suggested that in humid tropical forests, the average annual transpiration was 1045 mm (range 885–1285 mm). The resultant variation in annual tropical forests' evapotranspiration ranges from about 1000 to 1500 mm and the ratio of annual evapotranspiration to annual precipitation ranges mostly between 30 and 90% (Kume *et al.*, 2011). It is worth noting that Kume *et al.* (2011) found that apparently in tropical forests when the annual precipitation (P) is <2000 mm, the evapotranspiration increases with increases in P, and when P > 2000 mm, the evapotranspiration reaches the plateau of about 1500 mm.

There are more complex situations controlling tropical forests' evapotranspiration. Trees in the tropical monsoon (Am) climatic regions tend to cope with seasonal drought by stomatal control, leaf-fall, root water uptake at deeper soil depths and so forth (e.g. Igarashi *et al.*, 2015). Many pristine tropical forests have been cleared for conversion to plantations such as rubber trees and oil palms (e.g. Carlson *et al.*, 2012; Fox *et al.*, 2012). Such plantations' water use is comparable to or more than that of the original vegetations (e.g. Guardiola-Claramonte *et al.*, 2010; Kumagai *et al.*, 2015). Here, we should note that significant portions of the cleared tropical land revert quickly to secondary vegetation and, in terms of hydrological characteristics, increasingly resemble original forest with time (Giambelluca, 2002). Giambelluca *et al.* (2003) pointed out that evapotranspiration from fragmented forest tends to be enhanced by conditions in surrounding clearings. Tropical montane cloud forests, which are fog- and cloud-affected, need to be paid special attention on their hydrometeorology (Bruijnzeel *et al.*, 2011): small and large amounts of transpiration and canopy interception due to the small water vapour deficit and the cloud-water, respectively.

Despite the complexity in generalizing mechanisms of evapotranspiration, we attempt comparison of the tropical forests with other regions' evapotranspiration: annual evapotranspiration (E) from ground observations (e.g. watershed water balance, micrometeorological measurements and soil-water balance) were classified by annual mean temperature (T) and represented as a function of latitude (Fig. 6.1, adapted from Komatsu *et al.*, 2012). Despite a large variation in E at each latitude (~500 mm), which might be caused by altitude, forest type and local climate of the forest sites, differences in hydrological observation methods and so forth, there is a strong linear relationship between latitude and E. In humid climates of the ITCZ, the high P is expected to cause higher E (Fig. 6.1). Note that there is a large variation in E in the tropical forest regions, but many can be explained by the difference in the T, probably because of the altitude effects.

From this relationship between annual P and T, Komatsu *et al.* (2012) modified the Zhang *et al.* (2001) evapotranspiration model where:

$$E = P\frac{1+w(E_0/P)}{1+w(E_0/P)+(P/E_0)} \qquad (6.1)$$

in which

$$E_0 = 0.488T^2 + 27.5T + 412 \qquad (6.2)$$

where w is a coefficient representing plant water availability (= 2.0). E_0 is the potential evaporation, which was defined as a constant value calculated by Priestley and Taylor's (1972) equation in Zhang *et al.* (2001) but was modified in Komatsu *et al.* (2012) so that it can appropriately describe the temperature effect. Eqns 6.1 and 6.2 successfully reproduce E as a function of latitude (Fig. 6.1). In the tropical rainforest and monsoon climates it suggests that higher E eventuates through the conditions of a plentiful P and higher T.

Theoretical (Eqns 6.1 and 6.2) and observed relationships between annual P and E are

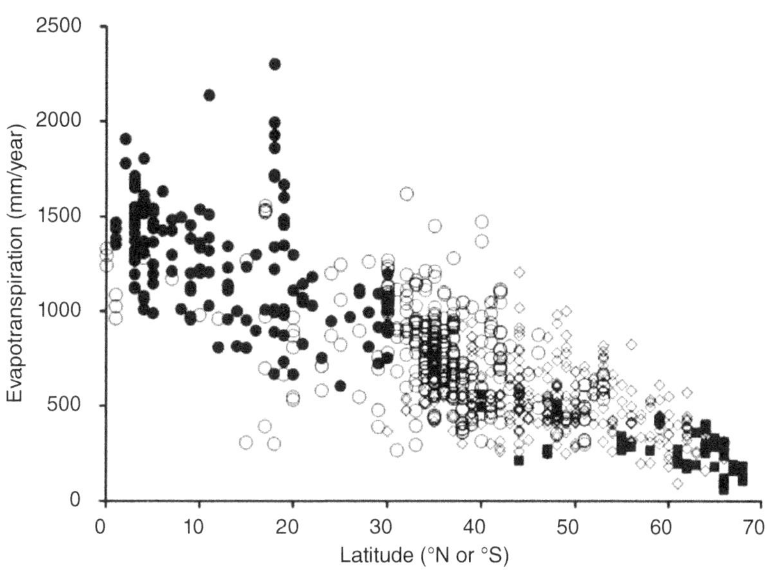

Fig. 6.1. Annual evapotranspiration as a function of latitude, classified by annual mean temperature (T; °C): closed circles = $T > 20$; open circles = $10 < T < 20$; open diamonds = $0 < T < 10$; closed squares = $T < 0$. (Data on evapotranspiration, latitude and temperature taken from Komatsu *et al.*, 2012.)

investigated in Fig. 6.2a1–c1. The shallower slope in the relationship between *E* and *P* in the neo-tropics (Fig. 6.2a1) and the Indo-Malay-Australasia tropics (Fig. 6.2c1) compared with the steeper relationship in the Afro-tropics (Fig. 6.2b1) can be explained by the saturation curve, Eqn 6.1. It is also surprising that such a simple formulation as Eqn 6.1 has an ability to describe the pan-tropical *E* characteristics because Fig. 6.2 contains data not only from natural seasonal and rain forests but also secondary forests and plantations. Further, it is interesting to note that from an interpretation of Budyko's vegetation categorization using the relationships among the radiation dryness index, net radiation and vegetation types (see Budyko and Miller, 1974), tropical forests can exist within the ranges of annual mean net radiation 110–133 W/m^2 and annual mean precipitation 1370–5000 mm. The lowest precipitation limit for the presence of tropical forest (i.e. 1370 mm/year) is found to be broadly comparable with the lowest value of *P* in each ecozone (Fig. 6.2a1–c1).

The local water-use ratio (LEUR) is defined as *E* divided by *P*, and represents the ratio of *P* water recycling from *E*. Figure 6.2a2–c2 shows these data for each tropical ecozone. The LEUR data show a significant decrease with *P* for all ecozones. This implies that in areas with smaller *P* (mainly in AFR), the *P* is effectively recycled

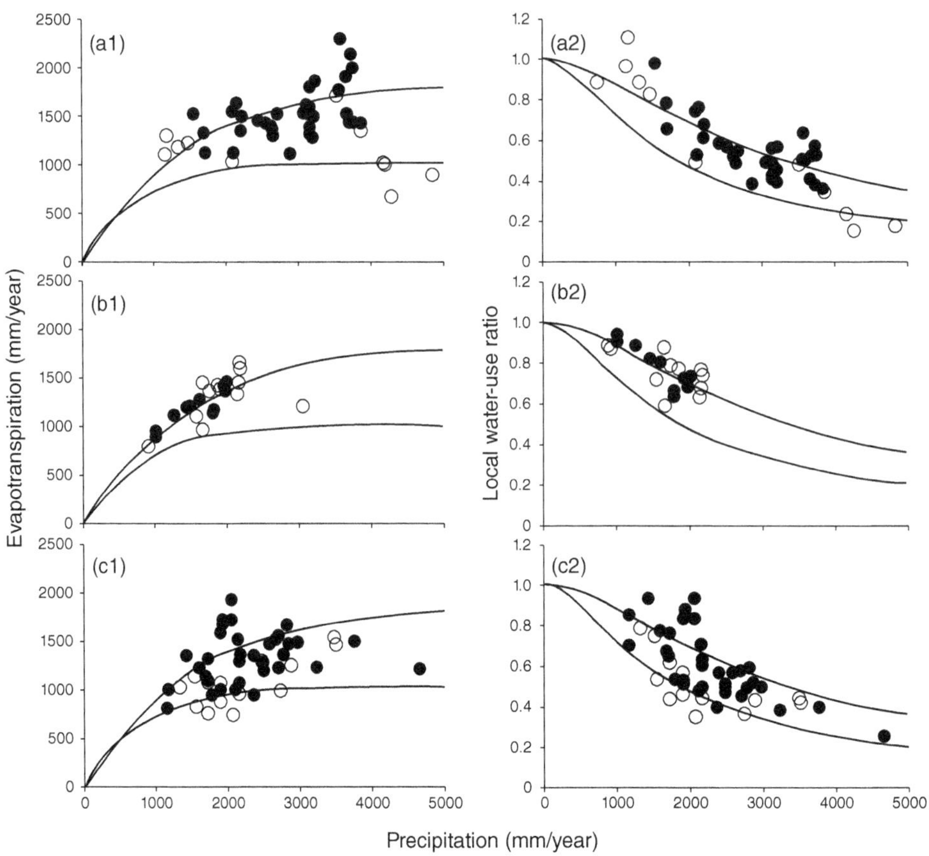

Fig. 6.2. Relationships between precipitation and annual evapotranspiration (a1, b1, c1) and local water-use ratio (a2, b2, c2) defined as evapotranspiration divided by precipitation, classified by annual mean temperature (*T*; °C): closed circles = *T* > 25; open circles = 20 < *T* < 25. Solid lines denote the theoretical relationships: upper and bottom lines for each panel assume *T* = 30 and 15, respectively. (a1, a2) neo-tropical ecozone; (b1, b2) Afro-tropical ecozone; and (c1, c2) Indo-Malay-Australasia tropical ecozone. (Data on evapotranspiration and precipitation taken from Komatsu *et al.*, 2012.)

from the E so that the vegetation is regulating its own regional water resources. In areas with larger P (such as NEO and IMA), P is also supplied by moist air from the surrounding terrain and oceans.

The atmosphere–land surface water balance method (see Oki *et al.*, 1995) with global data on P (Plate 5) and atmospheric water vapour divergence derived from ERA-Interim gridded four-dimensional meteorology data set (Dee *et al.*, 2011) produced a pan-tropical map of E and LEUR. Although E is well reproduced in areas with a relatively small amount of P, it is likely that large P in the method has induced an overestimation of E in such areas. Nevertheless, characteristics of spatial variation in E are well represented. Comparatively small E can be seen in the Afro-tropics, and a much larger E in the islands of Borneo and New Guinea in the Indo-Malay-Australasian tropics and in Amazonian headwaters in the neo-tropics. Areas with lower LEUR values tend to overlap areas with larger P (compare with Plate 5a) and this is consistent with the patterns shown in Fig. 6.2a2–c2. The exception is Borneo Island (IMA ecozone) where despite a plentiful P, higher values of LEUR are estimated due to near-zero annual atmospheric water vapour divergence/convergence (Kumagai *et al.*, 2013). Here, we should note that for considering the extent to which P relies on terrestrial E (i.e. moisture recycling), the role of global wind patterns, topography and land cover should be interpreted more in the context of continental moisture recycling (van der Ent *et al.*, 2010). Notably, Poveda *et al.* (2014) examined how 'aerial river' pathways modified by the effects of topography, orography and land cover types contribute to precipitation patterns in tropical South America.

6.4 Pan-Tropical Streamflow Generation

Figure 6.3 shows a plot of annual streamflow per unit drainage area (the strict definition of term 'runoff') against latitude (determined at the confluence with the ocean; adapted from Wohl *et al.*, 2012). The streamflow for December–February (winter in the northern hemisphere) and June–August (winter in the southern hemisphere) averaged per latitude was also estimated using the atmosphere–land surface water balance method with assumptions that deep groundwater flow all contributes to streamflow at the scale of large watersheds and changes in calculation-domain water storage can be neglected (Oki *et al.*, 1995) and added to Fig. 6.3. The negative values of computed 3-month streamflow are probably attributable to the role of dynamic subsurface storages. It is, however, apparent that the annual streamflow from tropical basins is typically much greater than that from temperate basins due primarily to the greater precipitation amounts in the humid tropics. Further, the range in annual observed streamflow is much larger within the latitudes 0 to 23.4° due to the presence of extensive areas with dry climates (BW and BS) in the tropics.

The greater streamflow present within the parts of the tropics with a rainforest climate (Af) and monsoon climate (Am) indicates that considerably more precipitation travels through watersheds towards streams than at other latitudes. The greater water flows within such tropical basins mean that the magnitudes of chemical and particle transport are likely to be greater and the watershed systems more sensitive to disturbance (Wohl *et al.*, 2012). Consequently, tropical forest hydrology has implications for the other scientific disciplines of biogeochemistry and geomorphology.

Some 89% of the channel network of the globe's streams/rivers comprises first-, second- and third-order channels (Table 2 in Downing *et al.*, 2012). This means the most streamflow is generated in the network of such low-order channels. Experimental watersheds typically comprise channels of first- to third-order size, and so are ideal locations for the study of the pathways of rainwater to the channel network via surface and/or subsurface pathways. These routes of water migration to channels are known as the pathways of 'streamflow generation' or simply 'runoff pathways' (Bonell, 2004; Burt and McDonnell, 2015). Most experimental watersheds used for the study of runoff pathways are located in temperate regions, with very few in tropical regions (Bonell, 2004; Burt and McDonnell, 2015). Figure 6.4a shows the locations of some key experimental watersheds with a long history of research on streamflow generation pathways that are located in the

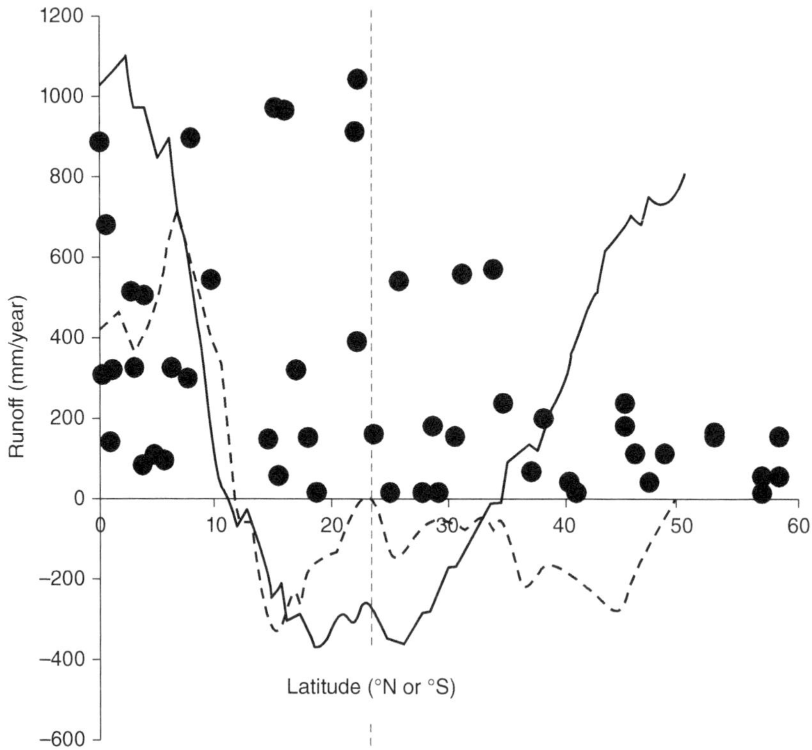

Fig. 6.3. Annual-scaled streamflow per unit drainage area as a function of latitude at the river mouth (circles) adapted from Wohl *et al.* (2012). Relationships between latitude and annually scaled and latitude-averaged streamflow computed from the atmosphere–land surface water balance method for December–February in the northern hemisphere (solid line) and June–August in the southern hemisphere (broken line) are also shown. Vertical dotted line denotes the latitude of the tropics north and south of the equator (23.4°).

tropics. Most of these watersheds are located beneath tropical forests, so the findings are most pertinent to tropical forest environments.

Streamflow generation pathways that have been observed within these tropical forest environments include:

1. *Infiltration-excess overland flow.* This water flow on slopes outside channels is caused by precipitation falling at a rate faster than the local coefficient of permeability at the ground surface (equivalent to the 'infiltration capacity' or surface 'saturated hydraulic conductivity').

2. *Saturation overland flow by direct precipitation.* Where rainfall falls on to ground at a rate less than that of the infiltration capacity, but where pores are already saturated, then no further infiltration can take place and new water travels over the surface.

3. *Subsurface flow.* Where rainfall infiltrates, some will evaporate from the soil or support transpiration from vegetation; the remainder will travel towards streams below the surface. Most of this water will enter the streams via the channel bed and banks, but some will return to the surface prior to reaching a channel (so-called *return flow*) and flow over the surface as *saturation overland flow*. Flow beneath the surface may be very shallow where lithomorphic soils overlie an impermeable geology, but may be over 100 m deep where permeable soils overlie permeable geology (whether unconsolidated materials or rock).

Some ambiguity in the definition of water pathways arises from the definition of the ground surface. Some scientists define overland flow as

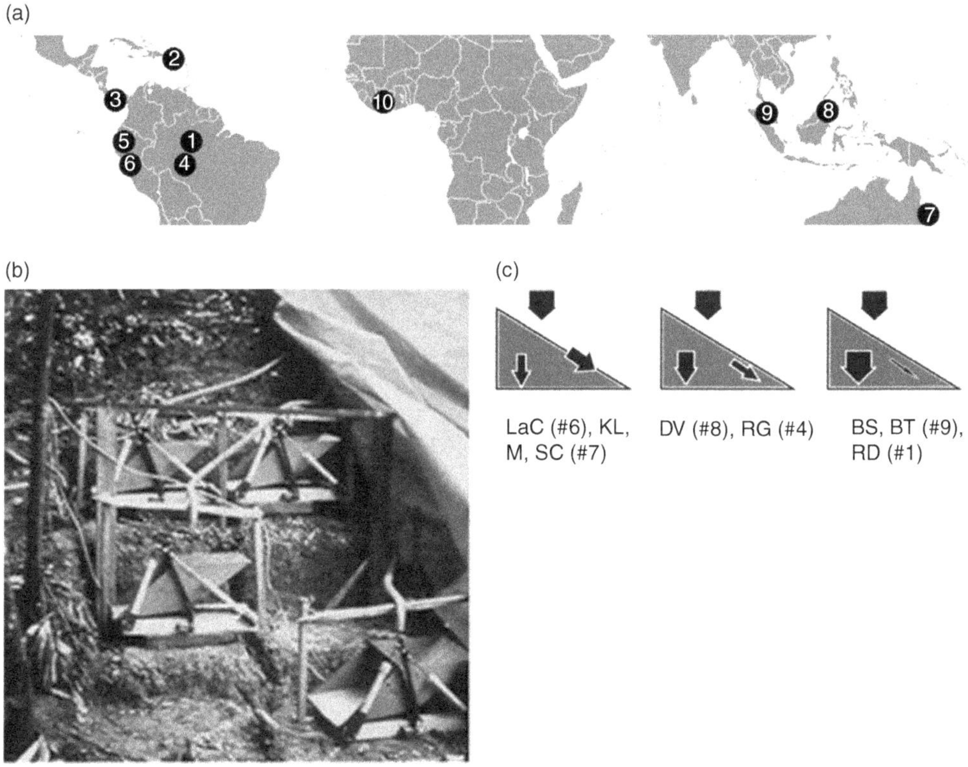

Fig. 6.4. (a) Location of ten experimental basins in the tropics with a long history of research into streamflow generation pathways: #1 = Reserva Ducke (Brazil); #2 = Bisley II (Puerto Rico); #3 = Barro Colorado & Lutzito (Panama); #4 = Rancho Grande & Jurena Ultisol (Brazil); #5 = Rio San Francisco (Ecuador); #6 = La Cuenca (Peru); #7 = South Creek (Australia); #8 = Danum Valley (Malaysia); #9 = Bukit Tarek & Bukit Berembun (Malaysia); #10 = Tai Forest (Ivory Coast) (for an extensive list of publications for these and other experimental sites in the tropics, see Elsenbeer, 2001; Bonell, 2004; Hugenschmidt *et al.*, 2014; Barthold and Woods, 2015). (b) The 'throughflow trough' system used by Bonell and Gilmour (1978) to demonstrate the presence of shallow lateral flows in the organic and mineral soils at the South Creek Experimental Watershed (#7), Queensland, Australia. (c) Inferred dominant streamflow generation pathways presented in Elsenbeer (2001). Experimental watershed LaC = La Cuenca (#6); KL = Kiani Lestari; M = Mendolong; SC = South Creek (#7); DV = Danum Valley (#8); RG = Rancho Grande (#4); BS = Bukit Soeharto; BT = Bukit Tarek (#9); RD = Reserva Ducke (#1) (see Elsenbeer, 2001 for details).

water moving over the surface of a mineral A (or E) soil horizon, while others use the definition of water moving laterally above the typically overlying organic horizons (i.e. L, H and/or O horizons).

Streamflow generation studies in the tropics that have directly observed the presence of overland flow where the rainfall intensity exceeds the infiltration capacity are very limited, but the Tai Forest basin is a good example (Bonell, 2004). Studies that have observed overland flow on permeable but saturated soils include the South Creek (Bonell and Gilmour, 1978) and La Cuenca basins (Elsenbeer and Vertessy, 2000). Several studies have direct evidence of lateral (downslope) flows within the subsurface generated during rainstorms. Bonell and Gilmour (1978) observed the presence of these flows using so-called 'throughflow troughs' (Fig. 6.4b), while Chappell and Sherlock (2005) observed the presence of these flows by monitoring tracer migration.

The more pertinent question is not whether a particular pathway is present or not, but whether it is the *dominant pathway* producing >50% of the observed total streamflow, particularly during storm events (i.e. without any manipulation of hydrographs by separation methods). Too often researchers infer the dominance of a particular pathway simply from the observed presence of the pathway, rather than relating measured *flows per unit basin area from that pathway* with those observed in the *stream per unit basin area*. For example, the measured lateral flows per unit basin area of Gilmour *et al.* (1980) are only a tiny fraction of the observed streamflow per unit basin area. Consequently, the importance of measured near-surface flows at this site (and many other sites) has been overemphasized in comparison to the unmeasured flows within deeper soil and unconsolidated rock strata. The resultant ambiguity and misinterpretation of the dominant pathway is amply illustrated by inconsistencies in the inferred dominant pathways presented in the reviews of Elsenbeer (2001), Bonell (2004) and Barthold and Woods (2015) as shown in Fig. 6.4c and Table 6.1.

A further area of concern is the focus of most studies on pathways in the solum (i.e. A and B soil horizons) alone, as highlighted by Bonell and Balek (1993) and Bonell (2004). There is an increasing awareness that at some tropical sites, soils may be developed on unconsolidated geological materials that are permeable and have deeper pathways within these strata. Examples of experimental watersheds developed on these deeper porous media in the tropics include the Lake Calado microbasin near Reserva Ducke basin in Brazil (Lesack, 1993); the Jungle Falls basin in Singapore (Chappell and Sherlock, 2005; Rahardjo *et al.*, 2010); the Bukit Timah (Noguchi *et al.*, 2005) and Bukit Berembun (Chappell *et al.*, 2004) basins in Malaysia; and the O Thom II basin in Cambodia (Shimizu *et al.*, 2007). Equally, other low-order basins in the tropics may be located on rock aquifers or rocks with fracture systems that produce very deep water pathways. The Arboleda basin near La Selva, Costa Rica (Genereux *et al.*, 2005) has such pathways. To alert researchers to the potential presence and role of these deeper streamflow generation pathways, Chappell *et al.* (2007) developed a perceptual model of runoff pathways, where a type III system has a dominance of pathways via unconsolidated geological materials and a type IV system has a dominance of pathways via rock aquifers or fracture systems (Fig. 6.5). Experimental basins lacking any evidence of major pathways through geological strata (solid rock or unconsolidated materials) are classified as either type II systems (where a B soil horizon is present, e.g. Chappell *et al.*, 1998) or type I systems (where a lithomorphic soil is developed on steep, impermeable mountain slopes). The potential presence of deeper pathways within existing experimental watersheds needs to be a key focus for new research. With more complete observational evidence, including information gained from both hydrometric and tracer studies (as recommended by Barthold and Woods, 2015), researchers may be closer to developing a unified, numerical model of the dominant pathways of streamflow generation applicable across all tropical forest environments.

Table 6.1. Inferred dominant streamflow generation pathways presented in Table 14.1 of Bonell (2004). See Bonell (2004) for further details of the experimental watersheds and pathway definitions.

Dominant pathway	Experimental watershed[a]
Predominantly vertical pathways	Reserva Ducke (#1), Bukit Tarek (#9), Fazenda Dimona, Mgera
Predominantly lateral pathways (infiltration-excess overland flow)	Tai Forest (#10)
Predominantly lateral pathways (saturation-excess overland flow)	South Creek (#7), Barro Colorado (#3), La Cuenca (#6), Maburae
Predominantly lateral pathways (subsurface stormflow within the soil)	Danum Valley (Sungai Baru Barat) (#8), Rancho Grande (#4), Bisley II (#2), Dodmane, Kannike, Ife, ECEREX
Predominantly lateral pathways (subsurface stormflow at the soil–bedrock interface)	Kuala Belalong

[a]Each # represents the location in Fig. 6.4a.

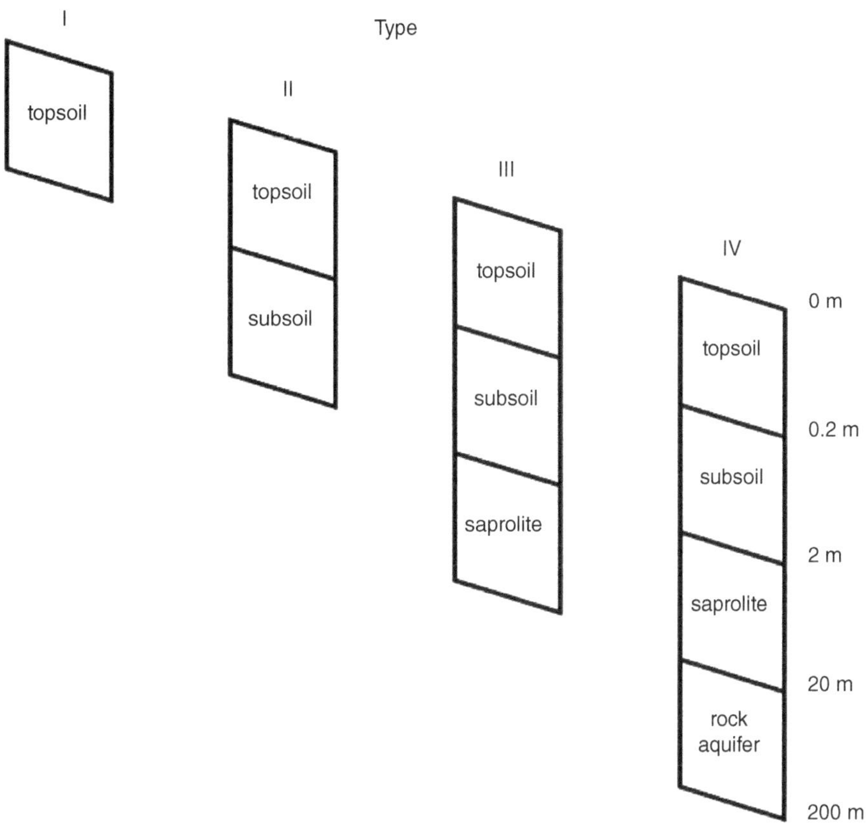

Fig. 6.5. Schematic representation of water pathway systems within tropical forests classified according to the presence of as many as four strata (soil horizon or topsoil (A); soil horizon or subsoil (B); unconsolidated geological materials; and solid rock) with the names type I to IV. Example depths in the logarithmic depth scale are given. (Adapted from Chappell *et al.*, 2007 and US DOE, 2012.)

6.5 Research Needs

The projected growth in atmospheric greenhouse gases within the coming century, as predicted by the Intergovernmental Panel on Climate Change's (IPCC) A1B (Balanced across all sources) scenario, will significantly increase tropical surface temperatures ranging from ~3 to 5°C for South-East Asia, the Amazon and West Africa (IPCC, 2007). Also, the A1B scenario predicts concomitant modifications to precipitation patterns: a general increase in precipitation for West Africa, intensification of seasonality of precipitation for South-East Asia (i.e. more and less precipitation in the wet and dry seasons, respectively) and a general decrease in precipitation for the southern and eastern Amazon. More dramatically, as an instance, the 1997–1998 El Niño was the strongest in the 20th century and its associated drought in Borneo Island in South-East Asia was the most severe (statistically, a drought such as in 1998 may occur once in ~360 years). Tree mortality rates during that drought were 6.37%/year, as compared with 0.89%/year during the pre-drought period (1993–1997) in the studied forest site in western Borneo (Nakagawa *et al.*, 2000). Global warming is likely to cause changes in Pacific regional climate that might alter ENSO activity in the future (e.g. Timmermann *et al.*, 1999). While it is not still clear how ENSO will be affected by global warming, it is possible that the frequency or amplitude of ENSO events could increase (Collins *et al.*, 2010).

In addition, the tropics are known to be a very active domain in terms of changes in land cover. According to Hansen *et al.*'s (2013) investigation using Earth observation satellite data from 2000 to 2012, the tropics were the only domain to exhibit a statistically significant increasing trend in annual forest loss, 2101 km^2/year, and tropical rainforests loss was 32 % of the global forest loss. South American tropical dry forests have been lost at the highest rate in the lost tropical forests. Furthermore, Hansen *et al.* (2013) confirmed that tropical forest loss rate in Brazil has decreased from 40,000 km^2 in 2003–2004 to 20,000 km^2 in 2010–2011, while forest loss in Indonesia increased drastically from 8000 km^2 in 2000–2001 to 20,000 km^2 in 2010–2011, indicating that an increase in Asian tropical forest loss 'compensates' a decrease in Amazonian tropical forest loss.

In the humid tropics, where the deforestation and land-use change are still ongoing and there is plenty of precipitation, a combination of climate change and drastic changes in land cover must induce pressure for freshwater resources and loss of soil via changes in local hydrological processes. Tropical forest evapotranspiration is generally greater than evapotranspiration from grasslands and thus changing land cover such as from forests to pastures would reduce the evapotranspiration and increase streamflow, resulting, in some cases, in increasing flood frequencies (see Bruijnzeel, 2001, 2004). The generation of localized infiltration-excess overland flow on compacted soil surfaces caused by forestry operations accelerates the geomorphological process of soil erosion (see Bruijnzeel, 2004; Sidle *et al.*, 2006; Sidle and Ziegler, 2012). Such land degradations would become increasingly worse by altered precipitation regimes through climate change (see Peña-Arancibia *et al.*, 2010; Wohl *et al.*, 2012). On the other hand, we should note that the hydrological properties of the secondary vegetation such as evapotranspiration and surface infiltration may quickly resemble those of the original forest again (see Giambelluca, 2002; Bruijnzeel, 2004).

From the global perspective, once more it should be emphasized that the tropical forests are a major source of global hydrological fluxes and thus changes in evapotranspiration rates could significantly impact both global and regional climates, in turn potentially affecting feedbacks to the atmosphere (see Bonan, 2008). Many previous research contributions to the knowledge on hydro-climate in the tropics suggest: forest cutting is more effective on local and global climate under maritime conditions than continental conditions (van der Molen *et al.*, 2006); rainwater is recycled earlier by wet canopy evaporation than via transpiration (van der Ent *et al.*, 2014); and in the maritime continent, changes in sea surface temperature influence precipitation regimes more than land cover changes (Bruijnzeel, 2004). Further, large-scale tropical deforestation and selective logging could result in warmer and drier conditions not only at the local scale, but also the teleconnections from converted tropical lands could pose a considerable risk to agriculture in other regions, due to impacts on precipitation against a background of warmer temperatures (see Lawrence and Vandecar, 2015).

As the first step for considering the future hydrological impact in the tropics, we need an understanding of the current and basic tropical hydrological cycling – and further, of the hydrological interactions among the earth surface and subsurface, vegetation and atmosphere – based on long-term and networked data acquisition and organization. As Wohl *et al.* (2012) pointed out, field-based hydrological measurements in many tropical countries have been less explored than those in the temperate regions and, to make matters worse, are waning. We should note that the lack of long-term observations homogenized throughout the tropics leads to a failure to validate hydro-climate models and thus the impossibility to extrapolate the future tropical forest hydrology by output of the models, which must be built referring to reliable observations and should use the observations as inputs. Besides making more effort for organizing a field-based hydrological observation network over the tropics, as shown in this chapter, remote sensing technologies and global-scale climate and geo-information databases can serve among the most promising tools to complement the lack of tropical field-observation sites.

6.6 Acknowledgements

The work was supported in part by Grant-in-Aid for Scientific Researches (#15H02645, #25281005) and the granted project 'Program for Risk Information on Climate Change' from the Ministry of Education, Science and Culture, Japan. We are deeply grateful to L.A. Bruijnzeel and the editors for their constructive comments on the manuscript.

References

Avissar, R. and Werth, D. (2005) Global teleconnections resulting from tropical deforestation. *Journal of Hydrometeorology* 6, 134–145.

Barthold, F.K. and Woods R.A. (2015) Stormflow generation: a meta-analysis of field evidence from small, forested catchments. *Water Resources Research* 51, 3730–3753.

Beer, C., Reichstein, M., Tomelleri, E., Ciais, P., Jung, M., Carvalhais, N., Rödenbeck, C., Arain, M.A., Baldocchi, D., Gordon, B., *et al.* (2010) Terrestrial gross carbon dioxide uptake: global distribution and covariation with climate. *Science* 329, 834–838.

Bonan, G.B. (2008) Forests and climate change: forcing, feedbacks, and the climate benefits of forests. *Science* 320, 1444–1449.

Bonell, M. (2004) Runoff generation in tropical forests. In: Bonell, M. and Bruijnzeel, L.A. (eds) *Forests, Water and People in the Humid Tropics: Past, Present and Future Hydrological Research for Integrated Land and Water Management*. UNESCO International Hydrology Series. Cambridge University Press, Cambridge, pp. 314–406.

Bonell, M. and Balek, J. (1993) Recent scientific developments and research needs in hydrological processes of the humid tropics. In: Bonell, M., Hufschmidt, M.M. and Gladwell, J.S. (eds) *Hydrology and Water Management in the Humid Tropics*. Cambridge University Press, Cambridge, pp. 167–260.

Bonell, M. and Gilmour, D.A. (1978) The development of overland flow in a tropical rainforest catchment. *Journal of Hydrology* 39, 365–382.

Boose, E.R., Foster, D.R. and Fluet, M. (1994) Hurricane impacts to tropical and temperate forest landscapes. *Ecological Monographs* 64, 369–400.

Boose, E.R., Serrano, M.I. and Foster, D.R. (2004) Landscape and regional impacts of hurricanes in Puerto Rico. *Ecological Monographs* 74, 335–352.

Bruijnzeel, L.A. (1990) *Hydrology of Moist Tropical Forests and Effects of Conversion: A State of Knowledge Review*. International Hydrological Programme, UNESCO, Paris.

Bruijnzeel, L.A. (2001) Forest hydrology. In: Evans, J. (ed.) *The Forests Handbook, Volume 1: An Overview of Forest Science*. Blackwell, Oxford, pp. 301–343.

Bruijnzeel, L.A. (2004) Hydrological functions of tropical forests: not seeing the soil for the trees? *Agriculture, Ecosystems & Environment* 104, 185–228.

Bruijnzeel, L.A., Mulligan, M. and Scatena, F.N. (2011) Hydrometeorology of tropical montane cloud forests: emerging patterns. *Hydrological Processes* 25, 465–498.

Budyko, M.I. and Miller, D.H. (1974) *Climate and Life*. Academic Press, New York.

Burt, T.P. and McDonnell, J.J. (2015) Whither field hydrology? The need for discovery science and outrageous hydrological hypotheses. *Water Resources Research* 51, 5919–5928.

Canadell, J.G., Le Quere, C., Raupach, M.R., Field, C.B., Buitenhuis, E.T., Ciais, P., Conway, T.J., Gillett, N.P., Houghton, R.A. and Marland, G. (2007) Contributions to accelerating atmospheric CO_2 growth from economic activity, carbon intensity, and efficiency of natural sinks. *Proceedings of the National Academy of Sciences USA* 104, 18866–18870.

Carlson, K.M., Curran, L.M., Ratnasari, D., Pittman, A.M., Soares-Filho, B.S., Asner, G.P., Trigg, S.N., Gaveau, D.A., Lawrence, D. and Rodrigues, H.O. (2012) Committed carbon emissions, deforestation, and community land conversion from oil palm plantation expansion in West Kalimantan, Indonesia. *Proceedings of the National Academy of Sciences USA* 109, 7559–7564.

Chappell, N.A. and Sherlock, M.D. (2005) Contrasting flow pathways within tropical forest slopes of Ultisol soil. *Earth Surface Processes and Landforms* 30, 735–753.

Chappell, N.A., Franks, S.W. and Larenus, J. (1998) Multi-scale permeability estimation in a tropical catchment. *Hydrological Processes* 12, 1507–1523.

Chappell, N.A., Tych, W., Yusop, Z., Rahim, N.A. and Kasran, B. (2004) Spatially-significant effects of selective tropical forestry on water, nutrient and sediment flows: a modelling-supported review. In: Bonell, M. and Bruijnzeel, L.A. (eds) *Forests, Water and People in the Humid Tropics: Past, Present and Future Hydrological Research for Integrated Land and Water Management*. UNESCO International Hydrology Series. Cambridge University Press, Cambridge, pp. 513–532.

Chappell, N.A., Sherlock, M., Bidin, K., Macdonald, R., Najman, Y. and Davies, G. (2007) Runoff processes in Southeast Asia: role of soil, regolith and rock type. In: Sawada, H., Araki, M., Chappell, N.A., LaFrankie, J.V. and Shimizu, A. (eds) *Forest Environments in the Mekong River Basin*. Springer, Tokyo, pp. 3–23.

Chen, M., Xie, P., Janowiak, J.E. and Arkin, P.A. (2002) Global land precipitation: a 50-yr monthly analysis based on gauge observations. *Journal of Hydrometeorology* 3, 249–266.

Cienciala, E., Kucera, J. and Malmer, A. (2000) Tree sap flow and stand transpiration of two *Acacia mangium* plantations in Sabah, Borneo. *Journal of Hydrology* 236, 109–120.

Collins, M., An, S.-I., Cai, W., Ganachaud, A., Guilyardi, E., Jin, F.-F., Jochum, M., Lengaigne, M., Power, S., Timmermann, A., *et al.* (2010) The impact of global warming on the tropical Pacific Ocean and El Niño. *Nature Geoscience* 3, 391–397.

Curran, L.M., Trigg, S.N., McDonald, A.K., Astiani, D., Hardiono, Y.M., Siregar, P., Caniago, I. and Kasischke, E. (2004) Lowland forest loss in protected areas of Indonesian Borneo. *Science* 303, 1000–1003.

Dee, D.P., Uppala, S.M., Simmons, A.J., Berrisford, P., Poli, P., Kobayashi, S., Andrae, U., Balmaseda, M.A., Balsamo, G., Bauer, P., *et al.* (2011) The ERA-Interim reanalysis: configuration and performance of the data assimilation system. *Quarterly Journal of the Royal Meteorological Society* 137, 553–597.

Downing, J.A., Cole, J.J., Duarte, C.M., Middelburg, J.J., Melack, J.M., Prairie, Y.T., Kortelainen, P., Striegl, R.G., McDowell, W.H. and Tranvik, L.J. (2012), Global abundance and size distribution of streams and rivers. *Inland Waters* 2, 229–236.

Elsenbeer, H. (2001) Hydrologic flowpaths in tropical soilscapes – a review. *Hydrological Processes* 15, 1751–1759.

Elsenbeer, H. and Vertessy, R.A. (2000) Stormflow generation and flowpath characteristics in an Amazonian rainforest catchment. *Hydrological Processes* 14, 2367–2381.

Feng, X., Porporato, A. and Rodriguez-Iturbe, I. (2013) Changes in rainfall seasonality in the tropics. *Nature Climate Change* 3, 811–815.

Fox, J., Vogler, J.B., Sen, O.L., Giambelluca, T.W. and Ziegler, A.D. (2012) Simulating land-cover change in montane mainland Southeast Asia. *Environmental Management* 49, 968–979.

Genereux, D.P., Jordan, M.T. and Carbonell, D. (2005) A paired-watershed budget study to quantify interbasin groundwater flow in a lowland rainforest, Costa Rica. *Water Resources Research* 41, W04011, doi: 10.1029/2004WR003635 (accessed 30 March 2016).

Giambelluca, T.W. (2002) Hydrology of altered tropical forest. *Hydrological Processes* 16, 1665–1669.

Giambelluca, T.W., Ziegler, A.D., Nullet, M.A., Truong, D.M. and Tran, L.T. (2003) Transpiration in a small tropical forest patch. *Agricultural and Forest Meteorology* 117, 1–22.

Gilmour, D.A., Bonell, M. and Sinclair, D.F. (1980) *An Investigation of Storm Drainage Processes in a Tropical Rainforest Catchment*. Technical Paper No. 56. Australian Water Resources Council, Canberra.

Guardiola-Claramonte, M., Troch, P.A., Ziegler, A.D., Giambelluca, T.W., Durcik, M., Vogler, J.B. and Nullet, M.A. (2010) Hydrologic effects of the expansion of rubber (*Hevea brasiliensis*) in a tropical catchment. *Ecohydrology* 3, 306–314.

Hansen, M.C., Potapov, P.V., Moore, R., Hancher, M., Turubanova, S.A., Tyukavina, A., Thau, D., Stehman, S.V., Goetz, S.J., Loveland, T.R., *et al.* (2013) High-resolution global maps of 21st-century forest cover change. *Science* 342, 850–853.

Hugenschmidt, C., Ingwersen, J., Sangchan, W., Sukvanachaikul, Y., Duffner, A., Uhlenbrook, S. and Streck, T. (2014) A three-component hydrograph separation based on geochemical tracers in a tropical mountainous headwater catchment in northern Thailand. *Hydrology and Earth System Science* 18, 525–537.

Igarashi, Y., Kumagai, T., Yoshifuji, N., Sato, T., Tanaka, N., Tanaka, K., Suzuki, M. and Tantasirin, C. (2015) Environmental control of canopy stomatal conductance in a tropical deciduous forest in northern Thailand. *Agricultural and Forest Meteorology* 202, 1–10.

IPCC (2007) *Climate Change 2007: Impacts, Adaptation and Vulnerability. Contribution of Working Group II to the Fourth Assessment Report of the Intergovernmental Panel on Climate Change*. Cambridge University Press, New York.

Kanamori, H., Yasunari, T. and Kuraji, K. (2013) Modulation of the diurnal cycle of rainfall associated with the MJO observed by a dense hourly rain gauge network at Sarawak, Borneo. *Journal of Climate* 26, 4858–4875.

Komatsu, H., Cho, J., Matsumoto, K. and Otsuki, K. (2012) Simple modeling of the global variation in annual forest evapotranspiration. *Journal of Hydrology* 420, 380–390.

Krusche, A.V., Ballester, M.V.R. and Leite, N.K. (2011) Hydrology and biogeochemistry of terra firme lowland tropical forests. In: Levia, D.F., Carlyle-Moses, D. and Tanaka, T. (eds) *Forest Hydrology and Biogeochemistry: Synthesis of Past Research and Future Directions*. Ecological Studies Vol. 216. Springer, New York, pp. 187–201.

Kumagai, T. and Kume, T. (2012) Influences of diurnal rainfall cycle on CO_2 exchange over Bornean tropical rainforests. *Ecological Modelling* 246, 91–98.

Kumagai, T. and Porporato, A. (2012a) Drought-induced mortality of a Bornean tropical rain forest amplified by climate change. *Journal of Geophysical Research–Biogeosciences* 117, G02032, doi: 10.1029/2011JG001835 (accessed 30 March 2016).

Kumagai, T. and Porporato, A. (2012b) Strategies of a Bornean tropical rainforest water use as a function of rainfall regime: isohydric or anisohydric? *Plant, Cell and Environment* 35, 61–71.

Kumagai, T., Saitoh, T.M., Sato, Y., Morooka, T., Manfroi, O.J., Kuraji, K. and Suzuki, M. (2004) Transpiration, canopy conductance and the decoupling coefficient of a lowland mixed dipterocarp forest in Sarawak, Borneo: dry spell effects. *Journal of Hydrology* 287, 237–251.

Kumagai, T., Saitoh, T.M., Sato, Y., Takahashi, H., Manfroi, O.J., Morooka, T., Kuraji, K., Suzuki, M., Yasunari, T. and Komatsu, H. (2005) Annual water balance and seasonality of evapotranspiration in a Bornean tropical rainforest. *Agricultural and Forest Meteorology* 128, 81–92.

Kumagai, T., Yoshifuji, N., Tanaka, N., Suzuki, M. and Kume, T. (2009) Comparison of soil moisture dynamics between a tropical rainforest and a tropical seasonal forest in Southeast Asia: impact of seasonal and year-to-year variations in rainfall. *Water Resources Research* 45, W04413, doi: 10.1029/2008WR007307 (accessed 30 March 2016).

Kumagai, T., Kanamori, H. and Yasunari, T. (2013) Deforestation-induced reduction in rainfall. *Hydrological Processes* 27, 3811–3814.

Kumagai, T., Mudd, R.G., Giambelluca, T.W., Kobayashi, N., Miyazawa, Y., Lim, T.K., Liu, W., Huang, M., Fox, J.M., Ziegler, A.D., *et al.* (2015) How do rubber (*Hevea brasiliensis*) plantations behave under seasonal water stress in northeastern Thailand and central Cambodia? *Agricultural and Forest Meteorology* 213, 10–22.

Kume, T., Tanaka, N., Kuraji, K., Komatsu, H., Yoshifuji, N., Saitoh, T.M., Suzuki, M. and Kumagai, T. (2011) Ten-year evapotranspiration estimates in a Bornean tropical rainforest. *Agricultural and Forest Meteorology* 151, 1183–1192.

Lawrence, D. and Vandecar, K. (2015) Effects of tropical deforestation on climate and agriculture. *Nature Climate Change* 5, 27–36.

Lesack, L.F.W. (1993) Water balance and hydrologic characteristics of a rain forest catchment in the Central Amazon Basin. *Water Resources Research* 29, 759–773.

Malhi, Y. and Wright, J. (2004) Spatial patterns and recent trends in the climate of tropical rainforest regions. *Philosophical Transactions of the Royal Society B: Biological Sciences* 359, 311–329.

Malhi, Y., Aragao, L.E.O.C., Galbraith, D., Huntingford, C., Fisher, R., Zelazowski, P., Sitch, S., McSweeney, C. and Meir, P. (2009) Exploring the likelihood and mechanism of a climate-change-induced dieback of the Amazon rainforest. *Proceedings of the National Academy of Sciences USA* 106, 20610–20615.

Meir, P., Cox, P.M. and Grace, J. (2006) The influence of terrestrial ecosystems on climate. *Trends in Ecology and Evolution* 21, 254–260.

Miyazawa, Y., Tateishi, M., Komatsu, H., Ma, V., Kajisa, T., Sokh, H., Mizoue, N. and Kumagai, T. (2014) Tropical tree water use under seasonal waterlogging and drought in central Cambodia. *Journal of Hydrology* 515, 81–89.

Mora, C., Frazier, A.G., Longman, R.J., Dacks, R.S., Walton, M.M., Tong, E.J., Sanchez, J.J., Kaiser, L.R., Stender, Y.O., Anderson, J.M., *et al.* (2013) The projected timing of climate departure from recent variability. *Nature* 502, 183–187.

Nakagawa, M., Tanaka, K., Nakashizuka, T., Ohkubo, T., Kato, T., Maeda, T., Sato, K., Miguchi, H., Nagamasu, H., Ogino, K., *et al.* (2000) Impact of severe drought associated with the 1997–1998 El Niño in a tropical forest in Sarawak. *Journal of Tropical Ecology* 16, 355–367.

Negrón-Juárez, R.I., Chambers, J.Q., Guimaraes, G., Zeng, H., Raupp, C.F., Marra, D.M., Ribeiro, G.H.P.M., Saatchi, S.S., Nelson, B.W. and Higuchi, N. (2010) Widespread Amazon forest tree mortality from a single cross-basin squall line event. *Geophysical Research Letters* 37, L16701, doi: 10.1029/2010GL043733 (accessed 30 March 2016).

Noguchi, S., Abdul Rahim, N. and Tani, M. (2005) Runoff characteristics in a tropical rain forest catchment. *Japan Agricultural Research Quarterly* 39, 215–219.

Oki, T., Musiake, K., Matsuyama, H. and Masuda, K. (1995) Global atmospheric water balance and runoff from large river basins. *Hydrological Processes* 9, 655–678.

Pan, Y., Birdsey, R.A., Fang, J., Houghton, R., Kauppi, P.E., Kurz, W.A., Phillips, O.L., Shvidenko, A., Lewis, S.L., Canadell, J.G., *et al.* (2011) A large and persistent carbon sink in the world's forests. *Science* 333, 988–993.

Peña-Arancibia, J.L., van Dijk, A.I.J.M., Mulligan, M. and Bruijnzeel, L.A. (2010) The role of climatic and terrain attributes in estimating baseflow recession in tropical catchments. *Hydrology and Earth System Sciences* 14, 2193–2205.

Phillips, O.L., Aragao, L.E.O.C., Lewis, S.L., Fisher, J.B., Lloyd, J., Lopez-Gonzales, G., Malhi, Y., Monteagudo, A., Peacock, J., Quesada, C.A., *et al.* (2009) Drought sensitivity of the Amazon rainforest. *Science* 323, 1344–1347.

Poveda, G., Jaramillo, L. and Vallejo, L.F. (2014) Seasonal precipitation patterns along pathways of South American low-level jets and aerial rivers. *Water Resources Research* 50, 98–118.

Priestley, C.H.B. and Taylor, R.J. (1972) On the assessment of surface heat flux and evaporation using large-scale parameters. *Monthly Weather Review* 100, 81–92.

Rahardjo, H., Nio, A., Leong, E. and Song, N. (2010) Effects of groundwater table position and soil properties on stability of slope during rainfall. *Journal of Geotechnical and Geoenvironmental Engineering* 136, 1555–1564.

Roberts, J., Cabral, O.M.R., Fisch, G., Molion, L.C.B., Moore, C.J. and Shuttleworth, W.J. (1993) Transpiration from an Amazonian rainforest calculated from stomatal conductance measurements. *Agricultural and Forest Meteorology* 65, 175–196.

Saatchi, S.S., Harris, N.L., Brown, S., Lefsky, M., Mitchard, E.T., Salas, W., Zutta, B.R., Buermann, W., Lewis, S.L., Hagen, S., *et al.* (2011) Benchmark map of forest carbon stocks in tropical regions across three continents. *Proceedings of the National Academy of Sciences USA* 108, 9899–9904.

Shimizu, A., Kabeya, N., Nobuhiro, T., Kubota, T., Tsuboyama, Y., Ito, E., Sano, M., Chann, S. and Keth, N. (2007) Runoff characteristics and observations on evapotranspiration in forest watersheds, central Cambodia. In: Sawada, H., Araki, M., Chappell, N.A., LaFrankie, J.V. and Shimizu, A. (eds) *Forest Environments in the Mekong River Basin*. Springer, Tokyo, pp. 135–146.

Shuttleworth, W.J., Gash, J.H.C., Lloyd, C.R., Moore, C.J., Roberts, J.M., Marques Filho, A.d.O., Fisch, G., Silva Filho, V.d.P., Ribeiro, M.d.N.G., Molion, L.C.B., *et al.* (1984) Eddy correlation measurements of energy partition for Amazonian forest. *Quarterly Journal of the Royal Meteorological Society* 110, 1143–1162.

Sidle, R.C. and Ziegler, A.D. (2012) The dilemma of mountain roads. *Nature Geoscience* 5, 437–438.

Sidle, R.C., Ziegler, A.D., Negishi, J.N., Nik, A.R., Siew, R. and Turkelboom, F. (2006) Erosion processes in steep terrain? Truths, myths, and uncertainties related to forest management in Southeast Asia. *Forest Ecology and Management* 224, 199–225.

Spracklen, D.V., Arnold, S.R. and Taylor, C.M. (2012) Observations of increased tropical rainfall preceded by air passage over forests. *Nature* 489, 282–285.

Tanaka, N., Kume, T., Yoshifuji, N., Tanaka, K., Takizawa, H., Shiraki, K., Tantasirin, C., Tangtham, N. and Suzuki, M. (2008) A review of evapotranspiration estimates from tropical forests in Thailand and adjacent regions. *Agricultural and Forest Meteorology* 148, 807–819.

Timmermann, A., Oberhuber, J., Bacher, A., Esch, M., Latif, M. and Roeckner, E. (1999) Increased El Niño frequency in a climate model forced by future greenhouse warming. *Nature* 398, 694–697.

US DOE (2012) *Research Priorities for Tropical Ecosystems Under Climate Change Workshop Report, DOE/SC-0153*. US Department of Energy, Office of Science, Washington, DC.

van der Ent, R.J., Savenije, H.H.G., Schaefli, B. and Steele-Dunne, S.C. (2010) Origin and fate of atmospheric moisture over continents. *Water Resources Research* 46, W09525, doi: 10.1029/2010WR009127 (accessed 30 March 2016).

van der Ent, R.J., Wang-Erlandsson, L., Keys, P.W. and Savenije, H.H.G. (2014) Contrasting roles of interception and transpiration in the hydrological cycle? Part 2: Moisture recycling. *Earth System Dynamics* 5, 471–489.

van der Molen, M.K., Dolman, A.J., Waterloo, M.J. and Bruijnzeel, L.A. (2006) Climate is affected more by maritime than by continental land use change: a multiple scale analysis. *Global and Planetary Change* 54, 128–149.

Vourlitis, G.L., de Souza Nogueira, J., de Almeida Lobo, F., Sendall, K.M., de Paulo, S.R., Antunes Dias, C.A., Pinto Jr, O.B. and de Andrade, N.L.R. (2008) Energy balance and canopy conductance of a tropical semi-deciduous forest of the southern Amazon Basin. *Water Resources Research* 44, W03412, doi: 10.1029/2006WR005526 (accessed 30 March 2016).

Walsh, R.P.D. (1996) Climate. In: Richards, P.W. (ed.) *The Tropical Rain Forest, An Ecological Study*, 2nd edn. Cambridge University Press, Cambridge, pp. 159–205.

Wohl, E., Barros, A., Brunsell, N., Chappell, N.A., Coe, M., Giambelluca, T., Goldsmith, S., Harmon, R., Hendrickx, J.M.H., Juvik, J., *et al.* (2012) The hydrology of the humid tropics. *Nature Climate Change* 2, 655–662.

Zhang, I., Dawes, W.R. and Walker, G.R. (2001) Response of mean annual evapotranspiration to vegetation changes at catchment scale. *Water Resources Research* 37, 701–708.

7 Hydrology of Flooded and Wetland Forests

T.M. Williams[1]*, K.W. Krauss[2] and T. Okruszko[3]
[1]Clemson University, Georgetown, South Carolina, USA; [2]US Geological Survey, Lafayette, Louisiana, USA; [3]Warsaw University of Life Sciences, Gdańsk, Poland

7.1 Introduction

In this chapter we examine the hydrology of forested areas that are subject to soil saturation by precipitation, groundwater or surface flooding. They include mangroves and other tidal forests, the forested portions of peatlands and tree-dominated wetlands defined by the Ramsar Convention (Mathews, 1993). They also include estuarine tidal forests, palustrine forested wetlands and the portions of palustrine scrub-shrub which are made up of immature tree species of the Cowardin *et al.* (1985) classification. A broad outline of the ecology of all wetlands is described in Mitsch and Gosselink (2015), wetlands specifically with tidal influence are described by Tiner (2013), while descriptions of northern and southern forested wetlands can be found in Trettin *et al.* (1996) and Messina and Conner (1998), respectively.

Since most forest species (except certain mangroves) cannot regenerate under continuous flooding (Lugo *et al.*, 1988), the wet limit of forested wetlands requires at least periodic drying of the soil surface. The dry limit is somewhat ambiguous but is generally associated with soil saturation that is of sufficient duration to produce visible soil features and limit the growth of plants not specifically adapted to saturated soils. In the USA, definition of the dry limit of wetlands is most particularly interplay of scientific, economic and political interests. Wetland activities are regulated by the Clean Water Act, administrated by the US Environmental Protection Agency. On-the-ground delineation and enforcement is done by the US Army Corps of Engineers (USACE). The USACE guidelines for delineation of wetlands relate to plant and soil indicators of saturation (USACE, 1987). National mapping of wetland vegetation is done by the US Fish and Wildlife Service (USFWS) in the Interior Department, while national mapping of soils is done by the Natural Resources Conservation Service (NRCS) of the US Department of Agriculture (USDA). Due to practical problems in delineation (damage to vegetation and/or soils), USACE published a practical method to relate measured water tables as a proxy to soil and vegetation indicators (USACE, 2005). That publication specified (in lieu of site-specific information) that soil saturation, sufficient for wetland designation, could be assumed to occur if the water table was within 12 inches (30 cm) of the soil surface continuously for 14 days of the growing season, defined by the NRCS as the period with soil temperatures above 28°F (–2.2°C) for 50% of years of record.

The depth criterion was chosen to reflect that soils are saturated a distance (ψ) above the

*Corresponding author; e-mail: tmwllms@clemson.edu

water table by capillary action. The distance ψ is the soil moisture tension (in cm) required for air to enter the soil and may also be called the bubbling pressure. Rawls *et al.* (1982) listed geometric mean values of ψ from 7.3 cm (sand) to 36.5 cm (clay) for 11 soil texture classes represented in the soil texture triangle. Comerford *et al.* (1996) examined high water table sandy soils (*Ultic Alaquods* – US system) in Florida and found evidence of soil saturation where water tables were within 15 cm of the surface. Their work suggested that in high water table forests the soil matrix also saturates a distance ψ from soil surface that agrees with values found by Rawls *et al.* (1982). The Rawls *et al.* (1982) estimates of ψ may be a reasonable method to estimate surface saturation.

Understanding the interaction of soils, hydrology and geomorphology may be critical to the study of slopes subject to mass movements, erosion and sediment transfer, runoff generation processes and human impacts on sediment processes (Sidle and Onda, 2004). Semeniuk and Semeniuk (1997) proposed to include geomorphical and hydroperiod descriptors into the Ramsar Convention classification. In northern Europe, peat accumulation and flooding depths are used to define mires (areas of peat accumulation), marshes (non-forested with little peat accumulation) and swamps (non-forested with deeper continuous flooding) (Okruszko *et al.*, 2011). These terms are counter to usage in the USA, where 'swamp' is used to refer almost exclusively to forested wetlands, and 'marsh' is used to refer to tidal fresh- and saltwater areas dominated by grasses and herbs that accumulate organic and inorganic sediments. In the USA, 'peatland' is used for all areas with organic soils, while in Europe only mires drained for agriculture or forestry are called peatlands. Brinson (1993) proposed classifying wetlands based on the hydrology of differing geomorphical configurations and differentiated the source of water in various wetland settings. Tiner (2011) produced dichotomous keys to differentiate and elaborate Cowardin *et al.* (1985) classes based on Brinson's ideas. In this chapter we focus on forested wetlands in regard to the source of water that causes soil saturation.

Soil saturation occurs when the soil moisture storage term (ΔSM) of the basic water balance equation is minimized (soil moisture maximized), or the balance of $P - E - T \pm Q_s \pm Q_g \geq \Delta SM$, where P is precipitation, E is evaporation of interception, open water evaporation and soil surface evaporation, T is transpiration, Q_s is surface water runoff and Q_g is subsurface (groundwater) runoff. This produces three basic types of forested wetland: (i) rain fed, where $P > E + T + Q_s + Q_g$; (ii) groundwater fed, where $P + Q_g > E + T + Q_s$; and (iii) surface fed, where $P + Q_s > E + T + Q_g$. Most surface-fed wetlands are in riverine settings with flooding due to river stage which is determined by upstream hydrology. Forests can also be flooded directly by action of tides and indirectly by tides altering the stage relationships of freshwater rivers.

7.2 Forested Wetlands Due to Excess Precipitation

Precipitation excess wetlands ($P > E + T + Q_s + Q_g$) are found in moist climatic regions. Annual P is likely to equal or exceed annual evapotranspiration (ET), and P *must* exceed ET for a significant portion of the year for wetland conditions to occur. Restricted surface and/or subsurface drainage are also generally present. These wetlands generally have shallow surface drainage, if any. In many cases these wetlands are found in relatively young geological formations where mature drainage patterns have not yet developed. Groundwater flows are also restricted vertically by a confining layer, or horizontally by slope, hydraulic conductivity (k), or the combination of the two. Brinson (1993) classified these wetlands as mineral or organic flats, Semeniuk and Semeniuk (1997) called them damplands and Tiner (2011) called them seasonally saturated wetlands.

7.2.1 Northern rain and snow region

The largest and most widespread concentrations of precipitation-fed wetlands are northern hemisphere bogs. They occur in Köppen's humid temperate–cool summer (Cf b,c) and subarctic (Df) climatic zones, including the north-eastern and north-central USA, eastern and most of northern Canada, north-eastern Europe, Fennoscandia, and Russia in Eurasia. In these zones P tends to exceed potential *ET* (*PET*) and a portion of annual precipitation falls as snow. Sphagnum mosses are the most common vegetation types found in wetlands across the entire zone. Common tree species are *Pinus* spp., or *Betula* spp. in Eurasia, and *Picea mariana*, *Larix laricina* and *Betula* spp. in North America. Most mires of these regions are composed of both bog (precipitation fed) and fen (groundwater fed), with raised bogs where peat accumulation produces slightly convex land forms.

The most important hydrological aspect of northern mires is the segregation of peat into acrotelm and catotelm layers (Ingram, 1978). The acrotelm is a surface accumulation layer of living moss and partially decomposed peat that is porous and has fairly low bulk density and high permeability. The catotelm is decomposed and compacted peat with both vertical and horizontal conductivity (k) three to five orders of magnitude lower than the acrotelm. Ingram (1983) suggests typical k values in the acrotelm of 10^{-1} cm/s and catotelm of 10^{-4} cm/s. Fraser *et al.* (2001) measured horizontal k in the acrotelm of 10^{-3} to 10^{-7} cm/s, with highest values near the surface and lowest at the top of the catotelm, and catotelm values of 10^{-6} to 10^{-8} cm/s. Devito *et al.* (1996) estimated k of 10^{-2} cm/s at 10 cm depth, 10^{-3} to 10^{-4} cm/s at 20–30 cm depth and values of 10^{-5} to 10^{-6} cm/s in deeper (>100 cm) peat. Boelter (1969) predated Ingram's terms but found k values of 1.8×10^{-3} cm/s for fibric (least decomposed) and 2×10^{-6} cm/s for sapric (most decomposed) peats. Chason and Siegel (1986) found higher k (2.5×10^{-4} to 2.6×10^{-2} cm/s) for peats from the catotelm in western Minnesota. The low k value of the catotelm and the variable connection of acrotelm to uplands are responsible for wide variation in wetland hydrology and watershed behaviour. Low k of the catotelm restricts lateral water movement to the upper portions of the acrotelm. The degree of upland connection to the mire is determined by the slope, thickness and conductivity of the upland material in contact with the acrotelm around the wetland margin.

Continental glaciation has produced a common landscape across much of the northern hemisphere. The northern regions are areas of glacial erosion with thin till deposits on bedrock from which all regolith has been eroded. Southern regions are areas of glacial deposition of thicker till deposits in moraines and drumlins, outwash sands and gravels in broad plains, and clay deposits in beds of former periglacial lakes. Scattered throughout the depositional till and outwash deposits are depressions formed by buried ice that slowly melted after glaciers retreated. Peat accumulations are found in various depressions throughout the glacially eroded landscape, in large wetlands of former glacial lakes, in depressions between drumlins and in many ice block depressions. Both glacial lakebeds and ice block depressions generally have a basal clay layer, deposited during the early post-glacial period, which can further restrict vertical groundwater exchange between the peat and mineral soil.

The hydrology of mires in the northern glacially eroded zone is dominated by flow associated with snowmelt (Woo and diCenzo, 1989). Exposed bedrock and areas of permafrost influence drainage pattern and runoff production north of 70°N in Canada (Hodgson and Young, 2001), and non-forested patterned 'appa' mires are generally found north of 63°N in Finland (Turunen *et al.*, 2002) and as far south as 60°N in Russia (Botch, 1990). Further south in Canada (61°N), St Amour *et al.* (2005) found flows in tributaries of the Mackenzie River showed surface water connections of most of the watersheds during snowmelt but flows from rain later in the season tended to occur only in watersheds with many fens and lakes. Discontinuous permafrost may be found as far south as 54°N in eastern Canada (Smith and Risenborough, 2002). Recent observation of permafrost thawing associated with a warming climate (Jorgeson *et al.*, 2006) suggests these limits may move northward in the coming decades.

South of permafrost, wetland hydrology is controlled by the thickness and type of outwash or till in the surrounding uplands (Buttle *et al.*, 2000). Where till is thin and discontinuous, overland flow from bedrock contributes directly to wetlands and streamflow. Where till or organic accumulations cover bedrock either continuously or as islands on the bedrock, flow occurs in the subsurface, generally at the soil–bedrock interface (Peters *et al.*, 1995). Coupling between slopes and wetlands occurs primarily with snowmelt. Summer rain produces stormflow only from wetlands connected to the stream network by flow within the acrotelm or saturated overland flow. The proportion of bog and fen in mires is related to till thickness and surrounding topography. Depressions in the area of thin till will have relatively narrow fen margins often with raised bog centres. Thicker till increases the proportion of groundwater entering the margin and the portion of the mire that is fen. The rate of groundwater recharge is also determined by the size of the source area and transmissivity ($T = k \times$ saturated thickness of the aquifer) of the upland aquifer.

Further south in the zone of glacial deposition the distribution of wetland types is more variable. The most ubiquitous but generally smaller wetlands are associated with ice block depressions. These depressions are more numerous near terminal moraines and vary from entirely vegetated mires to open lakes. The presence of a thin clay layer under the catotelm tends to further seal the depression, allowing perched water tables to exist in mires at elevations well above the general water table. Such mires may occur even in coarse textures of outwash plains. Devito *et al.* (1996) demonstrated that two similar depressions at 45°N were nearly entirely bog in an area of recharge with little connection of acrotelm to the upland water table aquifer, or entirely fen where a thick upland water table aquifer supplied flow through the acrotelm for most of the growing season. Bay (1967) described similar differences at watersheds S2 and S3 at the Marcell Experimental Forest in north-central Minnesota (47°N). Watershed S2 was perched above the regional water table and primarily a bog, while S3 was within an area of glacial outwash where the aquifer discharged through the acrotelm. A complete description of the hydrology of watershed S2 can be found in Verry *et al.* (2011) and is summarized in Chapter 14, this volume.

On drumlin-dominated watersheds (44°N), Todd *et al.* (2006) found few mires, but depressions between drumlins had mineral soil wetlands with vegetation characteristic of both wetlands and uplands (*Thuja occendentalis*, *Acer saccharum*, *Populus* spp., *Betula* spp. and *Typha* spp. near basin outlets). In these watersheds, runoff coefficients suggested both rain and snowmelt runoff originated only on the saturated wetlands. Contrary to the variable source area concept, wetlands at lower elevations did not produce flow until they received inflow from ephemeral streams originating from wetlands at higher elevations.

The hydrology of mires on periglacial lakebeds presents special problems associated with the size of the landscape. Studies of hydrology on former glacial lake wetlands have usually been done with piezometer nests placed in only small portions of the wetlands, with results interpreted from extrapolation of results of smaller wetlands. Fraser *et al.* (2001) examined a 28 km^2 wetland in Ontario (45.4°N, 75.4°W) with ten piezometer nests arranged along two 500 m transects. They found very limited recharge from regional groundwater, and flow from catotelm to acrotelm occurring only during drought. Even then flow was accompanied by a decline of head in the catotelm that would limit potential flow regardless of the extent of drought. Siegel and Glaser (1987) studied a 32 km^2 portion of the Lake Agassiz patterned wetlands in western Minnesota (48.1°N, 94.4°W), using three piezometer nests; one in a bog and the two others in fen areas. They found that regional groundwater was the source of much of the flow in the fen areas and resulted in great differences in pH (4 in bog, 7 in fen) and dissolved minerals (mainly Ca, Mg, Na). Heinselmann (1970), working on a 181 km^2 portion of the same Lake Agassiz wetland 100 km to the east (48.2°N, 93.5°W), found that high pH, Ca and Mg could only be found in a fen where upland discharge flowed into the acrotelm.

7.2.2 Southern rain-only region

Tropical peatlands represent 11% of the world's peat soils and 57% of those are found in South-East Asia (Andriesse, 1988). Nearly all are found in Köppen's tropical humid (Af, Am) climatic zones, with South-East Asia in the monsoonal Am zone. Although little work has been done on the hydrology, peat areas of Indonesia appear to resemble raised bogs of northern latitudes (Wösten *et al.*, 2008). Wösten *et al.* (2008) also found 10-year average annual rainfall of 2570 mm and estimated *ET* of 1500 mm, but found large differences in the amount of rainfall (range 1848–3788 mm) associated with the El Niño/Southern Oscillation. Chimer and Ewel (2005) found lower growth rates of forest trees, rapid decomposition of leaves, and peat accumulation due entirely to root growth and mortality within the peat layer in Micronesia, where rainfall rates were 5050 mm/year. Lähteenoja *et al.* (2009) measured peat accumulation rates similar to those of northern mires in the Peruvian section of the Amazonian lowlands, where average rainfall rates were 3100 mm/year.

Wösten *et al.* (2008) described rapid flow through the acrotelm during heavy rain, but rapid subsidence if water levels fell to the top of the catotelm at 70 cm. Areas of drained peatland were subject to rapid subsidence and fire during periods of rainfall below 2000 mm/year. Lähteenoja *et al.* (2009) found Amazonian wetlands to be much younger (600–3000 years) than northern mires and suggested that erosion by meander migration may limit the time a mire can exist in the Amazonian lowlands.

Wetlands, with excess rain as the primary source of saturation, can also be found in Köppen's humid subtropical climatic type (Cf), in North America (the south-eastern USA), in South America (southern Brazil, Uruguay and the pampas of Argentina), in South-East Asia (south-east China and southern Japanese Islands), in a small portion of South Africa, and in south-eastern Australia (Muller and Grymes, 1998). Forested areas in the south-eastern USA with this climate type have small deficits ($ET > P$) of 25–150 mm occurring in late spring to autumn (April–November) but larger surplus ($P > ET$) in winter to early spring (December–March). The seasonal surplus varies geographically from 430 mm in North Carolina decreasing to 150 mm in central Florida, and increasing westward to 600 mm in central Louisiana (Muller and Grymes, 1998).

In the south-eastern USA, much of the lower coastal plain is composed of late Pleistocene marine terraces of sands over heavy textured slack water deposits, with little erosional development and low drainage density. The resulting landscape has broad (1 km or more) flats between surface streams. Buol (1973) observed a general trend of increasing soil wetness with distance from streams in this region. Williams (1998) used a simple Darcy relationship to determine an equilibrium slope, based on rainfall surplus and lateral aquifer transmissivity, to predict how far from a drain a rainwater wetland will occur. Skaggs *et al.* (2005) showed how short-term daily water table data could be used to calibrate a model, DRAINMOD (Skaggs, 1978), to determine how far from a ditch wetland hydrology would be affected. They found standard agricultural drainage equations are equally applicable to forested areas, although k and drainable porosity (the quantity of water released by a small drop of the water table, also called specific yield) may differ substantially from values found on agricultural fields on the same soil series reported in the NRCS database (Skaggs *et al.*, 2011). DRAINMOD was also applied to compare and evaluate wetland hydrology for seven different criteria defining and identifying wetlands on three soils in the North Carolina Coastal Plain (Skaggs *et al.*, 1994).

7.3 Forested Wetlands due to Surplus Groundwater Flow

The most common form of forested wetland occurs where rainfall is supplemented by a positive flow of groundwater ($P + Q_g > E + T + Q_s$). They are primarily slope and basin wetlands in Brinson's (1993) classification, although groundwater subsidy can also be found in riverine types. Semeniuk and Semeniuk (1997) called them dampland, trough or paluslope, depending on the landform of basin, channel or side slope, respectively. This definition does not include all areas normally listed as groundwater-dependent ecosystems (Bertrand *et al.*, 2012). Groundwater-dependent ecosystems include all systems where transpiration is supplied or augmented by withdrawal from saturated soil regardless of the source of saturation. More about groundwater-dependent ecosystems inventory methods and field guides can be found from the USDA Forest Service (USDA, 2012a,b). In this section we consider wetland forests where soil saturation is maintained by a subsidy of inflow from subsurface sources.

Groundwater subsidy will generally occur in regions where stream channels generally have increasing baseflow in the downstream direction. Groundwater discharge and subsidy to the water balance will occur wherever the transmissivity (T) up gradient of the point of interest exceeds T down gradient. As T is defined as the product of hydraulic conductivity and aquifer thickness, declines of T are generally due to thinning of the aquifer. Aquifer thickness is determined by surface topography and topography of the upper bounds of a less permeable subsurface layer. In regions where the water table aquifer is bounded by impermeable bedrock or compacted glacial till, water table thickness will be determined by the difference between the subsurface topography of the restricting layer and surface topography. In regions where the aquiclude below the water table aquifer is a sedimentary layer, the thickness of the water table aquifer will be determined to a great degree by surface topography.

The preceding discussion of northern bogs and fens described the close association of many fens with groundwater discharge from adjacent slopes. In northern Europe groundwater subsidy to fens and in the upland edge of floodplains creates a characteristic vegetation of black alder (*Alnus glutinosa*) forest and *Carex* sedges often called alder carr. In the northern USA alder (*Alnus rugosa*) and *Carex* spp. occur in similar landscape positions with trees such as *L. laricina*, *Fraxinus nigra* or *Thuja occidentalis*. In Canada, the term 'lagg' has been used to encompass the transition between bog and upland on many mires. Verry *et al.* (2011) found water in this zone included runoff from the bog as well as subsurface stormflow from the adjacent uplands. This narrow lag zone produced most streamflow and also was the only area where the mire contributed to recharge of the deeper regional aquifer.

Williams (1998) outlined conditions in the south-eastern USA where the interaction of subsurface and surface topography tended to produce groundwater subsidy to the water balance. In the mountains and piedmont wetland occurrence was due primarily to positions of thin regolith (often actual rock outcrops), while in the coastal plain groundwater subsidy occurred at the toe of nearly all slopes either in depressions or on the upland margin of streams.

7.4 Example Hydrology of Rain Excess and Groundwater Excess Wetlands

Both precipitation excess and groundwater excess wetlands occur in humid regions and are found in close proximity. Both are controlled by the slope of the landscape and aquifer transmissivity, but precipitation-fed wetlands are generally more influenced by regional climate and local weather. An example from sites on the northern South Carolina coast near Georgetown (33.4°N, 79.2°W) demonstrates facets of precipitation and groundwater excess wetlands in a region where rain is the only form of precipitation in most years.

The example sites are all located on sandy soils found throughout the US Atlantic Coastal Plain. Water table elevation at 45 shallow wells, located within 4 km of the Georgetown, South Carolina, National Oceanic and Atmospheric Administration (NOAA) weather observation station, was measured for 14 years (Fig. 7.1). Four wells of that study are examined in this example. Wells #3, #8 and #11 are on broad flats on Leon soils (*Aeric Aleaquod*). The former beach ridges of wells #4, #8 and #11 are truncated north-east of well #2, which is located on Hobcaw soil (*Typic Umbraquults*) more common on riverine terraces (Colquhoun, 1974). Well #8 (at 4.64 m above mean sea level (amsl)) and well #11 (4.01 m amsl) are on a former beach ridge that is believed to be roughly 100,000 years old, while well #4 (1.81 m amsl) is on a younger ridge possibly 20,000–40,000 years old. Well #2 (0.59 m amsl) is on a slope between a small dune deposit and a surface drain located over 3 km from any tidal source. The water table aquifer at each well is bounded by a leaky aquiclude of silty clay, 1.5 m below sea level. The aquifer has a lateral conductivity (k) of 3.4×10^{-3} cm/s (Williams, 1981). Vegetation at all sites includes few obligate hydrophytes, due to land management that includes prescribed burning, and many facultative hydrophytes. Soils at wells #2 and #11 have distinct redoxomorphic indicators (indicators of distinct oxidation and reduction associated with soil saturation) within 30 cm of the surface, at well #4 indicators are less distinct and well #8 does not have redoxomorphic features in the upper 30 cm. Soil and vegetation (primarily soil) indicate that wells #2 and #11 are wetlands under the US system, while well #8 is not, and well #4 is ambiguous.

Figure 7.2a–d shows weekly water table depths at each of these wells from July 1975 through September 1989. In each graph a horizontal line at −15 cm indicates the point above which these sandy soils can be assumed to be saturated. The growing season in Georgetown (late February through November) for each year

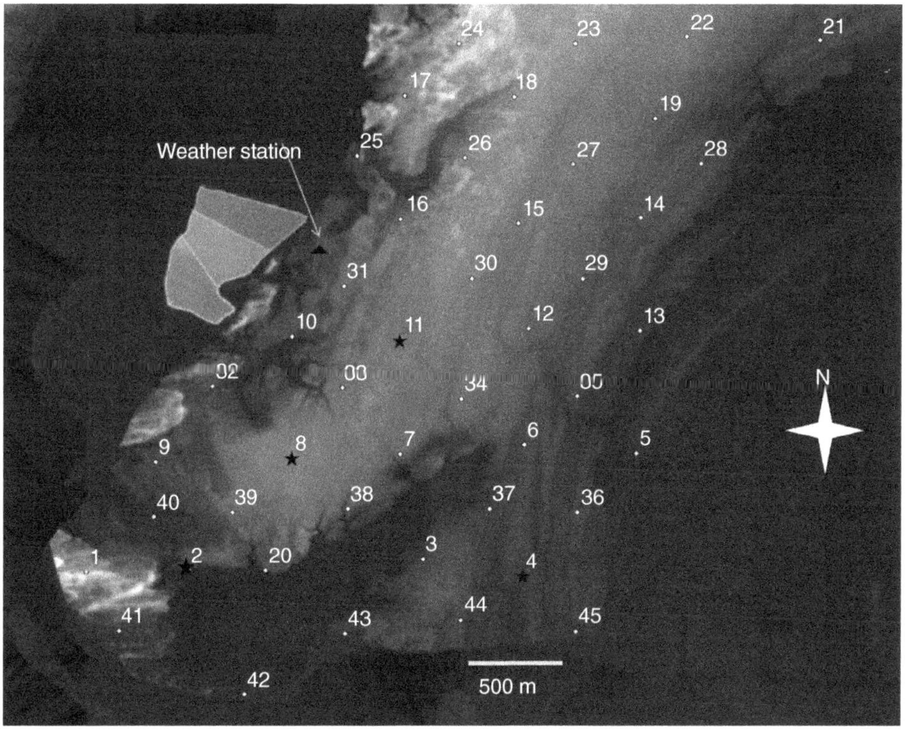

Fig. 7.1. Location of wells #2, #4, #8 and #11 near Georgetown, South Carolina, USA (33.4°N, 79.2°W). LiDAR DEM shades vary from black near sea level to white at 10 m. Tidal creeks on the eastern side and Winyah Bay to the west appear as various dark shades due to tidal change as LiDAR was being acquired. Light coloured polygons near weather station are active dredge spoil disposal sites (LiDAR = Light detection and ranging; DEM = digital elevation model).

is represented by the outline boxes. Since readings were at weekly intervals, three consecutive data points above the horizontal line within the growing season indicate soil saturation that influences vegetation and soils. Since data points are difficult to read the graph also has the words 'yes' or 'no', in each year, to indicate whether that criterion was met. Well #4 (Fig. 7.2a) met the saturation criterion for seven of the 14 years. Well #8 (Fig. 7.2b) was at a higher elevation and met the criterion in only five of the 14 years. Lest one jump to a hasty conclusion, well #11 (Fig. 7.2c) was also on the higher ridge and met the criterion in 12 of the 14 years. Well #2 (Fig. 7.2d) at the lowest elevation also had soil saturation in 13 of the 14 years.

Soil saturation is essentially the result of a point-based water balance that is due to variation in *P*, *E*, *T*, Q_s and Q_g. At all wells $Q_s = 0$, and we can assume *P* was nearly equal within the area of 670 ha. Different water table behaviour observed in the four wells is due to variation in *E*, *T* and Q_g, while differences between years are due to variation in *P*. For the 15 years of the study, mean precipitation was 1320 mm, with a maximum of 1714 mm and a minimum of 898 mm, compared with a 50-year (1950–2000) mean of 1335 mm, with a maximum of 1897 mm and a minimum of 768 mm (Fig. 7.3). Inspection of Figs 7.2 and 7.3 reveals that during years of high rainfall, as in 1982, all wells show soil saturation of wetlands. All were also wet in 1979, although 1979 rainfall was well below normal or the long-term average. The Georgetown area was hit by two hurricanes, David in 1979 and Hugo in 1989, both of which resulted in >300 mm rainfall in late September and saturation of the soil for the remainder of the growing season. In 1983 the situation was reversed, with no well meeting the wetland criterion with

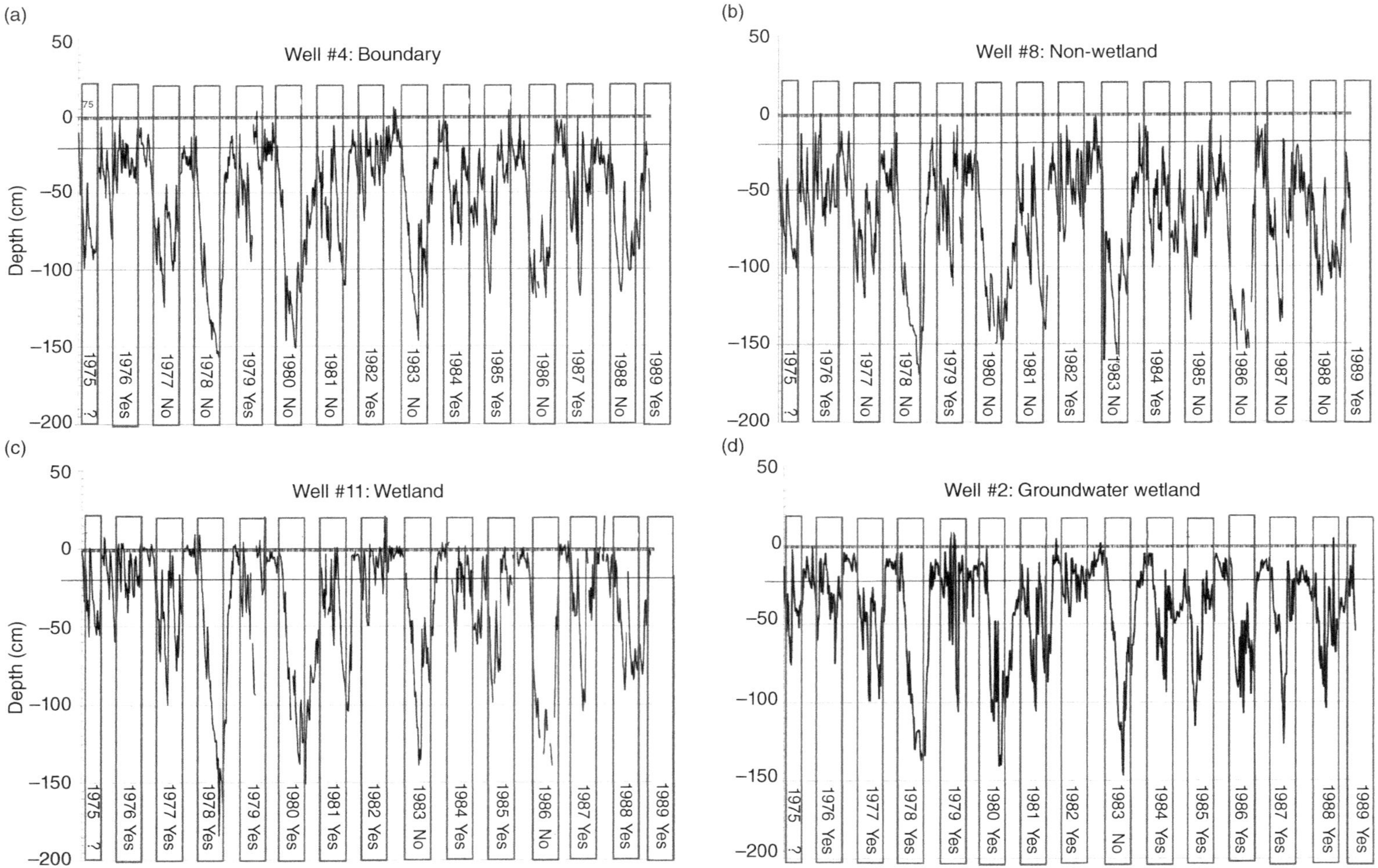

Fig. 7.2. Water table depths for wells #4 (a), #8 (b), #11 (c) and #2 (d) from July 1975 through July 1989. Line at 15 cm depth indicates water table depth at which soil is saturated. Boxed sections indicate the growing season for each year. 'Yes' or 'no' within box indicates whether soil was saturated more than 5% of the growing season for that well in that year.

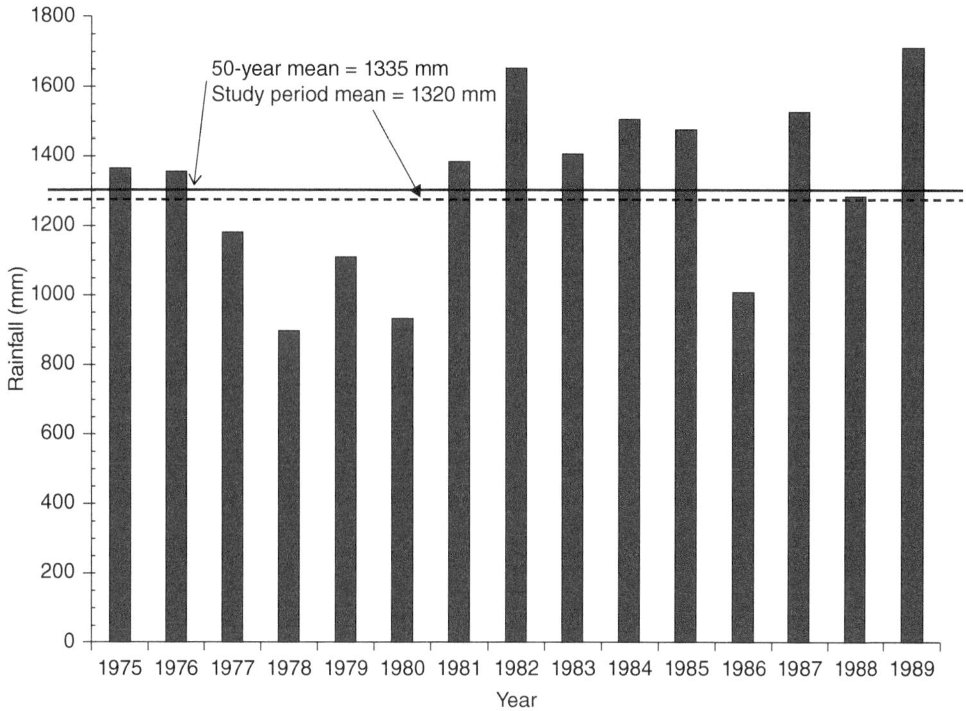

Fig. 7.3. Annual rainfall for Georgetown, South Carolina, USA from 1975 through 1989. Rainfall measured at the weather station shown in Fig. 7.1. (Data from SCDNR, 2015a.)

rainfall above normal. A dry spring and early summer lowered the water table below the capillary fringe and all soils had considerable unsaturated soil; with no large tropical system occurring that autumn, the water table did not recover until early 1984.

A rapid drop of the water table occurs in early spring of most years due to rapid leaf expansion in March, minimum precipitation in April and high potential *ET* (*PET*) from April to July (Fig. 7.4). Both interception and transpiration components of *ET* are related to leaf area which is, in turn, related to stand basal area. No cutting of tree stands was done near any of these wells throughout the study and stands surrounding the wells were inventoried in 1986. Basal area in those stands is listed in Table 7.1. Well #2 clearly has a subsidy that can sustain much higher interception and transpiration than any of the other wells. Comparisons of well #11 with well #2 for 1986 (a dry year) and for a 10-day period of no rain in 1977 are shown in Fig. 7.5. The yearly data (Fig. 7.5a) indicate that the water table at well #2 remained within 15 cm of the surface for most of the month of March while well #11 was dropping rapidly. Although the surfaces were not saturated later in the year, well #2 responses to late summer rainfall exceeded those at well #11. Well #2 displays attributes of a groundwater wetland due to recharge from the water table aquifer of the small (3.5 ha) dune north of the well. Roulet (1990) described the role of groundwater in fen hydrology as maintaining high water tables into the growing season in proportion to *T* and the storage capacity of the aquifer. For well #2 the aquifer storage is quite small and the effect was not long lasting.

During the 10-day period in 1977 (Fig. 7.5b) well #2 exhibits short-term water table fluctuations characteristic of groundwater recharge, which is common to all riparian forests when transpiration is high. By contrast well #11 shows no groundwater subsidy with a drop during the day but little, if any, rebound during the night. White (1932) proposed a method, well described in Mitsch and Gosselink (2015), to determine groundwater subsidy and *ET* from these fluctuations.

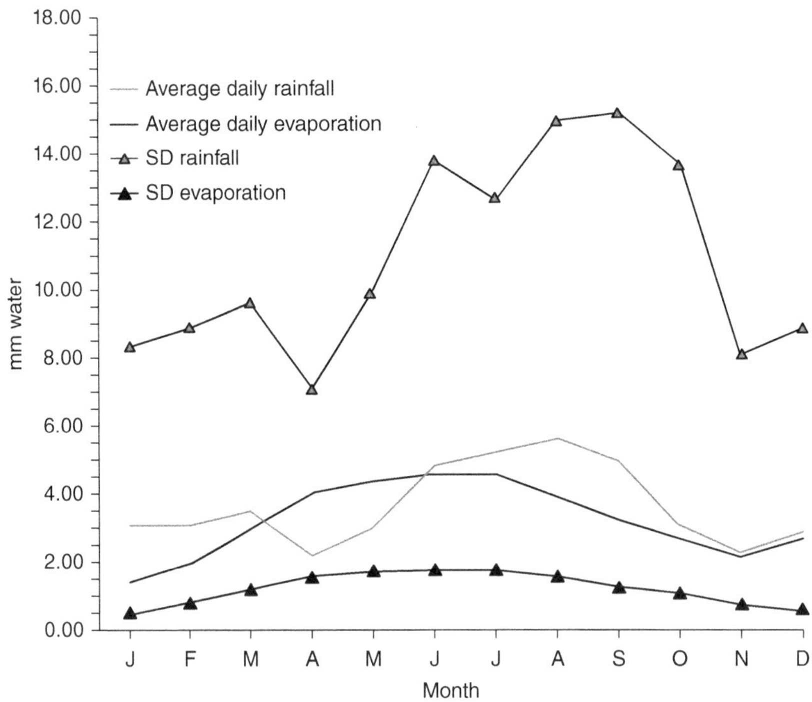

Fig. 7.4. Monthly average daily rainfall at Georgetown, South Carolina, USA and the standard deviation (SD) of that value, calculated from daily records from 1940 to 2000; and average daily potential evapotranspiration estimated as 70% of pan evaporation from the nearest Class A weather station in Charleston, South Carolina (32.95°N, 80.23°W) and the SD of that value.

Table 7.1. Land elevation, basal area and land slopes surrounding each well in the example. The aquiclude top is at an elevation of –1.5 m amsl, aquifer thickness varied from 2.09 to 5.14 m at soil saturation. Slopes were determined using ArcGIS 10 on a 1.5 m × 1.5 m DEM (Fig. 7.1). Cross-sections of surface elevations were measured 100 m in each of eight directions from each well.

Well number	Elevation (m amsl)	Basal area (m^2/ha)	Average land slope ($\times 10^{-4}$) 100 m from well (or slope as cm/100 m)								
			N	NE	E	SE	S	SW	W	NW	Ave.
#4	1.81	17.6	28	37	28	–20	–8.3	–8.9	1.5	38	12
#8	4.64	9.4	–9.3	–0.8	–19	–0.8	–8.6	–0.07	–22	–12	–20
#11	4.01	15.8	–2.4	7.5	–18	15	–26	–28	–27	1.2	–9.7
#2	0.59	32.4	101	4.1	2.5	25	12	–18	–21	165	56

amsl, above mean sea level; DEM, digital elevation model.

Loheide *et al.* (2005) showed that the specific yield term, assumed constant in White's method, actually varies with initial water table depth and duration of the drawdown period. Laio *et al.* (2009) used stochastic modelling to examine riparian *ET* and found water table response required estimating changes in the unsaturated region above the water table. Daily water table fluctuation indicates a groundwater subsidy at a site, but does not contain sufficient information to estimate either the groundwater subsidy or the rate of evapotranspiration.

Wells #4, #8 and #11 have the same soil series and are similarly situated on the same landscape topography, located on broad, flat, former beach deposits far from surface drainage. However, one might expect during high water table periods that short-range transfers would

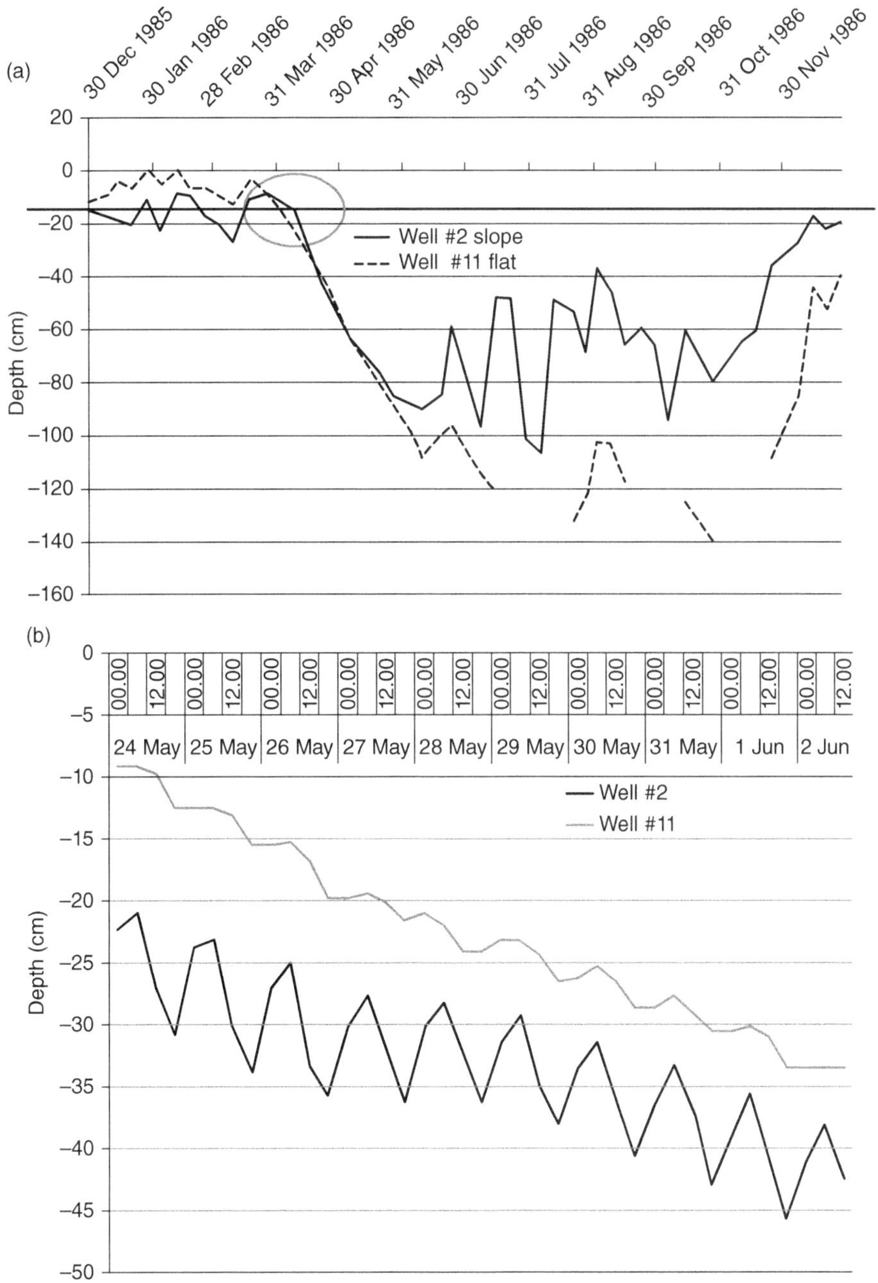

Fig. 7.5. Comparison of water table depth at wetlands with rain subsidy at well #11 and groundwater subsidy at well #2 for the dry year of 1986 (a) and for ten rain-free days in late spring of 1977 (b).

occur in the permeable surface (upper 5–10 cm) horizons. A 1.5 m × 1.5 m LiDAR (Light detection and Ranging) digital elevation model allowed measurement of slopes in the vicinity of each well (Table 7.1). Since micro-topography varied by 8–15 cm, slope was measured across each pair of pixels along the entire 100 m, resulting in 68 estimations of slope in that direction. Average surface slopes at each well clearly indicated the surface properties that led to differing water table behaviour. Well #8 is at a meso-topographic high elevation with negative slopes in all directions and an average slope of –0.002. Well #4 sits on a side slope with positive slope to the north and negative slope to the south, and an average slightly positive slope of 0.0012. Well #11 is on a very flat area with slightly positive slope to the north-east and south-east and negative slope to the west, with an average of –0.00097.

In the preceding example of the hydrology of rain-fed wetlands, soil saturation was most often found at the beginning of the growing season. Saturation during mid and later portions of the growing season was a result of high rainfall events generally associated with tropical cyclones. With average rainfall surplus of 380 mm and aquifer horizontal k of 3.4×10^{-3} cm/s, continuous soil saturation exceeded 5% of the growing season most of the time on a flat with an average negative slope of 9.7 cm/100 m, half the time on a side slope with a positive slope of 12 cm/100 m, and seldom on a slight mound with a negative slope of 20 cm/100 m. A small groundwater subsidy was found at a site with a positive slope of 56 cm/100 m and slopes on the recharge side over 1%. Changes in soil saturation are associated with topographic slopes that are best revealed with LiDAR sensing and GIS analysis.

7.5 Forested Wetlands due to Surplus Surface Water Flow

Forested wetlands where $P + Q_s > E + T + Q_g$ are classified as riverine, lacustrine fringe and tidal fringe by Brinson (1993) or lotic, lentic and estuarine by Tiner (2013). Lacustrine fringes are generally found adjacent to large lakes and are quite limited in extent. Tidal fringe forested wetlands include the large extent of mangroves in tropical regions and the tidal freshwater forested wetlands (TFFW) found at the landward edge of large estuaries or rivers feeding into coastal reaches (Conner *et al.*, 2007). Riverine forested wetlands are common around the world and can be found in the widest range of climatic regions. This definition will include riparian wetlands where phreatophytes withdraw transpiration from the water table aquifer and capillary fringe. Surface flows subsidize lacustrine fringe wetlands primarily by infiltration through the lakebed, and tidal fringe (see Section 7.5.1) and riverine forested wetlands by overbank flooding as well as infiltration into the bed.

Rivers can be thought of as a branched network of unbranched segments connected at nodes (Strahler, 1957). Individual segments can interact with groundwater in one of three ways: (i) groundwater flows into the stream and increases flow rate (gaining stream); (ii) water flows from the stream bottom into the regional groundwater and decreases flow (losing stream); or (iii) stream water flow is not connected to the regional groundwater system (Winter, 2001). A common situation occurs when lower-order streams are located in a region of higher rainfall (often mountains) and the river flows into a more arid region. Ivkovic (2009) examined the Namoi watershed in south-eastern Australia and found gaining streams in the eastern headwaters, variable gaining/losing streams downstream to the west, losing streams further west and a region of disconnected but losing streams furthest west. Average annual rainfall varied from 1100 mm at the eastern divide to 470 mm at the furthest west gauging station. Through the disconnected western section, transpiration of riparian phreatophytes utilized all water infiltrating from the streambed and a zone of unsaturated soil existed below the riparian water table aquifer. The Namoi River is like rivers in many parts of the world that flow from high mountains, or uplands in high rainfall regions, across arid regions. Over most of the world, riparian forests in these valleys have been replaced by agriculture.

Surface water subsidy due to overbank flooding has been the assumption of much of the research and literature discussing floodplain forests. Conversion of much of the lower Mississippi floodplain from forest to agriculture caused concern about the loss of natural functions of that system and resulted in intense research into floodplain forest ecology. Unfortunately, that research tended to focus on tree tolerance to flooding (saturated soils actually) and basic ecological relationships but seldom addressed hydrology beyond recording water level in the research plots (MacDonald *et al.*, 1979). Hook (1984) created a list of flood tolerance for most species found in south-eastern US floodplains. Variations of that list have been used repeatedly in graphics showing a cross-section of a simple stepped floodplain where most flood-tolerant species (*Taxodium distichum*, *Nyssa aquatica*, *Salix nigra*) were placed next to the river, slightly less tolerant (*Carya aquatica*) on a next higher step, moderately tolerant (*Acer rubrum*, *Acer sacharinum*) a step higher, all the way to flood-intolerant species on the upland. Such figures oversimplify geomorphology and hydrology of floodplains and are less useful than the original table. A more realistic depiction of the natural levee and slope explaining the real variation of floodplain geomorphology is presented by Hupp (2000), using US terminology.

LiDAR elevation data display the true complexity of floodplain topography as in Plate 6, an 11 km section of the Congaree River valley, near Columbia, South Carolina (33.8°N, 80.8°W). A portion of mature southern floodplain forest has been set aside in the Congaree National Park on a moderate sized (mean flow 245 m^3/s) river (USGS, 2015). Cross-sections (Fig. 7.6a–c) reveal that across the floodplain (natural levee to low point near terrace) the average slope is 0.000795 (Fig. 7.6a), the longitudinal slope of the floodplain is 0.000329 (Fig. 7.6b) and the water surface had an average slope of 0.000175. Water reflects LiDAR, producing an elevation for the water surface rather than the soil surface below it. However, for such a long reach (25 km) the bed slope will approximate the water surface.

All of the features described by Hupp (2000) can be found on the LiDAR representation (Plate 6) in their true complexity. Along the entire channel the natural levee is from 50 cm to 1 m higher than the floodplain behind it. The most striking floodplain feature is the scroll topography deposited as point bars in the past. Many of these are found in conjunction with an oxbow. The connection is obvious in the lower right where two meanders will soon form new oxbows. In both cases the outer edge of a meander is separated from a downstream meander by only the narrow natural levee. When the river exceeds bank full a channel may form to cut off one of these meanders and form a new oxbow. Three older oxbows are highlighted with standing water at elevations 28.79, 28.35 and 28.21 m.

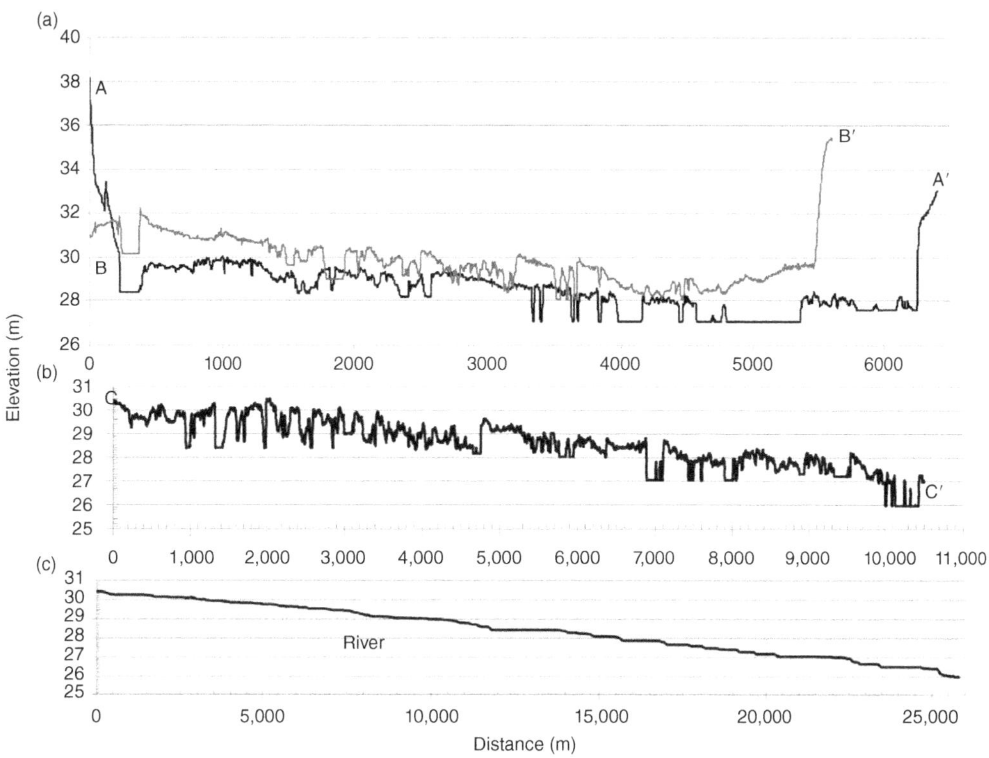

Fig. 7.6. Elevations of the Congaree River floodplain. Cross-sections A–A′, B–B′ and C–C′ are marked on Plate 6. Graph (c), 'River' shows water surface elevation from upstream to downstream end of the channel in Plate 6.

Hupp (2000) also acknowledges the floodplain is itself a watershed, approximately 55 km^2 as represented in Plate 6, where networks of smaller creeks will transport local storm water to the river. Where small streams cross the natural levee, what Hupp calls a crevasse splay, it will discharge to the river at low flow but can allow floodwaters into the floodplain even if the flow is not at bank full stage. Since the strongest slope is across the floodplain from levee to the toe of the bluff, most storm water, as well as overbank flooding, will tend to flow towards the bluff. As the floodplain also slopes downstream, water flowing along the bluff will eventually flow into the river at some point further downstream, where it will be called a slough. In Plate 6 small streams connect the series of oxbows highlighted, at elevations of 28.79, 28.25 and 28.21 m, to the highlighted slough where it enters the river at an elevation of 28.21 m. Often the outlet to a slough will be where the river meanders reach the opposite valley side.

Surface water subsidy to wetland forests on river floodplains is much more complex than simple overbank flooding. Alluvial channels adjust to flow and sediment load (Leopold *et al.*, 1964). The width, depth and slope of the channel are shaped by floods large enough to rearrange sediments and frequent enough to have a large cumulative impact, found to be a flood with a recurrence interval of about 1.5 years. Overbank flows drop sediment near the river, creating slopes towards the edge of the floodplain over time. Rain on the floodplain also produces stormflow that moves in a jumble of old oxbows and small creeks, some of which enter the river directly, allowing river flow into the floodplain during flows that are below bank full. Wetland forest species are most likely to occur in the low topography near the edge of the floodplain, where floodwaters, stormflow and upland groundwater subsidy are all concentrated. The floodplain also contains a variety of depressions associated with former channels that hold water, creating pockets of saturated soil at many of elevations across the floodplain. The back swamp is not a single feature but the vast mosaic of point deposits, oxbows and creeks that lies below the elevation of water in the river.

7.5.1 Forested wetlands subjected to tides

For many forested wetlands, the basic constituents of the water balance have been described. However, tidal forested wetlands comprise a significant area globally, and actually occupy coastlines disproportionately at subtropical and tropical latitudes in the form of mangrove forests. TFFW are also quite prevalent at some temperate latitudes, especially where coastal geomorphology interacts with lunar tides to extend tidal ranges up major rivers. Mangrove forests occupy 13.7 to 15.2 million ha globally (Spalding *et al.*, 2010; Giri *et al.*, 2011), while TFFW is conservatively estimated to be at least 200,000 ha in the south-eastern USA alone (Field *et al.*, 1991; Doyle *et al.*, 2007). For tidal forests, daily tidal fluctuations may overwhelm the mass balance of surface and groundwater flows, complicating broad application of rating curves to classic riverine mass balance determinations. River stage, baseflows and discharge are caused by tidal forcing as well as gravity-driven flows, such that flooding occurs from daily rising tides, seasonally rising and falling river water, and their interactions. That said, not all mangroves or TFFW are associated with rivers, making surface flood prediction a matter of trusting the local elevation datum and tide gauge, after accounting for local ponding.

Tidal forest designation typically refers to forests with visible surface water associated at least seasonally with lunar tides. Forests with root zone tidal fluctuations on barrier islands or far inland are often excluded. Indeed, tidal forests have historically been classified by the degree of inundation; intertidal vegetation establishes below an inundation threshold. Above that threshold, various species by productivity zones can be defined based upon a progressive decrease in tidal inundation upslope (Watson, 1928; Friess *et al.*, 2012).

The nature of surface water flooding even offers a classification system for mangroves (Lugo and Snedaker, 1974): *fringe* forests and *overwash island* forests are flooded with nearly every tide; *riverine* forests are flooded by tides for longer durations when river stage is high and nearer to the mouth; and *basin* forests occur in inland depressions that can retain water after being flooded by high rainfall events or by the highest tides. Position-based inundation gives rise to different geomorphical, geochemical and structural characteristics of mangrove soils (McKee *et al.*, 1988), which yield gradients in productivity. Inundation–productivity feedbacks have enamoured mangrove ecologists for decades, as much variation has been documented in mangrove forest structure coincident with the degree of surface flooding (Smith, 1992). In contrast, a hydrological understanding of TFFW is relatively new (Rheinhardt and Hershner, 1992; Day *et al.*, 2007); such forests are generally restricted to upper intertidal positions and supplanted by marsh at mid-to-lower intertidal positions.

Three general categories of surface inundation are displayed among most tidal forests, defined by characteristic hydrographs. The first hydrograph (Type 1) reveals regular cycles of surface water flooding followed by drainage during each tidal cycle, except during spring tides when surface water has little time to drain during ebb before the next flood cycle (Fig. 7.7a). Mean water levels are often above ground, giving rise to longer flood durations approaching 50% of the year for the most seaward mangrove forests. Local rainfall also affects the balance of flooding by forcing more water to back up during tidal flows, and even slowing drainage during tidal ebbs. In transitioning from lower to upper intertidal locations, tidal range decreases as sites gain surface elevation relative to sea level, affecting flood depth first before the frequency of tidal pulses is altered (Fig. 7.7, compare a with b). In general, Type 1 hydrographs are representative of many mangrove forests occurring along rivers, in deltas, as overwash islands or along the fringes of open estuarine waterbodies. As tides drain water levels below the soil surface, changes in the slope of the hydrograph identify the approximate soil surface of the wetland; water drains much slower through soil than through air.

The second type of hydrograph (Type 2) reveals tidal forests with basin characteristics, whereby high tides or major rainfall events strand water and force tidal fluctuations atop surface water (Fig. 7.7c). This hydrograph is also quite common; for example, approximately 52% of south Florida mangrove forests have basin/inland characteristics (Twilley *et al.*, 1986). Similar hydrographs can represent both basin mangroves, which are established in depressions, and TFFW, which are often depressional behind river levees or in back swamp settings (Duberstein, 2011). Dewatering of basin surface waters occurs over multiple days during an ebb tidal cycle because smaller tides do not extend to these forests to recharge surface waters. The influence of forest evapotranspiration is evident during this time frame, which serves to drain water levels below ground between spring tidal cycles (Fig. 7.7c).

The third type of hydrograph (Type 3) actually defines a different type of tidal forest; one driven by wind tides in lieu of lunar tides. Extremely common in some regions but rarely described, Day *et al.* (2007) detailed tidal water-level fluctuations within a microtidal (mean tide = 0.32 m) forest in Louisiana, USA where small lunar tidal fluctuations are superimposed upon wind tides. As offshore winds blow water inland, water levels rise above the surface of the tidal forest soil; only then are microtidal fluctuations readily observed (Fig. 7.7d). When frontal passages push water out of the tidal forest, tidal fluctuations disappear as water levels fall below ground. This is more common locally in places like the Louisiana Deltaic Plain or north-eastern North Carolina, where lunar tides are small, open waterbodies and estuaries are large and shallow, and anthropogenic landscape modifications or natural structures (e.g. barrier islands) restrict tidal flows regionally.

Typically, hydroperiods are described by flood frequency, flood duration and, sometimes, mean water table depth (Nuttle, 1997). Flood duration is most commonly used. For example, the majority of mangrove forests are inundated far less often than they are drained over annual cycles (Lewis, 2005). Accounts include 30% flooding for sites in Tampa Bay, Florida (Lewis, 2005); 29–53% flooding for basin and riverine sites in south-west Florida (Krauss *et al.*, 2006); 35% flooding for overwash island sites in Florida (Carlson *et al.*, 1983); and <35% flooding for some sites in Jamaica (Chapman, 1944). While it is uncertain exactly how long mangroves might survive higher flood durations, certain mangrove species can push higher inundation thresholds; for example, some *Rhizophora mangle* forests are

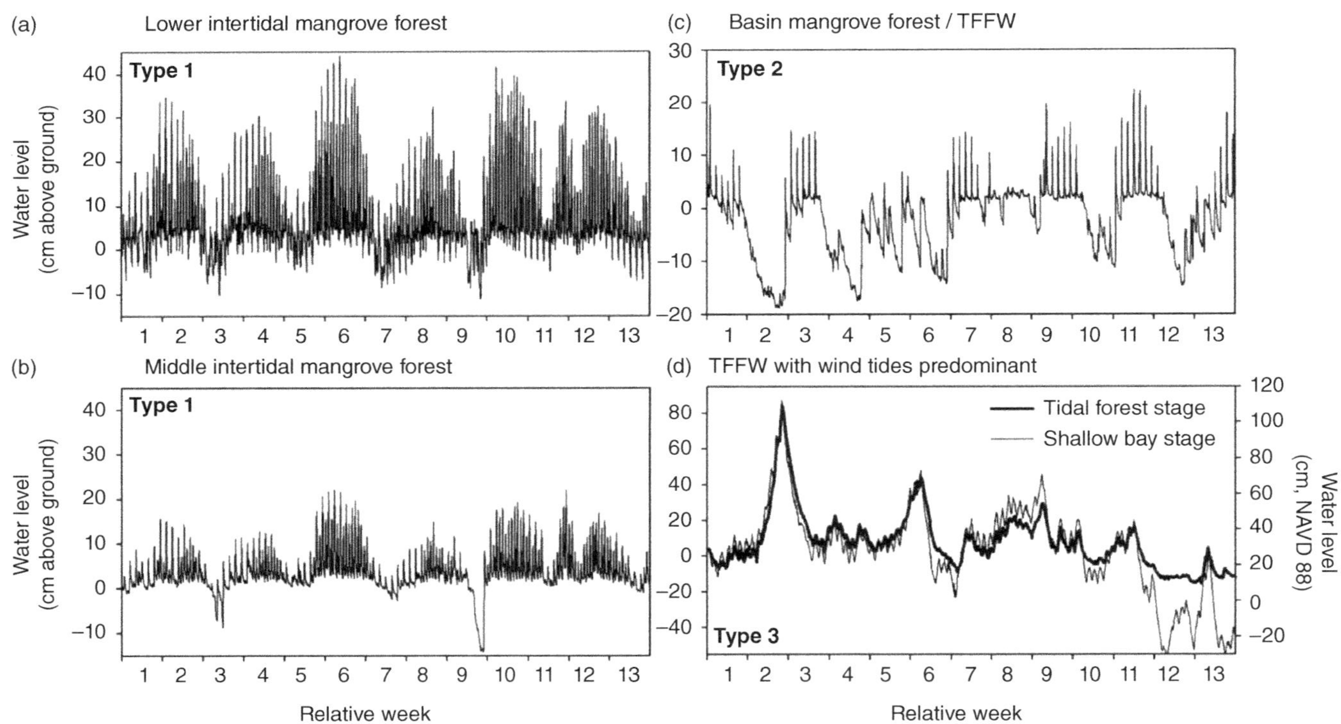

Fig. 7.7. Representative tidal forest hydrographs depicting Type 1 (a, b), Type 2 (c) and Type 3 (d) surface and shallow groundwater hydrological signatures of tidal forests. Hydrographs (a) and (b) are from riverine mangrove forests located along the Shark River in Everglades National Park (Florida, USA), with the location of the hydrograph in (b) approximately 5.8 km inland from that in (a). Both are aligned on the same time scale for flood depth and frequency comparison. Hydrograph (c) is from a basin mangrove forest located at Rookery Bay National Estuarine Research Reserve in Naples, Florida, USA, but is very similar to hydrographs from tidal freshwater forested wetlands (TFFW) located along major Atlantic coastal rivers (d) (NAVD 88 = North American Vertical Datum of 1988). Hydrograph (c) is re-drawn from Day *et al.* (2007) and represents a tidal forest in the Barataria Basin, Louisiana, USA, in comparison with local embayment stage. All hydrographs are from interior forest locations and are presented relative to the soil surface (0 on the left *y*-axis).

flooded 71–96% of the time (Chapman, 1944; Carlson *et al.*, 1983). Since TFFW are typically associated with upper intertidal positions, many are flooded for <20% of the year (Day *et al.*, 2007; Krauss *et al.*, 2009). This determination depends both on whether assessments are made in higher versus lower rainfall years and on what site elevation is used to make determinations (e.g. base of hollow versus top of hummock). For example, TFFW immediately adjacent to the Savannah River were inundated for 55% of the time relative to the bottom of a hollow during one high rainfall year (Duberstein and Conner, 2009).

Thus, two primary criteria need to be considered before hydroperiod metrics are standardized among sites and tidal forest types. First, a standard elevation needs to be established. In TFFW having hummocks and hollows, or in mangrove forests possessing mounds created through faunal excavation (e.g. mud lobster, *Thalassina anomala*), the lowest elevations between hummocks or mounds often do not drain completely during ebbing tides, creating nearly continuous saturation of hollows (Day *et al.*, 2007). Hydroperiod determinations would vary considerably when using that elevation in lieu of the top of hummocks, 15–20 cm higher in TFFW (Rheinhardt and Hershner, 1992; Duberstein *et al.*, 2013) but up to 1–2 m higher relative to mud lobster mounds in mangroves (Macnae, 1969). For mangroves and TFFW, hummocks or excavated mounds comprise 20–30% of the forest floor (Lindquist *et al.*, 2009; Duberstein, 2011).

Second, the definition of a flood event needs to be established. In characterizing tidal sites by the number of new flood pulses, Krauss *et al.* (2009) described no more than 170 independent tidal flooding events per year for TFFW along the Savannah River (Georgia, USA) and Waccamaw River (South Carolina, USA). These determinations were made relative to the bottom of the hollows; if hollows did not fully de-water between or among independent tide events, they were considered a single flood event. This might work well when the focus is on the role that flooding has on soil oxygenation. However, when considering the potential for sediment loading, for example, the absolute number of tidal pulses is more important regardless of whether hollows were de-watered. Thus, in reanalysing the same data and applying different assumptions, greater than 500 independent tidal flooding events per year were characterized for many of the same forests previously described along the Savannah and Waccamaw rivers (Ensign *et al.*, 2014).

7.6 Summary

Forested wetlands are found throughout the world and in a variety of landscape positions. They can be classified by the source of water which causes a surplus of inflow over losses. Precipitation may be the only cause of water surplus for sites on nearly level slopes. These sites are characterized by poor nutrition and slow growth, yet organic matter may accumulate on the surface due to slow decomposition. Drainage of these sites to improve tree growth is quite common in Scandinavia and the south-eastern USA.

Sites where the inflow surplus is due to groundwater additions to precipitation are generally located on concave landscape positions. The amount of subsidy and the length of flooding generally depend on the storage characteristics of the aquifer feeding the site, and ranges from small, short-lived subsidy for aquifers with small storage to nearly constant saturation when fed by aquifers with large storage. In regions where the aquiclude under the water table aquifer is uneven, these wetlands can occur on planar slopes due to thinning of the aquifer. The most common location occurs at the abrupt change in slope from hillside to valley bottom. Forested wetlands in this landscape position provide the important ecosystem service of nitrate reduction in agricultural areas.

Surface water subsidy occurs primarily along rivers and in coastal areas that are influenced by tides. Surface subsidy associated with rivers can expand wetland forests well into regions where $P \ll ET$. In such regions, riverine forested wetlands can be important sites of aquifer recharge. Forested wetlands associated with tidal forcing are the least well understood and may be most threatened by climate change.

References

Andriesse, J.P. (1988) *Nature and Management of Tropical Peat Soils*. FAO Soils Bulletin No. 59. Food and Agriculture Organization of the United Nations, Rome.

Bay, R.R. (1967) Ground water and vegetation in two peat bogs in northern Minnesota. *Ecology* 48, 308–310.

Bertrand, G., Goldscheider, N., Gobat, J.M. and Hunkler, D. (2012) Review: From multi-scale conceptualization to a classification system of groundwater dependent ecosystems. *Hydrogeology Journal* 20, 5–25.

Boelter, D.H. (1969) Physical properties of peats as related to degree of decomposition. *Soil Science Society of America Proceedings* 33, 606–609.

Botch, M. (1990) Aapa-mires near Leningrad at the southern limit of their distribution. *Annales Bontanici Fennici* 27, 281–286.

Brinson, M.M. (1993) *A Hydrogeomorphic Classification for Wetlands*. Technical Report WRP-DE-4. US Army Corps of Engineers, Waterways Experiment Station, Vicksburg, Mississippi.

Buol, S.W. (ed.) (1973) *Soils of the Southern United States and Puerto Rico*. Southern Cooperative Series Bulletin No. 174. Agricultural Experiment Stations of the Southern United States and Puerto Rico Land Grant Universities, Fort Worth, Texas.

Buttle, J.M., Creed, I.F. and Pomeroy, J.W. (2000) Advances in Canadian forest hydrology, 1995–1998. *Hydrological Processes* 14, 1551–1578.

Carlson, P.R., Yarboro, L.A., Zimmermann, C.F. and Montgomery, J.R. (1983) Pore water chemistry of an overwash mangrove island. *Florida Scientist* 46, 239–249.

Chapman, V.J. (1944) 1939 Cambridge University expedition to Jamaica – Part 1. A study of the botanical processes concerned in the development of the Jamaican shore-line. *Journal of the Linnean Society of London, Botany* 52, 407–447.

Chason, D.B. and Siegel, D.I. (1986) Hydraulic conductivity and related physical properties of peat, Lost River Peatlands, Northern Minnesota. *Soil Science* 142, 91–99.

Chimer, R.A. and Ewel, K.C. (2005) A tropical freshwater wetland: production, decomposition and peat formation. *Wetlands Ecology and Management* 13, 671–684.

Colquhoun, D.J. (1974) Cyclic surficial stratigraphic units of the Middle and Lower Coastal Plain, central South Carolina. In: Oats, R.O. and DuBar, J.R. (eds) *Post-Miocene Stratigraphy, Central and Southern Atlantic Coastal Plain*. Utah State University Press, Logan, Utah, pp. 179–190.

Comerford, N.B., Jerez, A., Freitas, J.A. and Montgomery, J. (1996) Soil water table, reducing conditions, and hydrologic regime in a Florida flatwood landscape. *Soil Science* 161, 194–199.

Conner, W.H., Doyle, T.W. and Krauss, K.W. (2007) *Ecology of Tidal Freshwater Forested Wetlands of the Southeastern United States*. Springer, New York.

Cowardin, L.M., Carter, V., Golet, F.C. and LaRoe, E.T. (1985) *Classification of Wetlands and Deepwater Habitats of the United States*. FBS/OBS 79/31. US Department of the Interior, Fish and Wildlife Service, Washington, DC.

Day, R.H., Williams, T.M. and Swarzenski, C.M. (2007) Hydrology of tidal freshwater forested wetlands of the Southeastern United States. In: Conner, W.H., Doyle, T.W. and Krauss, K.W. (eds) *Ecology of Tidal Freshwater Forested Wetlands of the Southeastern United States*. Springer, New York, pp. 29–63.

Devito, K.J., Hill, A.R. and Roulet, N. (1996) Groundwater–surface water interactions in headwater forested wetlands of the Canadian Shield. *Journal of Hydrology* 181, 127–147.

Doyle, T.W., O'Neil, C.P., Melder, M.P.V., From, A.S. and Palta, M.M. (2007) Tidal freshwater swamps of the Southeastern United States: effects of land use, hurricanes, sea-level rise, and climate change. In: Conner, W.H., Doyle, T.W. and Krauss, K.W. (eds) *Ecology of Tidal Freshwater Forested Wetlands of the Southeastern United States*. Springer, New York, pp. 1–28.

Duberstein, J.A. (2011) Composition and ecophysiological proficiency of tidal freshwater forested wetlands: investigating basin, landscape, and microtopographical scales. PhD dissertation, Clemson University, Clemson, South Carolina.

Duberstein, J.A. and Conner, W.H. (2009) Use of hummocks and hollows by trees in tidal freshwater forested wetlands along the Savannah River. *Forest Ecology and Management* 258, 1613–1618.

Duberstein, J.A., Krauss, K.W., Conner, W.H., Bridges, W.C. Jr and Shelburne, V.B. (2013) Do hummocks provide a physiological advantage to even the most flood tolerant of tidal freshwater trees? *Wetlands* 33, 399–408.

Ensign, S.H., Hupp, C.R., Noe, G.B., Krauss, K.W. and Stagg, C.L. (2014) Sediment accretion in tidal freshwater forests and oligohaline marshes of the Waccamaw and Savannah Rivers, USA. *Estuaries and Coasts* 37, 1107–1119.

Field, D.W., Reyer, A., Genovese, P. and Shearer, B. (1991) *Coastal Wetlands of the United States – An Accounting of a Valuable National Resource*. National Ocean Service, National Oceanic and Atmospheric Administration, Rockville, Maryland.

Fraser, C.D.J., Roulet, N.T. and Lafleur, M. (2001) Groundwater flow patterns in a large peatland. *Journal of Hydrology* 246, 142–154.

Friess, D.A., Krauss, K.W., Horstman, E.M., Balke, T., Bouma, T.J., Galli, D. and Webb, E.L. (2012) Are all intertidal wetlands naturally created equal? Bottlenecks, thresholds and knowledge gaps to mangrove and saltmarsh ecosystems. *Biological Reviews* 87, 346–366.

Giri, C., Ochieng, E., Tieszen, L.L., Zhu, Z., Singh, A., Loveland, T., Masek, J. and Duke, N. (2011) Status and distribution of mangrove forests of the world using earth observation satellite data. *Global Ecology and Biogeography* 20, 154–159.

Heinselmann, M.L. (1970) Landscape evolution, peatland types, and the environment in the Lake Agassiz Peatlands Natural Area, Minnesota. *Ecological Monographs* 40, 235–261.

Hodgson, R. and Young, K.I. (2001) Potential groundwater flow through a sorted net landscape, Arctic Canada. *Earth Surface Processes and Landforms* 26, 319–328.

Hook, D.D. (1984) Waterlogging tolerance of lowland tree species of the south. *Southern Journal of Applied Forestry* 8, 136–149.

Hupp, C. (2000) Hydrology, geomorphology and vegetation of Coastal Plain rivers in the southeastern United States. *Hydrological Processes* 14, 2991–3010.

Ingram, H.A.P. (1978) Soil layers in mires: function and terminology. *The Journal of Soil Science* 29, 224–227.

Ingram, H.A.P. (1983) Hydrology. In: Gore, A.J.P. (ed.) *Mires: Swamp, Bog, Fen and Moor*. Ecosystems of the World Series Vol. 4A. Elsevier, Oxford, pp. 67–158.

Ivkovic, K.M. (2009) A top-down approach to characterize aquifer-river interaction processes. *Journal of Hydrology* 365, 145–155.

Jorgeson, M.T., Shur, Y.L. and Pullman, E.R. (2006) Abrupt increase in permafrost degradation in Arctic Alaska. *Geophysical Research Letters* 33, L02503, doi: 10.1029/2005GL024960 (accessed 1 April 2016).

Krauss, K.W., Doyle, T.W., Twilley, R.R., Rivera-Monroy, V.H. and Sullivan, J.K. (2006) Evaluating the relative contributions of hydroperiod and soil fertility on growth of south Florida mangroves. *Hydrobiologia* 569, 311–324.

Krauss, K.W., Duberstein, J.A., Doyle, T.W., Conner, W.H., Day, R.H., Inabinette, L.W. and Whitbeck, J.L. (2009) Site condition, structure, and growth of baldcypress along tidal/non-tidal salinity gradients. *Wetlands* 29, 505–519.

Lähteenoja, O., Ruokolainen, K., Schulman, L. and Oinonen, M. (2009) Amazonian peatlands: an ignored C sink and potential source. *Global Change Biology* 15, 2311–2320.

Laio, F., Tamea, S., Ridolfi, L., D'Eodorico, P. and Rodriquiez-Iturbe, I. (2009) Ecohydrology of groundwater-dependent ecosystems 1. Stochastic water table dynamics. *Water Resources Research* 45, W05419, doi: 10.29/2008WR007292 (accessed 1 April 2016).

Leopold, L.B., Wolman, M.G. and Miller, J.P. (1964) *Fluvial Processes in Geomorphology*. Freeman, San Francisco, California.

Lewis, R.R. (2005) Ecological engineering for successful management and restoration of mangrove forests. *Ecological Engineering* 24, 403–418.

Lindquist, E.S., Krauss, K.W., Green, P.T., O'Dowd, D.J., Sherman, P.M. and Smith, T.J. III. (2009) Land crabs as key drivers in tropical coastal forest recruitment. *Biological Reviews* 84, 203–223.

Loheide, S.P. II, Butler, J.J. Jr and Gorelick, S.M. (2005) Estimation of groundwater consumption by phreatophytes using diurnal water table fluctuation: a saturated–unsaturated flow assessment. *Water Resources Research* 41, W07030. doi: 10.1029/2005WR003942 (accessed 1 April 2016).

Lugo, A.E. and Snedaker, S.C. (1974) The ecology of mangroves. *Annual Review of Ecology and Systematics* 5, 39–64.

Lugo, A.E., Brown, S. and Brinson, M.M. (1988) Forested wetlands in freshwater and salt-water environments. *Limnology and Oceanography* 33, 894–909.

MacDonald, P.O., Frayer, W.E. and Clauser, J.K. (1979) *Documentation, Chronology and Future Directions of Hardwood Habitat Losses in the Lower Mississippi Alluvial Plain, Vols 1 and 2*. US Department of the Interior, Fish and Wildlife Service, Washington, DC.

Macnae, W. (1969) A general account of the fauna and flora of mangrove swamps and forests in the Indo-West-Pacific region. *Advances in Marine Biology* 6, 73–270.

Mathews, G.V.T. (1993, reprinted 2003) *Ramsar Convention on Wetlands: Its History and Development*. Ramsar Convention Bureau, Gland, Switzerland.
McKee, K.L., Mendelssohn, I.A. and Hester, M.W. (1988) Reexamination of pore water sulfide concentrations and redox potentials near the aerial roots of *Rhizophora mangle* and *Avicennia germinans*. *American Journal of Botany* 75, 1352–1359.
Messina, M.G. and Conner, W.H. (1998) *Southern Forested Wetlands: Ecology and Management*. Lewis Publishers, Boca Raton, Florida.
Mitsch, W.J. and Gosselink, J.G. (2015) *Wetlands*. Wiley, Hoboken, New Jersey.
Muller, R.A. and Grymes, J.M. III (1998) Regional climates. In: Messina, M.G. and Conner, W.H. (eds) *Southern Forested Wetlands; Ecology and Management*. Lewis Publishers, Boca Raton, Florida, pp. 87–103.
Okruszko, T., Duel, H., Acreman, M., Grygoruk, M., Flörke, M. and Schneider, C. (2011) Broad-scale ecosystem services of European wetlands – overview of the current situation and future perspectives under different climate and water management scenarios. *Hydrological Sciences Journal* 56(8), 1–17.
Nuttle, W.K. (1997) Measurement of wetland hydroperiod using harmonic analysis. *Wetlands* 17, 82–89.
Peters, D.L., Buttle, J.M., Taylor, C.H. and LaZerte, B.D. (1995) Runoff production is a forested, shallow soil Canadian Shield basin. *Water Resources Research* 31, 1291–1304.
Rawls, R.J., Brakensiek, D.L. and Saxton, K.E. (1982) Estimation of soil water properties. *Transactions of American Society of Agricultural Engineers* 25, 1316–1320.
Rheinhardt, R.D. and Hershner, C. (1992) The relationship of below-ground hydrology to canopy composition in five tidal freshwater swamps. *Wetlands* 12, 208–216.
Roulet, N.T. (1990) Hydrology of a headwater basin wetland: groundwater discharge and wetland maintenance. *Hydrological Processes* 4, 387–400.
SCDNR (2015a) South Carolina State Climatology Office. Pan Evaporation Records for the South Carolina Area tables, table # 5 – Charleston, SC. Available at: http://www.dnr.sc.gov/climate/sco/Publications/pan_evap_tables.php#5 (accessed 12 August 2015).
SCDNR (2015b) South Carolina Department of Natural Resources, GIS Clearinghouse. Available at: https://www.dnr.state.sc.us/pls/gisdata/download_data.login (accessed 10 June 2015).
Semeniuk, V. and Semeniuk, C.A. (1997) A geomorphic approach to global classification for natural inland wetlands and rationalization of the system used by the Ramsar Convention – a discussion. *Wetlands Ecology and Management* 5, 145–158.
Sidle, R.C. and Onda, Y. (2004) Hydrogeomorphology: an overview of an emerging science. *Hydrological Processes* 18, 597–602.
Siegel, D.I. and Glaser, P.H. (1987) Groundwater flow in bog–fen complex, Lost River Peatland, northern Minnesota. *Journal of Ecology* 75, 743–754.
Skaggs, R.W. (1978) *A Water Management Model for Shallow Water Table Soils*. Report No. 134. Water Resources Research Institute of the University of North Carolina, NC State University, Raleigh, North Carolina.
Skaggs, R.W., Amatya, D.M., Evans, R.O. and Parsons, J.E. (1994) Characterization and evaluation of proposed hydrologic criteria for wetlands. *Journal of Soil and Water Conservation* 49, 354–363.
Skaggs, R.W., Chesheir, G.M. and Philips, B.D. (2005) Methods to determine lateral effect of a drainage ditch on wetland hydrology. *Transactions of the American Society of Agricultural Engineers* 48, 577–584.
Skaggs, R.W., Chescheir, G.M., Fernandez, G.P., Amatya, D.M. and Diggs, J. (2011) Effects of land use on soil properties and hydrology of drained coastal plain watersheds. *Transactions of the ASABE* 54, 1357–1365.
Smith, M.W. and Risenborough, D.W. (2002) Climate and the limits of permafrost: a zonal analysis. *Permafrost and Periglacial Processes* 13, 1–5.
Smith, T.J. III (1992) Forest structure. In: Robertson, A.I. and Alongi, D.M. (eds) *Tropical Mangrove Ecosystems*. American Geophysical Union, Washington, DC, pp. 101–136.
Spalding, M., Kainuma, M. and Collins, L. (2010) *World Atlas of Mangroves*. Earthscan, London.
Strahler, A.N. (1957) Quantitative analysis of watershed geomorphology. *Transactions of the American Geophysical Union* 38, 913–920.
St Amour, N.A., Gibson, J.J., Edwards, T.W.D., Prowse, T.D. and Pietroniro, A. (2005) Isotropic time-series partitioning of streamflow components in wetland-dominated catchments, lower Laird River basin, Northwest Territories, Canada. *Hydrological Processes* 19, 3357–3381.
Tiner, R.W. (2011) *Dichotomous Keys and Mapping Codes for Wetland Landscape Position, Landform, Water Flow Path, and Waterbody Type Descriptors: Version 2*. US Fish and Wildlife Service, National Wetlands Inventory Project, Northeast Region, Hadley, Massachusetts.

Tiner, R.W. (2013) *Tidal Wetlands Primer: An Introduction to their Ecology, Natural History, Status, and Conservation*. University of Massachusetts Press, Amherst, Massachusetts.
Todd, A.K., Buttle, J.M. and Taylor, C.H. (2006) Hydrologic dynamics and linkages in a wetland-dominated basin. *Journal of Hydrology* 319, 15–35.
Turunen, J., Tomppo, E., Tolonen, K. and Reinikainen, A. (2002) Estimating carbon accumulation rates of undrained mires in Finland – application to boreal and subarctic regions. *The Holocene* 12, 69–80.
Trettin, C.C., Jurgensen, M.F., Grigal, D.F., Gale, M.R. and Jeglum, J.K. (1996) *Northern Forested Wetlands Ecology and Management*. Lewis Publishers, Boca Raton, Florida.
Twilley, R.R., Lugo, A.E. and Patterson-Zucca, C. (1986) Litter production and turnover in basin mangrove forests in southwest Florida. *Ecology* 67, 670–683.
USACE (1987) *Corps of Engineers Wetland Delineation Manual*. Technical Report No. Y-87-1. US Army Corps of Engineers, Waterways Experiment Station, Vicksburg, Mississippi.
USACE (2005) *Technical Standard for Water-Table Monitoring of Potential Wetland Sites*. WRAP Technical Notes Collection (ERDC TN-WRAP-05-2). US Army Engineer Research and Development Center, Vicksburg, Mississippi.
USDA (2012a) *Groundwater-Dependent Ecosystems: Level I Inventory Field Guide – Inventory Methods for Assessment and Planning*. General Technical Report WO-86a. USDA Forest Service, Washington, DC.
USDA (2012b) *Groundwater-Dependent Ecosystems: Level II Inventory Field Guide – Inventory Methods for Project Design and Analysis*. General Technical Report WO-86b. USDA Forest Service, Washington, DC.
USGS (2015) National Water Information System. Columbia, South Carolina, Station 02169500. Available at: http://waterdata.usgs.gov/sc/nwis/uv?site_no=02169500 (accessed 27 September 2015).
Verry, E.S., Brooks, K.N., Nichols, D.S., Ferris, D.R. and Sebestyen, S.D. (2011) Watershed hydrology. In: Kolka, R.K., Sebestyen, S.D., Verry, E.S. and Brooks, K.N. (eds) *Peatland Biogeochemistry and Watershed Hydrology at the Marcell Experimental Forest*. CRC Press, Boca Raton, Florida, pp. 193–212.
Watson, J.G. (1928) Mangrove forests of the Malay Peninsula. *Malayan Forest Records* 6, 1–275.
White, W.N. (1932) *A Method of Investigating Ground-Water Supplies Based on Discharge by Plants and Evaporation from Soils: Results of Investigations in Escalante Valley, Utah*. Water Supply Paper 659-A. US Department of the Interior, Geological Survey, Washington, DC.
Williams, T.M. (1981) Determination of transmissivity of shallow aquifers on high water table forested areas by pumping tests. *Forest Science* 27, 124–127.
Williams, T.M. (1998) Hydrology. In: Messina, M.G. and Conner, W.H. (eds) *Southern Forested Wetlands; Ecology and Management*. Lewis Publishers, Boca Raton, Florida, pp. 103–122.
Winter, T.C. (2001) The concept of hydrologic landscapes. *Journal of the American Water Resources Association* 37, 335–349.
Woo, M.K. and diCenzo, P.D. (1989) Hydrology of small tributary streams in a subarctic wetland. *Canadian Journal of Earth Sciences* 26, 1557–1566.
Wösten, J.H.M, Clymans, E., Page, S.E., Rieley, J.O. and Limin, S.H. (2008) Peat–water interrelationships in a tropical peatland in Southeast Asia. *Catena* 73, 212–224.

Note: Any use of trade, product, or firm names is for descriptive purposes only and does not imply endorsement by the US government.

8 Forest Drainage

R.W. Skaggs[1]*, S. Tian[1], G.M. Chescheir[1], D.M. Amatya[2] and M.A. Youssef[1]
[1]*North Carolina State University, Raleigh, North Carolina, USA;*
[2]*USDA Forest Service, Cordesville, South Carolina, USA*

8.1 Introduction

Most of the world's 4030 million ha of forested lands are situated on hilly, mountainous or well-drained upland landscapes where improved drainage is not needed. However, there are millions of hectares of poorly drained forested lands where excessively wet soil conditions limit tree growth and access for harvesting and other management activities. Improved or artificial drainage has been used to improve forest productivity on such lands substantially. Drainage has increased timber growth in natural forests and, applied as a silvicultural practice, enabled harvesting, regeneration and increased production of plantation forests. Improved drainage is needed in regions where precipitation exceeds evapotranspiration (ET) on lands where natural drainage processes are not sufficient to remove the excess. Such conditions frequently occur in northern climates where ET is low and, in the absence of adequate natural drainage, soils remain saturated for long periods of time. Drainage may also be needed in lands that receive runoff and seepage from upslope, and in areas subjected to frequent flooding from adjacent streams. Peatlands, which form under very wet soil conditions, have been drained extensively to facilitate forest production in many parts of the world.

Paavilainen and Päivänen (1995) presented a detailed review of the history, methods and results of forest drainage of peatlands. They date reports of ditching of peatlands to promote tree growth to a 1773 Swedish publication and, based on a review of literature regarding drainage in Russia, the Baltic countries and Germany, noted that drainage to increase tree growth was well known in the region in the mid-19th century. While statistics documenting forest drainage go back to the mid-1800s in Sweden, Norway and Finland, the period of most intensive drainage activity started during the 1920s and 1930s, was inactive during World War II, and resumed in the 1950s and 1960s. In addition to northern and eastern Europe, drainage has been used in the British Isles, Canada and the USA as a relatively economical means of increasing forest productivity (Laine *et al.*, 1995; Paavilainen and Päivänen, 1995). Trottier (1991) concluded that, for poorly drained lands, few silvicultural practices can compete with drainage in terms of costs per unit increase of forest yield. By 1995 about 15 million ha of northern peatlands and other wetlands had been drained for forestry (Laine *et al.*, 1995). More than 90% of these lands are in Finland, Scandinavia and the former Soviet Union. The peak of forest drainage activity in Sweden was in the 1930s when drainage

*Corresponding author; e-mail: wayne_skaggs@ncsu.edu

was subsidized by the state to improve forest production while reducing unemployment (Paavilainen and Päivänen, 1995). Peltomaa (2007) reported that 5.5 million ha (more than 20%) of the 26 million ha of forest land in Finland is drained. About 4.5 million ha of the drained forest is peatlands. Beginning in the 1930s, with the greatest activity in the 1960s and 1970s, drainage was subsidized by the Finnish government in an effort to increase forest production. Peltomaa (2007) attributed the positive influence of drainage as one of the main reasons for a 40% increase in growing stock during the 30-year period 1970–2000. Forest products made up as much as 40% of Finnish exports in the 1970s and were still 20% of exports in 2005. Tomppo (1999) reported that drainage of forest lands in Finland had increased annual tree growth by 10.4 million m^3 since the beginning of the 1950s. While forest drainage has been applied on peat soils in Canada (Hillman, 1987), Quebec (Trottier, 1991), Ontario (Stanek, 1977) and Alberta (Hillman and Roberts, 2006), the area drained there and in northern USA states is a small fraction of that drained in northern Europe (Paavilainen and Päivänen, 1995).

The evolution of forest drainage in the US south started with a large-scale drainage project in the Hoffman Forest in eastern North Carolina in the 1930s (Fox *et al.*, 2007). Early observations of improved growth of pine adjacent to drainage ditches on both mineral and peat soils (Miller and Maki, 1957; Maki, 1960) led to field trials and more studies (Terry and Hughes, 1975, 1978), and finally to widespread drainage of forested wetlands. By the mid-1980s, drainage was used to provide access for harvest and regeneration, and to improve production on over 1 million ha of poorly drained forests in the coastal plains of states along the Atlantic and Gulf of Mexico (McCarthy and Skaggs, 1992). Expansion of drainage projects to establish new plantations on wetland forests ended by 1990 because of concern for their effect on jurisdictional wetlands and federal regulations for wetland protection. Government support for drainage was also reduced in other countries. Finland ceased subsidies for new forest drainage projects in 1992, due mostly to ecological concerns. However, in recognition of the economic importance of forest drainage it continued to subsidize maintenance and reclamation of old drainage systems (Peltomaa, 2007).

While forest drainage activity has been reduced substantially compared with 40 years ago, drainage is responsible for substantial increases in production on millions of hectares of natural and plantation forests, and the associated economic and social benefits. Optimum management and operation of existing drainage systems, as well as the design and construction of new systems, is complex since these systems need to address both production and environmental/ecological goals. An understanding of the methods and theory of drainage is needed to optimize drainage systems to achieve competing objectives. This chapter reviews the impacts of drainage on forest production and the hydrology of forested lands.

8.2 Purpose and Impact of Forest Drainage

The purpose and effects of drainage on forest production are well documented in the literature. There are two primary purposes: (i) to enable access and provide trafficable conditions such that planting, harvesting and other field operations can be conducted on time with minimum damage to soil and water resources; and (ii) to remove excess water from the soil profile to improve aeration status and promote tree growth. A related purpose/benefit of forest drainage in cold climates is to remove water from snowmelt and warm soils earlier in the season to promote growth (Peltomaa, 2007).

Both the need for and the effectiveness of improved subsurface drainage in providing trafficable soil conditions are depicted in Fig. 8.1, where soil on a poorly drained site (upper right of the picture, above the road) was severely puddled during the harvest operation. By contrast, the drainage ditch below and left of the road lowered the water table and significantly reduced compaction and puddling. Soil damage resulting from harvesting or site preparation during wet site conditions can reduce growth rates significantly and may be only partially offset by subsequent amelioration (Terry and Campbell, 1981). Terry and Hughes (1978) discussed the impacts of drainage on both natural stands and new pine plantations. They noted that drainage installation at least 1 year prior to

Fig. 8.1. Picture of soil conditions after harvesting on a plantation pine site, contrasting subsurface drainage (left of the road) with that without improved drainage (right of the road). Note severe puddling of soil in contrast with soil conditions on left of the road, where drainage had been improved. (Photo by Joe Hughes, 1981.)

harvest extends the logging season and minimizes soil damage, and concluded that about half of the cost of preharvest ditching was offset by reduced logging and site preparation costs, reduced site damage and increased site preparation effectiveness. Use of conventional equipment for both harvesting and site preparation on undrained sites is limited to dry seasons. Drainage extends the season for harvesting and makes it possible to conduct needed field operations in a timely fashion without damaging the soil.

While the impact of drainage on tree growth and yield has been studied by a number of researchers over the years (Miller and Maki, 1957; Graham and Rebuck, 1958; Maki, 1960; Klawitter *et al.*, 1970; White and Pritchett, 1970; Brightwell, 1973; Terry and Hughes, 1975; Trottier, 1991; Hillman and Roberts, 2006; Jutras *et al.*, 2007; Socha, 2012), published data on the subject are relatively limited. A number of articles by Weyerhaeuser scientists (Terry and Hughes, 1975, 1978; Campbell, 1976; Campbell and Hughes, 1991) reported results of a programme initiated in 1972 to improve drainage for the production of high-yield loblolly pine. Results originally summarized for pine by Terry and Hughes (1975) are given in Table 8.1, which has been expanded to include results published in recent years and for other regions. Results reported by Miller and Maki (1957), Klawitter *et al.* (1970), White and Pritchett (1970) and Terry and Hughes (1975) showed that drainage increased annual growth on very poorly drained mineral soils by 3.6 to 8.9 m^3/ha. These results are similar to those reported for peatlands in northern Europe and for bogs and poorly drained mineral soils in Quebec (Trottier, 1991). Annual increases in yield were typically more than 100% and in some cases much greater (Table 8.1). However, for cases where trees had negligible volume or rate of growth prior to drainage, large percentage increases may not be particularly meaningful (Payandeh, 1973). Growth responses to drainage not only differed among species, but also among stand ages. Socha (2012) determined that drainage

Table 8.1. Summary of results of studies to determine effects of drainage on tree growth and yield.

Study	Soil	Species	Age (years)	Units	Tree growth: Drained	Tree growth: Undrained	Increase (%)	Notes
Miller and Maki (1957)[a]	Portsmouth sl	Loblolly pine (*Pinus taeda*)	0–17	m^3/ha/year	4.7	0.34	4.4 (1300)	
Klawitter *et al.* (1970)[a]	Plummer Rains ls	Slash pine (*Pinus elliottii*)	19–22	m^3/ha/year	9.0	4.9	4.1 (84)	
White and Pritchett (1970)[a]	Leon s	Slash pine (*P. elliottii*)	0–5	m^3/ha/year	14.6	5.7	8.9 (160)	Water table depth (WTD) = 46 cm
					13.3	5.7	6.6 (120)	WTD = 92 cm
		Loblolly pine (*P. taeda*)			7.6	1.3	6.3 (490)	WTD = 46 cm
					4.9	1.3	3.6 (280)	WTD = 92 cm
Terry and Hughes (1975)[a]	Bayboro-Bladen	Loblolly pine (*P. taeda*)	0–13	m^3/ha/year	4.3	0.54	3.8 (700)	
Socha (2012)	Peatland	Scots pine (*Pinus sylvestris*)	0–100	m^3/ha/year	10	8	2 (25)	Modelled
Hillman and Roberts (2006)	Peatland	Black spruce (*Picea mariana*)	30–60	m^3/ha/year	0.59	0.11	0.48 (440)	Average from two sites
		Tamarack (*Larix laricina*)	50–60		0.87	0.74	0.13 (180)	Average from two sites
Trottier (1986) as cited in Hillman (1987)	Peatland	Black spruce (*P. mariana*)	30	m^3/ha/year	1.78	0.5	1.28 (250)	40 m drain spacing
Hökkä and Ojansuu (2004)	Peatland	Scots pine (*P. sylvestris*)	8–228	m^2/ha/year	12.5	5	7.5 (150)	Basal area

Continued

Table 8.1. Continued.

Study	Soil	Species	Age (years)	Units	Tree growth: Drained	Undrained	Increase (%)	Notes
Jutras *et al.* (2007)	Peatland	Black spruce (*P. mariana*)	45–75	mm/year (diameter)	1.37	0.44	0.93 (210)	Spacing = 20 m
					0.85	0.42	0.43 (100)	Spacing = 40 m
					0.65	0.45	0.2 (44)	Spacing = 60 m
Payandeh (1973)	Peatland	Black spruce (*P. mariana*)	79–119	m^3/m^2/year	0.005	0.00006	(8200)	Before and after drainage
Graham and Rebuck (1958)[a]		Pond pine (*Pinus serotina*)	0–22	m^2/ha/year	0.74	0.41	0.33 (80)	
Walker *et al.* (1961)	Bladen cl	Slash pine (*P. elliottii*)	0–2	m/year (height)	0.41	0.1	0.31 (310)	Seedlings
		Loblolly pine (*P. taeda*)			0.49	0.21	0.28 (140)	
Langdon (1976)	Myatt, Rains,	Loblolly pine	0–5	m/year	3.1	2.0	1.1 (55)	These three studies
Andrews (1993)	Lynchburg sl	(*P. taeda*)	0–21	(height)	14.1	11.7	2.4 (20)	conducted on the same site.
Kyle *et al.* (2005)			21–33	m^3/ha/year	334	341	–7 (–2)	Cumulative growth at three ages

[a]After Terry and Hughes (1975).

increased yields of Scots pine planted after drainage of a peatland in Poland by 25%, as compared with 15 and 6% increases for trees 30 and 40 years old, respectively, at the time of drainage. Langdon (1976) and Andrews (1993) reported significant increases of tree growth at ages 5 and 21 years for a drained loblolly pine stand in the coastal plain of Virginia, USA. However, Kyle *et al.* (2005) reported no significant tree volume increase for the same site at a stand age of 33 years. These results may have also been impacted by the natural drainage condition of the site. The soils on the site are classified as 'poorly drained' as opposed to the 'very poorly drained' soils of most of the other studies. Increased ET with stand age could have reduced the difference of water table depth between drained and undrained plots and hence the response to drainage (Kyle *et al.*, 2005). Hökkä and Ojansuu (2004) found that drainage increased site productivity by over 80% on a pine fen in northern Finland, but had only a moderate effect on another site with better natural drainage. In other cases tree growth responded well to drainage, but narrow ditch spacings were required to increase yields significantly. Jutras *et al.* (2007) found that drainage had little impact beyond 15 m from the ditch in a black spruce stand in Quebec.

8.3 Drainage Systems and Their Function

Most forest drainage systems can be characterized as one of two types or a combination of the two: (i) natural or systems that use and often enhance existing drainage patterns, branches, creeks and streams developed as a result of the watershed topography; and (ii) a grid system of parallel ditches such as that shown in Fig. 8.2. Where there is enough relief, natural drainage systems may provide a sufficient outlet for needed drainage. Additional ditches may be necessary to increase drainage intensity (DI) in some cases, but the basic drainage patterns are unchanged. The grid pattern is used in broad, poorly drained areas. Its regular pattern with relatively straight rows increases efficiency of site preparation, planting and harvesting. The drainage system for either a natural forest or a plantation will often be a combination of both types (Terry and Hughes, 1978). The drainage system may also be characterized as to whether it provides primarily surface drainage, subsurface drainage, or a combination of the two. The system shown in Fig. 8.2 provides primarily subsurface drainage through parallel ditches about 1 m deep and typically spaced 100 to 200 m apart. The tree seedlings are planted on beds, about 30 cm in height, which provide protection from flooded conditions and good soil–root contact. The water standing between the beds in Fig. 8.2 is the result of more than 150 mm of rainfall during a hurricane. In this case the furrows between the beds are not connected to the ditches; thus, the intensity of surface drainage is very low. Although there may be some runoff during extreme events, annual surface runoff is small and nearly all of the drainage water is removed by relatively slow subsurface flow. This has the advantage of reducing outflow rates from these watersheds during large storms and of preventing sediment and associated contaminants from moving into the ditches and on downstream. It also tends to keep water from intense runoff-producing rainfall events on the site making more of it available for ET.

For tight soils, subsurface drainage is slow and surface drainage may be the best option for removing excess water. In this case the furrows in Fig. 8.2 would be connected to the ditches such that most of the surface water would run off the site during the storm event. The beds would still provide protection from waterlogging and the water table would subsequently be drawn down by ET. Annual surface runoff would be greatly increased compared with a site with intensive subsurface drainage, as will be shown in a later example. The intensity or quality of surface drainage may be defined as the average depth of depression storage (i.e. the average depth of water stored on the surface at the time surface runoff ceases following a large rainfall event). For drainage of existing stands, beds are usually not an option. The intensity of surface drainage in that case is still dependent on depressional storage and is generally inversely proportional to ditch spacing, but is improved by distributing the ditch spoil such that entry of surface water is not impeded.

A schematic of the drainage system showing the evolution of the water table and its effect

Fig. 8.2. Drained pine plantation in the coastal plain of North Carolina, USA. Picture taken 1 day after rainfall of over 150 mm from hurricane Dennis, August 1981. Beds protect young seedlings from drowning. Note furrows between beds are not connected to ditches, virtually eliminating surface runoff in all but the most extreme rainfall events. (Photo by Joe Hughes, 1981.)

on drainage rates following a rainfall event is given in Fig. 8.3. Drainage rate is plotted as a function of water table elevation midway between the drains, *m*, in Fig. 8.3b. Drainage rates for specific water table positions 1–6 (Fig. 8.3a) are denoted in Fig. 8.3b. Exact solutions for this case may be obtained by numerically solving the governing equations for combined saturated and unsaturated flow (Skaggs and Tang, 1976). An approximate approach is to use a combination of methods as follows. When the profile is saturated and water is ponded on the surface (position 1), the drainage rate may be calculated by equations developed by Kirkham (1957) (denoted by DK in Fig. 8.3b). After the depth of surface water recedes due to drainage and evaporation to a depth below the top of the beds, water can no longer move across the surface to the vicinity of the drains (position 2) and the Kirkham equation is no longer applicable. Drainage rates continue to decline as the ponded water drains through the profile until the water table midway between the drains is just coincident with the surface (position 3). At this point the drainage rate can be estimated with the steady-state Hooghoudt equation (Bouwer and van Schilfgaarde, 1963), which may be written for ditches as:

$$q = 4K_e m(2d + m) / L^2, \qquad (8.1)$$

where q is drainage rate (cm/h), m is midpoint water table elevation above the drain, K_e is the equivalent lateral hydraulic conductivity of the profile (cm/h), d is the depth from the drain to the restrictive layer (cm) and L is the drain spacing (cm) (Fig. 8.3a). For drain tubes used in agricultural applications an equivalent depth d_e rather than the actual depth, d, is used to compensate for radial head losses near the drain. The drawdown process as the water table falls from position 3 to position 4 and finally to drain depth (position 5) is obviously not steady state but, in most cases, proceeds slowly, and the drainage rate can be estimated by the Hooghoudt equation. The water table may continue to recede (position 6) due to ET and/or seepage, but the drainage rate would be zero when the water

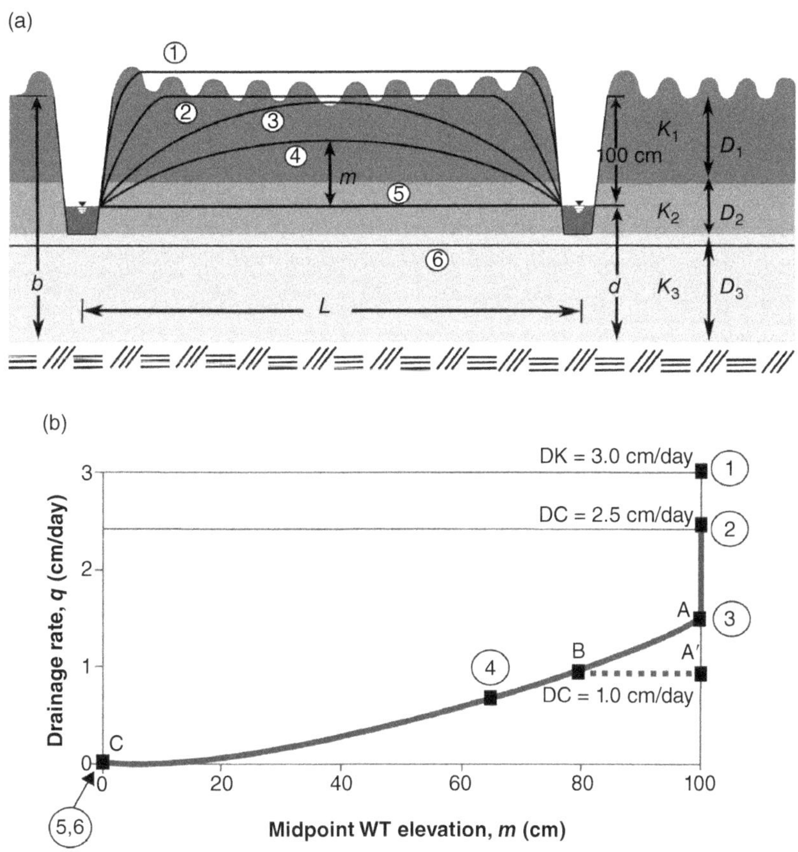

Fig. 8.3. (a) Schematic of a parallel subsurface drainage system with 1 m deep ditches spaced a distance *L* apart in a layered soil profile. (b) Drainage rates, *q*, corresponding to water table (WT) positions 1 to 6 shown in (a) are plotted as a function of the WT elevation, *m*, midway between the ditches. Drainage rates may be limited by the hydraulic capacity, DC, of the outlet, as shown.

table falls below drain depth. The drainage rate when the water table midway between the drains is at the surface (position 3) may be defined as the subsurface DI. DI is thus a function of the drain spacing and depth, and the thickness and hydraulic conductivity of the profile.

The values predicted by the Kirkham and Hooghoudt equations quantify the rate of water movement through the soil to the drains for given water table elevations. Most of the time, the water table is below the soil surface, drainage rates follow curve ABC in Fig. 8.3b and may be calculated with Eqn 8.1 above. The quality or intensity of subsurface drainage for a given site is typically quantified by the DI as defined above. However, in some cases the drainage rate may be limited not by the rate that water will move from the soil profile to the drain, but by the rate water will move through the ditch network to the drainage outlet; that is, by the hydraulic capacity of the system. The hydraulic capacity is called the drainage coefficient (DC) and depends on the size of the area being drained and the capacity of the outlet works, which is dependent on the size, slope and hydraulic roughness of the main drain or, in the case of pumped outlets, the pumping capacity. For example, let us assume the hydraulic capacity is 2.5 cm/day. When the profile is saturated and water is ponded on the surface such that it could theoretically drain through the soil at a rate of DK = 3.0 cm/day (Fig. 8.3b), the actual rate would be limited by the outlet capacity to DC = 2.5 cm/day. Once the water table falls to position 3 in Fig. 8.3b, q = DI = 1.5 cm/day which is smaller than DC, so the drainage event would follow curve ABC from there forward. In this example, DI < DC < DK, but DC may be greater than DK or less than DI, depending on the capacity of the outlet. For the case where DC is less than DI, say DC = 1.0 cm/day, the drainage rate for water table position 3 would be 1.0 cm/day and flow rates during a drainage event would follow A'BC in Fig. 8.3b. In this case water would back up in the ditches, the water table would become flatter and the flow rate from field to ditch would equilibrate with the DC.

Subsurface drainage rates may be reduced by obstructions in the ditches or downstream in outlet canals. In some cases obstructions such as weirs may be placed in the ditches to purposely reduce drainage rates. This practice is called controlled drainage (CD) and may be applied in both agricultural and forested lands to conserve water and reduce nutrient losses. Drainage rates for these cases may still be calculated with Eqn 8.1 as discussed above, with the value of m in Fig. 8.3 defined as the distance from the top of the weir in the ditch to the water table midway between drains. This approach could also be used to determine drainage rates when there is an obstruction or partial fill of the ditch.

The effect of drainage on water table depth is shown in Fig. 8.4 for a site in eastern North Carolina, USA. Measured water table depths for a drained loblolly pine plantation, an undrained forested wetland and a drained agricultural cropland are plotted for a 3-year period (1993–1995). Annual precipitation was 1004 mm (77% of average) in 1993, about average in 1994 (1284 mm) and 1368 mm in 1995. Very dry conditions during summer 1993 caused the water table to recede to depths greater than 1.2 m in the cropland site and to even greater depths in the deeper-rooted wetland and drained forested sites. The water table in the wetland was at or above the surface for extended periods in the winter and spring months of all three years and, except for the very dry summer 1993, well above the water tables in both the drained cropland and managed forest. Drainage from the wetland was mostly surface runoff, with minor subsurface drainage to widely spaced shallow natural drains. The average water table depth (1.54 m) on the drained forested site was much deeper and receded more rapidly than on the cropland site (0.75 m) or the undrained site (0.55 m). Drainage is the obvious reason for deeper water table depths in the forested compared with the wetland sites. Difference in ET is the reason for the greater water table depths on the drained forested versus the cropland site. Rooting depths are greater and the ET demand continues all year for the pine forest. The ditch depth was only 0.9 m but ET caused the water table to be drawn down to a maximum depth of more than 2 m for the drained forest, compared with only about 1.4 m for the agricultural site.

The response to drainage shown in Fig. 8.4 is in contrast to that reported for less permeable soils at other locations. For example, Jutras *et al.* (2007) reported that while drainage increased the annual growth rate of the diameter of black spruce close to the ditches in a peatland, it had only minor effects more than 15 m from the ditch. They concluded that narrow ditch spacing would be necessary to transform unproductive sites into productive ones. Such differences in response to drainage may be partly due to differences in climate, but are more likely due to differences in soil properties. The soil property having the greatest effect on drainage is the saturated hydraulic conductivity, K (Eqn 8.1). Paavilainen and Päivänen (1995) presented a summary of published measurements on a wide range of undisturbed peat soils. The K values varied from 4×10^{-2} to 9×10^{-8} cm/s (35 to 8×10^{-5} m/day), with magnitudes generally decreasing with increasing decomposition of peat. Published K values for mineral soils are roughly dependent on soil texture and vary from about 6×10^{-2} to 2×10^{-6} cm/s (Smedema *et al.*, 2005). The effect of K on response to drainage is shown in Table 8.2 for a 3 m deep homogeneous profile with parallel 0.9 m deep drainage ditches. Results show that profiles with K_e values less than 10^{-6} cm/s would have minimal response to subsurface drainage. Ditches spaced less than 2 m apart would be required for DI values of 5 mm/day. Depending on profile depth, K_e greater than 10^{-4} cm/s (0.36 cm/h) would be necessary for a typical forest drainage ditch spacing (40 m or greater) to result in a DI of just 1 mm/day. For soils with very low K_e values, the best alternative may be to provide drainage to remove surface water so that the water table can be lowered by ET. This will make runoff events flashier, but not have a great effect on annual catchment drainage (Robinson, 1986; Holden *et al.*, 2006).

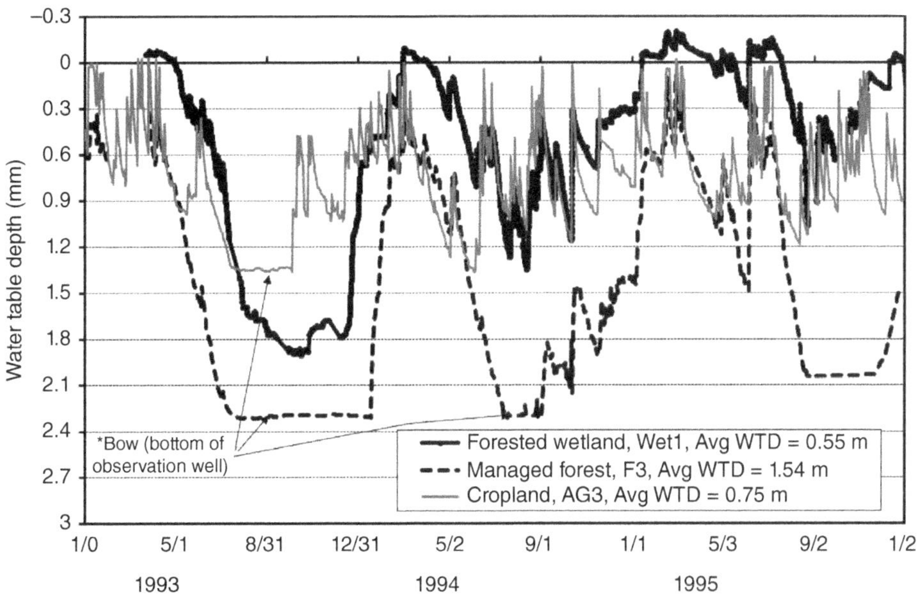

Fig. 8.4. Measured water table depths (WTD) for a drained loblolly pine plantation (F3), an undrained forested wetland (Wet1) and a drained agricultural field (AG3) near Plymouth, North Carolina, USA. Soils on the three sites (all located within a 3 km radius) are mineral (Portsmouth and Cape Fear sl and scl). (After Skaggs *et al.*, 2011.)

Table 8.2. Effect of hydraulic conductivity (K_e) on ditch spacing required for drainage intensity (DI) of 5 mm/day and on DI for a 40 m ditch spacing (L). Calculations made with Eqn 8.1 for ditch depth of 0.9 m and 3 m deep homogeneous soil profile (Fig. 8.3a).

K_e (cm/s)	Ditch spacing (m) for DI = 5 mm/day	DI (mm/d) for L = 40 m
10^{-8}	0.15	0 .0001
10^{-6}	1.5	0.01
10^{-4}	15	1
10^{-2}	150	100

8.4 Long-term Forest Drainage and Water Management Case Study

A long-duration watershed-scale study of forest drainage was conducted at the Carteret 7 site in Carteret County, North Carolina, USA. Initiated in 1986, the research site consists of three artificially drained experimental watersheds (D1, D2 and D3), each about 25 ha in size. Deloss fine sandy loam soil on the site is classified as very poorly drained with a shallow water table under natural conditions; the topography is nearly flat with slopes less than 0.1%. Each watershed is drained by four parallel lateral ditches about 1.2 m deep, spaced 100 m apart. A pump was installed on the outlet ditch to provide a reliable drainage outlet so that flow measurements could be made and water quality samples collected with minimum interference from elevated water levels in the outlet canal. The site was instrumented and water table and outflow data collection began in 1988 when the loblolly pine trees were 15 years old. Watershed D1 was maintained as the control with standard drainage practices from 1988 through 2009. Paired watershed studies were conducted to determine the hydrological and water quality impacts of several silvicultural and water management practices over the 21-year period 1988–2008.

After a 2-year calibration period, CD treatments were implemented on watersheds D2 and D3 to evaluate the impacts on water balance and storm event hydrology (Amatya *et al.*, 1996, 2000). In 1995, watershed D2 was harvested to study the impacts of harvesting, site preparation and regeneration on hydrology and water quality. At the same time, an orifice weir was installed on watershed D3 to study the performance of a weir arrangement that would extend drainage flow events and reduce peak flow rates (Amatya *et al.*, 2003). Watershed D3 was thinned in 2002 to study the impact of thinning on hydrology and drainage water quality (Amatya and Skaggs, 2008). The 21-year data set collected on the site was used to develop and test simulation models for predicting the hydrology of drained forested watersheds under the treatments referenced above.

8.4.1 General hydrology

Observations on the Carteret 7 watersheds indicated that the principal hydrological components for drained forested watersheds in the coastal plain are rainfall, ET and subsurface drainage. These processes are dominated by shallow water tables that result from the combination of very low relief, micro-topography that produces high surface detention storage, and aquitards within a few metres of the surface. A restrictive layer that begins at an average depth of about 2.8 m limits vertical seepage, which was estimated to be less than 3% of precipitation (Amatya *et al.*, 1996). Surface depressional storage is large on the bedded watersheds causing surface runoff to be small and, except for large tropical storms and hurricanes, negligible. Analysis of data for a 17-year period (1988–2005) on the control watershed (D1) showed that annual rainfall ranged from 852 to 2331 mm with an average of 1538 mm (Amatya and Skaggs, 2011). The large range in annual rainfall resulted from tropical storms and hurricanes in some years and drought in others. The annual runoff coefficient, defined as the ratio of outflow to rainfall, averaged 0.32 for the 17-year period. It ranged from 0.05 in the very dry year 2001 to 0.56 in the year of highest rainfall, 2003. Outflow on these watersheds is primarily subsurface flow to drainage ditches. Outflow rates were greater, more continuous and longer in duration in winter than in other seasons. Winter outflow was 59% of rainfall on average. The water table tended to be close to the surface during winter and early spring when ET demands are low, and during summer when hurricanes and tropical storms produced large outflows. However, it was drawn down to depths much deeper than the ditch depth during long dry periods in the summer and autumn. Annual ET, calculated as the difference between rainfall and outflow, averaged 1005 mm, which was close to the Penman–Monteith based average annual potential ET for a grass reference.

8.4.2 Effects of controlled drainage on hydrology of drained pine plantations

The DI needed for agricultural and silvicultural production varies with season and stage of the production cycle. For plantation forest the most critical stage is during harvest, site preparation for planting, and in the first years after planting when the seedlings require protection from high water table and excessive soil water conditions. ET is reduced during the seedling and early stage of growth, so drainage to lower the water table and provide suitable conditions for tree growth is more critical than later in the production cycle. For similar reasons, drainage is more critical in winter than in summer when the water table may be relatively deep due to ET alone (e.g. Fig. 8.4). Drainage in excess of that needed should be avoided as it removes water that could be used by the growing trees. Drainage can be reduced or managed on a temporal basis through the process of CD. CD may be applied in forested lands by the installation of a weir in the drainage outlet ditch such that the water level in the ditch must exceed the elevation of the weir for drainage water to leave the system.

Watershed D1 was maintained in conventional drainage with the weir level 1 m below the surface. CD to conserve water during the growing season was practised on D2 and CD to reduce drainage outflows during the spring was applied on D3. Results from the 3-year treatment period indicated that CD increased both ET and seepage and reduced outflows from D2 and D3 by 25 and 20 %, respectively, compared with the conventional drainage (Amatya *et al.*, 1996). The CD treatment on watershed D2 resulted in rises in water table elevations during the summer. But the rises were small and short-lived due to increased ET rates as compared with the spring treatment with lower ET demands. Spring-time CD on watershed D3 also reduced freshwater outflows substantially, minimizing off-site water quality impacts. CD significantly reduced storm outflows for all events, and peak outflow rates for most events. In some events, flows did not occur at all in watersheds with CD. When outflows occurred, duration of the event was reduced sharply because of reduced effective ditch depth. While sediment and nutrient transport from these flat forested watersheds is low compared with other land uses (Chescheir *et al.*, 2003), CD was effective in reducing those loads to surface waters (Amatya *et al.*, 1998).

8.4.3 Effect of harvesting, bedding and planting on hydrology

Watershed D2 was harvested in July 1995 at a stand age of 21 years. Continuous flow and water table records were analysed to determine the hydrological effects and their change with time after replanting. The biggest effect of harvesting is the removal of growing plants, which substantially reduces ET. This reduced water table depth and increased drainage outflow and runoff coefficient compared with the control (D1), which was not harvested. Harvesting and regeneration reduced annual ET by 28% and increased outflow by 49% during the 5-year period 1995–1999 (Skaggs *et al.*, 2006). The average runoff coefficient for the period was increased from 0.32 to 0.51. Analysis of the long-term flow and water table data indicated that differences in both drainage outflows and water table depths between D2 and the control (D1) had returned to normal by 2004, 7 years after replanting in 1997.

Hydrological data collected in the Carteret 7 studies provided clear evidence that land use and operations such as harvesting and site preparation for new planting may substantially impact soil properties that may also result in further hydrological changes. Recorded water table and drainage flow data were analysed to determine the relationship between q and m (e.g. Fig. 8.3b) for various stages of the production cycle. The measured $q(m)$ relationship was used with Eqn 8.1 to calculate the field effective saturated hydraulic conductivity for a range of water table elevations (m). Then the field effective saturated hydraulic conductivity was determined for the three principal layers of the soil profile above the restrictive horizon. Results are summarized in Table 8.3.

Results indicate that the field effective hydraulic conductivity (K) in the top 80 cm of the soil profile prior to harvest of the 21-year-old loblolly pine was 20 to 30 times greater than values given in the county soil survey for the Deloss soil series. The K value of 1.6 m/day for depths greater than 80 cm was apparently unaffected as it remained within the range given in the soil survey for Deloss throughout the preharvest to postharvest period. The high K values in the shallower layers are attributed to the presence of large pores that result from tree roots and biological activity that is uninterrupted for many years in a forest. Similar high K values were reported for an organic soil on the Parker tract in eastern North Carolina (Grace *et al.*, 2006) and for a mineral soil on the same tract (Skaggs *et al.*, 2011). All sites were in plantation forest. Hydraulic conductivity (K) determinations based on water table and drainage outflow measurements after harvest in 1995, but prior to site preparation for the new plantation in October 1996 (postharvest in Table 8.3), were the same as obtained for the preharvest

Table 8.3. Field effective hydraulic conductivity (m/day) by layer for the soil profile on Carteret 7 watershed D2 prior to and following harvest, bedding and planting. Transmissivity (*T*) of the soil profile is also shown. Values published in the county soil survey for Deloss soil series on the site are given in parentheses for reference and are considered typical for agricultural land uses. (After Skaggs *et al.*, 2006.)

		D2			
Depth (cm)	*K*, Deloss soil survey	Preharvest	Postharvest	Post bedding	7 years post planting
0–50	3.6 (1.2–3.6)	60	60	3.6	50
50–80	1.6 (0.36–1.6)	55	55	1.6	20
80–280	1.6 (0.36–1.6)	1.6	1.6	1.6	1.6
T (m^2/d)	5.5	50	50	5.5	34

condition. However, after bedding and planting, drainage rates were substantially reduced and the field effective *K* values determined from field data after bedding were in good agreement with the range of values published in the soil survey (Table 8.3). Apparently the bedding process destroyed the macropores in the surface layers such that the profile had effective *K* values similar to that expected for agricultural land use. These results indicate that *K* values needed for drainage design on plantations can be estimated from soil survey data. These values may be conservative as field effective *K* values in the top part of the profile may increase as the trees grow, only to return to original values after harvest and site preparation for new plantings. Other studies have found that drainage may change soil physical properties (possibly reducing *K*), through subsidence and compaction, and chemical properties, including decreased pH (Minkkinen *et al.*, 2008), decomposition rates of soil organic matter (Domisch *et al.*, 2000) and soil C stock (Minkkinen and Laine, 1998; Laiho, 2006).

8.5 Application of a Forest Drainage Simulation Model

Computer models can be applied for simulating hydrological processes and their interactions in drained forests. The models include DRAINMOD (Skaggs *et al.*, 2012), FLATWOODS (Sun *et al.*, 1998), SWAT (Arnold *et al.*, 1998) and MIKE SHE (Abbott *et al.*, 1986). As an example, an application of DRAINMOD is presented to illustrate impacts of subsurface DI on forest hydrology. DRAINMOD was developed in the 1970s to describe the performance of agricultural drainage systems. It is based on a water balance in the soil profile and uses the methods discussed in Section 8.3 to calculate drainage rates. Components of the water balance are simulated on an hourly basis for several years of weather record. The model used here is DRAINMOD-FOREST (Tian *et al.*, 2012, 2014), an enhanced version of DRAINMOD for forested landscapes; the model is briefly described in Chapter 9 (this volume). Simulations were conducted for the Deloss soil on the Carteret 7 site with mid-rotation (age 15 years) pine for DI ranging from 0.5 to 32 mm/day (corresponding to ditch spacing varying from 800 to 100 m), while other parameters were kept the same as in Tian *et al.* (2012). Results of the simulations show the effects of drainage system design on drainage objectives and the hydrology of drained watersheds. The effect of DI on average number of days with soil water and weather conditions suitable for field work is shown in Fig. 8.5a for three different periods of the year. A day was counted as a working day if the predicted water table depth was at least 0.6 m and the precipitation during the day was less than 10 mm. The number of working days increases sharply with increase of DI from 0.5 to 8 mm/day (Fig. 8.5a). Based on these results, a DI of between 5 and 8 mm/day would be recommended for this location. This would provide an average of 55 to 65 working days suitable for harvesting and site preparation during January–March, the wettest 90 days of the year. A DI of 5 to 8 mm/day is less than half of that required for agricultural production, which is about 15 mm/day for eastern North Carolina (Skaggs, 2007).

The effect of DI on average annual outflow for the 21-year simulation period (1988–2008) is shown in Fig. 8.5b. Results are shown for surface depression storages of 150 mm (characteristic of a bedded surface as shown in Fig. 8.2) and 25 mm, which is the minimum expected surface storage on either natural or non-bedded plantation forests on these nearly flat lands. For the bedded condition, average annual predicted subsurface drainage varied from a low of 420 mm for DI = 0.5 mm/day to 510 mm for DI = 32 mm/day. That is, the large majority of outflow from these bedded lands occurs as subsurface flow, even for wide ditch spacing and low DI. This is not the case when surface storage is small (25 mm). Increasing the DI from 0.5 to 32 mm/day for that case decreased predicted average annual surface runoff from 390 to 30 mm and increased annual subsurface drainage from 60 to 480 mm (Fig. 8.5b). Increasing the DI reduced predicted average annual ET and increased total outflow by about 60 mm (4% of annual precipitation) for both surface storage values considered. The 60 mm predicted increase in annual flow is about the same as reported by Robinson (1986) following the installation of a dense network of 0.5 m deep plough ditches on upland clay and peat soils in northern England. Drainage outflows accounted for about two-thirds of precipitation and ET one-third – almost exactly the reverse of the situation at Carteret where drainage accounts for about

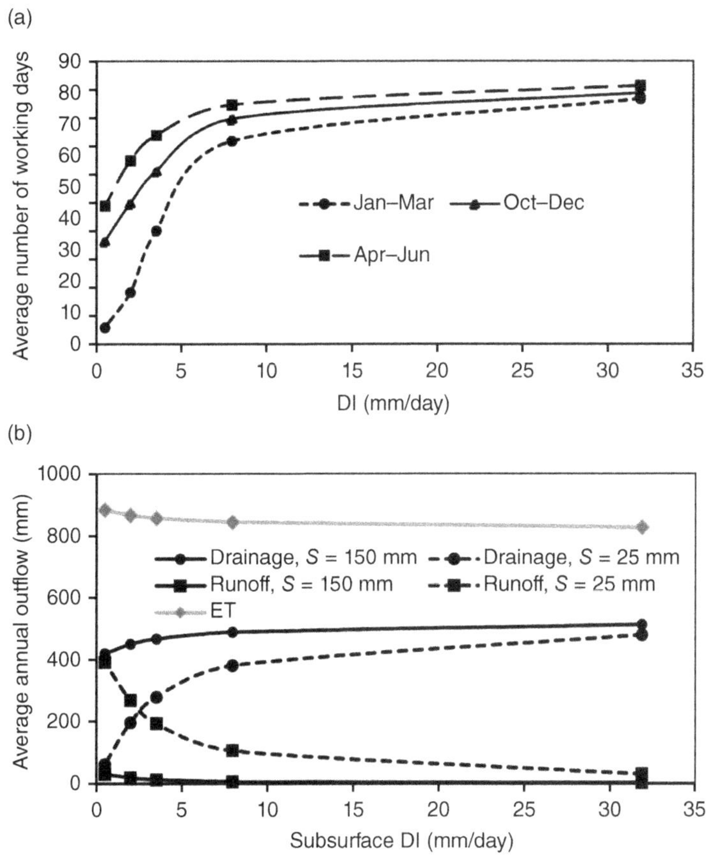

Fig. 8.5. (a) Effect of drainage intensity (DI) on predicted average number of days suitable for harvesting and site preparation during indicated periods. (b) Effect of subsurface DI on annual average drainage and surface runoff for average surface depression storages (*S*) of 15 cm and 2.5 cm. Results predicted by 21-year DRAINMOD-FOREST simulations for a Deloss sandy loam in Carteret County, North Carolina, USA (ET, evapotranspiration).

one-third of annual rainfall and ET roughly two-thirds. While the magnitude of increase in outflows was about the same as predicted for Carteret, the mechanisms were very different. The dense network of shallow ditches on the England site increased baseflows, quickly removed surface runoff, and increased peak flow rates and sediment loss. However, except for the zone very close to the ditch, drainage had limited effect on soil moisture (Robinson, 1986).

8.6 Summary

Drainage is used to improve access and yields on a small percentage of the world's forested lands. However, it has had a big impact on the millions of hectares on which it is applied. Drainage has increased timber yields on poorly drained peatlands and mineral soils in northern Europe, Canada and the southern USA. Substantial yield responses to drainage have been reported on both natural and plantation forests, with typical annual increases of 2 to 8 m^3/ha. In some cases yields have not responded to drainage due to climate, soil physical properties or fertility issues. First applied in the mid-1700s, forest drainage has a long history with the most active periods in the 1930s and from 1950 to about 1985. In recent years forest drainage has been de-emphasized because of concerns about its effects on ecology, biodiversity and related environmental issues.

Government programmes to subsidize forest drainage have been phased out in most countries, and new drainage projects to enhance forest production on wetland soils have been greatly reduced or effectively terminated by regulations to protect wetlands. In most countries exemptions to the regulations or special government programmes allow replanting on, and continued maintenance of, existing forest drainage systems. It is perhaps unreasonable to assume that the needs for wood and wood products for over 7 billion people can be provided without some ecological and environmental costs. Recognizing that, in spite of regulations limiting forest drainage, drained forests are here to stay, Lõhmus *et al.* (2015) suggested:

> Forest drainage can be seen as a scientifically exciting case for ecosystem management which must use novel approaches to reconcile timber production, water management and biodiversity conservation in functional forest–wetland mosaics and their hydrological networks.

Research has increased our understanding of the impacts of forest drainage and the response of hydrology, soils and tree growth to their design and management. For some cases, it is possible to control drainage outlets to conserve water during periods when drainage is not needed and remove excess water when it is. Simulation models have been developed for predicting, on a day-to-day basis, the effects of drainage management on hydrology, primary productivity, water quality and C stock. Their reliability and range of application will likely improve as we go forward. Future models may be run in real time to manage drainage on wetland forests to enhance both production and ecological objectives. While it may not be possible to economically produce timber and other forest products on forested wetlands without some impact on biodiversity and the environment, to do so in ways that create a sustainable balance between economic and environmental/ecological objectives appears to be a reasonable and achievable goal.

References

Abbott, M.B., Bathurst, J.C., Cunge, J.A., O'Connell, P.E. and Rasmussen, J. (1986) An introduction to the European Hydrological System – Systeme Hydrologique Europeen, 'SHE', 2: structure of a physically-based, distributed modelling system. *Journal of Hydrology* 87, 61–77.

Amatya, D.M. and Skaggs, R.W. (2008) Effects of thinning on hydrology and water quality of a drained pine forest in coastal North Carolina. In: Tollner, E.W. and Saleh, A. (eds) *21st Century Watershed Technology: Improving Water Quality and Environment Conference Proceedings, 29 March–3 April 2008, Concepcion, Chile*. ASABE Publication No. 701P0208cd. ASABE, St Joseph, Michigan.

Amatya, D.M. and Skaggs, R.W. (2011) Long-term hydrology and water quality of a drained pine plantation in North Carolina. *Transactions of the ASABE* 54, 2087–2098.

Amatya, D.M., Skaggs, R.W. and Gregory, J.D. (1996) Effects of controlled drainage on the hydrology of drained pine plantations in the North Carolina coastal plain. *Journal of Hydrology* 181, 211–232.

Amatya, D.M., Gilliam, J.W., Skaggs, R.W., Lebo, M.E. and Campbell, R.G. (1998) Effects of controlled drainage on forest water quality. *Journal of Environmental Quality* 27, 923–935.

Amatya, D.M., Gregory, J.D. and Skaggs, R.W. (2000) Effects of controlled drainage on storm event hydrology in a loblolly pine plantation. *Journal of the American Water Resources Association* 36, 175–190.

Amatya, D.M., Skaggs, R.W., Gilliam, J.W. and Hughes, J.H. (2003) Effects of orifice-weir outlet on hydrology and water quality of a drained forested watershed. *Southern Journal of Applied Forestry* 27, 130–142.

Andrews, L.M. (1993) Loblolly pine response to drainage and fertilization of hydric soils. MS thesis, Virginia Polytechnic Institute and State University, Blacksburg, Virginia.

Arnold, J.G., Srinivasan, R., Muttiah, R.S. and Williams, J.R. (1998) Large area hydrologic modeling and assessment part I: model development. *Journal of the American Water Resources Association* 34, 73–89.

Bouwer, H. and van Schilfgaarde, J. (1963) Simplified method of predicting fall of water table in drained land. *Transactions of the American Society of Agricultural Engineers* 6, 288–291.

Brightwell, C.S. (1973) *Slash Pine Plantations Respond to Water Control*. Woodland Research Note No. 26. Union Camp Corporation, Franklin, Virginia.

Campbell, R.G. (1976) Drainage of lower coastal plain soils. In: *Proceedings of the Sixth Southern Forest Workshop*. Southern Forest Soils Council, Charleston, South Carolina, pp. 17–27.

Campbell, R.G. and Hughes, J.H. (1991) Impact of forestry operations on pocosins and associated wetlands. *Wetlands* 11, 467–479.

Chescheir, G.M., Lebo, M.E., Amatya, D.M., Hughes, J., Gilliam, J.W., Skaggs, R.W. and Herrmann, R.B. (2003) *Hydrology and Water Quality of Forested Lands in Eastern North Carolina*. Research Bulletin No. 320. North Carolina Agricultural Research Service, Raleigh, North Carolina.

Domisch, T., Finer, L., Laiho, R., Karsisto, M. and Laine, J. (2000) Decomposition of Scots pine litter and the fate of released carbon in pristine and drained pine mires. *Soil Biology and Biochemistry* 32, 1571–1580.

Fox, T.R., Jokela, E.J. and Allen, H.L. (2007) The development of pine plantation silviculture in the southern United States. *Journal of Forestry* 105, 337–347.

Grace, J.M., Skaggs, R.W. and Chescheir, G.M. (2006) Hydrologic and water quality effects of thinning loblolly pine. *Transactions of the ASABE* 49, 645–654.

Graham, B.F. and Rebuck, A.L. (1958) The effect of drainage on the establishment and growth of pond pine (*Pinus serotina*). *Ecology* 39, 33–36.

Hillman, G.R. (1987) *Improving Wetlands for Forestry in Canada*. Information Report NOR-X-288. Northern Forestry Centre, Canadian Forestry Service, Edmonton, Alberta, Canada.

Hillman, G.R. and Roberts, J.J. (2006) Tamarack and black spruce growth on a boreal fen in central Alberta 9 years after drainage. *New Forests* 31, 225–243.

Holden, J., Evans, M.G., Burt, T.P. and Horton, M. (2006) Impact of land drainage on peatland hydrology. *Journal of Environmental Quality* 35, 1764–1778.

Hökkä, H. and Ojansuu, R. (2004) Height development of Scots pine on peatlands: describing change in site productivity with a site index model. *Canadian Journal of Forest Research–Revue Canadienne De Recherche Forestiere* 34, 1081–1092.

Jutras, S., Begin, J., Plamondon, A.P. and Hokka, H. (2007) Draining an unproductive black spruce peatland stand: 18-year post-treatment tree growth and stand productivity estimation. *Forestry Chronicle* 83, 723–732.

Kirkham, D. (1957) Theory of land drainage. In: Luthin, J.N. (ed.) *Drainage of Agricultural Lands*. American Society of Agronomy, Madison, Wisconsin, pp. 139–181.

Klawitter, R.A., Young, K.K. and Case, J.M. (1970) *Potential Site Index for Wet Pineland Soils of the Coastal Plain*. USDA Forest Service, State and Private Forestry, Southeast Area, Atlanta, Georgia.

Kyle, K.H., Andrews, L.J., Fox, T.R., Aust, W.M., Burger, J.A. and Hansen, G.H. (2005) Long-term effects of drainage, bedding, and fertilization on growth of loblolly pine (*Pinus taeda* L.) in the coastal plain of Virginia. *Southern Journal of Applied Forestry* 29, 205–214.

Laiho, R. (2006) Decomposition in peatlands: reconciling seemingly contrasting results on the impacts of lowered water levels. *Soil Biology and Biochemistry* 38, 2011–2024.

Laine, J., Vasander, H. and Laiho, R. (1995) Long-term effects of water level drawdown on the vegetation of drained pine mires in southern Finland. *Journal of Applied Ecology* 32, 785–802.

Langdon, O.G. (1976) *Study of Effects of Clearcutting, Drainage, and Bedding of Poorly Drained Sites on Water Table Levels and on Growth of Loblolly Pine Seedlings*. Progress Report. USDA Forest Service Government Printing Office, Washington, DC.

Lõhmus, A., Remm, L. and Rannap, R. (2015) Just a ditch in forest? Reconsidering draining in the context of sustainable forest management. *Bioscience* 65, 1066–1076.

Maki, T.E. (1960) Improving site quality by wetland drainage. In: Burns, P.Y. (ed.) *Southern Forest Soils, Proceedings of the Annual Forestry Symposium*. Louisiana State University Press, Baton Rouge, Louisiana, pp. 106–114.

McCarthy, E.J. and Skaggs, R.W. (1992) Simulation and evaluation of water management systems for a pine plantation watershed. *Southern Journal of Applied Forestry* 16, 48–56.

Miller, W.D. and Maki, T.E. (1957) Planting pines in pocosins. *Journal of Forestry* 55, 659–663.

Minkkinen, K. and Laine, J. (1998) Long-term effect of forest drainage on the peat carbon stores of pine mires in Finland. *Canadian Journal of Forest Research* 28, 1267–1275.

Minkkinen, K., Byrne, K.A. and Trettin, C. (2008) Climate impacts of peatland forestry. In: Strack, M. (ed.) *Peatlands and Climate Change*. International Peat Society, Jyväskylä, Finland. pp. 98–122.

Paavilainen, E. and Päivänen, J. (1995) *Peatland Forestry – Ecology and Principles*. Springer, New York.

Payandeh, B. (1973) Analysis of a forest drainage experiment in northern Ontario. I: Growth analysis. *Canadian Journal of Forest Research* 3, 387–398.

Peltomaa, R. (2007) Drainage of forests in Finland. *Irrigation and Drainage* 56, S151–S159.

Robinson, M. (1986) Changes in catchment runoff following drainage and afforestation. *Journal of Hydrology* 86, 71–84.

Skaggs, R.W. (2007) Criteria for calculating drain spacing and depth. *Transactions of the ASABE* 50, 1657–1662.

Skaggs, R.W. and Tang, Y.K. (1976) Saturated and unsaturated flow to parallel drains. *Journal of Irrigation and Drainage Division, ASCE* 102, 221–238.
Skaggs, R.W., Amatya, D., Chescheir, G.M., Blanton, C.D. and Gilliam, J.W. (2006) Effect of drainage and management practices on hydrology of pine plantation. In: Williams, T. (ed.) *Hydrology and Management of Forested Wetlands: Proceedings of the International Conference*. American Society of Agricultural and Biological Engineers, St Joseph, Michigan, pp. 3–14.
Skaggs, R.W., Chescheir, G.M., Fernandez, G.P., Amatya, D.M. and Diggs, J.D. (2011) Effects of land use of soil properties of drained coastal plains watersheds. *Transactions of ASABE* 54, 1357–1365.
Skaggs, R.W., Youssef, M.A. and Chescheir, G.M. (2012) DRAINMOD: model use, calibration, and validation. *Transactions of the ASABE* 55, 1509–1522.
Smedema, L.K., Vlotman, W.F. and Rycroft, D.W. (2005) *Modern Land Drainage: Planning, Design and Management of Agricultural Drainage Systems*, 2nd edn. Taylor & Francis, Delft, the Netherlands.
Socha, J. (2012) Long-term effect of wetland drainage on the productivity of Scots pine stands in Poland. *Forest Ecology and Management* 274, 172–180.
Stanek, W. (1977) Ontario clay belt peatlands – are they suitable for forest drainage? *Canadian Journal of Forest Research* 7, 656–665.
Sun, G., Riekerk, H. and Comerford, N.B. (1998) Modeling the forest hydrology of wetland-upland ecosystems in Florida. *Journal of the American Water Resources Association* 34, 827–841.
Terry, T.A. and Campbell, R.G. (1981) Soil management considerations in intensive forest management. In: *Proceedings of the Symposium on Engineering Systems for Forest Regeneration*. American Society of Agricultural Engineers, St Joseph, Michigan, pp. 98–106.
Terry, T.A. and Hughes, J.H. (1975) The effects of intensive management on planted loblolly pine (*Pinus taeda* L.) growth on poorly drained soils of the Atlantic Coastal Plain. In: Bernier, B. and Winget, C.H. (eds) *Forest Soils and Forest Land Management, Proceedings of the Fourth North American Forest Soils Conference*. Les Presses de l'Universite Laval, Quebec, Canada, pp. 351–377.
Terry, T.A. and Hughes, J.H. (1978) Drainage of excess water: why and how? In: Balmer, W.E. (ed.) *Proceedings of the Soil Moisture–Site Productivity Symposium*. USDA Forest Service, Southeastern Area, State and Private Forestry, Atlanta, Georgia, pp. 148–166.
Tian, S., Youssef, M.A., Skaggs, R.W., Amatya, D.M. and Chescheir, G.M. (2012) DRAINMOD-FOREST: integrated modeling of hydrology, soil carbon and nitrogen dynamics, and plant growth for drained forests. *Journal of Environmental Quality* 41, 764–782.
Tian, S., Youssef, M.A., Amatya, D.M. and Vance, E.D. (2014) Global sensitivity analysis of DRAINMOD-FOREST, an integrated forest ecosystem model. *Hydrological Processes* 28, 4389–4410.
Tomppo, E. (1999) Forest resources in Finnish peatlands in 1951–1994. *International Peat Journal* 9, 38–44.
Trottier, F. (1986) Accroissement de certains peuplements forestiers attribuable à la construction de cours d'eau artificiels. In: *Textes des Conférences Présentée au Colloque sur le Drainage Forestier*. Ordre des Ingenieurs Forestiers du Québec, Québec, Canada, pp. 66–84.
Trottier, F. (1991) Draining wooded peatlands: expected growth gains. In: Jeglum, J.K. and Overend, R.P. (eds) *Proceedings of the International Symposium on Peat and Peatlands Diversification and Innovation, Quebec City, Quebec, Canada*. Peatland Forestry, Vol. 1. Canadian Society for Peat and Peatlands, Dartmouth, Nova Scotia and Echo Bay, Ontario, Canada, pp. 6–10.
Walker, L.C., Daniels, J.M. and Daniels, R.L. (1961) Flooding and drainage effects on slash pine loblolly pine seedlings. Faculty Publication Paper No. 363. Available at: http://scholarworks.sfasu.edu/forestry/363 (accessed 3 April 2016).
White, E.H. and Pritchett, W.L. (1970) *Water Table Control and Fertilization for Pine Production in the Flatwoods*. Technical Bulletin No. 743. Agricultural Experiment Station. University of Florida, Gainesville, Florida.

Plate 1. Fire-induced erosion fills a weir at the Croppers Creek hydrological project in Victoria, Australia.

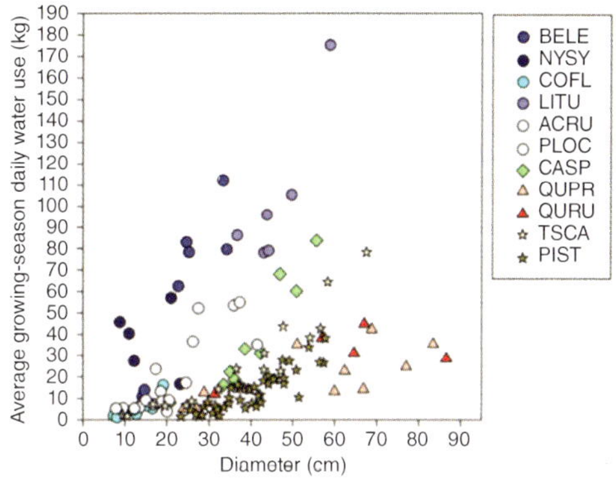

Plate 2. Observed daily water use estimated from sap flux density in trees of varying species (BELE, *Betula lenta*; NYSY, *Nyssa sylvatica*; COFL, *Cornus florida*; LITU, *Liriodendron tulipifera*; ACRU, *Acer rubrum*; PLOC, *Platnus occidentalis*; CASP, *Carya* spp.; QUPR, *Quercus prinus*; QURU, *Quercus rubra*; TSCA, *Tsuga canadensis*; PIST, *Pinus strobus*) in reference watersheds at the Coweeta Hydrological Laboratory in the south-eastern USA (except PIST). Symbols: circles are species with diffuse porous xylem anatomy; diamonds are species with semi-ring porous xylem anatomy; triangles are species with ring porous xylem anatomy; stars are species with tracheid xylem anatomy. (From Vose *et al*., 2011; *Ecohydrology* with permission.)

Plate 3. Forest encroachment due to land abandonment, Clarée Valley, Hautes Alpes, southern French Alps. (Photo: Carmen de Jong, 2008.)

Plate 4. Reduction of permeability and overall impacts of ski runs on forests, Queyras, Hautes Alpes, France, 2012. (Photo: Carmen de Jong, 2012.)

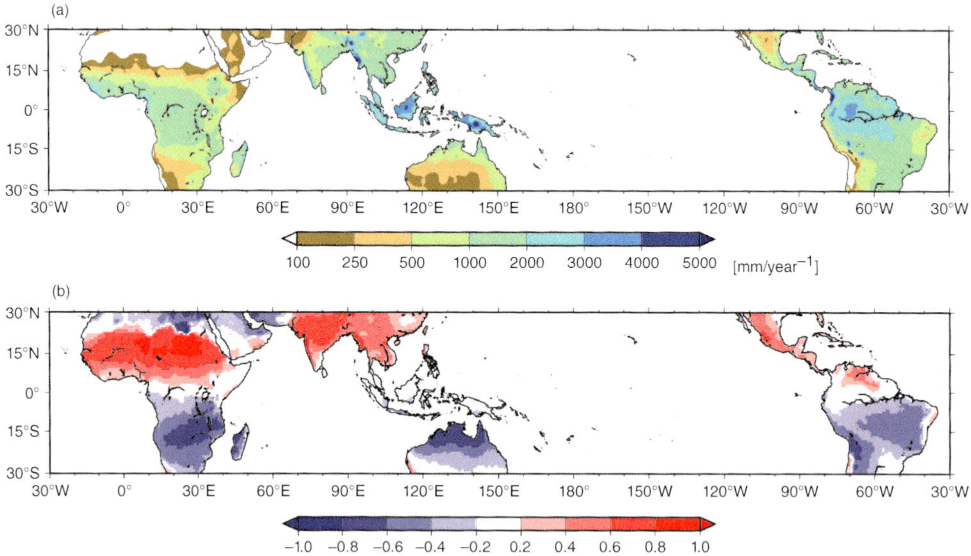

Plate 5. Pan-tropic maps of (a) annual mean precipitation and (b) precipitation seasonality represented by the ratio of difference in precipitation between June–August and December–February to annual precipitation. (Constructed using the National Oceanic and Atmospheric Association's (NOAA) PRECipitation REConstruction over Land (PREC/L) with values averaged over 1979–2012.)

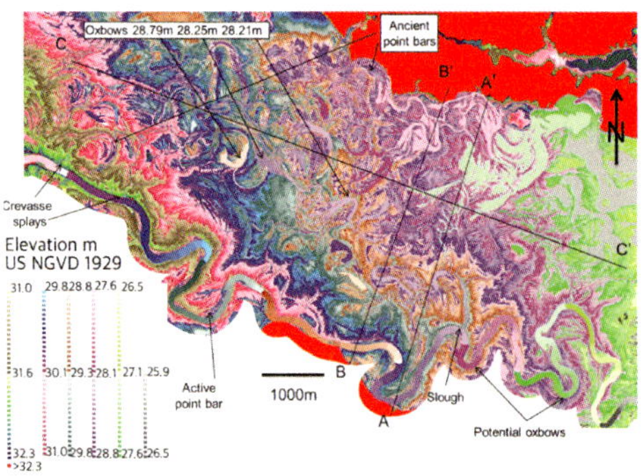

Plate 6. LiDAR DEM (10 ft ×10 ft, 3.05 m × 3.05 m) of a 11 km portion of the Congaree River floodplain in the Congaree Swamp National Park, Richland County, South Carolina, USA. Note areas of single colour are covered with water at the elevation depicted by that colour (DEM, digital elevation model). (LiDAR data collected and available from the South Carolina Department of Natural Resources; SCDNR, 2015b.)

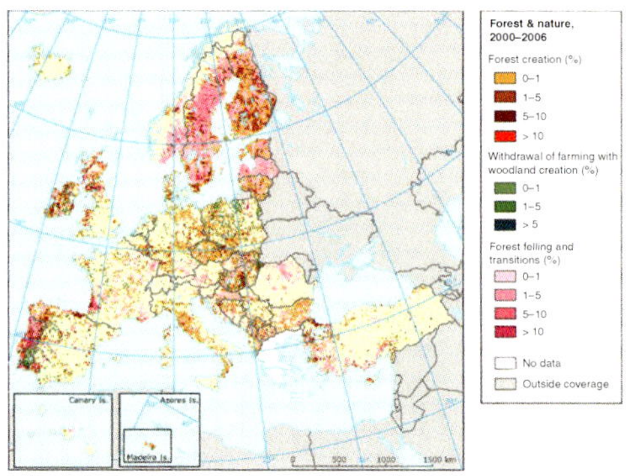

Plate 7. Forest cover change analysis for Europe between 2000 and 2006 using CORINE (Co-ORdinated INformation on the Environment) land cover data. The map shows the distribution and the intensity of selected land cover flows in forest area. (Produced with copyright from http://www.eea.europa.eu.)

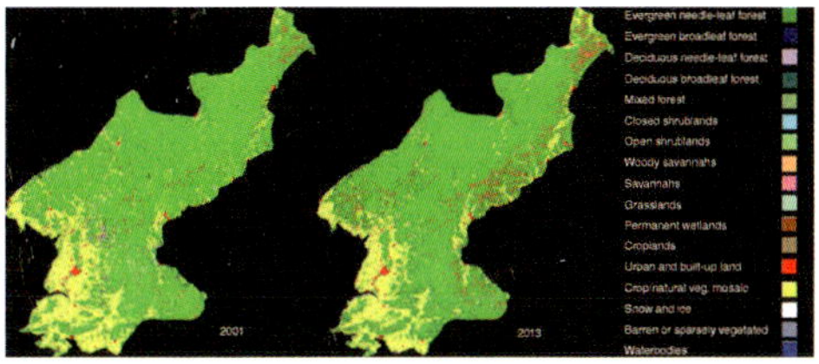

Plate 8. Land cover (including forest cover) change analysis maps (2001 and 2013) of North Korea completed using MODIS (Moderate Resolution Imaging Spectroradiometer) land cover product data. (From Kim *et al.*, 2015, 2016.)

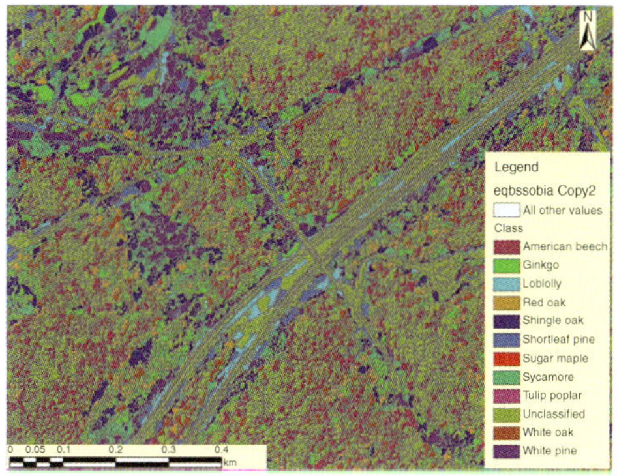

Plate 9. OBIA (object-based image analysis)-based forest tree speciation using high-resolution orthoimagery and VBA (Visual Basic for Application) coding in ArcObjects platform of ArcGIS 9.3.

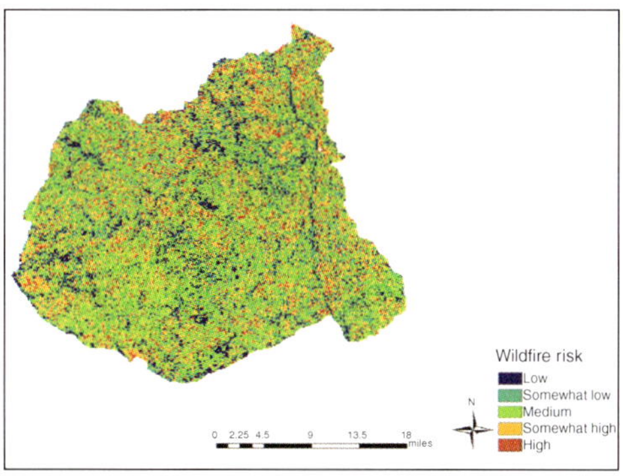

Plate 10. Wildfire susceptibility/vulnerability map of Sumter National Forest, South Carolina, USA.

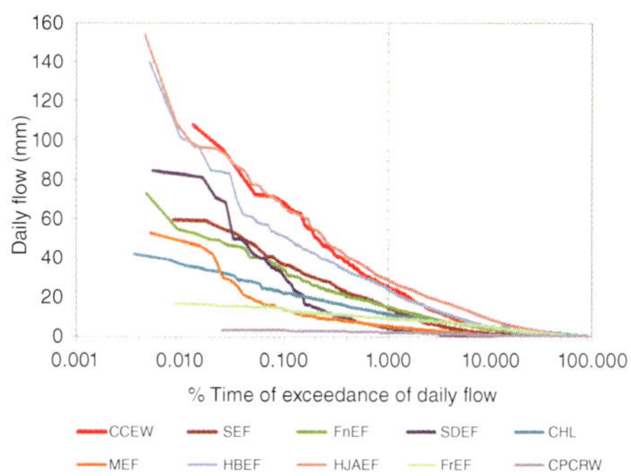

Plate 11. Daily flow duration curves for the entire range of percentage exceedance (CCEW, Caspar Creek Experimental Watershed; SEF, Santee Experimental Forest; FnEF, Fernow Experimental Forest; SDEF, San Dimas Experimental Forest; CHL, Coweeta Hydrologic Laboratory; MEF, Marcell Experimental Forest; HBEF, Hubbard Brook Experimental Forest; HJAEF, H.J. Andrews Experimental Forest; FrEF, Fraser Experimental Forest; CPCRW, Caribou-Poker Creek Research Watershed).

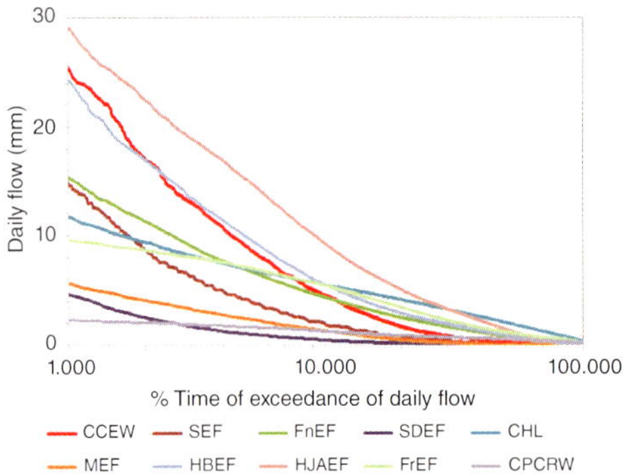

Plate 12. Daily flow duration curves for the range of percentage exceedance up to >1% only (CCEW, Caspar Creek Experimental Watershed; SEF, Santee Experimental Forest; FnEF, Fernow Experimental Forest; SDEF, San Dimas Experimental Forest; CHL, Coweeta Hydrologic Laboratory; MEF, Marcell Experimental Forest; HBEF, Hubbard Brook Experimental Forest; HJAEF, H.J. Andrews Experimental Forest; FrEF, Fraser Experimental Forest; CPCRW, Caribou-Poker Creek Research Watershed).

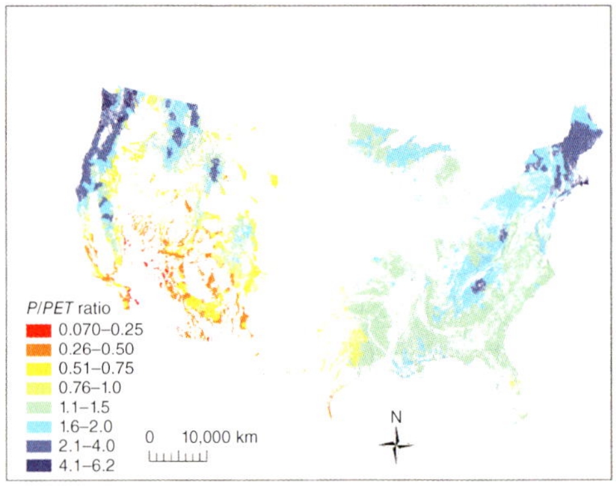

Plate 13. The ratio of long-term mean annual precipitation (*P*) to potential evapotranspiration (*PET*) over the conterminous USA. Potential evapotranspiration was calculated using the Hamon method.

Plate 14. The Bakhta River in the northern taiga zone. (Photograph: I.M. Danilin.)

Plate 15. Dark needle taiga on the western slopes of Yenisei Ridge in the Southern Taiga zone. (Photograph: A.S. Shishikin.)

9 Hydrological Modelling in Forested Systems

H.E. Golden[1], G.R. Evenson[2], S. Tian[3], D.M. Amatya[4] and G. Sun[5]
[1]US Environmental Protection Agency, Cincinnati, Ohio, USA; [2]Oak Ridge Institute of Science and Education, c/o US Environmental Protection Agency, Cincinnati, Ohio, USA; [3]North Carolina State University, Raleigh, North Carolina, USA; [4]USDA Forest Service, Cordesville, South Carolina, USA; [4]USDA Forest Service, Raleigh, North Carolina, USA

9.1 Introduction

Characterizing and quantifying interactions among components of the forest hydrological cycle is complex and usually requires a combination of field monitoring and modelling approaches (Weiler and McDonnell, 2004; National Research Council, 2008). Models are important tools for testing hypotheses, understanding hydrological processes and synthesizing experimental data (Sun *et al.*, 1998, 2011). A well-calibrated model that incorporates the general principles of forest hydrology can supplement field measurements (e.g. Hydrograph Separation Program, HYSEP; Sloto and Crouse, 1996; Barlow *et al.*, 2015) and, in turn, these measurements can provide data to improve a model and its performance. Forest hydrology models can also project water quantity and quality in catchments with limited recorded measurements, such as stream discharge (Sivapalan, 2003) and water balances at broad spatial scales (Sun *et al.*, 2011). Many forest hydrology models can also quantify forest biogeochemical cycling as well as surface water quality in catchments (DeWalle, 2003; Nelitz *et al.*, 2013).

There are increasing demands for improved hydrological models that project the hydrological responses to forest management practices (National Research Council, 2008; Amatya *et al.*, 2011; Vose *et al.*, 2011). This requires a suite of approaches that incorporate decades of research on the processes regulating transfers of water in forests and hydrological responses to forest management (Jones *et al.*, 2009; Buttle, 2011). Forest hydrology models are a necessity to project beyond current hydrological conditions from young stands to forests with full canopied catchments (e.g. quantifying the effects of forest site preparation, forest growth and silvicultural techniques on the hydrological cycle). Many forest hydrology models can also be applied to query how management, climate change and/or other land cover changes together affect the forest hydrological cycle and link physical and hydrological processes at the stand scale to that of the whole catchment. This involves projecting changes in components of the forest or forest catchment's water balances, including runoff, evapotranspiration, snow accumulation/melt, melting permafrost, and the cumulative effects of these changes on stream, river and lake processes (Beckers *et al.*, 2009; Sun *et al.*, 2011). Projecting such shifts in the forest hydrological cycle requires numerical modelling methods

*Corresponding author; e-mail: golden.heather@epamail.epa.gov

 Forest Hydrology: Processes, Management and Assessment (eds D.M. Amatya, T.M. Williams, L. Bren and C. de Jong)

because of their ability to conduct time-stepped simulations of specific hydrological processes and scale data to broader spatial extents using physically or process-based approaches (National Research Council, 2007).

This chapter provides a brief overview of forest hydrology modelling approaches for answering important global research and management questions. Many hundreds of hydrological models have been applied globally across multiple decades to represent and predict forest hydrological processes (Beckers *et al.*, 2009; Nelitz *et al.*, 2013; Amatya *et al.*, 2014). The focus of this chapter is on process-based models and approaches, specifically 'forest hydrology models'; that is, physically based simulation tools that quantify compartments of the forest hydrological cycle. Physically based models can be considered those that describe the conservation of mass, momentum and/or energy (Beckers *et al.*, 2009). While we provide minimal emphasis on empirical modelling methods, these approaches can be embedded within physically based models. For example, runoff from a parcel of land may be calculated using the USDA Natural Resources Conservation Service curve number method, an empirical approach for estimating rainfall–runoff responses based on combinations of soil, land cover and slope characteristics of a land parcel. While some modelling approaches we discuss are appropriate at the plot or stand scale, many are considered within the context of catchments. We consider the catchment scale to include multiple drainage areas ranging across various orders of magnitude (e.g. 0.1 km^2 to 1000 km^2), based on Golden *et al.* (2014), which is also consistent with Wei and Zhang (2011). Temporal scales of each model are associated with the time step the modeller selects to solve the governing equations within the model, typically hourly for streamflow hydrograph predictions, daily or monthly for large-scale ecosystem models and annually for the transient groundwater flow models.

9.2 Model Functionality and Complexity

Forest hydrology models can range in functionality and complexity. Each model's functionality results, in part, from the hydrological processes represented in the model, the mathematical equations expressed in these processes and how the spatial extent of the model domain is discretized (Beckers *et al.*, 2009). Most models simulate, at minimum, a basic water balance that includes moisture inputs (e.g. rainfall, snow and/or snowmelt) and outputs via evapotranspiration including canopy evaporation and runoff as a combination of surface and subsurface flows (Fig. 9.1). How the water balance is calculated varies widely based upon the complexity and spatial/temporal scale of the model. Simulated outputs are diverse across models as well, but generally include peak flow, low flow, total streamflow/water yield, evapotranspiration and/or changes in soil moisture over time.

9.2.1 Forest stand and soil moisture functions

Representation of forest hydrological processes is also diverse across different models. Typically, many models consider the forest ecosystem as mature (e.g. static) while other selected models (e.g. DRAINMOD-FOREST; Tian *et al.*, 2012) explicitly simulate forest physiological and phenological dynamics and how these dynamics affect a forest stand's water balance. Some models simulate the interactions among the soil, vegetation and atmosphere that affect the soil moisture dynamics and water-use efficiency of vegetation in forests (e.g. PnET-N-DNDC simulations of N_2O and NO emissions from forest soils; Li *et al.*, 2000; Stange *et al.*, 2000).

The majority of forest hydrology models require a numerical approach for estimating soil moisture dynamics, which are a key component of regulating evapotranspiration rates (often estimated by empirical methods such as the Priestly–Taylor, Hamon or Penman–Monteith approach for potential evapotranspiration (as the upper limit of evapotranspiration)) and rainfall–runoff processes at the catchment scale. Soil moisture conditions in forests reflect the water balances that are controlled by precipitation inputs (e.g. direct rainfall, throughfall, snow/snowmelt), evapotranspiration, the forest's soil water-storage capacity, other physical soil properties such as effective porosity, bulk density and

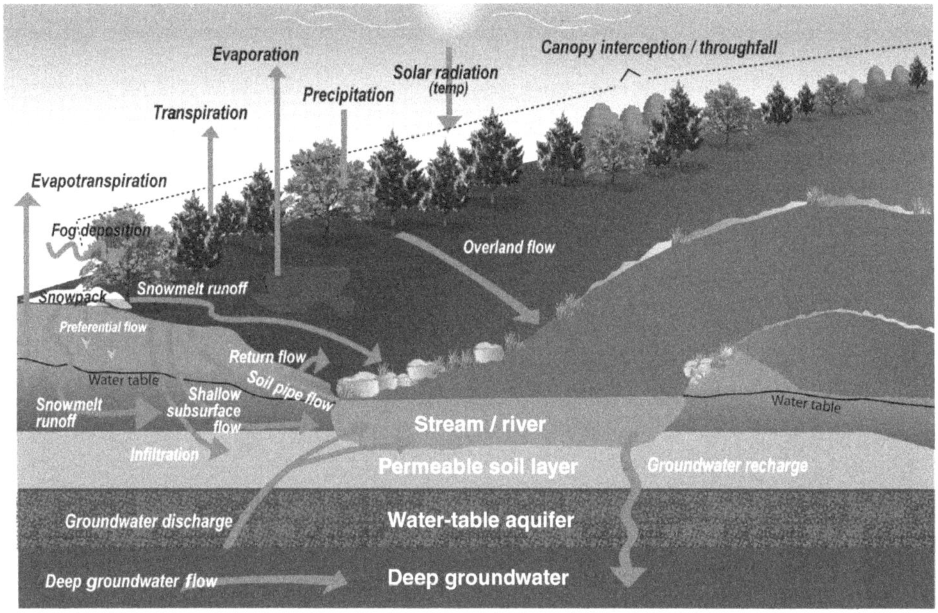

Fig. 9.1. Processes that can be modelled with forest hydrology simulation tools.

saturated hydraulic conductivity, and water table dynamics. Evaporation from the canopy interception and evapotranspiration rates are also controlled by leaf area index (LAI), canopy storage capacity and stomatal conductance, in addition to the soil moisture and climatic parameters. These time-varying soil moisture conditions and evapotranspiration rates are typically represented by a series of partial differential equations for different soil layers with variables (e.g. precipitation, canopy and soil/litter evaporation, evapotranspiration, flow inputs and outputs) that are calculated at the same time step.

9.2.2 Rainfall–runoff functions

The initiation of overland flow or subsurface flow in models that include rainfall–runoff dynamics (i.e. 'catchment models') occurs when a threshold soil moisture level, such as soil field capacity, is reached. Each model calculates a threshold value and runoff-generating processes differently. Forest hydrology models will typically represent one (but sometimes more than one) of these runoff-generating processes that might include variable source area (VSA) dynamics (e.g. TOPMODEL; Beven and Kirkby, 1979), the USDA Natural Resources Conservation Service curve number method (e.g. Soil and Water Assessment Tool; Neitsch *et al.*, 2011), Green–Ampt infiltration processes (e.g. HSPF), Hooghoudt's equation for shallow water table and drainage rates (e.g. DRAINMOD-FOREST; Tian *et al.*, 2012), soil moisture response function (e.g. VELMA; Abdelnour *et al.*, 2011) or soil moisture balance (WaSSI; Sun *et al.*, 2011) approaches, depending upon the temporal scale of simulation. Additional details on these processes are covered in Chapters 4, 6 and 8 (this volume).

For forest hydrology models that explicitly simulate rainfall–runoff processes, once runoff is initiated, several primary catchment-scale flowpath types could be represented in the model of interest. Surface runoff will likely include infiltration excess runoff (Horton, 1933), saturation-excess overland flow (Dunne and Black, 1970), including VSA dynamics, or a combination of both overland flow types. Subsurface stormflow (Hursh and Brater, 1941), including preferential flows, may also be implemented in the model's water balance routine, as well as return flows (i.e. flow that travels through the shallow subsurface before reissuing to the land surface). Surface depressional storage capacity, Manning's overland surface runoff coefficient, land slope, and other landscape and surface/litter vegetation characteristics are then used to route the excess rainfall after soil saturation to the nearest stream. Depending on the model's structure, deep groundwater flow, which produces baseflow in the study catchment's stream network, is calculated as part of the water balance (i.e. the surplus from water percolated to the deep bedrock and groundwater storage) or, in the cases where groundwater models or coupled surface–subsurface models are applied, the groundwater flow equation (i.e. the mathematical representation of groundwater flow through an aquifer) is solved explicitly using various soil hydraulic properties, particularly hydraulic conductivity. Several forest hydrology models calculate channel flow routing times to the catchment outlet once surface and subsurface runoff reaches the stream (e.g. SWAT). This is estimated using variables such as channel water levels, velocities, channel geometry and Manning's roughness coefficient.

9.2.3 Parameterization of functions

Depending upon the study or management objectives, complexity of the model and the forest composition, the number and breadth of parameters required to simulate key processes may vary substantially. In addition to several previously mentioned parameters related to soil and runoff, parameters related to simulation of forest evapotranspiration may include rooting depth and distribution, LAI, canopy density, canopy structure and interception capacity, stomatal or canopy conductance, and other biophysical characteristics including those that represent the understorey type/species (additional details are included in Chapter 3, this volume). Anthropogenic processes may also be incorporated into the functionality of many forest hydrological models. Some models can project variations in hydrological processes in response to climate change, management scenarios, wildfire, insect and disease outbreaks, and shifts in land cover/land use. For example, studies have applied modelling approaches to project the effects of forest harvesting and management on peak flows in British Columbia (Whitaker *et al.*, 2003; Schnorbus and Alila, 2004; Thyer *et al.*, 2004), portions of South America (Bathurst *et al.*, 2011; Birkinshaw *et al.*, 2011), China (Sun *et al.*, 2006), the north-western USA (Stednick, 1996, 2008; Schnorbus and Alila, 2004; Abdelnour *et al.*, 2011), among others. In a recent synthesis study, Amatya *et al.* (2014) outlined eight criteria for an ideal forest hydrology model that can describe impacts of forest fertilization on southern US forest landscapes. Model functionality may focus more strongly on glacial, tundra and permafrost processes, such as glacier melt (e.g. PREVAH; Viviroli *et al.*, 2009) and permafrost (e.g. Variable Infiltration Capacity (VIC) model; Liang *et al.*, 1994), and responses of these processes to anthropogenic changes. Further, some models have been developed and tested in mountainous systems where snowfall and snowmelt dynamics dominate (e.g. TOPMODEL; Hornberger *et al.*, 1994; Buytaert and Beven, 2011). Additional models better represent hydrological processes of low-gradient forest systems and/or where humid subtropical environments dominate (e.g. FLATWOODS, Sun *et al.*, 1998; DRAINMOD, Amatya and Skaggs, 2001; DRAINMOD-FOREST, Tian *et al.*, 2012).

9.2.4 Balancing model functionality and complexity

Model functionality and complexity go hand in hand: typically, the greater the number of functions the model simulates, the more complex the model. Forest hydrology models can vary considerably in complexity from simple empirical models (not discussed here) to process-based models that cover a range of low (ForHYM; Arp and Yin, 1992) to medium (VELMA; Abdelnour *et al.*, 2011) to highly complex (e.g. HydroGeoSphere; Brunner and Simmons, 2012) hydrological representations; that is, from simple bucket-type models to models that implement multiple water transport processes. Model complexity can also vary with spatial scale: highly complex and computationally intensive models often function best at finer spatial scales; as the spatial scale expands, resolution of the modelled system necessarily needs to coarsen to decrease computational demands. Forest hydrology process-based studies coupled with modelling approaches have most commonly been approached from a small catchment scale (National Research Council, 2008) using paired catchment approaches (von Stackelberg *et al.*, 2007) starting as early as 1909 to 1928 in North America (Bates and Henry, 1928). However, more recent studies have expanded their spatial scales towards the scale of management (e.g. large catchments, regions, nations, globally) and used generalizing principles derived from finer-scale studies. As such, based on the model structure, the spatial scale of interest and the management or research question, models can be discretized in different ways (e.g. by hydrological response units, sub-basins, finite difference grids) and parameters and processes can be spatially characterized as lumped (parameters and processes are generalized across space), semi-distributed (areas of the catchment are 'lumped' based on different physical characteristics such as land cover and soils) or distributed (parameters and processes vary spatially across the modelled system) (Kampf and Burges, 2007; Arnold *et al.*, 2015). Forest hydrology models can vary temporally, with some operating on a continuous time step (e.g. daily, monthly and/or annually).

9.3 Model Selection

A forest hydrology model is a simplification of reality. This is an important consideration when selecting the appropriate model for the forest hydrological management and/or research question. The current state-of-the-science remains limited on insights to choosing the most appropriate spatial resolution to represent hydrological processes of a specific system. Of utmost importance is developing a conceptual hydrological model of the study area based on spatial data (e.g. remote sensing (LiDAR), GIS), monitoring (e.g. streamflow, snowpack depths, temperature and humidity data, evapotranspiration, well- and piezometer-level measurements), and past modelling efforts and professional knowledge to: (i) determine the most important hydrological processes of the study area; (ii) select a model that can simulate these dominant processes; and (iii) determine whether the simplifying hydrological assumptions in the chosen model (e.g. spatial discretization and resolution) are valid for the system. For example, if a catchment's soils exhibit low infiltration capacity or precipitation rates exceed infiltration rates, a Hortonian rainfall–runoff model might be appropriate (Downer *et al.*, 2002). However, Hortonian flows rarely occur in fully forested conditions. Moreover, a lumped parameter model might be appropriate (compared with a spatially explicit model) where spatial heterogeneity is low, the spatial scale of the study area is broad (e.g. region, national), or a combination of the two. Model selection must also consider the management or research questions, the hydrological processes important to those questions and what future projections need to be simulated, such as climate change or forest management scenarios that vary in complexity.

Practical considerations for choosing a forest hydrology model include input data needs and parameter availability, computational time and cost–benefits of model complexity. For example, most catchment-based forest hydrology models require an accurate digital elevation model (DEM) and stream network layers as base data, in addition to measurements of hydrological processes (e.g. precipitation, temperature and relative humidity from meteorological station or modelled data); evapotranspiration (e.g. using water budget measurements, water vapour transfer methods, remote sensing, etc.); snowpack depths; water table variations); and downstream streamflow measurements (e.g. a stream gauge). Depending on the study question, water level data from groundwater wells, piezometers and other surface water features in the catchment (e.g. wetlands, lakes, dams) to better parameterize the model and quantify the full water balance are important. Further, model set-up, implementation and spin-up (the period taken for the model to equilibrate under the forcing, typically precipitation and temperature conditions) times – as well as the skill level required to execute the model – all increase with model complexity. Therefore, a consideration between the balance of benefits associated with minimizing model uncertainty versus the increased computational intensity costs associated with added model complexity is imperative (Freeze *et al.*, 1990).

9.4 Model Diagnostics and Evaluation

In order to determine whether a forest hydrology model is characterizing the system appropriately, model evaluation needs to be conducted. In the most general sense, model calibration is a process by which model parameters are adjusted within a predefined acceptable range so that the simulated model output matches an observed set of data. Traditionally with catchment models, the observed data are stream gauge records but can also include spatially distributed data on water table depths, soil moisture, evapotranspiration and other water balance components. In forest hydrology models that incorporate plant growth components, it may also be appropriate to use measured LAI, total biomass and other variables estimated by remote sensing for validation of productivity factors in addition to the hydrological variables. For parameter estimation programs (e.g. PEST, Doherty and Johnston, 2003; OSTRICH, Matott, 2005) an objective function (optimization) or, more accurately, multiple objective functions should be selected to generate the best-fit parameter set to match simulated results to observed data (Boyle *et al.*, 2003). A multi-objective framework reduces the problems associated with calibrating to local objective function minima. It also avoids subjectivity and information loss in model acceptability criteria by simultaneously minimizing observed and simulated differences of multiple functions (Doherty and Johnston, 2003; Flerchinger *et al.*, 2012). Most recently, Arnold *et al.* (2015) recommended a diagnostic approach that looks at signature patterns of behaviour in the model outputs to determine which processes, and thus parameters, need further adjustment during calibration. In a companion study, Malone *et al.* (2015) developed parameterization guidelines and considerations for hydrological models. Parameters are often fitted using measured data or calibrated within reasonable ranges, as determined by the system and/or literature values when data are lacking. However, careful consideration of equifinality (Beven, 1993) – which describes the process of arriving at the same simulated model output using a variety of different model parameter sets or structures, without knowing which one might be closest to 'reality' – is important.

Validation of the model traditionally suggests the successful testing of simulated outputs against observed data using an input data set different from the calibration data set. A split-sample approach (Klemes, 1986) is one popular example whereby calibration and validation are conducted during a sequential set of years: one continuous set is used for calibration, the other for validation. Such an approach to validation can be termed 'conditional validation', suggesting that the model has been validated using the calibrated model and separate data but can be updated with data that measure future conditions (e.g. changes in catchment factors or new state-of-the-science information) (Young, 2001).

9.4.1 Uncertainty and sensitivity analysis

Uncertainties in forest hydrology models must be accounted for in some capacity. Model uncertainties can take the form of parameter uncertainties, input data uncertainties, process uncertainties and predictive uncertainties, among others. Uncertainty analysis is conducted to quantify simulation output uncertainty by propagating uncertainties throughout the model and generating a probabilistic distribution of simulated outputs. How to handle uncertainties in hydrological modelling is a debate that has continued for decades (Matott *et al.*, 2009). Beven and Young (2013) suggest that uncertainties in hydrological models can be aleatory (irreducible) or epistemic (reducible) in nature. Aleatory uncertainties are random and can be treated probabilistically in the model, while epistemic errors are associated with current lack of knowledge of processes operating within the system. Whichever form of uncertainties exists in the model it is appropriate to detail the assumptions underlying these uncertainties and quantify them, where appropriate and feasible. Sensitivity analysis is one way of estimating the output uncertainties caused by changes in values of model parameters. Sensitivity analysis can determine which parameters assert the most quantifiable control over model outputs; that is, the analysis can quantify which model parameters produce a disproportionate change in simulated outputs based on a relatively small change in a parameter's value. For example,

Tian *et al.* (2014) and Dai *et al.* (2010) provide recent insights on global sensitivity analyses using the forest hydrological model DRAINMOD-FOREST and MIKE SHE, respectively.

9.5 Example Forest Hydrology Models

9.5.1 Watershed and plot models

PnET-BGC (Gbondo-Tugbawa *et al.*, 2001) and CENTURY (Parton *et al.*, 1993; Parton, 1996) are plot-scale models that simulate forest hydrological processes across a forest stand. PnET is a lumped-parameter, monthly or daily time-stepped and stand-level model that quantifies carbon and water dynamics in mature forests. Hydrological processes simulated by the model include canopy interception plant transpiration, macropore flow, lateral flow and deep percolation to the aquifer. CENTURY is a plot-scale terrestrial biogeochemical model that operates at a monthly time step (Parton *et al.*, 1993; Parton, 1996). The model is composed of linked sub-models representing forest production, grassland and crop production, soil organic matter and a water budget. The simplified water budget sub-model simulates monthly evaporation, transpiration, soil water content, snow water content and saturated flow between soil layers.

Catchment rainfall–runoff models refer to physically based models that simulate the forest hydrology water balance and predominant rainfall–runoff processes, including routing to a surface water system. These models use topographically defined catchments as boundaries and simulate surface and shallow subsurface processes. Unlike groundwater models, rainfall–runoff models quantify groundwater as part of the catchment water balance; the groundwater flow equation is not solved explicitly. Therefore, the deep groundwater system is considered a hydrological 'sink'. Several examples of catchment rainfall–runoff models that can be applied for forest hydrology include ForHyM (Arp and Yin, 1992; Meng *et al.*, 1995), TOPMODEL (Beven and Kirkby 1979), i-Tree Hydro (Wang *et al.*, 2008), VELMA (Abdelnour *et al.*, 2011, 2013), APEX (Williams and Izaurralde, 2005; Gassman *et al.*, 2007), PRMS (Leavesly *et al.*, 1983, 2005), DHSVM (Wigmosta *et al.*, 1994, 2002), BROOK90 (Federer *et al.*, 2003), VIC (Liang *et al.*, 1994, 1996) and INCA (Wade *et al.*, 2002) (Table 9.1).

ForHyM (Arp and Yin, 1992; Meng *et al.*, 1995) is a one-dimensional, empirical, lumped watershed hydrology model that operates at a daily time step and has been applied across multiple physiographical settings. The model includes a single vegetation layer and two soil layers. Hydrological processes simulated by the model include interception, throughfall, evapotranspiration, infiltration, vertical unsaturated water movement, streamflow, surface runoff, interflow, groundwater flow and snowmelt. TOPMODEL is a semi-physically based flexible mass balance modelling tool that simulates catchment-scale rainfall–runoff (Beven and Kirkby, 1979) and is particularly robust in forested catchments with shallow soils. Flow routing in TOPMODEL is driven by VSA dynamics and includes both saturated- and infiltration-excess overland flow. i-Tree Hydro (previously called UFORE-Hydro; Wang *et al.*, 2008) is a physically based, semi-distributed urban forestry hydrological model that simulates runoff volume and quality across different urban land covers. Simulations in iTree Hydro are at a daily time step and can operate at multiple watershed or plot (i.e. city, parcel) scales. A user can simulate the effects of various urban impervious and vegetation cover scenarios on the urban forest water balance, including interception, evapotranspiration, infiltration and runoff. The Visualizing Ecosystems for Land Management Assessment (VELMA) model is a spatially distributed ecohydrological model initially developed for forested catchments, particularly in the Pacific Northwest of the USA (Abdelnour *et al.*, 2011, 2013). VELMA can simulate multiple parts of the forest hydrological cycle (e.g. daily infiltration and redistribution, evapotranspiration, surface and subsurface runoff) using a four-layer soil column structure. The APEX model (Williams and Izaurralde, 2005; Gassman *et al.*, 2007) was developed to evaluate land management impacts of hydrology, water and soil quality, and vegetation growth and competition in upland watersheds. The forestry version includes rainfall interception by canopy/litter, silvicultural practices, and subsurface flow that includes deep percolation and lateral seepage using storage routing and pipeflow equations (Saleh *et al.*, 2004; Williams and Izaurralde, 2005).

Table 9.1. Example catchment models for forest hydrology applications.

Model	Hydrological approach	Time step(s)	Spatial scale(s)	Level of complexity	Appropriate regions of application	Model files publically accessible	Availability of online user manual and website
iTree-Hydro	Six main routines for rainfall–runoff processes: interception, impervious surface, soils, evaporation and transpiration, routing, and pollution. Uses time–area delay function or one-parameter diffusion-based exponential function for constructing downstream hydrograph	Daily	Multi-scale catchments and plots (i.e. city or parcel)	Medium	Multiple – can be applied to watersheds with different rainfall–runoff mechanisms; recent cold region module development (version 2; Yang *et al*., 2011)	Yes, by request	Yes; http://www.itreetools.org/hydro/
PnET (all)	Lumped-parameterized one-dimensional water balance model from canopy to soil	Daily to monthly	Plot to regional	Low	All forest ecosystems, both upland and lowland	Yes	Yes; http://www.pnet.sr.unh.edu/
CENTURY	Simplified water balance incorporating evapotranspiration, soil water content, saturated flow	Monthly	Plot	Low	Temperate and tropical forests	Yes	Yes; https://www.nrel.colostate.edu/projects/century/
ForHyM	One-dimensional process-based water balance model that also embodies some general empirical relationships for multiple model layers (forest canopy, snowpack, forest floor, soil and subsoil)	Daily, weekly	Catchment	Medium	Typically applied for northern forested watersheds with one or multiple biomes	Yes	Yes; http://watershed.for.unb.ca/research/forhym/
TOPMODEL	Semi-distributed rainfall–runoff model; variable source area dynamics but simulates both saturated- and infiltration-excess overland flows; assumes water table follows topography	Daily	Multi-scale catchments	Medium	Multiple – typically best in systems with moderate to steep topography and shallow soils	Yes, as an ArcGIS extension, R Statistical Package code or the original code	Yes; in the dynamic TOPMODEL R statistical version: http://cran.r-project.org/web/packages/dynatopmodel/index.html

VELMA	Spatially distributed ecohydrological model; simulates daily infiltration, evapotranspiration and surface/ subsurface runoff through four soil layers	Daily	Multi-scale catchments	Medium	Multiple – developed in small forested catchments of Pacific Northwest of USA but applied/tested elsewhere	No	Coming soon
APEX	Curve number method (five options) and Green–Ampt infiltration (four options), subsurface drainage, and five options for potential evapotranspiration for evapotranspiration estimates	Hourly, daily, monthly or annual	Field; catchment; grid-based	Medium	Upland agricultural and forested fields or watersheds	Yes	Yes; http://epicapex.tamu.edu/apex/
PRMS	Semi-distributed processed-based rainfall–runoff model for multiple spatial and temporal scales	Daily to centuries	Multi-scale catchments	Medium	Multiple	Yes	Yes; http://wwwbrr.cr.usgs.gov/prms/
DHSVM	Saturation excess infiltration; Darcy's law for unsaturated and kinematic for saturated subsurface	Sub-daily to annual	Catchment	Medium	Mountainous watersheds in Pacific Northwest of USA	Yes	Yes; http://www.hydro.washington.edu/Lettenmaier/Models/DHSVM/
BROOK90	One-dimensional water balance with canopy and multi-layer soil profile	Daily	Plot	Low	Applied worldwide although designed for forests within north-eastern USA	Yes	Yes; http://www.ecoshift.net/brook/brook90.htm
VIC	Grid-cell based water balance with canopy and three-layer soil profile	Daily to monthly	Regional, global	Medium	Any as intended to accompany large-scale general circulation models	Yes	Yes; http://www.hydro.washington.edu/Lettenmaier/Models/VIC/
INCA	Semi-distributed water balance with canopy, soil and riparian components	Daily	Catchment	Medium	No limitations; applied extensively in Europe	By request	Yes; http://www.reading.ac.uk/geographyandenvironmentalscience/research/INCA/

The Precipitation–Runoff Modeling System (PRMS) is a semi-distributed processed-based rainfall–runoff model that simulates components of the water balance, including evaporation, transpiration, runoff and infiltration, and quantifies interactions with forest/plant canopy, snowpack dynamics and soil hydrological processes (Leavesly *et al.*, 1983, 2005). PRMS has been applied across many landscape types and broad spatial scales. At broader spatial scales, PRMS is often calibrated in forested headwaters. BROOK90 is a one-dimensional process-based hydrological model that operates on a daily time step and was originally developed for forested catchments in the north-eastern USA (Federer *et al.*, 2003). The model includes components for interception by a single-layer canopy, snow accumulation and melt, direct evaporation from soil and snow, transpiration from a single-layer canopy and multi-layered soil, and multi-layered soil water movement. The Distributed Hydrology Soil Vegetation Model (DHSVM) is a watershed-scale hydrological model that operates at sub-daily to annual time steps (Wigmosta *et al.*, 1994, 2002). The model is composed of seven modules representing evapotranspiration, snowpack accumulation and melting, canopy snow interception and release, unsaturated subsurface flow, saturated subsurface flow, surface overland flow and channel flow. DHSVM is frequently applied to evaluate forest management hydrological effects across a variety of physiographical settings (Storck *et al.*, 1998; Bowling and Lettenmaier, 2001).

The Variable Infiltration Capacity (VIC) model is a macro-scale hydrological model that operates at daily to monthly time steps; it complements global-scale general circulation models (GCMs) used for climate simulations and weather prediction (Liang *et al.*, 1994, 1996). The model includes simulated forest evapotranspiration, canopy storage, surface and surface runoff, aerodynamic flux, and snow accumulation and melt. The Integrated Nitrogen Catchment Model (INCA) is a semi-distributed process-based watershed model that operates at a daily time step and is popularly used in Western European forested catchment studies (Whitehead *et al.*, 1998a,b; Wade *et al.*, 2002). The INCA hydrological module simulates soil moisture, storage and evaporation, topographic impacts on flow and streamflow, and can be applied to assess the effects of forest management on catchment-scale hydrology and biogeochemical cycling. Finally, a widely used watershed-scale distributed model, SWAT (Soil and Water Assessment Tool; Arnold *et al.*, 1998), originally developed for upland agricultural landscapes, has been tested, modified and updated for its application on large landscapes containing large portions of forest lands (von Stackelberg *et al.*, 2007; Watson *et al.*, 2009; Parajuli, 2010; Amatya and Jha, 2011).

9.5.2 Ecosystem models

Broad-scale ecosystem models are those that simulate combined terrestrial ecosystem processes with catchment rainfall–runoff and hydrological routing. These models can range in complexity from fully coupled physically based ecosystem dynamics and hydrological modelling systems to less mechanistic decision-making tools. Examples across this range of complexity include FOREST-BGC (Running and Gower, 1991), BIOME BGC (White *et al.*, 2000), RHESSys (Band *et al.*, 1993; Tague and Band, 2004) and WaSSI (Sun *et al.*, 2011, 2015; Caldwell *et al.*, 2012) (Table 9.2; Fig. 9.2 presents WaSSI as an example low-complexity ecosystem model). The FOREST-BGC model (Running and Gower, 1991) and its successors, such as BIOME-BGC model (White *et al.*, 2000) and other BGC family models, are process-based, stand-level ecosystem models that can be spatially aggregated and averaged to a per unit area basis. FOREST-BGC's water balance is simulated at a daily time step and includes evaporation, transpiration, rainfall interception, throughfall, soil moisture, snow water equivalent depth, and soil outflow of water. The Regional Hydro-Ecological Simulation System (RHESSys) is a semi-distributed hydrological model that operates at a daily time step and is used to simulate mountainous watersheds (Band *et al.*, 1993; Tague and Band, 2004). The hydrological component of the model simulates atmospheric processes, soil hydrological and transport processes including vertical seepage, soil evaporation and lateral flow, and canopy radiative and moisture processes. The WaSSI model is a relatively low-complexity, integrated, process-based model that

Table 9.2. Example ecosystem-scale models for forest hydrology applications.

Model	Hydrological approach	Simulation outputs: hydrology	Time step(s)	Spatial scale(s)	Level of complexity	Appropriate regions of application	Model files publically accessible	Availability of online user manual and website
FOREST-BGC	One-dimensional water balance with forest canopy and soil surface	Outflow, evapotranspiration	Daily	Catchment to regional	Medium	Any forest catchments	Yes	Yes; http://daac.ornl.gov/cgi-bin/dsviewer.pl?ds_id=36
BIOME BGC	Ecosystem model simulating hydrology across multiple scales based on one-dimensional water balance concepts	Evapotranspiration, soil water, snow depth, outflow	Daily	Multiple scale	Medium	Terrestrial ecosystem consisting of one or multiple biomes	Yes	Yes; http://www.ntsg.umt.edu/project/biome-bgc
RHESSys	Semi-distributed water balance incorporating TOPMODEL, CENTURY, DHSVM, BIOME BGC	Streamflow, carbon and nutrient dynamics at watershed outlet and within model components	Daily	Catchment	Medium	Mountainous catchments	Yes	Yes; http://fiesta.bren.ucsb.edu/~rhessys/
WaSSI	Distributed by HUC-12 (12-digit Hydrologic Unit Code) watersheds at the continental scale	Water yield, evapotranspiration, ecosystem productivity, water supply stress index	Monthly	Catchment	Low	Continental	Yes	Yes; http://www.fs.usda.gov/ccrc/tools/wassi

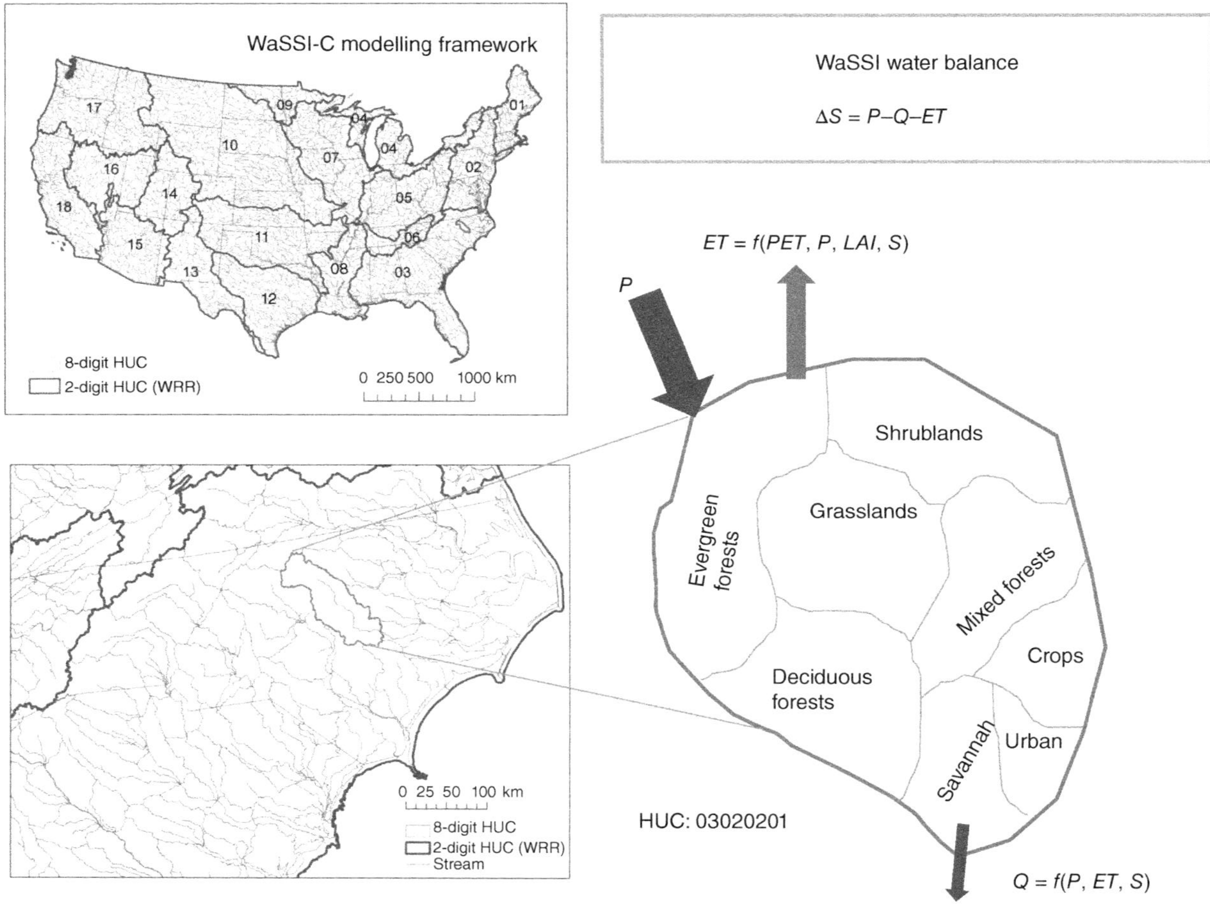

Fig. 9.2. WaSSI: an example relatively low-complexity, large-scale ecosystem model that can be applied for forest hydrology management and research questions. HUC is the Hydrologic Unit Code (a catchment identifier), WRR is water resource region, ΔS is the change in water storage at the catchment scale, P is precipitation, Q is catchment runoff, ET is evapotranspiration, PET is potential evapotranspiration and LAI is leaf area index.

describes key ecohydrological processes at broad spatial scales (Sun *et al.*, 2011, 2015; Caldwell *et al.*, 2012) (Fig. 9.2). It operates on a monthly time step and simulates the full monthly water balance (evapotranspiration, streamflow and soil moisture storage) for each land cover class at a user-defined watershed scale.

9.5.3 Groundwater models

Groundwater models can be applied to research and management questions related to forest hydrology to focus on the movement and transport of subsurface flows through saturated porous media. Groundwater models typically are bounded by deep subsurface flow networks that reach across multiple catchment boundaries and use Darcy's flow equation (i.e. the groundwater flow equation) to estimate deep groundwater transport, which is based on relationships among hydraulic conductivity, hydraulic gradient, fluid flow rates and the model domain contributing area. Surface water flows and features (e.g. lakes, ponds, wetlands, streams, rivers) are not modelled explicitly in groundwater simulations and are considered boundary conditions. Two example groundwater models that could be used for simulating forest hydrological systems with a strong groundwater component include MIKE SHE (Abbott, 1986a,b) and DRAINMOD-FOREST (Tian *et al.*, 2012) (Table 9.3). The MIKE SHE model (Abbott, 1986a,b) is a physically based, fully distributed hydrological modelling system that was designed to describe the full hydrological cycle in a watershed. The model simulates the hydrological processes of canopy interception, soil evaporation, transpiration, infiltration, overland flow, unsaturated flow in soils, groundwater flow in aquifers and channel flows in rivers. DRAINMOD-FOREST (Tian *et al.*, 2012) is a field-scale, process-based and integrated model for simulating hydrology, soil carbon and nitrogen cycles, and vegetation growth in lowland forests under various climate conditions and silvicultural practices. Hydrological processes in DRAINMOD-FOREST are simulated on a daily or hourly basis and include evapotranspiration, rainfall interception, infiltration, subsurface drainage, surface runoff, deep seepage, and soil water dynamics in the unsaturated zone.

9.5.4 Coupled surface–subsurface models

Coupled surface–subsurface models are highly complex modelling systems that link surface and groundwater models by dividing surface and subsurface flow into regions and solve the governing equations in each region using iterative solutions methods (e.g. Markstrom *et al.*, 2008) or simultaneously solve the governing equations for surface and subsurface flows (e.g. Panday and Huyakorn, 2004). These models consider feedback among various components of the surface and subsurface water balances (e.g. runoff, groundwater flows and evapotranspiration), and are thus extremely complex and computationally arduous. Two examples of such models that can be used to address forest hydrological management and research-related questions are HydroGeoSphere (Brunner and Simmons, 2012; Therrien *et al.*, 2010) and GSFLOW (Markstrom *et al.*, 2008) (Table 9.4). HydroGeoSphere is a physically based numerical model that simulates, at a variety of time steps, coupled surface (in two dimensions) and subsurface (in three dimensions) hydrological processes so that all primary components of the hydrological cycle are modelled (i.e. overland flow, streamflow, evaporation, transpiration, groundwater recharge, subsurface discharge into surface waterbodies) (Brunner and Simmons, 2012; Therrien *et al.*, 2010). GSFLOW is a high-complexity coupled surface–subsurface hydrological model that operates at a daily time step (Markstrom *et al.*, 2008; Fig. 9.3). The model integrates the surface-water Precipitation-Runoff Modeling System (PRMS) (Leavesley *et al.*, 1983, 1995) and the Modular Groundwater Flow Model (MODFLOW) (Harbaugh *et al.*, 2000; Harbaugh, 2005). PRMS simulates land-surface hydrological processes in evapotranspiration, runoff, infiltration and interflow, plant canopy interception and storage, and snowpack. MODFLOW simulates three-dimensional saturated groundwater flow and storage, one-dimensional unsaturated flow, and groundwater interaction with streams.

Table 9.3. Example groundwater models for forest hydrology applications.

Model	Hydrological approach	Simulation outputs: hydrology	Time step(s)	Spatial scale(s)	Level of complexity	Appropriate regions of application	Model files publically accessible	Availability of online user manual and website
MIKE SHE	Fully distributed	Evapotranspiration, streamflow, soil moisture content, water table level	Sub-daily, daily	Variable	High	All	No; this is a commercial product	Yes; http://www.mikepoweredbydhi.com/products/mike-she
DRAIN-MOD-FOREST	One-dimensional water balance of canopy, the soil surface and along a soil column	Evapotranspiration, outflow, soil water, seepage, water table level	Sub-daily, daily	Field scale	Medium	Lowland	Yes	No

Table 9.4. Example coupled surface–subsurface models for forest hydrology applications.

Model	Hydrological approach	Simulation outputs: hydrology	Time step(s)	Spatial scale(s)	Level of complexity	Appropriate regions of application	Model files publically accessible	Availability of online user manual and website
HydroGeoSphere	Surface domain represented as two-dimensional overland flow; subsurface represented as three-dimensional unsaturated/ saturated flow	All primary components of the hydrological cycle are modelled (i.e. overland flow, streamflow, evaporation, transpiration, groundwater recharge, subsurface discharge to surface waterbodies)	Daily to centuries	Multi-scale domains (e.g. plots, catchments)	Very high	Multiple	No, this is a commercial product	No; but information can be found at http://www. aquanty.com/ hydrogeosphere/
GSFLOW	Coupled surface and groundwater model; PRMS for surface; finite difference equation for subsurface	HRU (hydrological response unit) water balance, streamflow, groundwater dynamics	Daily	Watershed, regional	High	Any	Yes	Yes; http://wwwbrr. cr.usgs.gov/ projects/ SW_MoWS/ GSFLOW.html

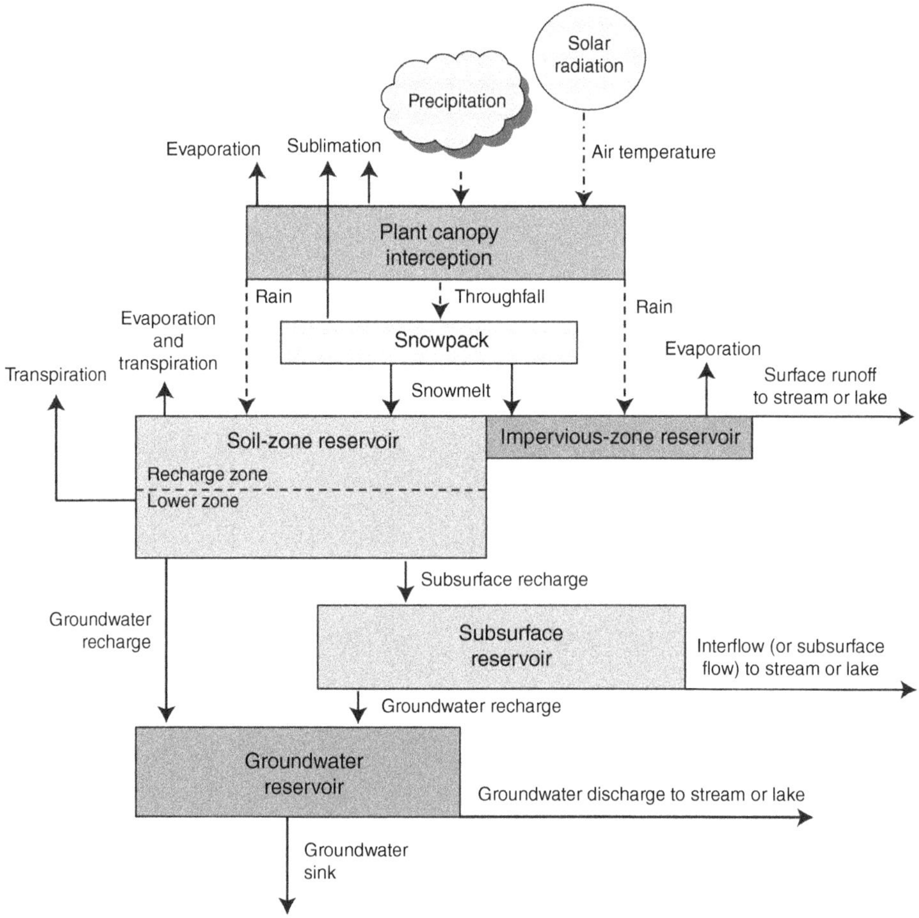

Fig. 9.3. GSFLOW model structure: an example complex, coupled surface–subsurface modelling system. (From Markstrom *et al.*, 2008, with permission; S. Markstrom, US Geological Survey, personal communication, 2015.)

9.6 Summary and Conclusions

Forest hydrology models are important tools for developing a clearer understanding of a forest stand or catchment's dominant hydrological processes and the process-based hydrological responses to future forest impacts, such as silvicultural practices, implementation of management activities and climate change, on water resources and other ecosystem services. These models can vary widely in complexity; therefore, clarity with regard to the research or management question in addition to the conceptual hydrological model of the forest stand or catchment is imperative for model selection. Model evaluation, including uncertainty and sensitivity analyses, is a primary approach to determine whether hydrological processes of interest and/or importance in the modelled system are well-characterized. With technological and high-speed computing developments in recent years, future forest hydrology modelling work will move further towards incorporating innovative remote sensing, geophysical and biogeochemical methods for improved parameterization and process understanding. Further, empirical methods (e.g. tracer and isotopic studies for hydrograph separation) and statistical approaches should continue to be integrated into mechanistic modelling structures. Finally, the development of new, simplified, yet physically based models might be most appropriate in some forested systems (Sidle *et al.*, 2011).

9.7 Disclaimer

The views expressed in this chapter are those of the authors and do not necessarily represent the views or policies of the US Environmental Protection Agency.

References

Abbott, M.B., Bathurst, J.C., Cunge, J.A., O'Connell, P.E. and Rasmussen, J. (1986a) An introduction to the European Hydrological System – Systeme Hydrologique European, 'SHE', 1: History and philosophy of a physically-based, distributed modelling system. *Journal of Hydrology* 87, 45–59.

Abbott, M.B., Bathurst, J.C., Cunge, J.A., O'Connell, P.E. and Rasmussen, J. (1986b) An introduction to the European Hydrological System – Systeme Hydrologique European, 'SHE', 2: Structure of a physically-based, distributed modelling system. *Journal of Hydrology* 87, 61–77.

Abdelnour, A., Stieglitz, M., Pan, F. and McKane, R. (2011) Catchment hydrological responses to forest harvest amount and spatial pattern. *Water Resources Research* 47, W09521, doi: 10.1029/2010WR010165 (accessed 3 April 2016).

Abdelnour, A., McKane, R., Stieglitz, M., Pan, F. and Cheng, Y. (2013) Effects of harvest on carbon and nitrogen dynamics in a Pacific Northwest forest catchment. *Water Resources Research* 49, 1292–1313.

Amatya, D.M. and Jha, M.K. (2011) Evaluating SWAT model for a low-gradient forested watershed in coastal South Carolina. *Transactions of the ASABE* 54, 2151–2163.

Amatya, D.M. and Skaggs, R.W. (2001) Hydrologic modeling of pine plantations on poorly drained soils. *Forest Science* 47, 103–114.

Amatya, D.M., Douglas-Mankin, K.R., Williams, T.M., Skaggs, R.W. and Nettles, J.E. (2011) Advances in forest hydrology: challenges and opportunities. *Transactions of the ASABE* 54, 2049–2056.

Amatya, D.M., Rossi, C.G., Dai, Z., Williams, R., Saleh, A., Youssef, M.A., Chescheir, G.M., Skaggs, R.W., Trettin, C.C., Vance, E. and Nettles, J.E. (2014) Modeling the fate of nitrogen applied to forest ecosystems – an assessment of model capabilities and potential applications. *Transactions of the ASABE* 56, 1731–1757.

Arnold, J.G., Srinivasan, R., Muttiah, R.S. and Williams, J.R. (1998) Large-area hydrological modeling and assessment: model development. *Journal of the American Water Resources Association* 34, 73–89.

Arnold, J.G., Youssef, M.A., Yen, H., White M.J., Sheshukov, A., Sadeghi A.M., Moriasi, D.N, Steiner, J.L., Amatya, D.M., Skaggs, R.W., *et al.* (2015) Hydrological processes and model representation: impact of soft data on calibration. *Transactions of the ASABE* 58, 1637–1660.

Arp, P.A. and Yin, X. (1992) Predicting water fluxes through forests from monthly precipitation and mean monthly air temperature records. *Canadian Journal of Forest Research* 22, 864–877.

Band, L.E., Patterson, P., Nemani, R. and Running, S.W. (1993) Forest ecosystem processes at the watershed scale: incorporating hillslope hydrology. *Agricultural and Forest Meteorology* 63, 93–126.

Barlow, P.M., Cunningham, W.L., Zhai, T. and Gray, M. (2015) *US Geological Survey Groundwater Toolbox, A Graphical and Mapping Interface for Analysis of Hydrologic Data (Version 1.0) – User Guide for Estimation of Base Flow, Runoff, and Groundwater Recharge from Streamflow Data. Techniques and Methods*, Book 3, Chapter B10. US Geological Survey, US Department of the Interior, Reston, Virginia.

Bates, C.G. and Henry, A.J. (1928) Forest and stream-flow experiment at Wagon Wheel Gap, Colorado. *US Weather Bureau Monthly Weather Review, Supplement No. 30*.

Bathurst, J.C., Birkinshaw, S.J., Cisneros, F., Fallas, J., Iroumé, A., Iturraspe, R., Gaviño Novillo, M., Urciuolo, A., Alvarado, A., Coello, C., *et al.* (2011) Forest impact on floods due to extreme rainfall and snowmelt in four Latin American environments 2: Model analysis. *Journal of Hydrology* 400, 292–304.

Beckers, J., Smerdon, B. and Wilson, M. (2009) Review of Hydrologic Forest Models for Forest Management and Climate Change Applications in British Columbia and Alberta. FORREX Forum for Research and Extension in Natural Resources Society, Kamloops, British Columbia, Canada. Available at: http://www.forrex.org/sites/default/files/forrex_series/FS25.pdf (accessed 26 June 2014).

Beven, K. (1993) Prophecy, reality and uncertainty in distributed hydrological modeling. *Advances in Water Resources* 16, 41–51.

Beven, K. and Kirkby, M. (1979) A physically-based, variable contributing area model of basin hydrology. *Hydrological Sciences Bulletin* 24, 43–69.

Beven, K. and Young, P. (2013) A guide to good practice in modeling semantics for authors and referees. *Water Resources Research* 49, 5092–5098.

Birkinshaw, S.J., Bathurst, J.C., Iroumé, A. and Palacios, H. (2011) The effect of forest cover on peak flow and sediment discharge – an integrated field and modelling study in central-southern Chile. *Hydrological Processes* 25, 1284–1297.

Bowling, L.C. and Lettenmaier, D.P. (2001) The effects of forest roads and harvest on catchment hydrology in a mountainous maritime environment. In: Wigmosta, M.S. and Burges, S.J. (eds) *Land Use and Watersheds: Human Influence on Hydrology and Geomorphology in Urban and Forest Areas*. American Geophysical Union, Washington, DC, pp. 145–164.

Boyle, D.P., Gupta, H.V. and Sorooshian, S. (2003) Multicriteria calibration of hydrologic models. In: Duan, Q., Gupta, H.V., Sorooshian, S., Rousseau, A.N. and Turcotte, R. (eds) *Calibration of Watershed Models*. American Geophysical Union, Washington, DC, pp. 185–196.

Brunner, P. and Simmons, C.T. (2012) HydroGeoSphere: a fully integrated, physically based hydrological model. *Ground Water* 50, 170–176.

Buttle, J.M. (2011) The effects of forest harvesting on forest hydrology and biogeochemistry. In: Levia, D.F., Tanaka, T. and Carlyle-Moses, D. (eds) *Forest Hydrology and Biogeochemistry: Synthesis of Past Research and Future Directions*. Ecological Studies Vol. 216. Springer, New York, pp. 659–677.

Buytaert, W.R, and Beven, K. (2011) Models as multiple working hypotheses: hydrological simulation of tropical alpine wetlands. *Hydrological Processes* 25, 1784–1799.

Caldwell, P.V., Sun, G., McNulty, S.G., Cohen, E.C. and Myers, M.J.A. (2012) Impacts of impervious cover, water withdrawals, and climate change on river flows in the conterminous US. *Hydrology and Earth System Sciences* 16, 2839–2857.

Dai, Z., Li, C., Trettin, C.C., Sun, G., Amatya, D.M. and Li, H. (2010) Bi-criteria evaluation of the MIKE SHE model for a forested watershed on the South Carolina coastal plain. *Hydrology and Earth System Sciences* 14, 1033–1046.

DeWalle, D.R. (2003) Forest hydrology revisited. *Hydrological Processes* 17, 1255–1256.

Doherty, J. and Johnston, J.M. (2003) Methodologies for calibration and predictive analysis of a watershed model. *Journal of the American Water Resources Association* 39, 251–265.

Downer, C.W., Ogden, F.L., Martin, W.D. and Harmon, R.S. (2002) Theory, development, and applicability of the surface water hydrologic model CASC2D. *Hydrological Processes* 16, 255–275.

Dunne, T. and Black, R.D. (1970) Partial-area contributions to storm runoff in a small New England watershed. *Water Resources Research* 6, 1296–1311.

Federer, C.A., Vorosmarty, C. and Fekete, B. (2003) Sensitivity of annual evaporation to soil and root properties in two models of contrasting complexity. *Journal of Hydrometeorology* 4, 1276–1290.

Flerchinger, G.N., Caldwell, T.G., Cho, J. and Hardegree, S.P. (2012) Simultaneous Heat and Water (SHAW) model: model use, calibration, and validation. *Transactions of the ASABE* 55, 1395–1411.

Freeze, R.A., Massmann, J., Smith, L., Sperling, T. and James, B. (1990) Hydrogeological decision analysis: 1. A framework. *Ground Water* 28, 738–766.

Gassman, P.W., Reyes, M.R., Green, C.H. and Arnold, J.G. (2007) The Soil and Water Assessment Tool: historical development, applications, and future directions. *Transactions of the ASABE* 50, 1211–1250.

Gbondo-Tugbawa, S.S., Driscoll, C.T., Aber, J.D. and Likens, G.E. (2001) Evaluation of an integrated biogeochemical model (PnET-BGC) at a northern hardwood forest ecosystem. *Water Resources Research* 37, 1057–1070.

Golden, H.E., Lane, C.R., Amatya, D.M., Bandilla, K.W., Raanan Kiperwas, H., Knightes, C.D. and Ssegane, H. (2014) Hydrologic connectivity between geographically isolated wetlands and surface water systems: a review of select modeling methods. *Environmental Modelling and Software* 53, 190–206.

Harbaugh, A.W. (2005) *MODFLOW-2005, The US Geological Survey Modular Ground-water Model: The Ground-water Flow Process*. US Department of the Interior, US Geological Survey, Reston, Virginia.

Harbaugh, A.W., Banta, E.R., Hill, M.C. and McDonald, M.G. (2000) *MODFLOW-2000, The US Geological Survey Modular Ground-water Model: User Guide to Modularization Concepts and the Ground-water Flow Process*. US Geological Survey, Reston, Virginia.

Hornberger, G.M., Bencala, K.E. and McKnight, D.M. (1994) Hydrological controls on dissolved organic carbon during snowmelt in the Snake River near Montezuma, Colorado. *Biogeochemistry* 25, 147–165.

Horton, R.E. (1933) The role of infiltration in the hydrologic cycle. *Transactions of the American Geophysical Union* 14, 446–460.

Hursh, C.R. and Brater, E. (1941) Separating storm hydrographs from small drainage areas into surface and subsurface-flow. *Transactions of the American Geophysical Union* 22, 863–871.

Jones, J.A., Achterman, G.L., Augustine, L.A., Creed, I.F., Folliott, P.F., MacDonald, L. and Wemple, B.C. (2009) Hydrologic effects of a changing forested landscape – challenges for the hydrological sciences. *Hydrological Processes* 23, 2699–2704.

Kampf, S.K. and Burges, S.J. (2007) A framework for classifying and comparing distributed hillslope and catchment hydrologic models. *Water Resources Research* 43, W05423, doi: 10.1029/2006WR005370 (accessed 3 April 2016).

Klemes, V. (1986) Operational testing of hydrological simulation-models. *Hydrological Sciences Journal – Journal des Sciences Hydrologiques* 31, 13–24.

Leavesly, G.H., Lichty, R.W., Troutman, B.M. and Saindon, L.G. (1983) Precipitation–Runoff Modeling System: User's Manual. US Geological Survey, Water-Resources Investigations Report 83-4238. Available at: http://pubs.usgs.gov/wri/1983/4238/report.pdf (accessed 5 February 2015).

Leavesly, G.H., Markstrom, S.L., Viger, R.J. and Hay, L.E. (2005) USGS Modular Modeling System (MMS)–Precipitation–Runoff Modeling System (PRMS) MMS-PRMS. In: Singh, V. and Frevert, D. (eds) *Watershed Models*. CRC Press, Boca Raton, Florida, pp. 159–177.

Li, C., Aber, J., Stange, F., Butterbach-Bahl, K. and Papen, H. (2000) A process-oriented model of N_2O and NO emissions from forest soils: 1. Model development. *Journal of Geophysical Research: Atmospheres* 105, 4369–4384.

Liang, X., Lettenmaier, D.P., Wood, E.F. and Burges, S.J. (1994) A simple hydrologically based model of land surface water and energy fluxes for general circulation models. *Journal of Geophysical Research* 99, 14415–14528.

Liang, X., Wood, E.F. and Lettenmaier, D.P. (1996) Surface soil moisture parameterization of the VIC-2L model: evaluation and modification. *Global and Planetary Change* 13, 195–206.

Malone, R.W., Yagow, G., Baffaut, C., Gitau, M., Qi, Z., Amatya, D., Parajuli, P., Bonta, J. and Green, T. (2015) Parameterization guidelines and considerations for hydrologic models. *Transactions of the ASABE* 58, 1681–1703.

Markstrom, S.L., Niswonger, R.G., Regan, R.S., Prudic, D.E. and Barlow, P.M. (2008) GSFLOW – coupled ground-water and surface-water flow model based on the integration of the Precipitation–Runoff Modeling System (PRMS) and the Modular Ground-Water Flow Model (MODFLOW-2005). Chapter 1 of Section D, Ground-Water/Surface-Water, Book 6, Modeling Techniques. Available at: http://pubs.usgs.gov/tm/tm6d1/pdf/tm6d1.pdf (accessed 3 April 2016).

Matott, L.S. (2005) OSTRICH: An Optimization Software Tool, Documentation and User's Guide, Version 1.6. University at Buffalo, Buffalo, New York. Available at: http://www.eng.buffalo.edu/~lsmatott/Ostrich/OstrichMain.html (accessed 6 February 2015).

Matott, L.S., Babendreier, J.E. and Purucker, S.T. (2009) Evaluating uncertainty in integrated environmental models: a review of concepts and tools. *Water Resources Research* 45, W06421, doi: 10.1029/2008WR007301 (accessed 3 April 2016).

Meng, F.R., Bourque, C.P.A., Jewett, K., Daugharty, D. and Arp, P.A. (1995) The Nashwaak Experimental Watershed Project – analyzing effects of clearcutting on soil-temperature, soil-moisture, snowpack, snowmelt and stream-flow. *Water, Air and Soil Pollution* 82, 363–374.

National Research Council (2007) *Models in Environmental Decision Making*. The National Academies Press, Washington, DC.

National Research Council (2008) *Hydrologic Effects of a Changing Forest Landscape*. The National Academies Press, Washington, DC.

Neitsch, S.L., Arnold, J.G., Kiniry, J.R. and Williams, J.R. (2011) *Soil and Water Assessment Tool: Theoretical Documentation, Version 2009*. Texas Water Resources Institute Technical Report No. 406. Texas A&M University System, College Station, Texas.

Nelitz, M., Boardley, S. and Smith, R. (2013) *Tools for Climate Change Vulnerability Assessments for Watersheds*. PN 1494. Prepared for Canadian Council of Ministers of the Environment by ESSA Technologies Limited, Vancouver, British Columbia, Canada.

Panday, S. and Huyakorn, P.S. (2004) A fully coupled physically-based spatially-distributed model for evaluating surface/subsurface flow. *Advances in Water Resources* 27, 361–382.

Parajuli, P.B. (2010) Assessing sensitivity of hydrologic responses to climate change from forested watershed in Mississippi. *Hydrological Processes* 24, 3785–3797.

Parton, W.J. (1996) The CENTURY model. In: Powlson, D., Smith, P. and Smith, J. (eds) *Evaluation of Soil Organic Matter Models*. Springer, Berlin/Heidelberg, pp. 283–291.

Parton, W.J., Scurlock, J.M.O., Ojima, D.S., Gilmanov, T.G., Scholes, R.J., Schimel, D.S., Kirchner, T., Menaut, J-C., Seastedt, T., Garcia Moya, E., *et al.* (1993) Observations and modeling of biomass and soil organic matter dynamics for the grassland biome worldwide. *Global Biogeochemical Cycles* 7, 785–809.

Running, S.W. and Gower, S.T. (1991) Forest-BGC, a general-model of forest ecosystem processes for regional applications. 2. Dynamic carbon allocation and nitrogen budgets. *Tree Physiology* 9, 147–160.

Saleh, A., Williams, J.R., Wood, J.C., Hauck, L.M. and Blackburn, W. (2004) Application of APEX for forestry. *Transactions of the American Society of Agricultural Engineers* 47, 751–765.

Schnorbus, M. and Alila, Y. (2004) Forest harvesting impacts on the peak flow regime in the Columbia Mountains of southeastern British Columbia: an investigation using long-term numerical modeling. *Water Resources Research* 40, W05205, doi: 10.1029/2003WR002918 (accessed 3 April 2016).

Sidle, R.C., Kim, K., Tsuboyama, Y. and Hosoda, I. (2011) Development and application of a simple hydrogeomorphic model for headwater catchments. *Water Resources Research* 47, W00H13, doi: 10.1029/2011WR010662 (accessed 3 April 2016).

Sivapalan, M. (2003) Prediction in ungauged basins: a grand challenge for theoretical hydrology. *Hydrological Processes* 17, 3163–3170.

Sloto, R.A. and Crouse, M.Y. (1996) HYSEP: A Computer Program for Streamflow Hydrograph Separation and Analysis. US Geological Survey, Water-Resources Investigations Report 96-4040. Available at: http://water.usgs.gov/software/HYSEP/ (accessed 3 April 2016).

Stange, F., Butterbach-Bahl, K., Papen, H., Zechmeister-Boltenstern, S., Li, C. and Aber, J. (2000) A process-oriented model of N_2O and NO emissions from forest soils: 2. Sensitivity analysis and validation. *Journal of Geophysical Research: Atmospheres* 105, 4385–4398.

Stednick, J. (1996) Monitoring the effects of timber harvest on annual water yield. *Journal of Hydrology* 176, 79–95.

Stednick, J. (2008) *Hydrological and Biological Responses to Forest Practices*. Ecological Studies Vol. 199. Springer, New York.

Storck, P., Bowling, L., Wetherbee, P. and Lettenmaier, D. (1998) Application of a GIS-based distributed hydrology model for prediction of forest harvest effects on peak stream flow in the Pacific Northwest. *Hydrological Processes* 12, 889–904.

Sun, G., Riekerk, H. and Comerford, N.B. (1998) Modeling the forest hydrology of wetland-upland ecosystems in Florida. *Journal of the American Water Resources Association* 34, 827–841.

Sun, G., Zhou, G., Zhang, Z., Wei, X., McNulty, S.G. and Vose, J.M. (2006) Potential water yield reduction due to forestation across China. *Journal of Hydrology* 328, 548–558.

Sun, G., Caldwell, P., Noormets, A., Cohen, E., McNulty, S., Treasure, E., Domec, J.C., Mu, Q., Xiao, J., John, R. and Chen, J. (2011) Upscaling key ecosystem functions across the conterminous United States by a water-centric ecosystem model. *Journal of Geophysical Research* 116, G00J05, doi: 10.1029/2010JG001573 (accessed 5 April 2016).

Sun, G., Caldwell, P.V. and McNulty, S.G. (2015) Modeling the potential role of forest thinning in maintaining water supplies under a changing climate across the conterminous United States. *Hydrological Processes* 29, 5016–5030.

Tague, C.L. and Band, L.E. (2004) RHESSys: Regional Hydro-Ecologic Simulation System – an object-oriented approach to spatially distributed modeling of carbon, water, and nutrient cycling. *Earth Interactions* 8, 1–42.

Therrien, R., McLaren, R.G., Sudicky, E.A. and Panday, S.M. (2010) *HydroGeoSphere: A Three-Dimensional Numerical Model describing Fully-Integrated Subsurface and Surface Flow and Solute Transport*. Groundwater Simulations Group, University of Waterloo, Waterloo, Ontario, Canada.

Tian, M.A., Youssef, M.A., Skaggs, R.W., Amatya, D.M. and Chescheir, G.M. (2012) DRAINMOD-FOREST: integrated modeling of hydrology, soil carbon, and nitrogen dynamics, and plant growth for drained forest. *Journal of Environmental Quality* 41, 764–782.

Tian, S., Youssef, M.A., Amatya, D.M. and Vance, E. (2014) Global sensitivity analysis of DRAINMOD-FOREST, an integrated forest ecosystem model. *Hydrological Processes* 28, 4389–4410.

Thyer, M., Beckers, J., Spittlehouse, D., Alila, Y. and Winkler, R. (2004) Diagnosing a distributed hydrologic model for two high-elevation forested catchments based on detailed stand- and basin-scale data. *Water Resources Research* 40, W01103, doi: 10.1029/2003WR002414 (accessed 5 April 2016).

Viviroli, D., Zappa, M., Gurtz, J. and Weingartner, R. (2009) An introduction to the hydrological modelling system PREVAH and its pre- and post-processing-tools. *Environmental Modelling and Software* 24, 1209–1222.

Von Stackelberg, N.O., Chescheir, G.M., Skaggs, R.W. and Amatya, D.M. (2007) Simulation of the hydrologic effect of afforestation in the Tacuarembo River basin, Uruguay. *Transactions of the ASABE* 50, 455–468.
Vose, J.M., Sun, G., Ford, C.R., Bredemeier, M., Otsuki, K., Wei, X.H., Zhang, Z.Q. and Zhang, L. (2011) Forest ecohydrological research in the 21st century: what are the critical needs? *Ecohydrology* 4, 146–158.
Wade, A., Durand, P., Beaujouan, V., Wessel, W., Raat, K., Whitehead, P., Butterfield, D., Rankinen, K. and Lepisto, A. (2002) A nitrogen model for European catchments: INCA, new model structure and equations. *Hydrology and Earth System Sciences* 6, 559–582.
Wang, J., Endreny, T.A. and Nowak, T.J. (2008) Mechanistic simulation of tree effects in an urban water balance model. *Journal of the American Water Resources Association* 44, 75–85.
Watson, B.M., McKeown, R.A., Putz, G. and MacDonald, J.D. (2009) Modification of SWAT for modelling streamflow from forested watersheds on Canadian boreal plain. *Journal of Environmental Engineering and Science* 7(S1), 145–159.
Wei, A. and Zhang, M. (2011) A review on research methods for assessing the impacts of forest disturbance on hydrology at large-scale watersheds. In: Li, C., Lafortezza, R. and Chen, J. (eds) *Landscape Ecology and Forest Management: Challenges and Solutions in a Changing Globe*. Springer, New York, pp. 119–147.
Weiler, M. and McDonnell, J. (2004) Virtual experiments: a new approach for improving process conceptualization in hillslope hydrology. *Journal of Hydrology* 285, 3–18.
Whitaker, A., Alila, Y., Beckers, J. and Toews, D. (2003) Application of the distributed hydrology soil vegetation model to Redfish Creek, British Columbia: model evaluation using internal catchment data. *Hydrological Processes* 17, 199–224.
White, M.A., Thornton, P.E., Running, S.W. and Nemani, R.R. (2000) Parameterization and sensitivity analysis of the BIOME-BGC terrestrial ecosystem model: net primary production controls. *Earth Interactions* 4, 1–85.
Whitehead, P., Wilson, E. and Butterfield, D. (1998a) A semi-distributed integrated nitrogen model for multiple source assessment in catchments (INCA): Part I – model structure and process equations. *Science of the Total Environment* 210, 547–558.
Whitehead, P., Wilson, E., Butterfield, D. and Seed, K. (1998b) A semi-distributed integrated flow and nitrogen model for multiple source assessment in catchments (INCA): Part II – application to large river basins in south Wales and eastern England. *Science of the Total Environment* 210, 559–583.
Wigmosta, M.S., Vail, L.W. and Lettenmaier, D.P. (1994) A distributed hydrology–vegetation model for complex terrain. *Water Resources Research* 30, 1665–1679.
Wigmosta, M.S., Nijssen, B., Storck, P. and Lettenmaier, D. (2002) The distributed hydrology soil vegetation model. In: Singh, V.P. and Frevert, D.K. (eds) *Mathematical Models of Small Watershed Hydrology and Applications*. Water Resources Publications, Littleton, Colorado, pp. 7–42.
Williams, J.R. and Izaurralde, R.C. (2005) The APEX model. In: Singh, V.P. and Frevert, D.K. (eds) *Watershed Models*. CRC Press, Boca Raton, Florida, pp. 437–482.
Yang, Y., Endreny, T.A. and Nowak, D.J. (2011) iTree-Hydro: snow hydrology update for the urban forest hydrology model. *Journal of the American Water Resources Association* 47, 1211–1218.
Young, P.C. (2001) Data-based mechanistic modelling and validation of rainfall–flow processes. In: Anderson, M.G. and Bates, P.D. (eds) *Model Validation: Perspectives in Hydrological Science*. Wiley, Chichester, UK, pp. 117–161.

10 Geospatial Technology Applications in Forest Hydrology

S.S. Panda[1]*, E. Masson[2], S. Sen[3], H.W. Kim[4] and D.M. Amatya[5]

[1]*University of North Georgia, Gainesville, Georgia, USA;* [2]*Université de Lille de Sciences et Technologies, Lille, France;* [3]*Indian Institute of Technology– Roorkee, Uttarakhand, India;* [4]*Anyang University, Anyang, Republic of Korea;* [5]*USDA Forest Service, Cordesville, South Carolina, USA*

10.1 Introduction

Two separate disciplines, hydrology and forestry, together constitute 'forest hydrology'. It is obvious that forestry and forest hydrology disciplines are spatial entities. Forestry is the science that seeks to understand the nature of forests through their life cycle and interactions with the surrounding environment. Forest hydrology includes forest soil water, streams and other small waterbodies encompassed by forest cover, and the hydrological cycle itself within a forested land cover. 'Forest' and (forest) 'Water' are two standardized land cover mapping classifications of the National Land Cover Database (NLCD) used by environmental planners in the USA (USGS LCI, 2015), CORINE (Co-ORdinated INformation on the Environment) data sets (EEA, 2006) established and used by the European Community, and other countries' national land cover mapping systems for developing an environmental management decision support system (DSS).

In Europe in general, and in France as an example, forest management (private and public) started earlier than in the USA. Since the 14th century, regulations and laws have been enacted in France to manage forests as a strategic resource (Morin, 2010) for timber production, for energy to sustain proto-industry's (steel production) needs and as a financial resource to raise funds for any purpose, including funding wars. Since the 18th century, forest management has been conducted in the USA as an ecosystem management approach while still including timber and fibre production as an important goal (Richmond, 2007).

Systemic forest management in the Indian subcontinent started late under the British colonial rule with establishment of the Imperial Forest Department in India in 1864 (Ramakrishnan *et al.*, 2012). An estimated 200+ million people in India depend on forests for their livelihoods in the form of fodder, fuelwood, increased agricultural growth through forest humus production and transportation to agricultural land with runoff and soil moisture conservation, and ecosystem services. In the Korean peninsula, however, unlike most of the Asian countries, forests are managed by private and public participation as done in the USA and Europe (Lee and Lee, 2005). Private and public participation in forest and forest hydrology management has its advantages (Lee and Lee, 2005). Therefore, due to its spatial discipline and recent proven management strategies, all together, forestry, and especially forest hydrology, could be managed well

*Corresponding author; e-mail: sudhanshu.panda@ung.edu

worldwide with the involvement of the respective governments and the availability of a sound DSS based on a geospatial technology (GT) application such as remote sensing (RS), geographic information systems (GIS), global navigation satellite systems (GNSS) and information technology (IT).

Forest hydrology can be managed by GT with the sound management decision support of water, soil, wildlife and environmental resources within the forest land cover. It is humanly impossible to deal with or analyse features of larger areas (like forests or their smaller fragments) for accurate management decision making, because site-specific forest management decision support (SSFMDS) based on scouting only would take years. Management of forests to support silviculture involves large-scale spatial and tabular (attribute) data (gigabytes or even terabytes) and numerous SSFMDS parameters including soil, climatologic, hydrological and crop growth attributes. This inherent data volume and intricacy related to SSFMDS, and especially forest hydrology phenomena, can be effectively and efficiently monitored using GT as conducted and well documented for site-specific crop management (von Gadow and Bredenkamp, 1992; Panda *et al.*, 2010). In fact, GT, especially GIS, has become a fundamental part of forestry management in many commercial forestry enterprises (Austin and Meyers, 1996). Currently, applications of advanced RS technologies such as ultra-high (<1 m) spatial resolution ortho- or satellite images, hyperspectral images and radio detection and ranging (RADAR) data have been extremely useful in the effective management of forest hydrology. Unmanned aerial vehicles (UAV) and unmanned aircraft systems (UAS) are making forest hydrology management more efficient through the acquisition of centimetre-scale spatial resolution images with user-specified bandwidths – thus helping in SSFMDS by mapping soil moisture, plant stomatal conductance, canopy temperature and leaf area index (LAI) to measure forest evapotranspiration (ET) and by monitoring forest fires (Grenzdörffer *et al.*, 2008).

GT, through raster imagery acquisition and mapping, has the ability to depict accurate pixel-based analysis of larger areas using several parameters such as land use/land cover (LULC), soil, elevation/topography, hydrology, transportation, population density and adjacency, and climate/weather, which directly or indirectly impact environmental management and especially forest hydrology management. RS technology helps in surveying the entire earth with unprecedented regularity; thus, major global forest cover change can be discovered or monitored efficiently to provide insight into forest hydrology management. Shuttle Radar Topography Mission (SRTM) satellites obtain global elevation data from which earth topographic changes can be monitored proficiently, suggesting changes to forest hydrology. In the current decade, with the introduction of LiDAR and UAV/UAS technology, earth elevation including tree heights in forests is being monitored with centimetre accuracy for forest biomass estimation and ET assessment (Zarco-Tejada *et al.*, 2014; Khosravipour *et al.*, 2015). Currently, weather satellites monitor global atmospheric conditions hourly, including water vapour in the atmosphere on a spatial basis (Panda *et al.*, 2015). RS imagery provides information on drought, vegetation vigour, flood damage, forest fires, deforestation and other natural disasters that are directly or indirectly influenced by forest hydrology (Panda *et al.*, 2015). D'urso and Minacapilli (2006) used a semi-empirical approach for forest surface soil water content estimation using radar data. Potential RS systems, such as colour infrared (CIR) aerial photography, most other multispectral scanners (MSS) (Landsat, QuickBird) and hyperspectral systems (AVIRIS, HyMap, CASI), bathymetric LiDAR, MISR, Hyperion, TOPEX/Poseidon, MERIS, AVHRR and CERES, are being used by scientists to remotely estimate the hydrological flux on the earth's surface, including forest land cover (Panda *et al.*, 2015).

GIS provides the tools to accurately map this information globally and locally, including development of automated geospatial models for precise and proficient forest hydrology management decision support (FHMDS). In recent times, most widely used global positioning system (GPS) technology (a part of GNSS) accurately tracks the position of environmental disasters such as forest fires, mudslides and other phenomena related to forest hydrology. IT helps improve the DSS development and popularize these fascinating but sometimes challenging-to-comprehend tools (Panda *et al.*, 2004b).

10.2 Geospatial Technology Application in Forest Hydrological Processes Management

The most important aspect of forest management is that forest cover provides a cleaner and more dependable supply of water compared with all other land covers on earth (Richmond, 2007). The basic forest hydrological concept explains that the first element of the hydrological process, interception of raindrops by the plant canopy or the forest canopy, occurs in abundance – about 25 to 30% of total precipitation (Zinke, 1967; see Chapters 1 and 3, this volume) – and with higher infiltration and lower runoff than any other land cover type except wetlands due to supportive forest soil texture and structure (Zinke, 1967).

People have been observing the link between forests and water for thousands of years (Amatya *et al.*, 2015). Before hydrology was recognized as a specialty or subfield of forestry, engineering, geography and other disciplines, the study of forests, water and climate was referred to as 'forest influences'. This is still a useful term and a meaningful concept (Barten, 2006). Globally, the forest flourishes when precipitation (P) is much greater than potential evapotranspiration (PET), the growing season is long, the climate is moderate and the frequency of natural disturbance is low (Barten, 2006).

Forest hydrology also influences natural disturbances, such as droughts in forests creating consequential wildfires, severe precipitation after prolonged drought and wildfire increasing chances of landslides, and unpredictable hydrological cycles in forest areas creating pest/disease infestation (see Chapter 1, this volume). The following subsections exemplify the importance of GT use in FHMDS.

10.2.1 Forest cover mapping and change analysis

The forest cover supports climate stabilization, biodiversity preservation, soil enrichment for agricultural lands, erosion control, clean water supply and its cycle regulation, bioenergy production, and fodder and timber supply for human and animal sustenance. Forest plant litters decompose and recycle nutrients through the shedding of leaves and seeds with the support of forest hydrological cycles, thus enriching the soil (Osman, 2013). These enriched soils from higher-elevation forest cover move to more flat topographic agricultural land and help in higher crop production (Osman, 2013). The tree roots and soil binding in the forest reduce excessive soil erosion (Kittredge, 1948). Forest cover regulates the water cycle by absorbing and redistributing rainwater equally to every species living within its range (Perry *et al.*, 2008). Moreover, riparian forest has proved its potential to clean surface water and reduce nitrate accumulation in soils and river flows (Lowrance, 1992; Pinay *et al.*, 1993). Additionally, mapping the forest plant canopy can help to quantify the first element of the hydrological cycle: interception and subsequent evaporation. Therefore, efficient mapping and analysing of the forest cover with GT supports better management DSS.

The Food and Agriculture Organization of the United Nations (FAO) monitors global forest cover with 250-m resolution MODIS data. The National Oceanic and Atmospheric Administration (NOAA) uses 1-km resolution AVHRR satellite imagery to constantly monitor the global vegetation change over time. Figure 10.1 represents the global forest cover density by climatic domain in 2010 as developed by FAO with MODIS data. Lepers *et al.* (2005) have used remotely sensed imagery to construct a temporal change analysis of global forest land cover between 1981 and 2000 (Fig. 10.2). The map and their study (Lepers *et al.*, 2005) provide quick insight into forest loss and its worldwide impact on forest hydrology, global climate, biodiversity and others. Areas in the map (Fig. 10.2) are defined as hotspots when deforestation rates exceed threshold values, as estimated from available deforestation data or from expert opinion.

The NLCD classifies three prominent forest land covers excluding forested wetland. Medium-resolution (30 m) Landsat (five MSS, seven ETM+) images are used to classify the land cover of the USA on a temporal basis (based on the satellites' fly-over cycle). The Anderson land cover classification scheme includes deciduous forest (#41), evergreen forest (#42) and mixed forest (#43) as forest categories. The National Aeronautics and Space Administration (NASA) GAP project develops US land cover maps at regular intervals,

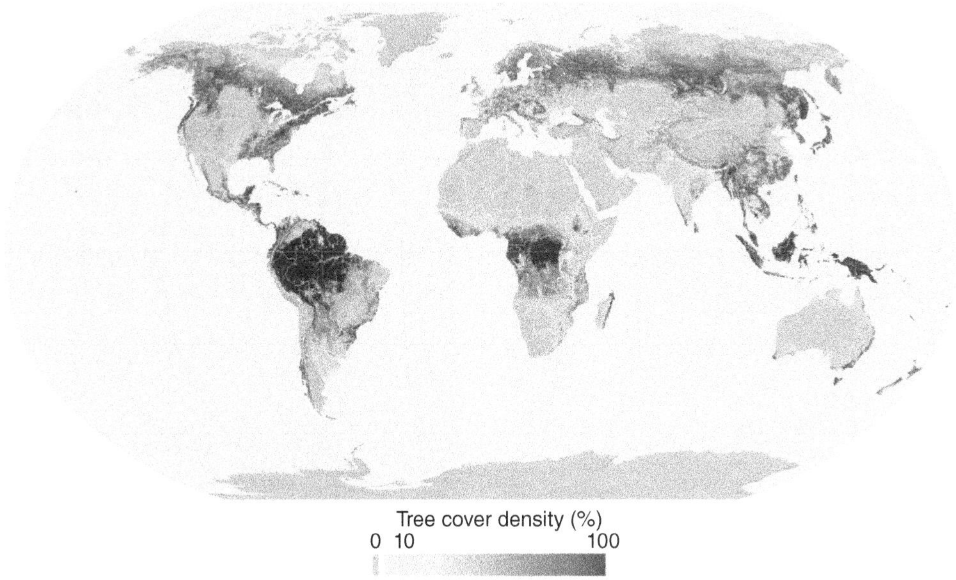

Fig. 10.1. Year 2010 world forest cover map by climatic domain developed by the FAO (Food and Agriculture Organization of the United Nations) using 250-m resolution MODIS (Moderate Resolution Imaging Spectroradiometer) satellite imagery. (From FAO, http://www.fao.org/forestry/fra/80298/en/; published with permission from FAO.)

i.e. 1974, 1985, 1992, 2001, 2005 (few states) and 2011. These NLCD data help in studying the US land cover change, especially forest cover change.

The conversion of natural land cover into human-dominated land-use types such as forest harvesting, deforestation, urbanization and agricultural intensification continues to be a change of global proportion with many environmentally unfriendly consequences for local climate, energy, hydrology and water balance, biogeochemistry and biodiversity (Potter *et al.*, 2007). Deforestation allows soil erosion, and nutrient-rich soils are lost into rivers, lakes and oceans (Panda *et al.*, 2004a). According to Sundquist (2007), the global tropical deforestation rate is about 8% of the current tropical forest inventory per decade. In the Indian subcontinent, Asia and Africa, shifting cultivation (e.g. slash-and-burn agriculture) is a prime example of mismanaging forest resources (forest soils and water) (Panda *et al.*, 2004a).

The global inventory of tropical land under shifting cultivation (including fallow) was 3 million km^2 by the 1980s (Sundquist, 2007; Fig. 10.2). Shifting cultivation in high-elevation forest lands adds to forest degradation by reducing the fertility level of the soil, which is accelerated by soil erosion due to land mass exposure. In the areas under shifting cultivation, nutrient losses occur through leaching, runoff and erosion, making the land uncultivable after two or three cropping seasons (Szott *et al.*, 1999; Panda *et al.*, 2005). Due to the changing dynamics of forest hydrology as a result of deforestation, it is almost impossible for the regeneration of the forest in the area under shifting cultivation (Szott *et al.*, 1999; Sundquist, 2007).

Potter *et al.* (2007) assessed land cover change detection in the majority of California, USA, using the MODIS 250-m resolution time series of enhanced vegetation index (EVI) data. The authors reported that areas affected by forest management and encroachment of residential development into natural vegetation zones should be prime locations for applications of land cover change detection. Goward *et al.* (2008) reported that a number of research projects within the North American Carbon Program (NACP) are combining RS and forest inventory data to map the extent and rate of forest disturbance in the conterminous USA. Given that disturbance processes vary in their extent, duration and

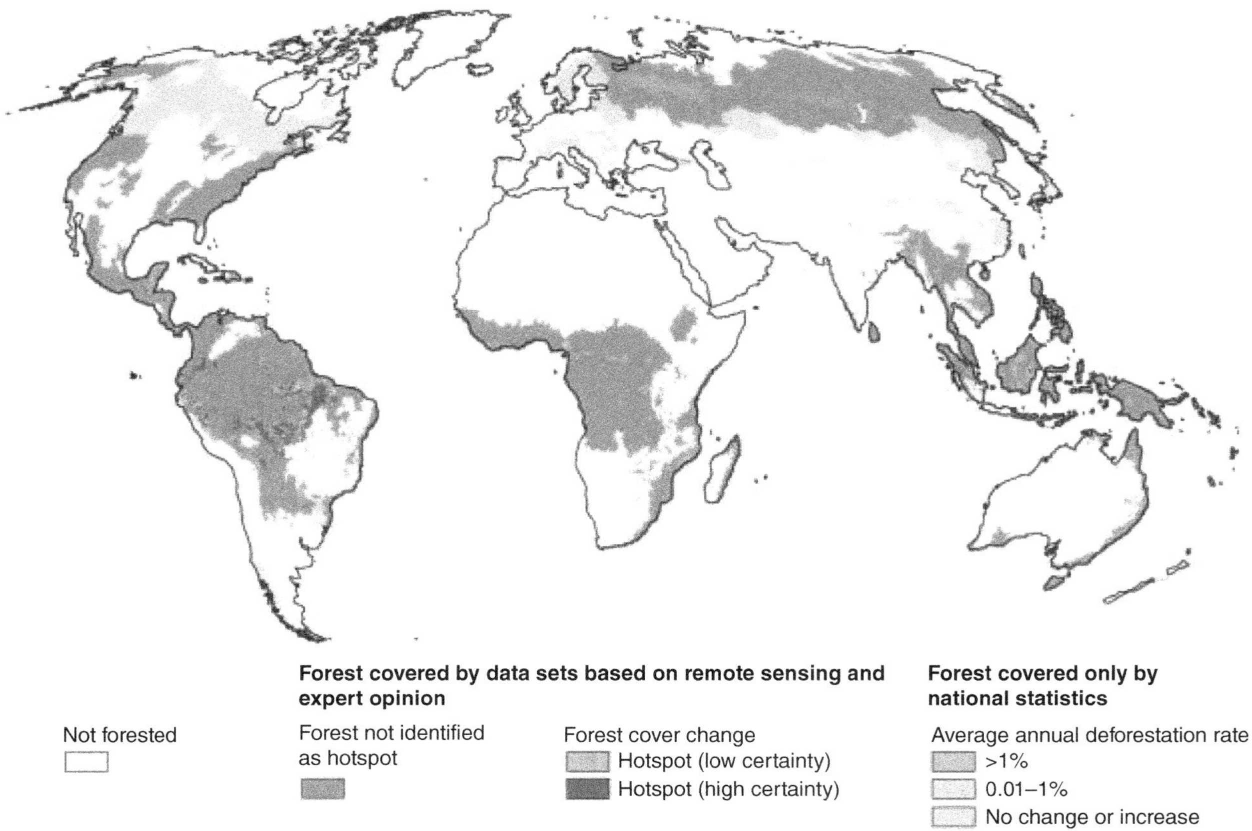

Fig. 10.2. Forest cover change between years 1981 and 2010 using remote sensing data and expert opinion. (From Lepers *et al.*, 2005, published with permission from the source.)

intensity, the authors suggested a multipronged approach with different satellite technologies targeted towards different space and time scales. For example, NOAA's AVHRR and NASA's MODIS are being used to map transient phenomenon occurring at the coarsest spatial scales, including insect outbreaks, drought stress and storm damage, and for estimating fire emissions and global mapping of active fires and burned areas.

Since 1994, the CORINE land cover data have provided land cover levels in Europe at a 25-ha minimum surface unit, and 5-ha change detection between each data version, according to four forest classes (EEA, 2006). The European programme used SPOT, MSS, TM, ETM+ and IRS P6 data for the 1994, 2000 and 2006 data sets. The 2012 update was released in September 2015 with significant improvement and reliability according to the RS data used (i.e. SPOT 4 and IRS P6 data) for a spatial coverage of 39 European countries. Plate 7 shows the forest and other natural cover changes in Europe between 2000 and 2006 as developed by the European Environmental Agency (EEA).

Kim *et al.* (2015, 2016) studied the land cover change in North Korea from 2001 to 2014. They found consistent decreases in normalized difference vegetation index (NDVI) values for 14 years, but interestingly observed a 4% increase in forest land covers that include evergreen needle-leaf forest, evergreen broadleaf forest, deciduous needle-leaf forest, deciduous broadleaf forest, mixed forest, closed shrublands, open shrublands and woody savannahs (Plate 8). Even though further study is required, this is a positive development by forest managers in North Korea. At a global scale, two aspects of climate change, namely temperature and precipitation, affect the photosynthesis of forest ecosystems. High-elevation tropical forests of five continents have been experiencing higher 'browning' (i.e. forests are losing foliage) and less photosynthetic activities (Krishnaswamy *et al.*, 2014).

Because the mangrove forests are declining in many parts of the world and even more rapidly than inland tropical forests, it is essential to determine the rate of change in cover and the causes behind it. Giri *et al.* (2007) used RS data along with geospatial mapping to understand the forest and its hydrodynamics through a multi-temporal analysis of Landsat satellite data from the 1970s, the 1990s and the 2000s in the mangrove forests of the Sundarbans of Bangladesh. They found that the mangrove forest and its intertidal zone hydrology processes are changing constantly due to erosion, aggradation, deforestation and mangrove rehabilitation programmes.

10.2.2 Forest soil water/moisture estimation and forested wetlands analysis

The soils horizon of forests consists of a prominent typical litter layer (O), a larger organic, nutrient-rich, mixed topsoil layer (A) and a mineral-rich layer (B and C). In general, forest soils naturally consist of high organic matter with high porosity and permeability, allowing high infiltration and low runoff (Pritchett, 1979; Osman, 2013). Soils in forested wetlands, found mostly in coastal and lower-elevation flat areas, in general are saturated in nature with abundant availability of soil water. Upland forests also hold high amounts of soil moisture (Jipp *et al.*, 1998). Therefore, forest soils are the nexus for many ecological processes, such as energy exchange, water storage and movement, nutrient cycling, plant growth, and carbon cycling at the base of the food web (Johnson *et al.*, 2000). Hence, forest soil and forest soil water are different from soil water in other land covers. According to the FAO soil map development process, forest soil is considered different from soils within other land cover types. For example, a lesser Himalayan overland flow study under forested versus degraded land cover showed that although Hortonian overland flow generation is dominant in both systems, hydrological characteristics vary in terms of runoff coefficient and soil physical properties. GTs, including ground-penetrating radar (GPR), can help study the soil water phenomena of the forest in a non-intrusive and efficient manner. For soil map development in the areas where no maps are available/developed, forest land cover (using RS data) can be used as the area of a specific type of soil (Panda *et al.*, 2004a). Figure 10.3 shows a GT-based procedure to create soil maps using forest hydrology information and FAO-suggested processes.

Understanding the dynamics of soil moisture and its measurements and modelling is critical for broad environmental areas such as agricultural and silvicultural crop management, water

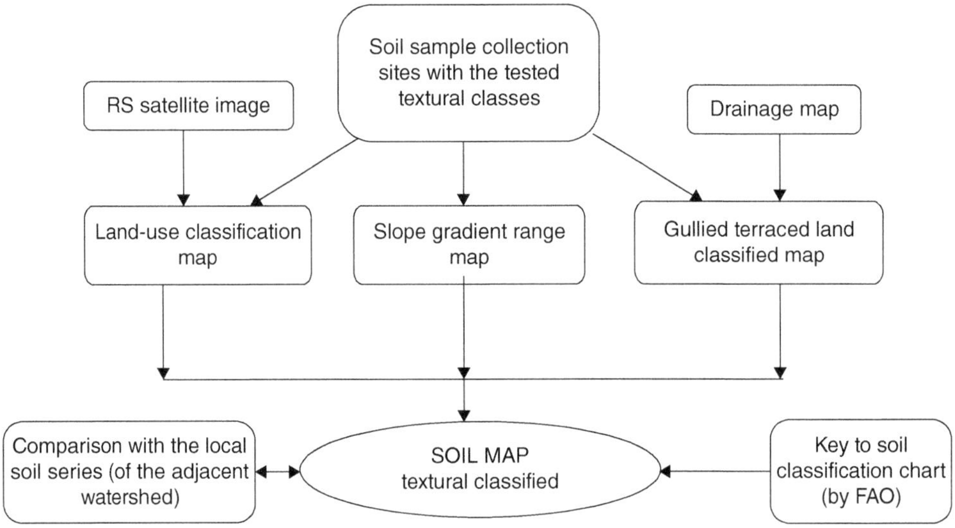

Fig. 10.3. Forested watershed soil map development procedure using geospatial technology (RS, remote sensing) and the soil classification key of the FAO (Food and Agriculture Organization of the United Nations). (From Panda *et al.*, 2004a.)

cycle and climate dynamics, flooding and forest fires, including hydrological processes. Although many methods are available to measure soil moisture, *in situ* measurement of the spatial distribution of soil moisture on a watershed/landscape scale is not typically possible. International efforts have been underway for decades to reliably measure soil moisture with an acceptable spatial resolution using a satellite-based RS technique.

Active microwave RS observations of backscattering, such as C-band vertically polarized synthetic aperture radar (SAR) observations from the second European Remote Sensing (ERS-2) satellite, have the potential to measure moisture content in a near-surface layer of soil (Walker *et al.*, 2004). However, SAR backscattering observations are highly dependent on topography, soil texture, surface roughness and soil moisture, meaning that soil moisture inversion from single-frequency and polarization SAR observations is difficult. The authors reported some improvements in measurements of near-surface soil moisture with the ERS-2 satellite over Landsat. Microwave RS-based soil moisture estimates are limited to bare soil or low to moderate amounts of vegetation cover. Passive microwave sensors have the advantage of collecting soil moisture remote data in areas with high vegetation cover like forest land cover, but with a trade-off in the spectral resolution range. While the most useful frequency range for soil moisture sensing is 1 to 5 GHz, passive microwave RS is in a range of 10 to 20 km (Njoku and Entekhabi, 1996). Njoku and Entekhabi (1996) outlined the basic principles of the passive microwave technique for soil moisture sensing and how to optimally assimilate passive microwave data into hydrological models. Schmugge *et al.* (2002) remotely estimated forest surface soil moisture from passive microwave data.

Nolan and Fatland (2003) reported that recent advancements in making soil moisture models may act as the Rosetta stone that allows for the InSAR (Interferometric Synthetic Aperture Radar) measurement of soil moisture using existing satellites. Lu *et al.* (2005) demonstrated the feasibility of measuring changes in water level beneath tree cover more accurately using C-band InSAR images from ERS-1 and ERS-2 satellites than the L-band for swamp forests in Louisiana, USA. This capability to measure water level changes in wetlands, and consequently in water storage capacity, using RS may provide a required input for hydrological models and flood hazard assessments. Panda *et al.* (2015) in their recent study used Band 5 (near-infrared) and Band 7 (mid-infrared) to estimate plant moisture (stomatal conductance) and soil moisture in the forest cover with mature and young pine, switchgrass and pine understorey with more than 70% accuracy.

10.2.3 Forest vegetation and biomass mapping

Forest vegetation has an apparent influence on microclimate (air temperature, humidity and wind speed) under the canopy compared with open area land covers. It is the 'active' surface for the absorption of solar energy and carbon dioxide and the release of oxygen and water vapour through evapotranspiration, and has a localized effect. In the context of global warming and a climate change scenario, understanding the forest microclimate with respect to forest vegetation or forest biomass is a necessity. High-resolution orthoimagery along with an advanced image processing approach is successful in forest vegetation speciation. Plate 9 depicts the advantage of an object-based image analysis (OBIA)-image segmentation approach in forest tree speciation in the Elachee Nature Center in Georgia, USA with the use of very high resolution (30 cm) orthoimagery and LiDAR data (to determine tree height) along with Visual Basic for Application (VBA) coding in ArcObjects platform.

In Europe in general, and in France specifically, forest mapping and species identification were developed using RS data such as aerial infrared photography in the 1970s (Touzet and Lecordix, 2010) and more recently (since the 1990s) SPOT imagery at 10- to 20-m resolution (i.e. panchromatic and visible and near-infrared (VNIR) bands). The Soil and Water Assessment Tool (SWAT) model ArcGIS add-in is being used to conduct ecohydrological modelling. Bärlund *et al.* (2007) assessed SWAT model performance in the evaluation of hydrology management actions for implementation of the Water Framework Directive in a Finnish forested catchment. The study suggests that GT is an efficient tool to differentiate individual trees in the forest, determine the forest biomass, and subsequently define the loss of water from the forest land via evapotranspiration. Panda *et al.* (2015) developed a procedure to assess the water loss through evapotranspiration in plots with pine only, pine plus understories, pine and switchgrass intercropping, and switchgrass only using 30 cm LiDAR, 15 cm orthoimagery and expert knowledge of the field. Riegel (2012) used LiDAR data to develop a forest biomass quantification model which provided insight into overall forest health and helped in forest ET modelling.

Forest net primary production (NPP) is a field of RS research that includes hyperspectral data from airborne or satellite platforms like AVIRIS or Hyperion (Ollinger and Smith, 2005). 'Primary production' is the accumulation of organic material produced by a plant (biomass). 'Net primary production' is the remaining biomass after subtracting energy used (respiration) for plant growth and development. In the coming years, the HYPXIM project (Michel *et al.*, 2011) aims at providing researchers, including forest RS topics, with hyperspectral satellite data, including VNIR and shortwave infrared sensors, at a high resolution (8 to 15 m). This project will be of great interest for forest research that needs the spatial resolution of airborne data with the global coverage facility of a satellite mission.

10.2.4 Forest evapotranspiration estimation

Different plant species compete for water at different amounts in a forest due to the dense and complex composition of the vegetation. However, the temporal water uptake or evapotranspiration (ET) rate and amount for each species of forest vegetation are poorly understood in a watershed/landscape. The forest ET rate depends upon many factors such as forest soils, vegetation, and climatic conditions such as air and canopy temperature, solar radiation, vapour pressure, wind velocity, and the nature and type of the evaporating surface in the forest range (Viessman and Lewis, 2002). Plant evaporation occurs mostly from the above canopy interception and understorey/litter evaporation. Transpiration encompasses the withdrawal and transport of water from the soil/aquifer system from plant roots and stems, and eventually from plant leaves into the atmosphere (Senay *et al.*, 2013). According to Viessman and Lewis (2002), available heat energy (radiation and air temperature), capacity to transport vapour away from the evaporative surface by wind and humidity, and soil water-content availability are the guiding factors for ET. LAI, canopy temperature (T_c), canopy (G_c) or stomatal conductance (g_s), wind velocity, and soil moisture or volumetric water content are the most important parameters of ET estimation (Panda *et al.*, 2014; see also Chapter 3, this volume, for more details on forest ET processes and controlling factors).

In recent years, RS-based GT has been increasingly used for development and application of ET models for determining and assessing the ET rates compared with field measured data for agricultural and irrigated crop ecosystems (Cammalleri *et al.*, 2014). These novel approaches have been tested recently for individual forest species (see Chapter 3, this volume; Panda *et al.*, 2014, 2016). These ET-related parameters (albedo, conductance, canopy temperature, soil moisture, LAI) are estimated with RS imagery data (Narasimhan *et al.*, 2003; Mu *et al.*, 2007; Chen *et al.*, 2014; Panda *et al.*, 2016). Thus, forest hydrologists could make decisions on forest vegetation species to grow or not grow. The RS-based spectral information is particularly useful in applications dealing with mapping and modelling biophysical properties of ecosystems such as water quality, plant vigour and soil nutrients (i.e. Landsat individual bands cater to very specific earth observation applications) (Panda *et al.*, 2016). As shown in Fig. 10.4, Landsat individual bands or a combination of bands through ratio development can estimate the ecohydrological parameters. Panda *et al.* (2016) have used free Landsat 7 and Landsat 8 images to develop ET and ET parameter models for homogeneous pine forest in coastal North Carolina, USA. Table 10.1 provides a geospatial-based input and ET/ET parameters output correlation chart. *Remote Sensing and Hydrology 2000* (Owe *et al.*, 2001) includes many individual research articles describing the use of RS in ET and ET parameter estimation along with other hydrological processes.

10.2.5 Forest hydrology attributed geohazards analysis

Different geohazards are directly or indirectly related to forests and forest hydrology. Geohazards such as wildfires, landslides, drought and flooding are very harmful for humans and the biodiversity directly associated with forest cover. All of these hazards can be monitored, managed and even pre-warned with the use of GT. The following presents a few applications analysing the susceptibility or vulnerability of such geohazards related to forest hydrology.

Forest fires

Forest fire management is a very big issue today. A persistent La Niña effect in the last four years (from 2011 to 2015) created severe drought conditions in the US west coast states (Lenihan and Bachelet, 2015). The drought in California and other west coast forests led to increased wildfires in 2015. Forest fires or wildfires are regulated by many environmental features of forests, including soil water content, forest topography, forest infrastructure, forest cover microclimate and especially forest species. A combined understanding of these spatial features would help manage wildfires better. Dudley *et al.* (2015) developed a geospatial model for determining locations of forest fire susceptibility in Sumter National Forest in South Carolina, USA. The authors used slope, aspect, slope rate of spread, slope suppression difficulty, NDVI, road buffer, fuel biomass density, urban fuel load and lightning strike frequency rasters to develop a comprehensive and fully automated geospatial model that predicts wildfire-vulnerable locations on a scale of low to high. Plate 10 provides the wildfire vulnerability map of Sumter National Forest (see also Chapter 13, this volume, for more about hydrology of forests after wildfire/prescribed fire).

RS applications for studying watershed-scale fires, their remote measurement techniques, their effects on biogeochemistry and the atmosphere, and their ecohydrological effects have been studied extensively by Riggan *et al.* (2004, 2009). Riggan's group also led development of the FireMapper thermal-imaging radiometer and its application to measurement and monitoring of large wildland fires and forest drought stress and mortality in mixed conifer forest (Riggan *et al.*, 2003). A study on tracking the MODIS NDVI time series to estimate fuel accumulation was conducted by Uyeda *et al.* (2015).

Forest fires significantly affect the hydrological cycle and thus rainfall–runoff modelling (Eisenbies *et al.*, 2007; Folton *et al.*, 2015). Recently, Chen *et al.* (2013) analysed satellite observations of terrestrial water storage from

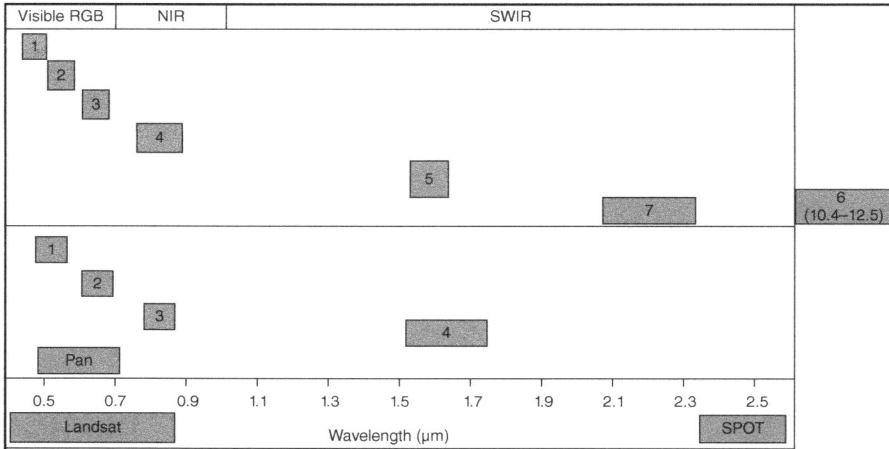

Landsat 7 ETM+ band (μm)	Applications	SPOT 4 band (μm)
Band 1 – Blue IR (0.45–0.52)	Coastal water mapping, soil/vegetation differentiation, deciduous/coniferous differentiation	
Band 2 – Green (0.52–0.60)	Green reflectance from healthy vegetation, iron content in rocks and soil	Band 1 (0.50–0.59) (Green)
Band 3 – Red (0.63–0.69)	Chlorophyll absorption for plant differentiation	Band 2 (0.61–0.68) (Red)
Band 4 – NIR (0.76–0.90)	Biomass survey, waterbody delineation	Band 3 (0.79–0.89) NIR
Band 5 – MIR (1.55–1.75)	Plant moisture content, cloud/snow differentiation	Band 4 (1.58–1.75) MIR
Band 6 – TIR (10.4– 2.5)	Thermal mapping, soil moisture	
Band 7 – MIR (2.08–2.35)	Soil analysis	

Landsat 8 OLI and TIRS bands (μm)	Applications
Band 1 – Coastal aerosol (0.43–0.45)	Coastal and aerosol studies
Band 2 – Blue (0.45–0.51)	Bathymetric mapping, distinguishing soil from vegetation and deciduous from coniferous vegetation
Band 3 – Green (0.53–0.59)	Emphasizes peak vegetation, which is useful for assessing plant vigour
Band 4 – Red (0.64–0.67)	Discriminates vegetation slopes
Band 5 – NIR (0.88–0.85)	Emphasizes biomass content and shorelines
Band 6 – SWIR 1 (1.57–1.65)	Discriminates moisture content of soil and vegetation; penetrates thin clouds
Band 7 – SWIR 2 (2.11–2.29)	Improved moisture content of soil and vegetation and thin cloud penetration
Band 8 – Panchromatic (0.50–0.68)	15 m resolution, sharper image definition
Band 9 – Cirrus (1.36–1.38)	Improved detection of cirrus cloud contamination
Band 10 – TIRS 1 (10.60–11.19)	100 m resolution, thermal mapping and estimated soil moisture
Band 11 – TIRS 2 (11.5–12.51)	100 m resolution, improved thermal mapping and estimated soil moisture

Fig. 10.4. General comparison of Landsat and SPOT spectral bands for earth observation application (R, red; G, green; B, blue; NIR, near-infrared; SWIR, shortwave infrared; MIR, mid-infrared; TIR, thermal infrared; ETM+, Enhanced Thematic Mapper Plus; OLI, Operational Land Imager; TIRS, Thermal Infrared Sensor).

Table 10.1. Input–output correlation relationship for model development.

Models (with 2006–2012 data)	Input parameters (remote sensing Landsat 7 ETM+ based)	Output parameters (field data)
ET	Plot SAVI means Plot NDVI means Plot VVI means Individual Band 5, 6 and 7 DN value averages	Plot averages of calculated ET values from FLUX instrument (average of 12.00–14.00 hours) (in W/m^2)
Soil moisture	Band 7 means	Plot averages of 30 cm depth soil moisture value (in %)
Canopy temperature	Band 6 means	Plot averages of 12.00–14.00 hours (in °C)
Canopy conductance	Band 5 means	Plot averages of 12.00–14.00 hours (in m/s)

ET, evapotranspiration; SAVI, soil-adjusted vegetation index; NDVI, normalized difference vegetation index; VVI, vegetation vigour index; DN, digital number.

the Gravity Recovery and Climate Experiment (GRACE) mission, along with satellite observations of fire activity from the MODIS mission for the Amazon region. Based on the contrasting analysis of data for high- and low-fire years from 2002 to 2011, the authors suggested that, at least qualitatively, water storage as measured by GRACE can provide information to help predict the severity of a fire season in the region several months in advance.

Landslides

Landslides are attributed to drought and large wildfires. Large wildfires after a persistent drought decrease the forest plant density, and hence the plant root and soil-binding power diminishes. Forest soils are looser due to drought conditions and, hence, are more vulnerable to erosion. Therefore, with immediately succeeding precipitation, a large mass of soil from the steep slope forest area slides down, creating life- and resource-threatening landslides. The geology of the forest area plays a greater role in landslides. Nolan *et al.* (2011), in their award-winning presentation in the 2011 Georgia Urban and Regional Information Systems Association (GA-URISA) conference, showed the advantage of GT to determine the susceptibility of landslides in the Coosawhatchee watershed in the Chattahoochee National Forest of north Georgia, USA. They used geospatial data such as soil texture, soil drainage, maximum water capacity, bulk density, lithology, basement depth, slope, storm surge and LULC.

Floods

Forest hydrology plays a bigger role in determining flooding susceptibility due to the distinct topography, soil composition and hydrological parameters in forest cover compared with other spatial locations. Forest cover is a low-contributing land cover towards flooding due to its soil composition (Booth *et al.*, 2002; van Dijk and Keenan, 2007). However, as discussed earlier, deforestation or forest degradation generally would change the soil dynamics and lead to higher runoff from the forest cover. In general, steeper topography is part of forest land cover and flooding vulnerability increases in those spatial locations. Several methods have been used to model the flood potential sites throughout the world, but GT usage is preferred, because all flooding parameters are considered to be spatial in nature. Choi and Liang (2010) in South Korea used the DEM (digital elevation model) hydrological soil group in their models to study the mostly mountainous watershed for flood vulnerability analysis. Ramsey *et al.* (2013) reported that SAR inundation mapping could provide an improved representation of coastal flooding, including flooding in the mangrove areas.

10.2.6 Forest stream water quality management

Stream water quality is the consequence of forest hydrology management. The riparian forest cover along streams is the transition or ecotone between terrestrial and aquatic ecosystems.

It supports a host of essential functions (Naiman and Décamps, 1997), like filtering runoff nutrients, providing shade that influences water temperature and dissolved oxygen concentration in waterbodies, putting leaf litter into the water as a carbon source for microbes and invertebrates at the base of the food web, supporting the stream banks structurally, supporting channels with large woody debris, diversifying stream habitats, and providing essential cover for flood flows and sediment transport. RS technology is efficiently being used to delineate the riparian forest cover, or the lack of it, along streams. The Watershed Habitat Evaluation and Biotic Integrity Protocol (WHEBIP) developed by Dr Reuben Goforth (Carlsen, 2004) and a similar protocol developed with the USDA Forest Service use stream riparian forest cover and stream channel attributes as major parameters to determine stream health. The lead author has developed an online estimation tool (https://web.ung.edu/gis/water/calculator.aspx) for calculating stream faecal coliform load from non-point and point sources, including forest land cover.

Zhang and Barten (2008) developed the Watershed Forest Management Information System (WFMIS) to help protect water resources from watershed/forest degradation. The WFMIS was developed as an extension of ArcGIS with three sub-modules to address non-point source pollution mitigation, road system management and silvicultural operations (Zhang and Barten, 2008). Panda *et al.* (2004b) developed a GIS-based watershed management DSS for determining water quality and quantity variability due to annual land cover changes. The study area, the 12-digit HUC (Hydrologic Unit Code) Beaver Lake watershed, was a forested watershed with more than 61% forest cover. This DSS is very important for FHMDS in water quality monitoring of forested streams (Panda *et al.*, 2004b). Zhang and Barten (2008) have also developed a stand-alone interface in VBA. A user can input the forest cover loss area in acres and the software will predict the water quality change (total P, total N, PO_4^{3-}, NO_3^- and total suspended solid) in kg/ha/year. The GT-based forest biomass studies discussed earlier would help quantify the water quality dynamics of forest streams. Forest trails, nutrient-rich forest soils and unique forest hydrological cycles are the causes of different forest stream water quality dynamics (Lowrance *et al.*, 1997). The Water Erosion Prediction Project (WEPP) is a process-based model that allows continuous simulation in small watersheds and hillslope profiles to estimate soil erosion and subsequent water quality dynamics in forests (Flanagan *et al.*, 1995). Geospatial interface for WEPP (GeoWEPP) has the potential to predict soil- and water erosion-based forest stream water quality monitoring and management using PRISM climate data, burn severity data, distributed WEPP land-use data, distributed WEPP soil parameters and DEM. The model would accurately and efficiently predict the forest soil erosion rate to support forest managers.

10.3 Modelling Forest Hydrological Processes with Geospatial Technology Support

Distributed models like MIKE Système Hydrologique Européen (SHE), SWAT, TOPMODEL (topographic model) and others are widely used to simulate ecohydrological processes in a large watershed-scale landscape, which generally contains the forest land use (Amatya *et al.*, 2011). Such distributed models use parameters directly related to the physical characteristics of the catchment (watershed), namely topography, soil, LULC and geology; and spatial variability in physical characteristics and meteorological conditions (Pietroniro and Leconte, 2000). Therefore, these models provide the possibility of deriving their inputs from remotely sensed data (Gupta *et al.*, 2008). The RS technique is useful in deriving high-resolution information in spatial and temporal domains about the hydrological parameters and thus provides a new means for calibration and validation of distributed hydrological models (Fortin *et al.*, 2001).

A decade in hydrological research on ungauged basins (Hrachowitz *et al.*, 2013) has demonstrated the interest of using RS in collecting data to predict water flows including topographical (i.e. DEM) and land cover layers for spatially distributed hydrological models (Doten *et al.*, 2006; Khan *et al.*, 2011). In France, impacts of the Mediterranean forest basin have been studied (Cosandey, 1993; Cosandey *et al.*, 2005) using IFN (National Forest Inventory, France) forest land cover data derived from aerial photography mapping. Using an existing hydrological model that includes lateral groundwater flow, Sutanudjaja *et al.* (2014) showed that remotely sensed soil moisture data can be valuable for accurately predicting groundwater dynamics at a local level and could be scaled up to provide more accurate information about groundwater variability, availability and reserves across the globe.

Troch *et al.* (2007) investigated the potential use of GRACE data to detect the monthly changes in terrestrial water storage in the Colorado River basin using *in situ* data from 2003 to 2006 and comparing those data against the basin-scale water balance (BSWB)-based models. The authors found that the GRACE results agree with the BSWB model that winter 2005 was generally wet, but the GRACE results disagree with the exact timing of this event. With respect to BSWB, GRACE underestimates the severity of the subsequent dry period. Scanlon *et al.* (2012) reported that general correspondence between GRACE and groundwater level data found in the California Central Valley validates the methodology and increases confidence in the use of GRACE satellites to monitor groundwater storage changes. Van Griensven *et al.* (2012) evaluated LAI and ET simulated by the SWAT model with corresponding values obtained using remotely sensed data. The authors' evaluation showed that values for ET tend to be slightly underestimated, while those for LAI were visibly overpredicted. At the same time, the satellite images clearly followed the land-use pattern of the basin and showed uniform values for the different types of vegetation. This suggests that the SWAT model's forest species input parameter development process needs updating to provide correct results in forest ET estimation.

10.4 New Technology in Forest Hydrology Management

Higgins *et al.* (2014) used a satellite-based interferometry technique to map the subsidence of the Ganges–Brahmaputra river delta covering 10,000 km^2 area over 4 years. The authors found that the delta is subsiding at a rate of about 10 mm/year around Dhaka, Bangladesh's capital, and at about 18 mm/year outside the city, and indicated that satellite interferometry can be a useful method in accurately gauging the subsidence in deltas. Such techniques may be useful in large deltas with mangrove forests in Asia and Africa. Groundwater is the last component of the hydrological cycle to realize the benefits of RS (Becker, 2006). The author explored the potential for RS of groundwater in the context of active and planned satellite-based sensors. Again, these methods may well be applicable for large groundwater-dominated forested landscapes around the world.

Ongoing efforts under the planned NASA/Center National d'Etudes Spatiales (CNES) Surface Water and Ocean Topography (SWOT) satellite mission, including the planned new algorithm using AirSWOT (an airborne platform approximating SWOT's capabilities), will provide an enhanced tool to accurately characterize river discharge from space by providing concurrent observations of water surface elevation, surface slope and inundated area for wide rivers (Pavelsky, 2012). Efforts are also underway to develop and expand space techniques to measure changes in terrestrial waters (Alsdorf *et al.*, 2003; Cazenave *et al.*, 2004). Such techniques will be useful for large forest landscapes like the Amazon River basin and streams/rivers draining long-term USDA Forest Service experimental forests and ranges in the conterminous USA.

10.5 Conclusions

This chapter provides a detailed discussion on the GT applications in forest hydrological processes management that includes: (i) forest cover mapping and change analysis; (ii) forest soil water/moisture estimation and forested wetlands analysis; (iii) forest vegetation and biomass mapping; (iv) forest ET estimation; (v) forest hydrology attributed geohazard analysis, such as forest fires, landslides and flooding; and (vi) forest stream water quality management. The chapter also provides insight on modelling forest hydrological processes with GT support. The last section of the chapter discusses new technology applications in forest hydrology management and provides suggestions on future studies.

As discussed in the chapter, GTs including RS, GIS, GNSS and IT have tremendous potential for better decision support in forest management and especially forest hydrology management. More and more hydrology models/software, such as GeoWEPP (http://geowepp.geog.buffalo.edu/versions/arcgis-10-x/), the Automated Geospatial Watershed Assessment (AGWA) tool (http://www.epa.gov/esd/land-sci/agwa/) and the USDA Forest Service database tools, Natural Resource Manager (NRM) (http://www.fs.fed.us/nrm/index.shtml), are being developed for forest hydrology management that use GT. As mentioned in the chapter, comprehensive complete automated geospatial models are being developed in ArcGIS ModelBuilder platform that can use any type of RS and GIS data to analyse forest hydrological behaviour. Most importantly, GPS technology is getting better and more efficient with the introduction of more satellites into space by Europe, Russia, India and China. The GNSS – the advanced version of GPS – is being used as a major tool in fighting forest fires, landslides and other forest-related geohazards in all parts of the world. Image spatial and spectral resolutions are getting better, in part due to the participation of private entrepreneurs in real-time image data collection, and also with the introduction of large-scale hyperspectral imaging.

The future of forest hydrology management lies in the hands of every stakeholder, but reliance on trained forest managers may not be enough to keep the global forest cover in good shape and health. Therefore, everyone has a responsibility towards global forest upkeep, as it was found that the forest cover flourishes when private and public entities collaborate. Erratic weather conditions due to global warming and climate change, and the consequential El Niño and La Niña effects, are creating severe disruption in forest management. Therefore, freely available MODIS and Landsat 8 data and the subsequently generated NDVI and EVI, along with open-source (free) GIS software like Map Window (http://www.mapwindow.org/), QGIS (http://www.qgis.org/en/site/) and GRASS (https://grass.osgeo.org/), would help develop FHMDS to save forest land cover from degradation. Above all, it is expected that with the advent of UAVs and UASs, which will be in the hands of many stakeholders in the near future, forest management could be easier. GT is getting easier, and the working procedures are becoming available in the public domain for the layman's use. Stakeholders should take advantage of these advanced technologies to take prudent steps towards FHMDS.

References

Alsdorf, D., Lettenmaier, D.D. and Vörösmarty, C. (2003) The need for global, satellite-based observations of terrestrial surface waters. *Eos, Transactions, American Geophysical Union* 84(29), 269–280.

Amatya, D.M., Douglas-Mankin, K.R., Williams, T.M., Skaggs, R.W. and Nettles, J.E. (2011) Advances in forest hydrology: challenges and opportunities. *Transactions of the ASABE* 54, 2049–2056.

Amatya, D.M., Sun, G., Green Rossi, C., Ssegane, H.S., Nettles, J.E. and Panda, S. (2015) Forests, land use change, and water. In: Zolin, C.A. and Rodrigues, R.deA.R. (eds) *Impact of Climate Change on Water Resources in Agriculture*. CRC Press, Boca Raton, Florida, pp. 116–153.

Austin, M.P. and Meyers, J.A. (1996) Current approaches to modelling the environmental niche of eucalypts: implication for management of forest biodiversity. *Forest Ecology and Management* 85, 95–106.

Bärlund, I., Kirkkala, T., Malve, O. and Kämäri, J. (2007) Assessing SWAT model performance in the evaluation of management actions for the implementation of the Water Framework Directive in a Finnish catchment. *Environmental Modelling & Software* 22, 719–724.

Barten, P.K. (2006) Overview of forest hydrology and forest management effects. Available at: http://www.maenvirothon.org/2010%20Overview_Barten_UMass_5Oct09-1.pdf (accessed 15 June 2015).

Becker, M.W. (2006) Potential for satellite remote sensing of ground water. *Ground Water* 44, 306–318.

Booth, D.B., Hartley, D. and Jackson, R. (2002) Forest cover, impervious-surface area, and the mitigation of stormwater impacts. *Journal of the American Water Resources Association* 38, 835–845.

Cammalleri, C., Anderson, M.C., Gao, F., Hain, C.R. and Kustas, W.P. (2014) Mapping daily evapotranspiration at field scales over rainfed and irrigated agricultural areas using remote sensing data fusion. *Agricultural and Forest Meteorology* 186, 1–11.

Carlsen, W.S., Trautmann, N.M., Cunningham, C.M., Krasny, M.E., Welman, A., Beck, H., Canning, H., Johnson, M., Barnaba, E., Goforth, R. and Hoskins, S. (2004) *Watershed Dynamics*. Cornell Scientific Inquiry Series. NSTA Press, Arlington, Virginia.

Cazenave, A., Milly, P.C.D., Douville, H., Benvineste, J., Kosuth, P. and Lettenmaier, D. (2004) Space techniques used to measure change in terrestrial waters. *Eos, Transactions, American Geophysical Union* 85(6), 106–107.

Chen, Y., Velicogna, I., Famiglietti, J.S. and Randerson, J.T. (2013) Satellite observations of terrestrial water storage provide early warning information about drought and fire season severity in the Amazon. *Journal of Geophysical Research – Biogeosciences* 118, 495–504.

Chen, Y., Xia, J., Liang, S., Feng, J., Fisher, J.B., Li, X., Li, X., Liu, S., Ma, Z., Miyata, A., *et al.* (2014) Comparison of satellite-based evapotranspiration models over terrestrial ecosystems in China. *Remote Sensing of Environment* 140, 279–293.

Choi, H.I. and Liang, X.Z. (2010) Improved terrestrial hydrologic representation in mesoscale land surface models. *Journal of Hydrometeorology* 11, 797–809.

Cosandey, C. (1993) Conséquences hydrologiques d'une coupe forestière. Le cas du bassin de la Latte (Mont-Lozère, France). In: Gresilen, M. (ed.) *L'eau, la terre et les hommes, Hommage à René Frécaut*. Presses Universitaires de Nancy, Nancy, France, pp. 355–363.

Cosandey C., Andréassian V., Martin, C., Didon-Lescot, J., Lavabre, J., Folton, N., Mathys, N. and Richard, D. (2005) The hydrological impact of the Mediterranean forest: a review of French research. *Journal of Hydrology* 301, 235–249.

Doten, C.O., Bowling, L.C., Lanini, J.S., Maurer, E.P. and Lettenmaier, D.P. (2006) A spatially distributed model for the dynamic prediction of sediment erosion and transport in mountainous forested watersheds. *Water Resources Research* 42, W04417, doi: 10.1029/2004WR003829 (accessed 6 April 2016).

Dudley, A., Panda, S.S., Amatya, D. and Kim, Y. (2015) Hydro-climatology based wildfire susceptibility automated geospatial model development for forest management. Presented at: *2015 Georgia Water Resources Conference,* 28–30 April 2015, Athens, Georgia, USA.

D'urso, G. and Minacapilli, M. (2006) Semi-empirical approach for surface soil water content estimation from radar data without a-priori information on surface roughness. *Journal of Hydrology* 321, 297–310.

EEA (European Environmental Agency) (2006) *CLC2006 Technical Guidelines*. Technical Report No. 17/2007. Office for Official Publications of the European Communities, Luxembourg.

Eisenbies, M.A., Aust, W.M., Burger, J.A. and Adams, M.B. (2007) Forest operations, extreme flooding events, and considerations for hydrologic modeling in the Appalachians – a review. *Forest Ecology and Management* 242, 77–98.

Flanagan, D.C., Ascough, J.C., Nicks, A.D., Nearing, M.A. and Laflen, J.M. (1995) Overview of the WEPP erosion prediction model. In: *USDA–Water Erosion Prediction Project: Hillslope Profile and Watershed Model Documentation*. US Department of Agriculture/National Soil Erosion Research Laboratory, West Lafayette, Indiana, pp. 1.1–1.12.

Folton, N., Andréassian, V. and Duperray, R. (2015) Hydrological impact of forest-fire from paired-catchment and rainfall–runoff modelling perspectives. Special issue: Modelling temporally-variable catchments. *Hydrological Sciences Journal* 60, 1213–1224.

Fortin, J.P., Turcotte, R., Massicotte, S., Moussa, R., Fitzback, J. and Villeneuve, J.P. (2001) Distributed watershed model compatible with remote sensing and GIS data. I: Description of model. *Journal of Hydrologic Engineering* 6, 91–99.

Giri, C., Pengra, B., Zhu, Z., Singh, A. and Tieszen, L.L. (2007) Monitoring mangrove forest dynamics of the Sundarbans in Bangladesh and India using multi-temporal satellite data from 1973 to 2000. *Estuarine, Coastal and Shelf Science* 73, 91–100.

Goward, S.N., Masek, J.G., Cohen, W., Moisen, G., Collatz, G.J., Healey, S., Houghton, R.A., Huang, C., Kennedy, R., Law, B., *et al.* (2008) Forest disturbance and North American carbon flux. *Eos, Transactions, American Geophysical Union* 89(11), 106–107.

Grenzdörffer, G.J., Engel, A. and Teichert, B. (2008) The photogrammetric potential of low-cost UAVs in forestry and agriculture. *International Archives of the Photogrammetry, Remote Sensing and Spatial Information Sciences* 31(B3), 1207–1214.

Gupta, H.V., Wagener, T. and Liu, Y. (2008) Reconciling theory with observations: elements of a diagnostic approach to model evaluation. *Hydrological Processes* 22, 3802–3813.

Higgins, S.A., Overeem, I., Steckler, M.S., Syvitski, J.P.M., Seeber, L. and Akhter, S.H. (2014) InSAR measurements of compaction and subsidence in the Ganges–Brahmaputra Delta, Bangladesh. *Journal of Geophysical Research, Earth Surface* 119, 1768–1781.

Hrachowitz, M., Savenije, H.H.G., Blöschl, G., McDonnell, J.J., Sivapalan, M., Pomeroy, J.W., Arheimer, B., Blume, T., Clark, M.P., Ehret, U., *et al.* (2013) A decade of predictions in ungauged basins (PUB) – a review. *Hydrological Sciences Journal* 58, 1198–1255.

Jipp, P.H., Nepstad, D.C., Cassel, D.K. and De Carvalho, C.R. (1998) Deep soil moisture storage and transpiration in forests and pastures of seasonally-dry Amazonia. In: Markham, A. (ed.) *Potential Impacts of Climate Change on Tropical Forest Ecosystems*. Springer, Dordrecht, the Netherlands, pp. 255–272.

Johnson, D.W., Cheng, W. and Burke, I.C. (2000) Biotic and abiotic nitrogen retention in a variety of forest soils. *Soil Science Society of America Journal* 64, 1503–1514.

Khan, S., Hong, Y., Wang, J., Yilmaz, K.K., Gourley, J.J., Adler, R.F., Brakenridge, G.R., Policelli, F., Habib, S. and Irwin, D. (2011) Satellite remote sensing and hydrologic modeling for flood inundation mapping in Lake Victoria Basin: implications for hydrologic prediction in ungauged basins. *IEEE Transactions on Geoscience and Remote Sensing* 49, 85–95.

Khosravipour, A., Skidmore, A.K., Wang, T., Isenburg, M. and Khoshelham, K. (2015) Effect of slope on treetop detection using a LiDAR canopy height model. *ISPRS Journal of Photogrammetry and Remote Sensing* 104, 44–52.

Kim, R.H., Kwon, J.H., Amatya, D.M. and Kim, H.W. (2015) Monitoring forest degradation in North Korea using satellite imagery. Presented at: *70th Annual Meeting of the Korean Association of Biological Sciences*, August 12–13, 2015, Seoul National University, Seoul, Republic of Korea.

Kim, R.H., Kim, H.W., Lee, J.-H., Kim, Y.S., Kim, A.L., Lee, S.H. and Koo, Y.S. (2016) Analysis of land cover and vegetation change in North Korea using MODIS data for the 14 years (2001–2014) (in Korean). *Journal of the Korean Forest Society* (in press).

Kittredge, J. (1948) *Forest Influences: The Effects of Woody Vegetation on Climate, Water, and Soil, With Applications to the Conservation of Water and the Control of Floods and Erosion*. McGraw-Hill, New York.

Krishnaswamy, J., Robert, J. and Shijo, J. (2014) Consistent response of vegetation dynamics to recent climate change in tropical mountain regions. *Global Change Biology* 20, 203–215.

Lee, D.K. and Lee, Y.K. (2005) Roles of Saemaul Undong in reforestation and NGO activities for sustainable forest management in Korea. *Journal of Sustainable Forestry* 20(4), 1–16.

Lenihan, J.M. and Bachelet, D. (2015) Historical climate and suppression effects on simulated fire and carbon dynamics in the conterminous United States. In: Bachelet, D. and Turner, D. (eds) *Global Vegetation Dynamics: Concepts and Applications in the MC1 Model*. Wiley, New York, pp. 17–30.

Lepers, E., Lambin, E.F., Janetos, A.C., DeFries, R., Achard, R., Ramankutty, N. and Scholes, R.J. (2005) A synthesis of information on rapid land-cover change for the period 1981–2000. *BioScience* 55, 115–124.

Lowrance R. (1992) Groundwater nitrate and denitrification in a coastal plain riparian forest. *Journal of Environmental Quality* 21, 401–405.

Lowrance, R., Altier, L.S., Newbold, J.D., Schnabel, R.R., Groffman, P.M., Denver, J.M., Correll, D.L., Gilliam, J.W., Robinson, J.L., Brinsfield, R.G., *et al.* (1997) Water quality functions of riparian forest buffers in Chesapeake Bay watersheds. *Environmental Management* 21, 687–712.

Lu, Z., Crane, M., Kwoun, O.-I., Wells, C., Swarzenski, C. and Rykhus, R. (2005) C-band radar observes water level change in swamp forests. *Eos, Transactions, American Geophysical Union* 86(14), 141–144.

Michel, S., Gamet, P. and Lefevre-Fonollosa, M.-J. (2011) HYPXIM – a hyperspectral satellite defined for science, security and defence users. In: *Proceedings of 2011 3rd Workshop on Hyperspectral Image and Signal Processing: Evolution in Remote Sensing (WHISPERS), Lisbon, 6–9 June 2011*. Institute of Electrical and Electronics Engineers, New York, doi: 10.1109/WHISPERS.2011.6080864 (accessed 6 April 2016).

Morin, G.-A. (2010) La continuité de la gestion des forêts françaises de l'ancien régime à nos jours, ou comment l'Etat a-t-il pris en compte le long terme. *Revue Française d'Administration Publique* (134), pp. 233–248.

Mu, Q., Heinsch, F.A., Zhao, M. and Running, S.W. (2007) Development of a global evapotranspiration algorithm based on MODIS and global meteorology data. *Remote Sensing of Environment* 111, 519–536.

Naiman, R.J. and Décamps, H. (1997) The ecology of interfaces: riparian zones. *Annual Review of Ecology and Systematics* 28, 621–658.

Narasimhan, B., Srinivasan, R. and Whittaker, A.D. (2003) Estimation of potential evapotranspiration from NOAA-AVHRR satellite. *Applied Engineering in Agriculture* 19, 309–318.

Njoku, E.G. and Entekhabi, D. (1996) Passive microwave remote sensing of soil moisture. *Journal of Hydrology* 184, 101–129.

Nolan, J., Panda, S.S. and Mobasher, K. (2011) *Landslide Probability Study of Coosawhatchee 8-digit HUC Watershed using Geospatial Technology*. Georgia Urban and Regional Information Systems Association (GA-URISA), Atlanta, Georgia.

Nolan, M. and Fatland, D.R. (2003) New DEMs may stimulate significant advancements in remote sensing of soil moisture. *Eos, Transactions, American Geophysical Union* 84(25), 233–236.

Ollinger, S.V. and Smith, M.L. (2005) Net primary production and canopy nitrogen in a temperate forest landscape: an analysis using imaging spectroscopy, modeling and field data. *Ecosystems* 8, 760–778.

Osman, K.T. (2013) Forest soils. In: Osman, K.T. (ed.) *Soils: Principles, Properties and Management*. Springer, Dordrecht, the Netherlands, pp. 229–251.

Owe, M., Brubaker, K., Ritchie, J. and Rango, A. (eds) (2001) *Remote Sensing and Hydrology 2000*. International Association of Hydrological Sciences, Wallingford, UK.

Panda, S.S., Andrianasolo, H., Murty, V.V.N. and Nualchawee, K. (2004a) Forest management planning for soil conservation using satellite images, GIS mapping, and soil erosion modeling. *Journal of Environmental Hydrology* 12(13), 1–16.

Panda, S.S., Chaubey, I., Matlock, M.D., Haggard, B.E. and White, K.L. (2004b) Development of a GIS-based decision support system for Beaver Lake watershed management. In: *Proceedings of the American Water Resources Association (AWRA) Spring Specialty Conference, May 17–19, 2004, Nashville, Tennessee, USA*. Available at: http://www.awra.org/proceedings/0405pro_toc.html (accessed 21 April 2016).

Panda, S.S., Andrianasolo, H. and Steele, D. (2005) Application of geotechnology to watershed soil conservation planning at the field scale. *Journal of Environmental Hydrology* 13, 1–22.

Panda, S.S., Ames, D.P. and Panigrahi, S. (2010) Application of vegetation indices for agricultural crop yield prediction using neural network. *Remote Sensing* 2, 673–696.

Panda, S.S., Amatya, D.M. and Hoogenboom, G. (2014) Stomatal conductance, canopy temperature, and leaf area index estimation using remote sensing and OBIA techniques. *Journal of Spatial Hydrology* 12(1), 24 p.

Panda, S.S., Rao, M., Fitzgerald, J. and Thenkabail, P.S. (2015) Remote sensing satellites and sensors: optical, radar, LiDAR, microwave, hyperspectral, and UAVs. In: Thenkabail, P.S. (ed.) *Remote Sensing Handbook*, Vol. I. CRC Press, New York, pp. 3–57.

Panda, S.S., Amatya, D.M., Sun, G. and Bowman, A. (2016) Remote estimation of a managed pine forest evapotranspiration with geospatial technology. *Transactions of the ASABE* (in press).

Pavelsky, T. (2012) Developing new algorithms for estimating river discharge from space. A meeting note. *Eos, Transactions, American Geophysical Union* 93(45), 457.

Perry, D.A., Oren, R. and Hart, S.C. (2008) *Forest Ecosystems*, 2nd edn. Johns Hopkins University Press, Baltimore, Maryland.

Pietroniro, A. and Leconte, R. (2000) A review of Canadian remote sensing applications in hydrology, 1995–1999. *Hydrological Processes* 14, 1641–1666.

Pinay, G., Roques, L. and Fabre, A. (1993) Spatial and temporal patterns of denitrification in a riparian forest. *Journal of Applied Ecology* 30, 581–591.

Potter, C., Genovese, V., Gross, P., Boriah, S., Steinbach, M. and Kumar, V. (2007) Revealing land cover change in California with satellite data. *Eos, Transactions, American Geophysical Union* 88(26), 269–274.

Pritchett, W.L. (1979) *Properties and Management of Forest Soils*. Wiley, New York.

Ramakrishnan, P.S., Rao, K.S., Chandrasekhara, U.M., Chhetri, N., Gupta, H.K., Patnaik, S., Saxena, K.G. and Sharma, E. (2012) South Asia. In: Parrotta, J.A. and Trosper, R.L. (eds) *Traditional Forest-Related Knowledge: Sustaining Communities, Ecosystems and Biocultural Diversity*. World Forests Vol. 12. Springer, Dordrecht, the Netherlands, pp. 315–356.

Ramsey, E. III, Rangoonwala, A. and Bannister, T. (2013) Coastal flood inundation monitoring with satellite C-band and L-band synthetic aperture radar data. *Journal of the American Water Resources Association* 49, 1239–1260.

Richmond, E. (2007) Forest hydrology. In: Fierro, P. Jr and Nyer, E.K. (eds) *The Water Encyclopedia: Science and Issues*, 3rd edn. CRC Press, Boca Raton, Florida.

Riegel, J.B. (2012) A comparison of remote sensing methods for estimating above-ground carbon biomass at a wetland restoration area in the southeastern coastal plain. MS thesis, Duke University, Durham, North Carolina, USA.

Riggan, P.J. and Tissell, R.G. (2009) Airborne remote sensing of wildland fires. In: Bytnerowicz, A., Arbaugh, M., Anderson, C. and Riebau, A. (eds) *Wildland Fires and Air Pollution*. Developments in Environmental Science Vol. 8. Elsevier Publishers, Amsterdam, pp. 139–168.

Riggan, P.J., Tissell, R.G. and Hoffman, J.W. (2003) Application of the FireMapper™ thermal-imaging radiometer for wildfire suppression. *2003 IEEE Aerospace Conference Proceedings* 4, 1863–1872.

Riggan, P.J., Tissell, R.G., Lockwood, R.N., Brass, J.A., Pereira, J.A.R., Miranda, H.S., Miranda, A.C., Campos, T. and Higgins, R. (2004) Remote measurement of energy and carbon flux from wildfires in Brazil. *Ecological Applications* 14, 855–872.

Scanlon, B.R., Longuevergne, L. and Long, D. (2012) Ground referencing GRACE satellite estimates of groundwater storage changes in the California Central Valley, USA. *Water Resources Research* 48, W04520, doi: 10.1029/2011WR011312 (accessed 6 April 2016).

Schmugge, T.J., Kustas, W.P., Ritchie, J.C., Jackson, T.J. and Rango, A. (2002) Remote sensing in hydrology. *Advances in Water Resources* 25, 1367–1385.

Senay, G.B., Bohms, S., Singh, R.K., Gowda, P.H., Velpuri, N.M., Alemu, H. and Verdin, J.P. (2013) Operational evapotranspiration mapping using remote sensing and weather datasets: a new parameterization for the SSEB approach. *Journal of the American Water Resources Association* 49, 577–597.

Sundquist, B. (2007) Forest land degradation: a global perspective, 6th edn. Available at: http://www.civilizationsfuture.com/bsundquist/df0.html (accessed 21 April 2016).

Sutanudjaja, E.H., van Beek, L.P.H., de Jong, S.M., van Geer, V.C. and Bierkens, M.R.P. (2014) Calibrating a large-extent high-resolution coupled groundwater–land surface model using soil moisture and discharge data. *Water Resources Research* 50, 687–705.

Szott, L.T., Palm, C.A. and Buresh, R.J. (1999) Ecosystem fertility and fallow function in the humid and subhumid tropics. *Agroforestry Systems* 47, 163–196.

Touzet, T. and Lecordix, F. (2010) La carte forestière sans papier. *Comité français de Cartographie* 206, 53–62.

Troch, P., Durcik, M., Seneviratne, S., Hirschi, M., Teuling, A., Hurkmans, R. and Hasan, S. (2007) New data sets to estimate terrestrial water storage change. *Eos, Transactions, American Geophysical Union* 88(45), 469–470.

USGS LCI (US Geological Survey, Land Cover Institute) (2015) NLCD 92 Land Cover Class Definitions. Available at: http://landcover.usgs.gov/classes.php (accessed 15 August 2015).

Uyeda, K.A., Stow, D.A. and Riggan, P.J. (2015) Tracking MODIS NDVI time series to estimate fuel accumulation. *Remote Sensing Letters* 6, 587–596.

Van Dijk, A.I. and Keenan, R.J. (2007) Planted forests and water in perspective. *Forest Ecology and Management* 251, 1–9.

Van Griensven, A., Maskey, S. and Stefanova, A. (2012) The use of satellite images for evaluating a SWAT model: application on the Vit Basin, Bulgaria. In: Seppelt, R., Voinov, A.A., Lange, S. and Bankamp, D. (eds) *International Environmental Modelling and Software Society (iEMSs) 2012 International Congress on Environmental Modelling and Software. Managing Resources of a Limited Planet: Pathways and Visions under Uncertainty, Sixth Biennial Meeting, Leipzig, Germany*. Available at: http://www.iemss.org/society/index.php/iemss-2012-proceedings (accessed 6 April 2016).

Viessman, W. and Lewis, G.L. (2002) *Introduction to Hydrology*. Prentice-Hall, Upper Saddle River, New Jersey.

Von Gadow, K. and Bredenkamp, B. (1992) *Forest Management*. Academica, Pretoria, South Africa.

Walker, J.P., Houser, P.R. and Willgoose, G.R. (2004) Active microwave remote sensing for soil moisture measurement: a field evaluation using ERS-2. *Hydrological Processes* 18, 1975–1997.

Zarco-Tejada, P.J., Diaz-Varela, R., Angileri, V. and Loudjani, P. (2014) Tree height quantification using very high resolution imagery acquired from an unmanned aerial vehicle (UAV) and automatic 3D photo-reconstruction methods. *European Journal of Agronomy* 55, 89–99.

Zhang, Y. and Barten, P.K. (2008) Watershed Forest Management Information System (WFMIS). *Environmental Modelling & Software* 24, 569–575.

Zinke, P.J. (1967) Forest interception studies in the United States. In: Sopper, W.E. and Lull, H.W. (eds) *International Symposium on Forest Hydrology*. Pergamon Press, New York, pp. 137–161.

11 Forest Cover Changes and Hydrology in Large Watersheds

X. Wei[1]*, Q. Li[1], M. Zhang[2], W. Liu[3] and H. Fan[3]
[1]University of British Columbia, Kelowna, British Columbia, Canada; [2]University of Electronic Science and Technology of China, Chengdu, People's Republic of China; [3]Nanchang Institute of Technology, Nanchang, People's Republic of China

11.1 Introduction

Forests play an important role in the water cycle by influencing rainfall interception, evapotranspiration, soil infiltration and storage, and streamflow. The impacts of forest changes caused by either natural or human forces (e.g. wildfire, deforestation, reforestation, urbanization) on hydrology have been studied for a century, either by the traditional experimental paired watershed approach or hydrological modelling (see Chapter 9, this volume). A general understanding is that deforestation can substantially increase annual streamflow, magnify peak flows and alter baseflows (Stednick, 1996; Moore and Wondzell, 2005; Creed *et al.*, 2014), while reforestation can decrease annual streamflow and reduce peak flows. However, these results are drawn mainly from experimental watershed studies conducted at small spatial scales (<100 km^2, most of which are less than 10 km^2) and they cannot be simply extrapolated to large watersheds (>1000 km^2) (Shuttleworth, 1988; Shaman *et al.*, 2004) because of more complexities of land forms (e.g. managed and natural forests, wetlands, lakes, open lands) and their interactions. This highlights a critical need for conducting separate research on the impacts of forest cover changes and water in large watersheds. The objectives of this chapter are to: (i) briefly summarize impacts of forest cover changes on hydrology in large watersheds; (ii) describe various existing research methods in evaluating forest cover change effects on hydrology in large watersheds; and (iii) identify research challenges and future research priorities.

Studying the impacts of forest cover changes and water in large watersheds is challenging. The first challenge is the lack of an efficient, commonly accepted methodology. The greatest difficulty in a large watershed study lies in separating the effects of forest changes (e.g. disturbances) and climate variability on hydrology (Zheng *et al.*, 2009; Wei and Zhang, 2011). Forest cover changes and climatic variability are generally viewed as two major drivers interactively influencing streamflow in large forested watersheds (Buttle and Metcalfe, 2000; Sharma *et al.*, 2000). It is commonly accepted that the effects of climate variability on hydrology must be excluded in order to quantify the hydrological impacts of forest cover changes in large watersheds. The experimental paired watersheds or physically based hydrological models, commonly used to study the hydrological effects of forest cover changes in small watersheds, however, have limitations when applied to large watersheds (Tuteja *et al.*, 2007; Scott and Prinsloo, 2008;

*Corresponding author; e-mail: adam.wei@ubc.ca

Zhao *et al.*, 2010). The experimental paired watershed approach is generally infeasible for large watersheds given the great difficulty in locating suitable control watersheds (Fohrer *et al.*, 2005). Similarly, physically based hydrological models, such as the Distributed Hydrology–Soils–Vegetation Model (DHSVM), MIKE Système Hydrologique Européen (SHE), the Variable Infiltration Capacity (VIC), and other similar models are applicable only for the watersheds that are well monitored with extensive, long-term available data on vegetation, soil, topography, land use, hydrology and climate (Stednick, 2008; Kirchner, 2009; Wei and Zhang, 2010). Moreover, the empirical relationships between different watershed processes and components used in hydrological models are drawn mainly from small watershed studies and may be problematic when transferred to large watersheds (Kirchner, 2006). Therefore, the most commonly used methods in small-scale paired watershed studies have limited utility in forest hydrological studies on large watersheds.

Second, the lack of a suitable indicator for representing and integrating various types of forest cover changes or disturbances is another challenge in large watershed studies (Wei and Zhang, 2010; Zhang, 2013). For example, in a large watershed, different types of forest disturbances (both natural and anthropogenic) are accumulated over space and time. To quantitatively represent cumulative forest disturbances over time at a watershed scale, an integrated indicator other than a simple indicator such as total disturbed area or forest cover rate is needed. A suitable forest disturbance indicator for a large watershed should not only represent all types of disturbances and intensity ranges, but also include their cumulative forest disturbance histories and subsequent recovery processes following disturbances over space and time (Wei and Zhang, 2010). Equivalent clearcut area (ECA) is defined as the area that has been harvested, cleared or burned with a reduction factor to account for hydrological recovery due to forest regeneration after disturbances (BC Ministry of Forests and Rangeland, 1999). The indicator of cumulative equivalent clearcut area (the sum of annual equivalent clearcut area, hereinafter referred to as CECA) has been successfully used in the Pacific Northwest to test watershed-scale forest logging or wildfire and their effects on various watershed processes including aquatic habitat, hydrology and aquatic biology (Whitaker *et al.*, 2002; Chen and Wei, 2008; Lin and Wei, 2008). For example, the annual ECA and CECA of all forest disturbances including logging, wildfire and mountain pine beetle infestation accounted for 13.4% and 31.2% of the watershed area in 2004, respectively, in Baker Creek watershed (1570 km^2) located in British Columbia, Canada (Zhang and Wei, 2012). Other indicators such as remote sensing-based NDVI (normalized difference vegetation index) (Yang *et al.*, 2014) and total watershed sapwood area (Jaskierniak *et al.*, 2015) have also been applied. However, no full comparisons have been made yet to determine which indicators or indices are more suitable than others.

Finally, the lack of suitable study watersheds can also constrain forest hydrological studies in large watersheds. In order to detect the effects of cumulative forest changes on hydrology, a large watershed must experience significant forest changes or disturbances (e.g. CECA of >20–30%) and must also include a sufficiently long period without forest disturbances (or with limited forest disturbances) as a comparable reference or control period. Long-term data on forest cover change history, climate and hydrology must also be available. Moreover, large watersheds are more prone to anthropogenic activities (e.g. channelization, reservoir or dam and/or road constructions including legacy water management structures such as levees and impoundments) and their hydrological impacts (Magilligan and Nislow, 2005). Given the fact that the majority of large watersheds are poorly regulated or monitored, it is rather challenging to find suitable study watersheds to assess forest changes and their effects on hydrology.

Despite a limited number of studies to date, the topic of forest cover changes and hydrology in large watersheds has received growing attention mainly because many practices and policies of natural resource management are operated on large landscape, watershed or even regional scales. Scientific information on large watersheds is critically needed to support the design of natural resource management strategies, especially given the fact that climate change (e.g. global warming) and anthropogenic activities (e.g. logging, urbanization and land conversion) are altering watershed processes and ecosystem functions dramatically and extensively, and are leading to more frequent and catastrophic forest disturbances (e.g. insect infestation and wildfire) (Schindler, 2001). A comprehensive understanding of the impacts of forest cover changes on water in large watersheds is essential for the sustainability of long-term water supply and the protection of watershed ecosystem functions under a changing environment.

11.2 Forest Cover Changes and Water in Large Watersheds

Forest cover changes and water in large watersheds have received growing attention in the past few decades mainly because of increasing demand for scientific information on large-scale watersheds or landscapes to support sustainable natural resources management. In spite of limited studies, significant progress employing different methods such as statistics (Wei *et al.*, 2013; Zhang and Wei, 2014a) and modelling (Christiaens and Feyen, 2001; Chen *et al.*, 2005) has been made on this subject.

We synthesized 160 global published case studies on forest cover changes (deforestation and reforestation) and annual water yield (AWY) in large watersheds (>1000 km^2). We found that deforestation increases AWY, while reforestation decreases it, which is consistent with the results from small paired experimental watershed studies. Our meta-data analysis also shows that greater areal forest cover changes cause larger AWY responses regardless of change directions (deforestation or reforestation impacts) (Fig. 11.1).

The forest cover changes not only alter annual mean flow substantially, but also change peak flows. However, rare studies have been conducted on assessing forest cover changes and peak flows in large watersheds. In addition, the results on peak flow response to forest cover changes are inconsistent, with large variations. Many studies showed that hydrological responses to alteration of forest covers are not significant in large-scale basins. For instance, the study in north-eastern Ontario, Canada by Buttle and Metcalfe (2000) found limited streamflow responses in some large-sized watersheds (ranging from 401 to 11,900 km^2) to land cover changes (5–25%) and no definitive changes in annual peak flows. Wilk *et al.* (2001) did not find any significant hydrological change in the Nam Pong River basin (12,100 km^2) in north-east Thailand after a reduction of forest cover from 80% in 1957 to 27% in 1995, which may be due to shaded trees left in the agriculture area and secondary growth in the abandoned plots.

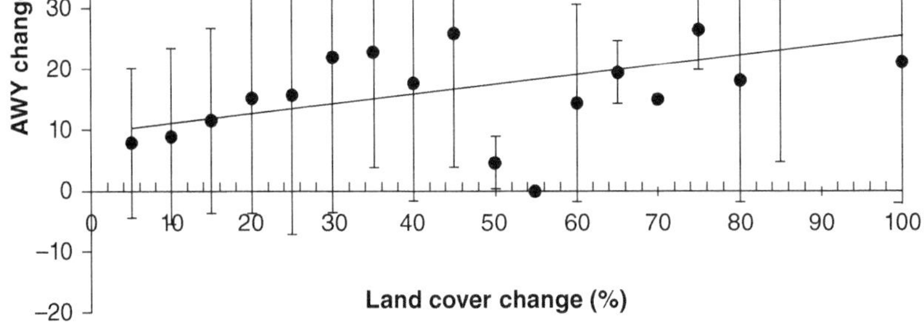

Fig. 11.1. Percentage change in annual water yield (AWY) with each 5% forest cover change based on 160 global published case studies (with standard deviations represented by error bars).

In contrast, some studies showed that peak or high flows were increased significantly by deforestation in large-scale watersheds. For example, peak flows or high flow regimes were increased dramatically in several large watersheds located in the interior of British Columbia, Canada, including the Willow River watershed (Lin and Wei, 2008), Tulameen River watershed (Zhang, 2013) and Baker River watershed (Zhang and Wei, 2012).

An interesting case study on the comparison of peak flow responses to forest disturbance between two neighbouring large watersheds (Bowron River and Willow River watersheds, located in the interior of British Columbia, Canada) is worth mentioning here (Zhang and Wei, 2014b). Both watersheds experienced similar forest disturbance levels (ECA of 25–30%). Their results showed that forest harvesting in the Willow watershed dramatically increased annual and spring mean flows as well as annual and spring peak flows, whereas it caused an insignificant change in those hydrological variables in the Bowron watershed. The contrasted differences in hydrological responses are due to the differences in topography, spatial heterogeneity, forest harvesting characteristics and climate between the two watersheds. The relative uniform topography and climate in the Willow watershed may promote hydrological synchronization effects, whereas larger variation in elevations, together with forest harvesting that occurred at lower elevations, may cause hydrological de-synchronization effects in the Bowron watershed. The contrasted results demonstrate that the effects of forest disturbance on hydrology in large watersheds are likely watershed-specific and any attempt to generalize hydrological responses to forest changes must be carried out with caution.

The studies on low or base flow responses to forest cover changes in large watersheds are even rarer. The results from small watershed studies showed that the responses of low flows to logging could be positive, negative or even negligible (Calder and Maidment, 1992; Moore and Wondzell, 2005), while reforestation generally decreased low flow (Andreassian, 2004). Due to more complexities in land forms, channel morphology and topographies in large watersheds, it is generally expected that the responses of low flows to forest changes in large watersheds are more varied. For example, in a large, severely disturbed watershed, the Baker River watershed (ECA of about 60%), low flows were significantly increased (Zhang and Wei, 2014a). Interestingly, Zhou *et al.* (2010) also found that large-scale reforestation (forest recovery) plays a positive role in redistributing water from the wet season to the dry season and, consequently, in increasing water yield in the dry season. Nevertheless, more case studies are needed before any meaningful conclusions on forest cover changes and low flows in large watersheds can be provided.

11.3 Research Methods

Current approaches on hydrological responses associated with forest changes in large watersheds can be classified into two general categories: hydrological modelling and non-modelling (Wei and Zhang, 2011; Zhang, 2013). Selection of a suitable research approach depends mainly upon the purpose of the research, data availability and the number of available watersheds. Below are six methods commonly applied in this subject.

11.3.1 Hydrological modelling

Hydrological models are frequently used in large watershed hydrological research. Hydrological models can be divided into lumped, semi-distributed and fully distributed models in light of their spatial representations (Zhang, 2013). Lumped hydrological models treat a watershed as a whole system or entity, and do not consider the detailed spatial representations of watershed elements and processes. Semi-distributed models divide a watershed into several sub-basins. However, the spatial heterogeneity is expressed only to some extent, not in great detail. Unlike lumped or semi-distributed models, distributed models can well represent a watershed by assigning input data and physical characteristics to grids or elements within the delineated sub-basins. Physically based distributed models are able to provide distributed approximations or predictions of hydrological variables across watersheds, and thus have a better representation of reality. However, a physical-based fully distributed model requires a large data set and input parameters on various processes, components and their interactions. For large-scale watershed research, a semi-distributed model is commonly used because of a general absence of detailed data and input parameters at large scales.

The one-factor-at-a-time approach (OFAT), commonly used in sensitivity analysis (Wilson *et al.*, 1987a; Pitman, 1994; Gao *et al.*, 1996), is also used in association with hydrological models to distinguish the impact of climate factors and land cover change on watershed hydrology (Wilson *et al.*, 1987b; Karvonen *et al.*, 1999). In a hypothetical example, the impacts of climatic variability and land use change on streamflow are assessed with the available data in the period of 1960 to 2000. First, we keep the land cover in 1960 unchanged over the simulation period while climate change is allowed from 1960 to 2000. Then, we simulate the streamflow change (ΔQ_C), which can be treated as the impact of climatic variability on hydrology. Second, keeping the climate of 1960 unchanged while land cover is changed, we then calculate the streamflow change (ΔQ_L) as the impact of land cover change. Finally, we assume the changes of both climate and land cover, and then calculate the streamflow change (ΔQ_{L+C}). In this way, the relative contributions of forest and land cover changes and climatic variability to hydrology can be computed.

Various distributed hydrological models have been used successfully to quantitatively study the effects of climate change and forest change/land cover change on hydrology, such as the Soil and Water Assessment Tool (SWAT) (Chen *et al.*, 2005; Zhang, A.J., *et al.*, 2012), DHSVM (Sun and Bosilovich, 1996; Stonesifer, 2007) and MIKE SHE (Christiaens and Feyen, 2001), etc. In spite of increased applications, hydrological models are still based on our current theories that are deeply rooted in the physics of small-scale processes. This gives rise to difficulties in representing non-linear hydrological processes and their interactions at all scales across heterogeneous landscapes. In addition, calibrating and testing a model may not always assure its validity, since there are some inherent drawbacks in the approaches of parameter calibration and validation (Kirchner, 2006). We often over-parameterize our models to meet high accuracy levels, ignoring the equifinality problem that different parameter sets for a model might yield the same result during calibration, but distinctly different predictions when conditions are altered (Kirchner, 2006).

11.3.2 Breakpoints and double mass curves

The double mass curve (DMC) is a simple and intuitive method, widely used in long-term trend analysis of hydrometeorological elements. DMC draws a curve between two cumulative hydrometeorological variables to test the consistency of the two variables or to analyse the trend change and its strength (Buttle and Metcalfe, 2000; Siriwardena *et al.*, 2006; Yao *et al.*, 2012). The DMC method can be also used to separate the relative influences of forest change and climatic effects on hydrology (Koster and Suarez, 1999). For example, a modified DMC (MDMC) between cumulative annual streamflow and cumulative effective precipitation (the difference between total precipitation and evapotranspiration) is constructed for a large forested watershed (Wei and Zhang, 2010; Zhang, M., *et al.*, 2012; Zhang, 2013) (Fig. 11.2). In this way, climatic effect on annual streamflow can be eliminated. In the period of no forest disturbance, the curve should produce a straight line, a baseline that describes the linear relationship between annual streamflow and annual effective precipitation, and a break in this curve (e.g. year 1986 in Fig. 11.2) would suggest the change of annual streamflow caused by forest disturbance. In other words, a step change or regime shift occurs in the slope of MDMC and the slope before the break is different from that afterwards. However, this visually detected breakpoint needs confirmation of its statistical significance by a non-parametric test or application of an autoregressive integrated moving average (ARIMA) model (Box and Pierce, 1970). The difference between actual observations and the predicted line after the change point can be calculated, and is regarded as the cumulative impact of forest changes.

11.3.3 Sensitivity-based approach

The sensitivity-based approach is similar to the elasticity method (Dooge *et al.*, 1999) and is

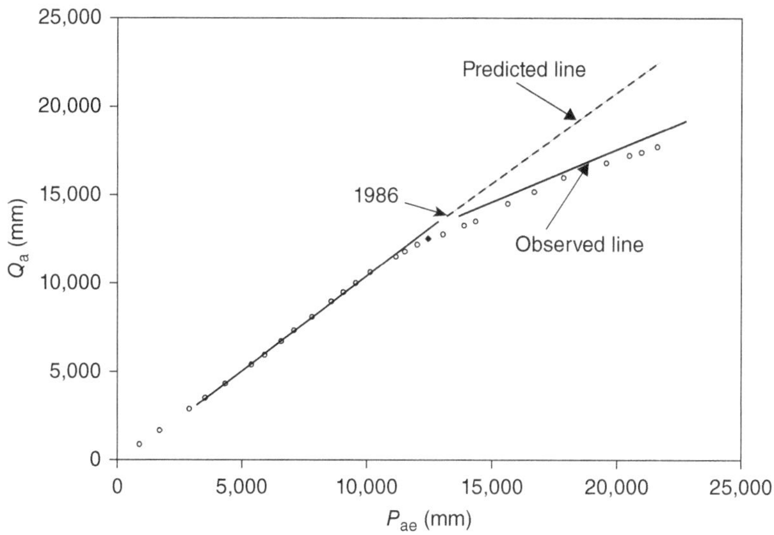

Fig. 11.2. A hypothetical example of application of the MDMC (modified double mass curve) between cumulative annual streamflow (Q_a) and cumulative annual effective precipitation (P_{ae}) for quantifying the effects of forest changes on annual mean flow.

used to calculate the effect of climate variability on streamflow. Perturbations in both precipitation (P) and potential evapotranspiration (PET) can lead to changes of water balance. It can be assumed that a change in mean annual streamflow can be determined using the following expression (Koster and Suarez, 1999; Jones *et al.*, 2006):

$$\Delta Q_{clim} = \beta \Delta P + \gamma \Delta PET \quad (11.1)$$

where ΔQ_{clim}, ΔP and ΔPET are changes in streamflow, precipitation and potential evapotranspiration, respectively. β and γ are the sensitivity coefficients of streamflow to precipitation and potential evapotranspiration, expressed as:

$$\beta = \frac{1+2x+3wx^2}{\left(1+w+wx^2\right)^2} \quad (11.2)$$

and

$$\gamma = \frac{1+2x}{\left(1+x+wx^2\right)^2} \quad (11.3)$$

where x is the mean annual index of dryness (equal to PET/P) and the values of vegetation factor w for forest, grassland and shrub land are 2, 0.5 and 1, respectively (Zhang *et al.*, 2001).

This method is suitable for the analysis of a single basin and for quantitative calculation of the impact of climate variables on streamflow. Once the effects of climatic variability on flow are estimated, the effects of forest disturbance or land-use changes can be deducted from total streamflow variations. The method has been used successfully for several case studies (Dooge *et al.*, 1999; Zhang *et al.*, 2001; Jones *et al.*, 2006). There may be two challenges in this method. First, it is not easy to determine w values for specific forest vegetation types. Where there are always different types of forests in a large watershed, how to select a specific w value remains challenging. Second, the effect of forest disturbance or land-use changes on hydrology is estimated indirectly from total hydrological variations and the effects of climatic variability. Thus, its reliability is dependent on the accuracy of the other two terms.

11.3.4 Simple water balance

The water balance methods provide a framework to determine changes in the water balance components (Liu *et al.*, 2009). A simple water balance model can be used to determine the influence of climate and vegetation on streamflow at a watershed scale:

$$P = ET + Q + \Delta S \tag{11.4}$$

where P is precipitation, ET is actual evapotranspiration, Q is streamflow and ΔS is change in catchment water storage. When averaged over a long period, deep percolation (recharge) and change in soil moisture storage is often only 5 to 10% of the annual water balance, and therefore the change in catchment water storage (ΔS) can be neglected (Ponce and Shetty, 1995; Zhang *et al.*, 1999).

Precipitation and actual evapotranspiration constitute the most important variables to influence streamflow change at the watershed scale. Precipitation, which varies both in temporal trend and spatial distribution, is regarded as independent of vegetation types (Zhang *et al.*, 2001), which mainly reflect changes of climate. However, actual evapotranspiration is a complex process. There are various ways to estimate watershed-scale evapotranspiration. For example, following the Budyko hypothesis, the simple two-parameter model for estimating the actual evapotranspiration was developed (Budyko, 1961). The model is consistent with the previous theoretical work and shows good agreement with more than 250 catchment-scale measurements from around the world (Zhang *et al.*, 2001, 2004):

$$\frac{ET}{P} = \frac{1 + w(PET/P)}{1 + w(PET/P) + (PET/P)^{-1}} \tag{11.5}$$

where ET is the actual evapotranspiration and PET is reference evapotranspiration, a substitute for potential evapotranspiration calculated by the Penman–Monteith method (Allen *et al.*, 1998). w is the plant-available water coefficient estimated in the same way as in the sensitivity-based approach (Zhang *et al.*, 2001).

The following steps describe how a simple water balance method is implemented to estimate the effects of forest cover change on hydrology (Liu *et al.*, 2009). First, according to Eqn 11.5, calculate the actual evapotranspiration using the original data including precipitation, air temperature, relative humidity, wind speeds and sunshine hours after calculating PET by the Penman–Monteith method. Second, estimate annual streamflow using Eqn 11.4. In this step, the change of annual streamflow is influenced by both climate variability and vegetation changes and, thus, the calculated annual streamflow can be defined as Q_{v+c} (i.e. $Q_{v+c} = P - ET$). Third, removing the decreasing or increasing trend of precipitation, air temperature, relative humidity, wind speeds and sunshine hours in data series to make them as stationary time series (Xu *et al.*, 2006), recalculate PET and ET using the de-trended climate variables and estimate annual streamflow according to Eqns 11.4 and 11.5 (i.e. $Q_v = P - ET_{new}$). In this step, the change of annual streamflow reflects mainly the influence of vegetation changes and, thus, the recalculated annual streamflow can be defined as Q_v. Finally, calculate the difference between Q_{v+c} and Q_v; thus the change of annual streamflow caused by the climate variability ($Q_c = Q_{v+c} - Q_v$) can then be estimated (Liu *et al.*, 2009).

A simple water balance method provides a new way to distinguish the impact of climatic variables and vegetation factors on hydrological change. However, the choice of w values and the difficulty associated with removal of the decreasing or increasing trends in climate data may introduce some errors in this method.

11.3.5 Time trend method

In the time trend method, a relationship is established between streamflow and climatic variables before the basin's vegetation perturbation occurs and is then used to predict the streamflow response post-perturbation assuming undisturbed basin conditions. The typical time trend approach is to divide the whole study period into a calibration period and the prediction period. The model accuracy depends on the length of the calibration or pre-perturbation periods (Zhao *et al.*, 2010).

During the calibration period the streamflow is calculated as:

$$Q_1 = aP_1 + b \tag{11.6}$$

During the prediction period the expected streamflow is calculated as:

$$Q_2' = aP_2 + b \tag{11.7}$$

and

$$\Delta Q_v = \overline{Q_2} - \overline{Q_2'} \tag{11.8}$$

where P is precipitation, Q is streamflow, Q' is the predicted streamflow for the catchment after treatment (from using Eqn 11.7 developed during the calibration period), ΔQ_v represents the change in mean annual streamflow because of vegetation change, subscripts 1 and 2 represent respectively the calibration period and the prediction period, and b is the fitted regression coefficient. $\overline{Q_2}$ is the average observed streamflow in the prediction period and $\overline{Q_2'}$ is the average predicted streamflow calculated by Eqn 11.7 using the regression coefficients from the calibration period.

This method uses a simple regression to express the relationship between precipitation and streamflow both before and after forest cover changes. The method only requires data of precipitation, streamflow and other meteorological variables, and the requirement on the detailed forest cover change can be ignored to some extent. The method may accept discontinuous data. Depending on different hydrometeorological characteristics in a study basin, time trend method performance at yearly hydrological variables is better than on the variables at monthly or daily intervals. This is because the rainfall–runoff relationship in a watershed at an annual interval is much stronger than those at daily and monthly ones. Guardiola-Claramonte *et al.* (2011) proposed to consider the impact of temperature on the relationship between precipitation and streamflow.

11.3.6 Tomer–Schilling framework

Tomer and Schilling developed a coupled water–energy balance framework that requires long-term annual time series of precipitation (P), streamflow (Q) and potential evapotranspiration (PET) to assess if unused available energy ($PET/P > 1$) and water ($PET/P < 1$) were related to climate and/or to land management in agricultural catchments (Tomer and Schilling, 2009; Peña-Arancibia *et al.*, 2012). The conceptual framework qualitatively discriminates whether the dominant drivers of observed changes are related to land cover change and/or climate. The framework relating changes in land cover and/or climate to the observed changes in the excess amounts of water (P_{ex}) and excess amounts of energy (E_{ex}) as fractions is illustrated in Peña-Arancibia *et al.* (2012).

The Tomer–Schilling framework assumes that land cover change will affect actual evapotranspiration (ET) but not P or PET, acknowledging that effects of land cover change on P and PET can be considered indirectly at this scale and possibly would be of second order compared with changes in ET in the woodland environment. Thus, land cover change will cause ecohydrological shifts towards increased P_{ex} and E_{ex}, or towards decreased P_{ex} and E_{ex}. Changes in climate are required to cause increased P_{ex} and decreased E_{ex}, due to the temporal increase in the P/PET ratio and vice versa (Peña-Arancibia *et al.*, 2012):

$$\overline{P_{ex}} = \frac{(\overline{P} - \overline{ET})}{\overline{P}} \tag{11.9}$$

and

$$E_{ex} = \frac{(\overline{PET} - \overline{ET})}{\overline{PET}} \tag{11.10}$$

The Tomer–Schilling framework is an effective and qualitative analysis tool of hydrological processes. The method can not only analyse long-term impacts of climate and forest cover on hydrology, but also explains the main factors of hydrological responses in different time periods. However, as with all qualitative methods, the impact of each variable of climate or vegetation cannot be quantitatively evaluated, and this points to the same difficulties in studying the common law of hydrological responses under different topographical and vegetative conditions.

11.4 Future Research Priorities

Here, we propose the following future research priorities in this subject of large-scale studies of forest cover changes and water. First, more case studies are needed. Assessing the relative contributions of forest or land cover changes and climatic variability to hydrology is rather limited in large watersheds. Zhang and Wei (2014b) compared two adjacent large watersheds located in the interior of British Columbia, Canada and found contrasting conclusions under similar forest disturbance levels. They further concluded that the effects of forest change on hydrology in large watersheds are likely watershed-specific. This clearly demonstrates that more case studies are needed before general conclusions can be derived. Second, a research priority should be given to further development and improvement of existing research methods. Although quite a few research methods are currently available for studying the impacts of forest cover change and climate change on hydrology, there is not a single commonly accepted method. The lack of commonly accepted methods may limit our ability to compare the results of different studies. Third, in large watershed studies, analytical results are based largely on data quality and spatial coverage. Inherent spatial variabilities in precipitation, temperature, solar radiation, humidity, wind speed, surface albedo, canopy characteristics, etc. in these large watersheds and uncertainties in parameter estimates constrain our ability to derive robust conclusions. For instance, only a few climatic stations are located in large watersheds. In addition, the existing climatic stations are often located in easily accessible places. Thus the spatial variability of precipitation cannot easily be addressed in large watersheds. Future studies should be designed to specifically address such uncertainties. Finally, more research should be focused on mechanisms, processes and their interactions. It is difficult to study the mechanisms and processes in large watersheds, mainly due to lack of a suitable methodology and data for assessing the complicated interactions and cumulative behaviours across various spatial scales. However, such research is critical for explaining and verifying the findings obtained through statistical and modelling approaches.

References

Allen, R.G., Pereira, L.S., Raes, D. and Smith, M. (1998) *Crop Evapotranspiration – Guidelines for Computing Crop Water Requirements.* FAO Irrigation and Drainage Paper No. 56. Food and Agriculture Organization of the United Nations, Rome.

Andreassian, V. (2004) Waters and forests: from historical controversy to scientific debate. *Journal of Hydrology* 291, 1–27.

BC Ministry of Forests and Rangeland (1999) Appendix 8. In: *Interior Watershed Assessment Guidebook*, 2nd edn. Ministry of Forests and Rangeland, Victoria, British Columbia, Canada.

Box, G.E. and Pierce, D.A. (1970) Distribution of residual autocorrelations in autoregressive-integrated moving average time series models. *Journal of the American Statistical Association* 65, 1509–1526.

Budyko, M.I. (1961) The heat balance of the earth's surface. *Soviet Geography* 2, 3–13.

Buttle, J. and Metcalfe, R. (2000) Boreal forest disturbance and streamflow response, northeastern Ontario. *Canadian Journal of Fisheries and Aquatic Sciences* 57, 5–18.

Calder, I.R. and Maidment, D. (1992) *Hydrologic Effects of Land-Use Change*. McGraw-Hill, New York.

Chen, J.F., Li, X.B. and Zhang, M. (2005) Simulating the impacts of climate variation and land-cover changes on basin hydrology: a case study of the Suomo basin. *Science in China Series D – Earth Sciences* 48, 1501–1509.

Chen, W. and Wei, X. (2008) Assessing the relations between aquatic habitat indicators and forest harvesting at watershed scale in the interior of British Columbia. *Forest Ecology and Management* 256, 152–160.

Christiaens, K. and Feyen, J. (2001) Analysis of uncertainties associated with different methods to determine soil hydraulic properties and their propagation in the distributed hydrological MIKE SHE model. *Journal of Hydrology* 246, 63–81.

Creed, I.F., Spargo, A.T., Jones, J.A., Buttle, J.M., Adams, M.B., Beall, F.D., Booth, E.G., Campbell, J.L., Clow, D., Elder, K., *et al.* (2014) Changing forest water yields in response to climate warming: results from long-term experimental watershed sites across North America. *Global Change Biology* 20, 3191–3208.

Dooge, J.C.I., Bruen, M. and Parmentier, B. (1999) A simple model for estimating the sensitivity of runoff to long-term changes in precipitation without a change in vegetation. *Advances in Water Resources* 23, 153–163.

Fohrer, N., Haverkamp, S. and Frede, H.G. (2005) Assessment of the effects of land use patterns on hydrologic landscape functions: development of sustainable land use concepts for low mountain range areas. *Hydrological Processes* 19, 659–672.

Gao, X.G., Sorooshian, S. and Gupta, H.V. (1996) Sensitivity analysis of the biosphere–atmosphere transfer scheme. *Journal of Geophysical Research: Atmospheres* 101, 7279–7289.

Guardiola-Claramonte, M., Troch, P.A., Breshears, D.D., Huxman, T.E., Switanek, M.B., Durcik, M. and Cobb, N.S. (2011) Decreased streamflow in semi-arid basins following drought-induced tree die-off: a counter-intuitive and indirect climate impact on hydrology. *Journal of Hydrology* 406, 225–233.
Jaskierniak, D., Kuczera, G., Benyon, R. and Wallace, L. (2015) Using tree detection algorithms to predict stand sapwood area, basal area and stocking density in *Eucalyptus regnans* forest. *Remote Sensing* 7, 7298–7323.
Jones, R.N., Chiew, F.H.S., Boughton, W.C. and Zhang L. (2006) Estimating the sensitivity of mean annual runoff to climate change using selected hydrological models. *Advances in Water Resources* 29, 1419–1429.
Karvonen, T., Koivusalo, H., Jauhiainen, M., Palko, J. and Weppling, K. (1999) A hydrological model for predicting runoff from different land use areas. *Journal of Hydrology* 217, 253–265.
Kirchner, J.W. (2006) Getting the right answers for the right reasons: linking measurements, analyses, and models to advance the science of hydrology. *Water Resources Research* 42, W03S04, doi: 10.1029/2005WR004362 (accessed 8 April 2016).
Kirchner, J.W. (2009) Catchments as simple dynamical systems: catchment characterization, rainfall-runoff modeling, and doing hydrology backward. *Water Resources Research* 45, W02409, doi: 10.1029/2008WR006912 (accessed 8 April 2016).
Koster, R.D. and Suarez, M.J. (1999) A simple framework for examining the interannual variability of land surface moisture fluxes. *Journal of Climate* 12, 1911–1917.
Lin, Y. and Wei, X. (2008) The impact of large-scale forest harvesting on hydrology in the Willow watershed of Central British Columbia. *Journal of Hydrology* 359, 141–149.
Liu, Q., Yang, Z.F., Cui, B.S. and Sun, T. (2009) Temporal trends of hydro-climatic variables and runoff response to climatic variability and vegetation changes in the Yiluo River basin, China. *Hydrological Processes* 23, 3030–3039.
Magilligan, F.J. and Nislow, K.H. (2005) Changes in hydrologic regime by dams. *Geomorphology* 71(1), 61–78.
Moore, R.D. and Wondzell, S.M. (2005) Physical hydrology and the effects of forest harvesting in the Pacific Northwest: a review. *Journal of the American Water Resources Association* 41, 763–784.
Peña-Arancibia, J.L., van Dijk, A.I., Guerschman, J.P., Mulligan, M., Bruijnzeel, L.A.S. and McVicar, T.R. (2012) Detecting changes in streamflow after partial woodland clearing in two large catchments in the seasonal tropics. *Journal of Hydrology* 416, 60–71.
Pitman, A.J. (1994) Assessing the sensitivity of a land-surface scheme to the parameter values using a single-column model. *Journal of Climate* 7, 1856–1869.
Ponce, V.M. and Shetty, A.V. (1995) A conceptual-model of catchment water-balance. 1. Formulation and calibration. *Journal of Hydrology* 173, 27–40.
Schindler, D.W. (2001) The cumulative effects of climate warming and other human stresses on Canadian freshwaters in the new millennium. *Canadian Journal of Fisheries and Aquatic Sciences* 58, 18–29.
Scott, D.F. and Prinsloo, F.W. (2008) Longer-term effects of pine and eucalypt plantations on streamflow. *Water Resources Research* 44, W00A08 doi: 10.1029/2007WR006781 (accessed 8 April 2016).
Shaman, J., Stieglitz, M. and Burns, D. (2004) Are big basins just the sum of small catchments? *Hydrological Processes* 18, 3195–3206.
Sharma, K.P., Vorosmarty, C.J. and Moore, B. (2000) Sensitivity of the Himalayan hydrology to land-use and climatic changes. *Climatic Change* 47, 117–139.
Shuttleworth, W.J. (1988) Macrohydrology – the new challenge for process hydrology. *Journal of Hydrology* 100, 31–56.
Siriwardena, L., Finlayson, B.L. and McMahon, T.A. (2006) The impact of land use change on catchment hydrology in large catchments: the Comet River, Central Queensland, Australia. *Journal of Hydrology* 326, 199–214.
Stednick, J.D. (1996) Monitoring the effects of timber harvest on annual water yield. *Journal of Hydrology* 176, 79–95.
Stednick, J.D. (2008) Long-term streamflow changes following timber harvesting. In: Stednick, J.D. (ed.) *Hydrological and Biological Responses to Forest Practices*. Springer, New York, pp. 139–155.
Stonesifer, C.S. (2007) *Modeling the cumulative effects of forest fire on watershed hydrology: a post-fire application of the Distributed Hydrology-Soil-Vegetation Model (DHSVM)*. Msc. thesis, The University of Montana, Missoula, Montana.
Sun, W.Y. and Bosilovich, M.G. (1996) Planetary boundary layer and surface layer sensitivity to land surface parameters. *Boundary-Layer Meteorology* 77, 353-378.

Tomer, M.D. and Schilling, K.E. (2009) A simple approach to distinguish land-use and climate-change effects on watershed hydrology. *Journal of Hydrology* 376, 24–33.

Tuteja, N.K., Vaze, J., Teng, J. and Mutendeudzi, M. (2007) Partitioning the effects of pine plantations and climate variability on runoff from a large catchment in southeastern Australia. *Water Resources Research* 43, W08415, doi: 10.1029/2006WR005016 (accessed 8 April 2016).

Wei, X. and Zhang, M. (2010) Quantifying streamflow change caused by forest disturbance at a large spatial scale: a single watershed study. *Water Resources Research* 46, W12525, doi: 10.1029/2010WR009250 (accessed 8 April 2016).

Wei, X. and Zhang, M. (2011) Research methods for assessing the impacts of forest disturbance on hydrology at large-scale watersheds. In: Li, C., Lafortezza, R. and Chen, J (eds) *Landscape Ecology in Forest Management and Conservation: Challenges and Solutions for Global Change*. Higher Education Press, Beijing and Springer, Berlin/Heidelberg, pp. 119–147.

Wei, X., Liu, W.F. and Zhou, P.C. (2013) Quantifying the relative contributions of forest change and climatic variability to hydrology in large watersheds: a critical review of research methods. *Water* 5, 728–746.

Whitaker, A., Alila, Y., Beckers, J. and Toews, D. (2002) Evaluating peak flow sensitivity to clear-cutting in different elevation bands of a snowmelt-dominated mountainous catchment. *Water Resources Research* 38, 1172, doi: 10.1029/2001WR000514 (accessed 8 April 2016).

Wilk, J., Andersson, L. and Plermkamon, V. (2001) Hydrological impacts of forest conversion to agriculture in a large river basin in northeast Thailand. *Hydrological Processes* 15, 2729–2748.

Wilson, M., Henderson-Sellers, A., Dickinson, R. and Kennedy, P. (1987a) Sensitivity of the biosphere–atmosphere transfer scheme (BATS) to the inclusion of variable soil characteristics. *Journal of Climate and Applied Meteorology* 26, 341–362.

Wilson, M.F., Henderson-Sellers, A., Dickinson, R. and Kennedy, P. (1987b) Investigation of the sensitivity of the land surface parameterization of the NCAR community climate model in regions of tundra vegetation. *Journal of Climatology* 7, 319–343.

Xu, C.Y., Gong, L.B., Jiang, T., Chen, D.L. and Singh, V.P. (2006) Analysis of spatial distribution and temporal trend of reference evapotranspiration and pan evaporation in Changjiang (Yangtze River) catchment. *Journal of Hydrology* 327, 81–93.

Yang, H., Qi, J., Xu, X., Yang, D. and Lv, H. (2014) The regional variation in climate elasticity and climate contribution to runoff across China. *Journal of Hydrology* 517, 607–616.

Yao, Y.F., Cai T.J., Wei, X.H., Zhang, M.F. and Ju, C.Y. (2012) Effect of forest recovery on summer streamflow in small forested watersheds, Northeastern China. *Hydrological Processes* 26, 1208–1214.

Zhang, A.J., Zhang, C., Fu, G.B., Wang, B.D., Bao, Z.X. and Zheng, H.X. (2012) Assessments of impacts of climate change and human activities on runoff with SWAT for the Huifa River Basin, Northeast China. *Water Resources Management* 26, 2199–2217.

Zhang, L., Walker, G.R. and Dawes, W. (1999) *Predicting the Effect of Vegetation Changes on Catchment Average Water Balance*. Technical Report No. 99/12. Monash Univeristy, Melbourne, Victoria, Australia.

Zhang, L., Dawes, W.R. and Walker, G.R. (2001) Response of mean annual evapotranspiration to vegetation changes at catchment scale. *Water Resources Research* 37, 701–708.

Zhang, L., Hickel, K., Dawes, W.R., Chiew, F.H.S., Western, A.W. and Briggs, P.R. (2004) A rational function approach for estimating mean annual evapotranspiration. *Water Resources Research* 40, W02502, doi: 10.1029/2003WR002710 (accessed 8 April 2016).

Zhang, M. (2013) The effects of cumulative forest disturbances on hydrology in the interior of British Columbia, Canada. PhD Thesis, The University of British Columbia, Okanagan, British Columbia, Canada.

Zhang, M. and Wei, X. (2012) The effects of cumulative forest disturbance on streamflow in a large watershed in the central interior of British Columbia, Canada. *Hydrology and Earth System Sciences* 16, 2021–2034.

Zhang, M. and Wei, X. (2014a) Alteration of flow regimes caused by large- scale forest disturbance: a case study from a large watershed in the interior of British Columbia, Canada. *Ecohydrology* 7, 544–556.

Zhang, M. and Wei, X. (2014b) Contrasted hydrological responses to forest harvesting in two large neighbouring watersheds in snow hydrology dominant environment: implications for forest management and future forest hydrology studies. *Hydrological Processes* 28, 6183–6195.

Zhang, M., Wei, X., Sun, P.S. and Liu, S.R. (2012) The effect of forest harvesting and climatic variability on runoff in a large watershed: the case study in the Upper Minjiang River of Yangtze River basin. *Journal of Hydrology* 464, 1–11.

Zhao, F.F., Zhang, L., Xu, Z.X. and Scott, D.F. (2010) Evaluation of methods for estimating the effects of vegetation change and climate variability on streamflow. *Water Resources Research* 46, W03505, doi: 10.1029/2009WR007702 (accessed 8 April 2016).

Zheng, H.X., Zhang, L., Zhu, R., Liu, C.M., Sato, Y. and Fukushima, Y. (2009) Responses of streamflow to climate and land surface change in the headwaters of the Yellow River Basin. *Water Resources Research* 45, W00A19, doi: 10.1029/2007WR006665 (accessed 8 April 2016).

Zhou, G.Y., Wei, X., Luo, H.Y., Zhang, M., Li, Y.L., Qiao, Y.N., Liu, H.G. and Wang, C.L. (2010) Forest recovery and river discharge at the regional scale of Guangdong Province, China. *Water Resources Research* 46, W09503, doi: 10.1029/2009WR008829 (accessed 8 April 2016).

12 Hydrological Effects of Forest Management

J.D. Stednick[1]* and C.A. Troendle[2]
[1]*Colorado State University, Fort Collins, Colorado, USA;*
[2]*Retired, USDA Forest Service, Fort Collins, Colorado, USA*

12.1 Introduction

Forested catchments provide a range of ecosystem services, including delivery of clean waters suitable for many uses: low flow maintenance, peak flow regulation (flood attenuation) groundwater recharge and soil conservation (Colman, 1953; Hamilton, 2008). High infiltration rates in undisturbed forests produce little overland flow, meaning precipitation generally passes through the soil before reaching the stream network, resulting in low erosion and sedimentation rates and high water quality (Anderson *et al.*, 1976; Beschta, 1990; Bruijnzeel, 2004; Calder, 2007). Forest health has a direct correlation to stream health (de la Cretaz and Barten, 2007).

The integral relationship between forests and water resources begs the question, 'what is the hydrological effect of forest management activities?' Forest management is the practical application of biological, physical, economic and social principles to the growth, regeneration, utilization and conservation of forests to meet specified goals and objectives while maintaining the productivity of the forest – forest management includes management for aesthetics, fish, recreation, urban values, water, wilderness, wildlife, wood products and other forest resource values (SAF, 2008). Forest management activities may include road construction, timber harvesting, use of prescribed fire and chemical (fertilizer, insecticide and herbicide) applications. Forest management activities that disturb or remove vegetation potentially affect hydrological processes. Soil disturbance from tree felling is generally minor, but movement of logs or whole trees to a landing or collection point may disturb the soil surface significantly. These disturbances are often not connected to the hydrological network, which minimizes catchment impacts. Soil surface disturbances related to collection and haul roads can be more damaging because of the connectivity to the stream network. Best Management Practices (BMPs) for road design, layout and maintenance minimize the damage (Adams and Ringer, 1994). Stand improvement may include selective harvesting of trees in either dominant or subordinate crown positions. Forest stand thinning may increase water and nutrient availability, but these resources are utilized quickly by remaining vegetation. This chapter reviews the potential effects of forest management activities, particularly timber harvesting, on streamflows.

Paired catchment experiments are the most common approach used to assess effects of forest management activities on streamflow. This approach uses two or more catchments, one

*Corresponding author; e-mail: jds@cnr.colostate.edu

designated as control and at least one other as treatment. Paired catchments are either adjacent or very close to one another geographically so as to be affected by the same climatic factors. The success of paired catchment studies initially depends on how similar control and treatment catchments are with respect to their geology, soils, topography and vegetation (Moore and Wondzell, 2005). Prior to disturbance of the treatment catchment, there is a calibration period to allow quantification of differences in flow between the two catchments that are attributable to differences in their geology and topography (Whitehead and Robinson, 1993). An understanding of the catchment hydrology is required when interpreting results from such studies in order to distinguish harvesting-related streamflow changes from those attributable to other factors (Fuller *et al.*, 1988).

The earliest catchment studies were designed to determine the balance between precipitation and streamflow and how this balance was affected by land cover and land-use practices. The effects of timber harvesting on water yield in particular, but also water quality, were first investigated as a paired catchment study at Wagon Wheel Gap, near Creede, Colorado, USA beginning in 1908 (Bates and Henry, 1928; Ice and Stednick, 2004). It was soon recognized that catchment studies could also be used to understand how forestry practices affect stream water quality (Swank and Johnson, 1994).

A large number of small field-scale experimental studies using a paired catchment approach have been conducted in Australia, New Zealand, South Africa, South America, Great Britain, China, Japan and the USA to better understand forest hydrological processes, their interactions with the environment and their ecohydrological impacts (Hibbert, 1967; Swank and Douglass, 1974; US EPA, 1980; Bosch and Hewlett, 1982; Sahin and Hall, 1996; Stednick, 1996; Sun *et al.*, 2001; Andreassian, 2004; Jackson *et al.*, 2004; Brown *et al.*, 2005; Edwards and Troendle, 2008; NRC, 2008; Chescheir *et al.*, 2009; Bren and McGuire, 2012; Bren and Lane, 2014). The paired catchment approach allows separation of climatic effects from vegetative effects. Differences in streamflow are quantified and used to assess the effect of forest management by comparing observed flows in the treatment catchment versus predicted values calculated from the pre-treatment relationship with control, had it not been disturbed. These studies generally fall into one of four categories including afforestation, deforestation, regrowth or vegetation type conversion (Brown *et al.*, 2005). We have elected not to repeat those results here, but rather illustrate the processes involved and categorically describe responses.

Many ongoing paired catchment studies have changed emphasis from the effects of forest management practices to long-term changes in water resources as related to changing atmospheric inputs or climate variability in temperature and precipitation. Nonetheless, even these studies use the common metrics of annual water yield, peak flow and low flow as discussed below.

12.1.1 Annual water yield

The reduction of forest canopy decreases interception and evapotranspiration losses and increases runoff proportionally, but non-linearly. There are two thresholds that must be overcome to increase water yield. The first threshold is annual precipitation. Paired catchment studies show that sustainable runoff is produced only when annual precipitation exceeds 450–500 mm annually (Bosch and Hewlett, 1982; MacDonald and Stednick, 2003; Scherer and Pike, 2003). In regions that receive less than 500 mm, the amount of precipitation, on average, is inadequate to exceed evaporative demand. These areas are often water limited and a decrease in forest cover will not necessarily produce increased runoff, but increase soil evaporation or use by residual vegetation. One way to increase runoff in these areas is to decrease infiltration rates (soil compaction for rainfall harvesting, for example).

Areas that receive more than 500 mm of annual precipitation are less likely to be water limited. Vegetation removal, through timber harvesting for example, has to reduce vegetative cover below the point where residual vegetation can still use all the water or an increase in water yield will not be detected. A minimum of 20% forest cover or basal area needs to be removed for a detectable increase in annual water yield (Bosch and Hewlett, 1982), but varies by biogeoclimatic area or hydrological region area (US EPA, 1980;

Stednick, 1996). Water yield increases are non-linearly proportional to the degree of vegetation removal. The greatest increase in water yield generally occurs the first full year after treatment.

Annual water yield and the change in water yield following timber harvest, both within and between catchments, are dependent on climatic variability and antecedent moisture conditions. Water yield response to a given precipitation event is a reflection of the antecedent soil moisture conditions on the catchment at the time of the event. Precipitation falling on wetter soils will generally result in greater water yield than will be generated from the same event falling on drier soils and soils are generally wetter on harvested catchments. In more arid environments, or during drier portions of the year, the difference in antecedent moisture content between forested and harvested catchments might be minimal as will be the water yield response and the difference in water yield response. Under more humid conditions, the differences in antecedent conditions between forested and harvested catchment are usually greater as is the difference in water yield that will occur in response to a given precipitation event. On average, water yield and changes in water yield following timber harvest will increase with increasing precipitation both within and between catchments. This generality applies to changes in water yield following timber harvest primarily when differences in antecedent soil moisture exist. In areas of high rainfall or during periods of low evapotranspiration (winter), differences in antecedent conditions between forested and harvested catchments may be negated, as will be the difference in water yield response.

In more humid environments, or areas of high rainfall, water yield increases are usually the greatest but effects of harvesting are the shortest due to rapid forest regrowth. In drier regions, changes in water yield are not as pronounced but are more persistent since the vegetative regrowth takes more time. Increases in water yield after forest harvesting are not equally distributed over the water year but they do reflect the antecedent soil moisture conditions that exist at the time precipitation is made available to the soil. Precipitation tends to fall as rain in lower elevations and latitudes and coastal regions, while snow is deposited in higher elevations or cold regions during the winter months. In rain-dominated areas, increased water yields are observed during the late autumn and winter months, when the soil mantle moisture deficit is being recharged. In snow-dominated regions, the greatest increases in water yield following logging are usually observed during the late spring to early summer months when snowmelt recharges the soil. While snow water equivalent (SWE) has been shown to increase with decreasing vegetation, the degree and timing of increased water yield depend on the timing of soil moisture recharge.

In addition to soil moisture recharge, solar energy (a function of slope and aspect) is an important factor in the amount and timing of runoff. For example, south-facing slopes (in the northern hemisphere) generally have less dense vegetation and receive more sunlight than north-facing slopes. This results in lower interception losses but higher evaporative rates that persists following timber harvest. The opportunity for increases in water yield following timber harvest is reduced on south slopes relative to north slopes.

The increase in water yield decreases as the forest regrows. Within even-aged stands without significant understorey, these effects include: increases in annual water yield, increases in late summer and autumn low flows, variable responses (no change or increases) in peak flows and possibly earlier timing of peak flows. Uneven-aged forest stands usually have less response to harvesting since increased water and nutrients fluxes are utilized by the remaining vegetation. Afforestation is the conversion of land from non-forest to forest cover. Increased evapotranspiration and increased interception from the forest as compared with the non-forest may decrease annual water yields. Depending on previous site conditions, peak flows may decrease due to improved infiltration and low flows may increase due to increases in soil moisture storage. Finally, there is the question of linkages between forest cover and precipitation which will not be addressed in this chapter.

Attempting to quantify the effect of forest harvesting on annual water yield is time consuming and expensive. It requires a long-term commitment from researchers, which necessitates sufficient funding and requires a commitment from the landowner to maintain a catchment in an undisturbed condition. Catchment-specific study results are often difficult to extrapolate, especially given the variability in response even for a specific hydrological region (Stednick, 1996), coupled with climatic variability that is now much more appreciated. Furthermore, differences in catchment characteristics, as well as forest type, composition and harvesting method, compound extrapolation. Extrapolating results to other catchments must be done with care and model efforts can only give suggestions of water resource responses.

12.1.2 Peak flows

Peak flows are the maximum flow rate that occurs within a specified period of time, usually on an annual basis, and occur between May and June due to spring snowmelt or from long-duration rain events in rain-dominated environments. The literature showed mixed responses to peak flow increases (Hamilton, 1985; Austin, 1999; Scherer, 2001) and it is a contentious topic (van Dijk and Keenan, 2007).

Effects of harvesting on peak flows are often examined in catchment studies because of flooding concerns. In addition, increases in peak flows can cause increases in stream scouring and bank undercutting, which in turn can affect water quality and aquatic habitats through the transport of sediment (Stednick, 2000). Roads constructed to facilitate timber harvesting and forest management can also affect the magnitude and timing of peak flow (Reiter and Beschta, 1995; Wemple *et al.*, 1996; Gucinski *et al.*, 2001). Compacted road surfaces limit water infiltration; road cut banks can intercept slower subsurface flows and transform them into more rapid surface flows; and road ditches and culverts can reroute water directly into streams (Scherer and Pike, 2003). Road BMPs are used to minimize hydrological effects.

In snow-dominated environments, the timing of peak flows may be advanced by timber harvest operations due to faster and earlier snowmelt rates (as cited by Scherer, 2001) that enter wetter soils, causing an earlier recharge and requiring less meltwater to be retained on site. A literature review showed a range of advancement from zero to 18 days (Austin, 1999). An early public concern was the potential synchronization of peak flows, which means that peak flows from various portions of each sub-catchment would combine to form a higher flood peak. Forest harvesting can alter peak flows by de-synchronizing snowmelt over a catchment and reducing the total peak flow. The resulting hydrograph usually shows two relatively lower peaks rather than a larger one. Such responses are attributed to early snowmelt in logged areas, followed later by snowmelt in forested areas; often seen in a bimodal snowmelt hydrograph.

Rain-on-snow events can occur in the warm snow zone or transitional snow zone. Here the precipitation event as rain falling on snow results in some snowmelt and increased runoff from the usually larger than normal precipitation depth. Such events usually occur over large areas and the effects of forest management activities on runoff cannot be separated.

In rain-dominated systems harvesting does not result in an increase in peak flow in situations where the soil is recharged in both the forested and harvested areas, beyond the effects of interception reduction on rainfall input. The only time harvesting significantly alters peak discharge, beyond the interception influence, is when there are differences in antecedent conditions between forested and harvested areas and the harvested area is more responsive. Thus most changes in peak flow occur during the growing season when soil moisture differences are present and the effects of both reduced interception and reduced soil water retention occur. The primary difference between rain-dominated and snow-dominated is largely a function of the timing of when precipitation is available to infiltrate (individual rainfall events or accumulated snowpack, magnitude of the input event and antecedent soil moisture conditions).

The effects of timber harvesting on peak flows has received discussion and debate in the literature (Jones and Grant, 1996, 2001; Thomas and Megahan, 1998, 2001; Beschta *et al.*, 2000; van Dijk and Keenan, 2007). Much of the debate stems from the variability in results from various studies, coupled with different statistical approaches in interpreting the changes in peak flow.

No single variable (e.g. amount of forest removed, harvesting method, silvicultural treatment) can account or predict peak flow changes. It appears that peak flow increases occur less often under contemporary forest practice in most settings, probably attributable to smaller harvest areas as a portion of the catchment, minimization of road lengths and streamside channel vegetation left undisturbed. Similarly, no studies were found that measured increased peak flows in any forest stand for harvesting practices other than clearcutting (i.e. shelterwood, patchcut or various thinning prescriptions).

12.1.3 Low flows

Another common public misconception is that timber harvesting decreases low flows (e.g. Chang, 2005 among others). During the summer months, the water savings resulting from reduced interception and evapotranspiration can increase low flows. Of a review of 350 worldwide studies on the effects of forest harvesting on water resources, only 28 addressed low flows. Of these, 16 had increased low flows, ten had no change and two had a decrease (Austin, 1999). The last two studies were in coastal Oregon, USA. The first study hypothesized that canopy drip or fog drip beneath the trees from fog and cloud interception was reduced and reduced low flows (Harr, 1982). This occult precipitation added to the total net precipitations. Under such circumstances, forest removal could decrease annual water increases if the amount of added precipitation is significant. The second case study showed that changes in species composition during forest regeneration or succession affected catchment hydrology. A change in riparian vegetation from conifers to deciduous species after harvest reduced dry-weather streamflow (low flows) (Hicks *et al.*, 1991). A similar decrease in flows can occur when climax mixed hardwoods are replaced by pioneer hardwood species (Swank and Johnson, 1994).

During low flows (i.e. baseflow), the removal of forest (or other vegetative) cover in the riparian area can increase streamflow on smaller streams from evapotranspiration savings. Contemporary forest harvesting practices usually exclude the riparian area from activity to protect water resources, so such a change may not be common. Interest in the physiographical influences on low flow generation is increasing (i.e. Tague and Grant, 2004). Low flows often increase after harvesting, but increases are variable and difficult to analyse statistically. The longevity of low flow changes is generally not addressed in the literature (Reiter and Beschta, 1995; Gucinski *et al.*, 2001; Stednick, 2008). Low flow increases have been reported following fire and insect disturbance (Scherer, 2001). Low flow changes seem to return to pre-treatment conditions in a matter of a few years (Austin, 1999; Stednick, 2008).

For this review chapter, it was seen that the definition of low flows varies in the literature; ranging from an instantaneous flow rate, to number of days below a certain threshold (Stednick, 2008), to actual flow recurrence intervals such as 3-, 7-, 10- or 30-day low flow. Expression of low flows as a change percentage may be misleading since a small change in low flow would be expressed as a large percentage. Furthermore, quantification of low flow rates, even with artificial control sections, may be problematic.

12.2 Effects of Forest Fire

Prescribed fire for vegetation removal, slash removal or site preparation is a common forest management activity in many forest types. The seriousness of a fire depends on the size and intensity of the fire, soil characteristics, slope steepness, the amount and character of the precipitation to which the burned area is subject after the fire, the vegetation type present before the fire, the length of time the soil is bare before revegetation, the type and amount of vegetation that comes back after the fire, the proportion of the catchment burned, characteristics of the unburned catchment area, and the channel stability and condition to carry increased streamflows. The hydrological response of a catchment to a fire is an extremely complex interaction of many variables. Many different processes at different scales of time make comparison of results from different catchments difficult. Catchment or catchment-level responses to fire, either controlled fire or wildfire, may result in changes in streamflow as measured by total yield, peak flows and low flows (see Chapter 13, this volume, for more detail).

The amount of erosion and sedimentation that occur after such a fire is highly variable. A plausible explanation is that convective storms are driving high erosion rates. Convective storms during the summer months, which are characterized by high-intensity precipitation, result in the soil surface receiving a large amount of precipitation over a short timespan. As a result, most of the precipitation cannot infiltrate the soil and is converted to runoff as overland flow, resulting in downstream flooding with sediment and debris-laden waters. Conversely, wildfires that are of lower fire severity may not result in soil hydrophobicity and not alter streamflow generation mechanisms, and streamflow (and suspended sediment) would not increase (Troendle and Bevenger, 1996).

12.3 Effects of Insects and Disease

In 1939 a wind storm in Colorado, USA created ideal breeding conditions for an Engelmann spruce beetle epidemic (Love, 1955). By 1946, the beetle had killed up to 80% of the forest trees. Using a paired catchment approach, average water yield increased for a 15-year post-epidemic period. Maximum annual instantaneous streamflows increased from 0 to 27%. Overall, the increased water yield was attributed to greater accumulations of snow in the killed areas (Love, 1955). This was the first study to document increased streamflow from an insect defoliation event and adds to the Colorado history in forest hydrology. Later analysis of streamflow records revealed that the smallest increases on both catchments occurred during the first 5-year period (when the beetle population was multiplying to epidemic proportions) and the largest increases occurred 15 years later (Bethlahmy, 1974, 1975).

A mountain pine beetle outbreak in the mid-1970s killed an estimated 35% of the trees in Jack Creek in south-west Montana, USA. Data analysis indicated an increase in water yield, an advance of 2–3 weeks in the annual hydrograph snowmelt peak and an increase in low flows. Because of de-synchronization of streamflow peaks, the increased annual water yields did not produce a large difference in peak flows (Potts, 1984).

The paired catchment technique was also used to assess streamflow changes of Camp Creek in interior British Columbia, Canada after clearcut logging occurred over 30% of its catchment area. Existing hydrometric data for Camp Creek (beetle infested) and those of an adjacent control, Greata Creek, were analysed for both the pre-logging and post-logging periods. Post-logging Camp Creek streamflow changes included increased annual yield and peak flows, as well as earlier annual peak flow and half-flow volume occurrence dates (Cheng, 1989).

Other study results on the effects of insect outbreaks on streamflow were occasionally accompanied by timber harvesting, resulting in variable findings. Retrospective models were used to model the forest type and age over time to assess the hydrological effect of pine beetle activity and predicted water yield increases (Troendle and Nankervis, 2014). Conversely, a physically based model suggested no water yield response (Mikkelson *et al.*, 2013).

No study was found examining the effect of forest disease on streamflow. Studies documented changes in water quality but not water quantity. Paired catchment studies to assess the effects of insect or disease are difficult to conduct, since control or undisturbed catchments are difficult to maintain and be kept in an undisturbed condition. Paired catchments would have similar vegetation and be subject to the same disturbance agent.

12.4 Future Investigative Methods

The paired catchment approach has been the traditional approach in determining the effects of forest management practices on streamflow. The following guidelines for paired catchment studies have been proposed: (i) hydrological similarities between catchments should be assessed throughout the pre-treatment data collection period; (ii) catchments should be 1000 ha or less in size (larger catchments appear to integrate things better and error terms are lower and calibration tighter) (Troendle *et al.*, 2001); (iii) treatment should be executed during a single event and percentage harvested should be extensive (>20%); and (iv) pre- and post-treatment streamflow data should be sufficient for a high power of detecting a change if one exists (>10 years for both pre- and post-treatment) (McFarlane, 2001; Buttle, 2011) or use modelling approaches (Zhang *et al.*, 2001; Bren and Hopmans, 2007).

However, the finding and maintenance of an undisturbed catchment has become difficult and expensive, thus alternative approaches are being used to answer the effects of contemporary forest practices on water resources, several of which are described below.

12.4.1 Single catchment studies

This method examines a single catchment during calibration and treatment periods. During the calibration period, streamflow data are related statistically to weather data to develop a hydroclimatic model (often a simple regression). During the treatment period, the model is used to estimate what streamflow would have been in the absence of treatment. Effects of treatment on streamflow are calculated as differences between observed and estimated values. Uncertainty in model estimates can obscure treatment effects (NCASI, 2009). Increasing the length of the calibration period can improve model estimates but cannot overcome some inherent limitations of the single catchment approach. For example, if weather data are collected from a single station, model estimates of streamflow are based on weather data that are most likely not representative of conditions in the entire catchment (Chang, 2005). The popularity of the paired catchment method is due in part to its generally greater power to detect treatment effects (Loftis and MacDonald, 2000).

12.4.2 Retrospective studies

Another alternative is to use previously collected streamflow and precipitation data (NCASI, 2009). Retrospective studies involve an after-the-fact pairing of harvested catchments with undisturbed catchments for which pre-harvesting data exist (Moore and Wondzell, 2005). As control catchments become less available, and the additional question of data stationarity with precipitation and streamflow, retrospective studies will no doubt increase (i.e. McFarlane, 2001; Webb *et al.*, 2012).

12.4.3 Nested catchment studies

Nested catchment studies can provide insights into hydrological processes across spatial scales by measuring treatment effects in large catchments and sub-catchments of those catchments (NCASI, 2009). When coupled with process modelling, nested catchment studies can measure treatment effects and provide insight to causal mechanisms (Alila and Beckers, 2001; Alila *et al.*, 2005). Some examples from the USA are Caspar Creek in California (Ziemer, 2001; Keppeler, 2007), Mica Creek in Idaho (Hubbart *et al.*, 2007), Alto Catchment Study in Texas (McBroom *et al.*, 2008), Hinkle Creek in Oregon (Zegre *et al.*, 2010), Alsea Catchment Study Revisited (e.g. Stednick, 2008) and Deadhorse Creek in Colorado (Troendle, 1987); and Bowron Catchment in Canada (Wei and Davidson, 1998).

12.4.4 Statistical approaches

The paired catchment approach typically uses an analysis of covariance to determine the significance of post-treatment water hydrological responses. Responses are various water quantity metrics such as annual water yield, instantaneous peak flow or low flows. The utility of this approach is limited by the variability between the catchments, type II error and the question of control catchment stationarity. Prediction residuals are used to determine if a significant change occurred between the pre- and post-treatment periods. Earlier studies used the 95% confidence level (i.e. Moring, 1975).

Paired catchment studies are used for determining the changes in water yield resulting from changes in vegetation at various time scales including the annual yield, the seasonal pattern of flows, and changes in both annual and seasonal flow duration curves. Comparisons between paired catchment results and a mean annual water balance model showed good agreement (Brown *et al.*, 2005). Analysis of annual water yield changes from afforestation, deforestation and regrowth experiments demonstrates that the time taken to reach a new equilibrium under permanent land-use change varies considerably. Deforestation experiments reach a new equilibrium more quickly than afforestation experiments. Seasonal changes in water yield highlight the proportionally larger impact on low flows (Brown *et al.*, 2005; van Dijk and Keenan, 2007).

Change-point analysis is conducted with a non-parametric test for homogeneity. The Pettitt test was developed to identify change points in hydrological time series when the exact time of change is unknown (Pettitt, 1979). This approach determines significant changes in mean values of a series, pinpointing abrupt changes in the record. The test counts the number of times a member of the first sample exceeds a member of the second sample. If a change point is detected, the time series is divided into two parts around the timing of the change point. The Pettitt test is frequently used in combination with statistical trend tests to assess the effects of catchment changes on hydrological time series data (Ma *et al.*, 2008; Zhang *et al.*, 2008; Salarijazi *et al.*, 2012). A change point can distinguish streamflow changes due to natural disturbance or land-use changes from streamflow changes due to climate variability.

Precipitation–runoff models have been developed to look at the effects of harvesting on water resources (Whitaker *et al.*, 2003; Seibert and McDonnell, 2010; Seibert *et al.*, 2010). Monte Carlo simulations reduce model parameter error. A change-detection method using daily streamflow values was used to assess streamflow changes after harvesting (Seibert and McDonnell, 2010; Zegre *et al.*, 2010). Inter-catchment variability was quantified before and after treatment to better identify catchment response to timber harvesting. Numerical modelling using long-term data and classes of data has been developed (Schnorbus and Alila, 2004).

More recently in paired catchment studies, a change-detection technique using moving sums of recursive residuals (MOSUM) can select calibration periods for each control–treatment catchment pair to reduce regression model uncertainty, which may mask treatment effects (Ssegane *et al.*, 2015). Better separation of evaporation changes from transpiration changes after hurricane damage used a moving-window type temporal analysis of streamflow data to capture decadal-long hydrological processes (Jayakaran *et al.*, 2014).

12.5 Summary

Various reviews have been conducted on the effects of timber harvesting on hydrology and responses are variable for annual water yield, low flows, peak flows and timing of peak flows. A threshold of 20% of the catchment needs to be harvested to have a measurable water yield response. Silvicultural practices other than clearcutting can exceed that threshold; however, few studies have been conducted that demonstrate a measurable response to streamflow for other than clearcutting in operational situations. Results from small study catchments cannot be extrapolated to larger catchments. A water yield increase in a small catchment following timber harvesting cannot be quantified or measured at a larger scale given common stream gauging practices.

Given the difficulty of maintaining a control catchment, it is our opinion that more retrospective studies will be done with existing hydrometeorological data and new statistical methods. Increased applications of statistics for non-stationary data and change-detection methods, coupled with data collection platforms with finer time steps, will allow for more rigorous interpretation of the hydrological effects of forest management activities.

References

Adams, P.W. and Ringer, J.O. (1994) *The Effects of Timber Harvesting and Forest Roads on Water Quantity and Quality in the Pacific Northwest: Summary and Annotated Bibliography.* Forest Engineering Department, Oregon State University, Corvallis, Oregon.

Alila, Y. and Beckers, J. (2001) Using numerical modeling to address hydrologic forest management issues in British Columbia. *Hydrological Processes* 15, 3371–3387.

Alila, Y., Weiler, M., Green, K., Gluns, D. and Jost, G. (2005) Designing experimental catchments to understand and quantify the influence of land-use management and natural variability on the hydrological response at multiple scales. In: Predictions in Ungauged Basins (PUB), Workshop Proceedings, 28–30 November 2005. Available at: https://www.for.gov.bc.ca/hfd/library/fia/2006/FSP_Y062294c.pdf (accessed 14 April 2016).

Anderson, H.W., Hoover, M.D. and Reinhart, K.G. (1976) *Forests and Water: Effects of Forest Management on Floods, Sedimentation and Water Supply*. General Technical Report PSW-GTR-18. USDA Forest Service, Pacific Southwest Research Station, Albany, California.

Andreassian, V. (2004) Water and forests: from historical controversy to scientific debate. *Journal of Hydrology* 291, 1–27.

Austin, S.A. (1999) Streamflow response to forest management: a meta-analysis using published data and flow duration curves. MSc. thesis, Colorado State University, Fort Collins, Colorado.

Bates, C.G. and Henry, A.J. (1928) Forest and streamflow experiment at Wagon Wheel Gap, Colorado. Final report upon completion of the second phase of the experiment. *USDA Monthly Weather Revue, Supplement 30*.

Beschta, R.L. (1990) Effects of forests on water quantity and quality. In: Walstad, J.D., Radosevich, S.R. and Sandberg, D.V. (eds) *Natural and Prescribed Fire in Pacific Northwest Forests*. Oregon State University Press, Corvallis, Oregon, pp. 219–232.

Beschta, R., Pyles, M.R., Skaugset, A.E. and Surfleet, C.G. (2000) Peak flow responses to forest practices in the western cascades of Oregon, USA. *Journal of Hydrology* 233, 102–120.

Bethlahmy, N. (1974) More streamflow after a bark beetle epidemic. *Journal of Hydrology* 23, 185–189.

Bethlahmy, N. (1975) A Colorado episode: beetle epidemic, ghost forests, more streamflow. *Northwest Science* 49, 95–105.

Bosch, J.M. and Hewlett, J.D. (1982) A review of catchment experiments to determine the effect of vegetation changes on water yield and evapotranspiration. *Journal of Hydrology* 55, 3–23.

Bren, L. and Hopmans, P. (2007) Paired catchments observations on the water yield of mature eucalypt and immature radiate pine plantations in Victoria, Australia. *Journal of Hydrology* 336, 416–429.

Bren, L. and Lane, P. (2014) Optimal development of calibration equations for paired catchment projects. *Journal of Hydrology* 510, 720–731.

Bren, L. and McGuire, D. (2012) Paired catchment experiments and forestry politics in Australia. In: Webb, A.A., Bonell, M., Bren, L., Lane, P.J.N., McGuire, D., Neary, D.J., Nettles, J., Scott, D.F., Stednick, J. and Wang, Y. (eds) *Revisiting Experimental Catchment Studies in Forest Hydrology (Proceedings of a Workshop held during the XXV IUGG General Assembly in Melbourne, June–July 2011)*. IAHS Publication No. 353. International Association of Hydrological Sciences, Wallingford, UK, pp. 106–116.

Brown, A.E., Zhang, L., McMahon, T.A., Western, A.W. and Vertessy, R. (2005) A review of paired catchment studies for determining changes in water yield resulting from alterations in vegetation. *Journal of Hydrology* 310, 28–61.

Bruijnzeel, L.A. (2004) Hydrological functions of tropical forests: not seeing the soil for the trees? *Agricultural Ecosystems and Environment* 104, 185–228.

Buttle, J.M. (2011) The effects of forest harvesting on forest hydrology and biogeochemistry. In: Levia, D.F., Carlyle-Moses, D. and Tanaka, T. (eds) *Forest Hydrology and Biogeochemistry: Synthesis of Past Research and Future Directions*. Ecological Studies Vol. 216. Springer, New York, pp. 659–678.

Calder, I.R. (2007) Forests and water – ensuring forest benefits outweigh water costs. *Forest Ecology and Management* 251, 110–120.

Chang, M. (2005) *Forest Hydrology: An Introduction to Water and Forests*. Taylor & Francis, New York.

Cheng, J.D. (1989) Streamflow changes after clear-cut logging of a pine beetle-infested catchment in southern British Columbia, Canada. *Water Resources Research* 25, 449–456.

Chescheir, G.M., Skaggs, R.W. and Amatya, D.M. (2009) Quantifying the hydrologic impacts of afforestation in Uruguay: a paired watershed study. In: *Proceedings of the XIII World Forestry Congress, Buenos Aires, Argentina, 18–25 October 2009*. World Forestry Congress, Buenos Aires, Argentina, pp. 18–23.

Colman, E.A. (1953) *Vegetation and Catchment Management*. Ronald Press, New York.

de la Cretaz, A. and Barten, P.K. (2007) *Land Use Effects on Streamflow and Water Quality in Northeastern United States*. CRC Press, Boca Raton, Florida.

Edwards, P. and Troendle, C.A. (2008) Water yield and hydrology. In: Audin, L.J. (ed.) Cumulative Watershed Effects of Fuels Management. USDA Forest Service, Northeastern Area. Available at: http://www.na.fs.fed.us/fire/cwedocs/12_Water_Yield_Hydrology_Edwards_Troendle.pdf (accessed 8 April 2016).

Fuller, R.D., Simone, D.M. and Driscoll, C.T. (1988) Forest clearcutting effects on trace metal concentrations: spatial patterns in soil solutions and streams. *Water, Air, and Soil Pollution* 40, 185–195.

Gucinski, H., Furniss, M.J., Ziemer, R.R. and Brookes, M.H. (2001) *Forest Roads: A Synthesis of Scientific Information*. General Technical Report PNW-GTR-509. USDA Forest Service, Pacific Northwest Research Station, Portland, Oregon.

Hamilton, L. (1985) Overcoming myths about soil and water impacts of tropical forest land uses. In: El-Swaify, S.A., Moldenhauer, W.C. and Lo, A. (eds) *Soil Erosion and Conservation*. Soil Conservation Society of America, Ankeny, Iowa, pp. 680–690.

Hamilton, L. (2008) *Forests and Water. Thematic Study for the Global Forest Resources Assessment*. FAO Forestry Paper No. 155. Food and Agriculture Organization of the United Nations, Rome.

Harr, R.D. (1982) Fog drip in the Bull Run Municipal Catchment, Oregon. *Water Resources Bulletin* 18, 785–789.

Hibbert, A.R. (1967) Forest treatment effects on water yield. In: Sopper, W.E. and Lull, H.W. (eds) *International Symposium of Forest Hydrology*. Pergamon Press, Oxford, pp. 527–543.

Hicks, B.J., Beschta, R.L. and Harr, R.D. (1991) Long-term changes in streamflow following logging in Western Oregon and associated fisheries implications. *Water Resources Bulletin* 27, 217–226.

Hubbart, J.A., Link, T.E., Gravelle, J.A. and Elliot, W.J. (2007) Timber harvest impacts on water yield in the Continental/Maritime hydroclimatic region of the United States. *Forest Science* 53, 169–180.

Ice, G.G. and Stednick, J.D. (2004) Introduction. In: Ice, G.G. and Stednick, J.D. (eds) *A Century of Forest and Wildland Watershed Lessons*. Society of American Foresters, Bethesda, Maryland, pp. vii–x.

Jackson, C.R., Sun, G., Amatya, D.M., Swank, W.T., Riedel, M., Patric, J., Williams, T., Vose, J.M., Trettin, K., Aust, W.M., *et al.* (2004) Fifty years of forest hydrology research in the southeast: some lessons learned. In: Ice, G.G. and Stednick, J.D. (eds) *A Century of Forest and Wildland Watershed Lessons*. Society of American Foresters, Bethesda, Maryland, pp. 33–112.

Jayakaran, A.D., Williams, T.M., Ssegane, H., Amatya, D.M., Song, B. and Trettin, C.C. (2014) Hurricane impacts on a pair of coastal watersheds: implications of selective hurricane damage to forest structure and streamflow dynamics. *Hydrology and Earth System Sciences* 18, 1151–1164.

Jones, J.A. and Grant, G.E. (1996) Peak flow responses to clear-cutting and roads in small and large basins, western Cascades, Oregon. *Water Resources Research* 32, 959–974.

Jones, J.A. and Grant, G.E. (2001) Comment on ‘Peak flow responses to clear-cutting and roads in small and large basins, western Cascades, Oregon: A second opinion’ by R. B. Thomas and W. F. Megahan. *Water Resources Research* 37, 175–178.

Keppeler, E. (2007) Effects of timber harvest on fog drip and streamflow, Caspar Creek experimental catchments, Mendocino County, California. In: Standiford, R.B., Giusti, G.A., Valachovic, Y., Zielinski, W.J. and Furniss, M.J. (eds) *Proceedings of the Redwood Region Forest Science Symposium: What Does the Future Hold?* General Technical Report PSW-GTR-194. USDA Forest Service, Pacific Southwest Research Station, Albany, California, pp. 85–93.

Loftis, J.C. and MacDonald, L.H. (2000) Exploring the benefits of paired catchments for detecting cumulative effects. In: *Catchment Management and Catchment Operations 2000*. American Society of Civil Engineers, New York, pp. 1–9.

Love, L.D. (1955) The effect on streamflow of the killing of spruce and pine by the Englemann spruce beetle. *Transactions of the American Geophysical Union* 36, 113–118.

Ma, Z., Kang, S., Zhang, L., Tong, L. and Su, X. (2008) Analysis of impacts of climate variability and human activity on streamflow for a river basin in arid region of northwest China. *Journal of Hydrology* 352, 239–249.

MacDonald, L.H. and Stednick, J.D. (2003) *Forest and Water: A State of the Art Review for Colorado*. Completion Report No. 196. Colorado Water Resources Research Institute, Colorado State University, Fort Collins, Colorado.

McBroom, M.W., Beasley, R.S., Chang, M. and Ice, G.G. (2008) Storm runoff and sediment losses from forest clearcutting and stand reestablishment with best management practices in the Southeastern United States. *Hydrological Processes* 22, 1509–1522.

McFarlane, B.E. (2001) Retrospective analysis of the effects of harvesting on peak flow in southeastern British Columbia. In: Toews, D.A.A. and. Chatwin, S. (eds) *Catchment Assessment in the Southern Interior of British Columbia, Workshop Proceedings. Working Paper No. 57*. Research Branch, British Columbia Ministry of Forests, Victoria, British Columbia, Canada, pp. 81–93. Available at: http://www.for.gov.bc.ca/hfd/pubs/Docs/Wp/Wp57.htm (accessed 10 August 2015).

Mikkelson, K., Bearup, L.A., Maxwell, R., Stednick, J.D., McCray, J.E. and Sharp, J.O. (2013) Bark beetle infestation impacts on nutrient cycling, water quality and interdependent hydrological effects. *Biogeochemistry* 115, 1–21.

Moore, R.D. and Wondzell, S.M. (2005) Physical hydrology and the effects of forest harvesting in the Pacific Northwest: a review. *Journal of the American Water Resources Association* 41, 763–784.

Moring, J.R. (1975) *The Alsea Watershed Study: Effects of Logging on the Aquatic Resources of Three Headwater Streams of the Alsea River, Oregon: Part II – Changes in Environmental Conditions*. Fisheries Research Report No. 9. Oregon Department of Fish and Wildlife, Corvallis, Oregon.

NCASI (2009) *Effects of Forest Management on Water Resources in Canada: A Research Review.* Technical Bulletin No. 969. National Council for Air and Stream Improvement, Inc., Research Triangle Park, North Carolina.

NRC (2008) *Hydrologic Effects of a Changing Forest Landscape.* Committee on Hydrologic Impacts from Forest Management Activities, National Research Council, Washington, DC.

Pettitt, A. (1979) A nonparametric approach to the change-point problem. *Applied Statistics* 28, 126–135.

Potts, D. (1984) Hydrologic impacts of a large-scale mountain pine beetle (*Dendroctonus ponderosae* Hopkins) epidemic. *Journal of the American Water Resources Association* 20, 373–377.

Reiter, M.L. and Beschta, R.L. (1995) The effects of forest practices on water. In: *Cumulative Effects of Forest Practices in Oregon*. Oregon Department of Forestry, Salem, Oregon, Chapter 7.

SAF (2008) *The Dictionary of Forestry*. Society of American Foresters, Bethesda, Maryland.

Sahin, V. and Hall, M.J. (1996) The effects of afforestation and deforestation on water yields. *Journal of Hydrology* 178, 293–309.

Salarijazi, M., Akhond-Ali, A., Adib, A. and Daneshkhah, A. (2012) Trend and change-point detection for the annual stream-flow series of the Karun River at the Ahvaz hydrometric station. *African Journal of Agricultural Research* 7, 4540–4552.

Scherer, R. (2001) Effect of changes in forest cover on streamflow: a literature review. In: Toews, D.A.A. and Chatwin, S. (eds) *Catchment Assessment in the Southern Interior of British Columbia, Workshop Proceedings*. Working Paper No. 57. Research Branch, British Columbia Ministry of Forests, Victoria, British Columbia, Canada, pp. 44–55.

Scherer, R. and Pike, R. (2003) *Effects of Forest Management Activities on Streamflow in the Okanagan Basin, British Columbia: Outcomes of a Literature Review and Workshop*. FORREX Series No. 9. FORREX–Forest Research and Extension Partnership, Kamloops, British Columbia, Canada.

Schnorbus, M. and Alila, Y. (2004) Forest harvesting impacts on the peak flow regime in the Columbia Mountains of southeastern British Columbia: an investigation using long-term numerical modeling. *Water Resources Research* 40, W05205, doi: 10.1029/2003WR002918 (accessed 8 April 2016).

Seibert, J. and McDonnell, J. (2010) Land-cover impacts on streamflow: change-detection modelling approach that incorporates parameter uncertainty. *Hydrological Sciences Journal* 55, 316–332.

Seibert, J., McDonnell, J. and Woodsmith, R. (2010) Effects of wildfire on catchment runoff responses: a modelling approach to detect changes in snow-dominated forested catchments. *Hydrological Research* 41, 378–390.

Ssegane, H., Amatya, D.M., Muwamba, A., Chescheir, G.M., Appelboom, T., Tollner, E.W., Nettles, J.E., Youssef, M.A., Birgand, F. and Skaggs, R.W. (2015) Hydrologic calibration of paired watersheds using a MOSUM approach. *Hydrology and Earth Systems Science* 19, 1–35.

Stednick, J.D. (1996) Monitoring the effects of timber harvest on annual water yield. *Journal of Hydrology* 176, 79–95.

Stednick, J.D. (2000) Effects of vegetation management on water quality: timber management. In: Dissmeyer, G. (ed.) *Drinking Water from Forests and Grasslands*. General Technical Report SRS-039. USDA Forest Service, Southern Research Station, Asheville, North Carolina, pp. 147–167.

Stednick, J.D. (ed.) (2008) *Hydrological and Biological Responses to Temperate Coniferous Forest Practices*. Springer, New York.

Sun, G., McNulty, S.G., Shepard, J.P., Amatya, D.M., Riekerk, H., Comerford, N.B., Skaggs, W. and Swift, L. Jr (2001) Effects of timber management on the hydrology of wetland forests in the southern United States. *Forest Ecology and Management* 143, 227–236.

Swank, W.T. and Douglass, J.E. (1974) Streamflow greatly reduced by converting deciduous hardwood stands to pine. *Science* 185, 857–859.

Swank, W.T. and Johnson, C.E. (1994) Small catchment research in the evaluation and development of forest management practices. In: Moldan, B. and Cerny, J. (eds) *Biogeochemistry of Small Catchments: A Tool for Environmental Research*. SCOPE 51. Wiley, Chichester, UK, pp. 383–408.

Tague, C. and Grant, G.E. (2004) A geological framework for interpreting the low-flow regimes of Cascade streams, Willamette River Basin, Oregon. *Water Resources Research* 40, W04303, doi: 10.1029/2003WR002629 (accessed 8 April 2016).

Thomas, R.B. and Megahan, W.F. (1998) Peak flow responses to clear-cutting and roads in small and large basins, western Cascades, Oregon: a second opinion. *Water Resources Research* 34, 3393–3403.

Thomas, R.B. and Megahan W.F. (2001) Reply to 'Comment on "Peak flow responses to clear-cutting and roads in small and large basins, western Cascades, Oregon: a second opinion" by R. B. Thomas and W. F. Megahan'. *Water Resources Research* 37, 181–183.

Troendle, C.A. (1987) The effect of partial and clearcutting on streamflow at Deadhorse Creek, Colorado. *Journal of Hydrology* 90, 145–157.

Troendle, C.A. and Bevenger, G.S. (1996) Effect of fire on streamflow and sediment transport, Shoshone National Forest, Wyoming. In: Greenlee, J. (ed.) *Proceedings of the Second Biennial Conference on the Greater Yellowstone Ecosystem, The Ecological Implications of Fire in Greater Yellowstone National Park, Wyoming*. International Association of Wildland Fire, Fairfield, Washington, pp. 43–52.

Troendle, C.A. and Nankervis, J.M. (2014) *The Effects of Insect Mortality and Other Disturbances on Water Yield in the North Platte Basin*. Final Report. Prepared for the Wyoming Water Development Council, Cheyenne, Wyoming.

Troendle, C.A., Wilcox, M.S., Bevenger, G.S. and Porth, L. (2001) The Coon Creek Water Yield Augmentation Project: implementation of timber harvesting technology to increase streamflow. *Forest Ecology and Management* 143, 179–187.

US EPA (1980) Hydrology. In: *An Approach to Water Resources Evaluation of Non-point Silvicultural Sources*. EPA-600/8-80-012. US Environmental Protection Agency, Athens, Georgia, Chapter III.

Van Dijk, A.I.J.M. and Keenan, R. (2007) Planted forests and water in perspective. *Forest Ecology and Management* 251, 1–9.

Webb, A.A., Bonell, M., Bren, L., Lane, P.N.J., McGuire, D., Neary, D.G., Nettles, J., Scott, D.F., Stednick, J.D. and Wang, Y. (eds) (2012) *Revisiting Experimental Catchment Studies in Forest Hydrology (Proceedings of a Workshop held during the XXV IUGG General Assembly in Melbourne, June–July 2011)*. IAHS Publication No. 353. International Association of Hydrological Sciences, Wallingford, UK.

Wei, X. and Davidson, G.W. (1998) Impact of large-scale timber harvesting on the hydrology of the Bowron River catchment. In: Alila, Y. (ed.) *Mountains to Sea: Human Interaction with the Hydrologic Cycle, 51st Annual Conference Proceedings, 10–12 June 1998*. Canadian Water Resources Association, Victoria, British Columbia, Canada, pp. 45–52.

Wemple, B.C., Jones, J.A. and Grant, G.E. (1996) Channel network extension by logging roads in two basins in western Cascades, Oregon. *Water Resources Bulletin* 32, 1195–1207.

Whitaker, A., Alila, Y., Beckers, J. and Toews, D. (2003) Application of the distributed hydrology soil vegetation model to Redfish Creek, British Columbia: model evaluation using internal catchment data. *Hydrological Processes* 17, 199–224.

Whitehead, P.G. and Robinson, M. (1993) Experimental basin studies – an international and historical perspective of forest impacts. *Journal of Hydrology* 145, 217–230.

Zegre, N., Skaugset, A.E., Som, N., McDonnell, J.J. and Ganio, L.M. (2010) In lieu of the paired catchment approach: hydrologic model change detection at the catchment scale. *Water Resources Research* 46, W11544, doi: 10.1029/2009WR008601 (accessed 8 April 2016).

Zhang, L., Dawes, W.R. and Walker, D.R. (2001) Response of a mean annual evapotranspiration to vegetation changes at a catchment scale. *Water Resources Research* 37, 701–708.

Zhang, X., Zhang, L., Zhao, J., Rustomji, P. and Hairsine, P. (2008) Responses of streamflow to changes in climate and land use/cover in the Loess Plateau, China. *Water Resources Research* 44, W00A19, doi: 10.1029/2007WR006711 (accessed 8 April 2016).

Ziemer, R.R. (2001) Caspar Creek. In: Marutani, T., Brierley, G.J., Trustrum, N.A. and Page, M. (eds) *Source-to-Sink Sedimentary Cascades in Pacific Rim Geo-Systems*. Matsumoto Sabo Work Office, Ministry of Land, Infrastructure and Transport, Nagano, Japan, pp. 78–85.

13 Hydrology of Forests after Wildfire

P.R. Robichaud*
USDA Forest Service, Moscow, Idaho, USA

13.1 Introduction

The hydrological response from burned forest watersheds can be some of the most dramatic responses that occur from forested catchments (Bren, 2014). High-severity fires may lead to extremely high peak flows which often strip away easily erodible soil; conversely, low-severity fires may have minimal effect on the watershed response. Most forest watersheds with good hydrological conditions and adequate rainfall sustain stream baseflow conditions throughout the year and produce little erosion (DeBano *et al.*, 1998). Wildfire impacts these stable conditions by consuming accumulated forest floor material, forest vegetation and understorey vegetation (Table 13.1). This vegetation and forest floor material protects the soil from raindrop impact and overland flow, and promotes infiltration.

13.2 Fire Effects on Soil

13.2.1 Soil infiltration

Water infiltrating into the soil is highly dependent upon the surface conditions. Runoff from wildfire-burned hillslopes generally increases by one or two orders of magnitude (Moody *et al.*, 2013). The expanse of disturbance of the surface material or consumption of forest floor material during combustion is a major determining factor in the degree of disturbance to the surface material. This is usually the consumption of organic debris (commonly referred to as 'duff' or 'forest floor') and the fine organic matter that holds soil particles together. The amount of duff consumed during the combustion process is a function of the severity of the fire, including the temperature reached and the duration of heating. The post-fire hydrological response is directly related to the effect of the fire on the soil and duff layers (Robichaud, 1996; Parsons *et al.*, 2010). Post-fire condition of the mineral surface horizons is important because they determine the amount of mineral soil exposed to raindrop splash, overland flow and the development of water-repellent soil conditions (DeBano, 1981).

13.2.2 Soil water repellency

Soil water repellency has been well documented in burned and unburned soils in forests (Robichaud, 2000; Huffman *et al.*, 2001; Doerr *et al.*, 2006; Butzen *et al.*, 2015). Wildfires have often been associated with the formation of water-repellent soil conditions. These are thought to

*E-mail: probichaud@fs.fed.us

 Forest Hydrology: Processes, Management and Assessment (eds D.M. Amatya, T.M. Williams, L. Bren and C. de Jong)

Table 13.1. Hydrological processes affected by wildfire. Specific factors influencing hydrological changes include: soil type and structure; soil cover; vegetation type and regeneration rate; precipitation intensity and frequency; understorey and canopy vegetation cover; micro- and macro-topography features. (From Neary *et al.*, 2005.)

Hydrological process	Consequence of high burn severity	
Infiltration	↑ Overland flow	
	↑ Stormflow	
	↑ Water repellency	
Soil water storage	↑ Evaporation	↓ Infiltration
	↑ Water repellency	↓ Litter absorption
Forest floor/duff storage	↑ Evaporation	↓ Interception
	↑ Runoff	
	↑ Splash erosion	
	↑ Snow sublimation	
Interception/evapotranspiration	↑ Water yield	↓ Storage
	↑ Snowpack	↓ Evaporation
		↓ Transpiration
Surface runoff/overland flow	↑ Sediment yield	
	↑ Erosion	
	↑ Debris flow	
Streamflow	↑ Surface runoff	↓ Evaporation
	↑ Snowmelt rate	↓ Transpiration
	↑ Erosion	
Peak flow	↑ Volume	
	↑ Flash flood frequency	
	↑ Flood levels	
Baseflow	↑ Evaporation	↓ Infiltration
Water quality	↑ Suspended sediment	
	↑ Ash nutrients: Ca^{2+}, NO_3^-, NH_4^+, PO_4^{3-}	

decrease infiltration and increase runoff and soil erosion (DeBano *et al.*, 1998; Robichaud, 2000) (Fig. 13.1). Hydrophobic organic compounds which are in the litter and topsoil are volatized during combustion and released upwards to the atmosphere and downwards into the soil profile along a temperature gradient. Translocated hydrophobic compounds condense on cooler soil particles below the surface, leading to water-repellent conditions (DeBano *et al.*, 1976).

Sometimes, natural water-repellent soil conditions also occur in unburned forests due to coating of soil particles with hydrophobic compounds leached from organic matter accumulations, by-products of microbial activity, or fungal growth under thick layers of litter and duff material (Savage *et al.*, 1972; DeBano, 2000; Doerr *et al.*, 2000; Butzen *et al.*, 2015). Under unburned conditions, litter and vegetation cover promote water storage and mitigate water repellency impacts on infiltration and erosion. Fire removes protective organic layers (litter and duff), exposing the soil to raindrop impact and removing barriers to overland flow (Moffet *et al.*, 2007; Pierson *et al.*, 2008).

Seasonal variability in the presence and strength of soil water repellency under burned and unburned conditions has been observed (Doerr and Thomas, 2000; Dekker and Ritsema, 2000; Huffman *et al.*, 2001). Dekker *et al.* (2001) demonstrated that soil water repellency is a function of soil water content, that critical soil water thresholds demarcate wettable and water-repellent soil conditions, and that the relationship between moisture content and soil water repellency is affected by the drying regime. Time since burning was not a significant predictor of soil water repellency in pine forests of the Colorado Front Range (Huffman *et al.*, 2001) as the water-repellent soils became wettable when soil moisture levels exceeded 12 to 25%. These studies indicate that seasonal variability in site characteristics that influence soil water repellency can confound assessment of long-term soil water repellency persistence (Doerr *et al.*, 2000).

Fig. 13.1. Water-repellent soil conditions after a wildfire and a 30-minute rainfall event showing saturated surface soil and dry soil a few centimetres below the surface. Circles indicate water drops on the soil surface, rectangles indicate water drops that have infiltrated.

13.2.3 Soil water storage

Wildfires in a given region often occur after a drought cycle of several years (Westerling *et al.*, 2006). This cycle can affect the soil water storage even before the wildfire starts with reduced soil water in the soil profile. The same drought cycle that caused the region's wildfire season may persist for several years after the fire. Without the protective layer of duff and forest litter, it is often difficult to recover the soil water deficiency because winter snow may melt but without the 'sponge' holding effect of the duff, little water remains to replenish the soil profile. It may take several years before antecedent (pre-fire) soil water conditions are reached.

Vegetation recovery after wildfire depends on many factors (commonly including soil burn severity, distance to seed source and fire tolerance of native species), but especially on the precipitation and snowpack (and subsequent melting) in the first post-fire year. There is typically a surge of vegetation growth immediately after wildfire, with growth increasing at a non-linear rate over the next decade. The magnitude of growth depends on landscape dynamics. Vegetation depletes the soil water profile via root uptake and transpiration, but also provides important soil stabilization. The stabilizing influence of vegetation is likely to be more beneficial than any detriment caused by soil water depletion. Both fine and larger roots can also provide infiltration pathways within burned soil profiles that were rendered water repellent from the fire.

A forest wildfire in central Washington, USA caused an increase in soil water storage as transpiration was reduced from the dead burnt trees (Klock and Helvey, 1976). Conversely, in Arizona, a burned-over ponderosa (*Pinus ponderosa*) forest had decreased soil water storage due to increase in overland flow and drying of the bare soil surface (Campbell *et al.*, 1977). Thus, soil water storage is a function of soil and site conditions as well as local climate.

13.2.4 Forest floor/duff

The hydrological response of the forest soil is influenced by the effects of the wildfire on the organic material found above the mineral soil in the forest floor (Fosberg, 1977; Brown *et al.*, 1985). This organic material commonly has three distinct layers. The top ('litter') layer is the undecomposed, unconsolidated material consisting of debris such as twigs, grasses, leaves and needles. Below the litter is the fermentation layer, which consists of partially decomposed organic material, often bound with fungus. Humus, the third and deepest organic layer, is extensively decomposed material found just above the A horizon of the mineral soil. In the field, it can be difficult to discern the physical separation between the fermentation and humus layers because humus is usually mixed in varying proportions with partially decomposed organic materials. Forest scientists and fire managers commonly use the term 'duff' to refer collectively to the fermentation and humus layers, while the term 'forest floor' is used to refer to all the surface organic horizons (litter and duff) overlying the mineral soil (DeBano *et al.*, 1998). Although there is usually a clear division between the mineral soil and overlying duff, site disturbances may mix varying amounts of mineral soil into the duff.

The ground-level effects of wildfires can range from removal of litter to total consumption of the forest floor and alteration of the mineral soil structure below (Wells *et al.*, 1979; Brown *et al.*, 1985; DeBano *et al.*, 1998; Ryan, 2002). Mineral soil that becomes exposed when the forest floor is completely consumed is highly susceptible to erosion (Wells *et al.*, 1979; Soto *et al.*, 1994), thereby increasing the sediment available for transport (Nyman *et al.*, 2013). Additionally, infiltration and water storage capacity of the mineral soil are significantly reduced because the 'sponge' effect of the organic forest floor material is gone and the mineral soil cannot absorb short-duration, high-intensity rainfall (Baker, 1990).

Any remaining unburned duff layer below the ash layer can behave as water-repellent patches when dry and water-absorbent patches when moist. This patchiness increases the spatial variability of the soil properties and adds complexity to understanding post-wildfire runoff and erosion responses. Even when water repellency is extreme, prolonged rainfall can cause the soil to be transformed to a 'normal' wettable state (Doerr *et al.*, 2000; Stoof *et al.*, 2011), but soil can regain its repellent state once dry conditions return (Shakesby and Doerr, 2006).

13.2.5 Soil and spatial variability

Soil properties are naturally highly variable. Soil erosion experiments generally find standard deviations in erodibility values are similar to the mean value and coefficients of variation greater than 30% are common (Elliot *et al.*, 1989). Soils near the tops of ridges tend to be coarser grained and shallower, whereas soils at the bottoms of hillslopes may be finer grained. The disturbance from fire (high soil burn severity versus low burn severity) rather than inherent soil properties often dominates the erodibility of the soils. Nyman *et al.* (2013) suggest that sandy soils which are naturally highly erodible are likely to become more erodible after fire, whereas clay loam soils quickly stabilize after the initial loss of loose particles. The distribution of the disturbance and the subsequent secondary effects are seldom uniform (Robichaud *et al.*, 2007).

The combined effects of a mosaic in fire severity and soil variability result in spatial variability of soil erodibility that has some degree of predictability, but a great deal of natural variation. For instance, the effect of water repellency decreases with an increase in spatial scale (Larsen *et al.*, 2009), because water will often find infiltration pathways via natural hillslope or landscape features. There will be areas following wildfire where the fire burned at a higher soil burn severity (as defined by Parsons *et al.*, 2010), leading to a complete loss of surface cover. These are likely to show the development or enhancement of a water-repellent soil condition. There will be other areas where the fire burned at low soil burn severity, resulting in an area of minimal erosion risk; a large percentage of any fire will generally exhibit a combination of these characteristics. Spatial variability analyses have shown that following some wildfires, there are definite trends in degree of fire severity, whereas the variability is evenly distributed on a hillslope or watershed following other fires (Robichaud and Miller, 1999).

13.3 Fire Effects on Vegetation

13.3.1 Interception and evapotranspiration

Wildfire can have a significant effect on the vegetation, ranging from complete combustion of the canopy for hundreds of square kilometres to little charring of needles or leaves. Forests experience reduction in evaporative losses through interception and evapotranspiration, thereby increasing rain and snow reaching the ground and increasing soil moisture, runoff and streamflow (Neary *et al.*, 2005). The combustion of forest canopies has been shown to have a significant effect on interception by decreasing stand rainfall-intercepting capacity. For example, duff and vegetation canopy combustion has been found to decrease water storage or 'hydrologic buffering' capacity, especially on north aspect slopes in dry conifer forests (Ebel, 2013).

Removal of forest canopy by wildfire can also increase accumulation of the snowpack; the difference may be a function of reduced interception from the tree canopy (Burles and Boon, 2011; Gleason *et al.*, 2013). For example, it has been found that burned forest canopy produced a 4–11% increase in snow water equivalent accumulation compared with that produced by the mature forest stand. This is similar to other disturbance that removes canopy, for example clearcut logging (Winkler *et al.*, 2010). These same burned areas experienced a greater ablation rate compared with mature forest stands that was attributed to earlier and more rapid snowmelt from increased solar radiation (Burles and Boon, 2011; Gleason *et al.*, 2013).

Decreased canopy cover also generally drives the reduction of evapotranspiration. For example, Dore *et al.* (2012) found evapotranspiration to decrease immediately after the fire in a semi-arid pine and mixed eucalypt forest. This was attributed to the low or non-existent vegetation cover leading to a high ratio of evaporation compared with transpiration. Transpiration decreased the most in dry conifer forest followed by wet conifer forest, deciduous forest and grassland. Fire severity also influences evapotranspiration as the eucalypt forest experienced 41% lower evapotranspiration after a high-severity burn compared with unburned forest, whereas moderate severity of burning resulted in only 3% lower evapotranspiration in the first and second years following the fire. The lower evapotranspiration was offset by regenerating seedlings in addition to forest floor evapotranspiration and interception loss (Nolan *et al.*, 2014). Recovery from wildfire for evapotranspiration and interception will occur between 3 and 4 years after burning (Soto and Diaz-Fierros, 1997; Nolan *et al.*, 2014) as these processes are correlated with increases in leaf area, canopy density or basal area (Soto and Diaz-Fierros, 1997).

13.4 Fire Effects on Watershed Response

13.4.1 Precipitation

Precipitation patterns in the years following the disturbance are crucial in determining the hydrological response. If precipitation is minimal, there will be little erosion, but there will also be little natural or seeded vegetation regrowth and little soil recovery from water-repellent conditions, meaning that the site can remain susceptible to erosion for another year or two.

If the precipitation comes as short-duration, high-intensity storms then erosion can be severe. If the weather is very wet, and the soils are water repellent, there is a high likelihood of severe soil erosion, but there will also be rapid vegetation recovery. Runoff and erosion from rainfall or rain-on-snow events will be much greater than runoff from melting snow. Once a site has recovered, rainfall rates in excess of 50 mm/h, or total rainfall amounts greater than 100 mm within a day, are necessary before any significant upland erosion will occur. Such rainfall intensities seldom occur in many forested areas.

13.4.2 Surface runoff/overland flow

When high-severity fire results in poor hydrological conditions, most precipitation does not infiltrate into the soil and streamflow response to precipitation is rapid. In such a case, runoff and peak flows can increase by several orders of magnitude and can cause extreme hydrological impacts (Neary *et al.*, 2005; Moody and Martin, 2009; Robichaud *et al.*, 2010). These increased watershed responses are typically caused by infiltration-excess and sometimes by saturation-excess overland flow or a combination of both (Sheridan *et al.*, 2007; Moody *et al.*, 2013). Increased runoff is often attributed to a combination of: development of soil water repellency, the increase in amount of bare soil, the decrease in canopy interception and the lack of surface water storage. Convective rainfall events are the primary cause of increased runoff.

13.4.3 Streamflow

In general, rainfall runoff methods assume temporally and spatially uniform rainfall (which is usually not applicable to burned areas in mountainous terrain) and runoff contributions to channel flow from the entire drainage basin area. Depending on the post-wildfire response, hillslope runoff-generating processes may switch between infiltration-excess and saturation-excess overland flow (Ebel *et al.*, 2012; Moody *et al.*, 2013). Runoff generation by infiltration-excess has been found to be more sensitive to the uncertainty associated with precipitation than saturation-excess.

Annual streamflow discharge from a 560 ha burned-over watershed in the Cascade Range of central Washington, USA increased five times relative to a pre-fire streamflow. Differences between the pre- and post-fire streamflow discharge varied from nearly 110 mm in a dry year to about 477 mm in a wet year as summarized by Neary *et al.* (2005). Campbell *et al.* (1977) observed a 3.5 times increase of 20 mm in average annual stormflow discharge from a small (8 ha) severely burned watershed following the occurrence of a wildfire in a south-western US ponderosa pine forest. Average annual stormflow discharge from a smaller 4 ha moderately burned watershed increased 2.3 times to almost 15 mm in relation to an unburned (control) watershed. The average runoff efficiency on the severely burned watershed was 357% greater when the precipitation input was rain and 51% less in snowmelt periods. The observed differences during rainfall events were largely due to the lower tree density, a greater reduction in litter cover and a more extensive formation of water-repellent soil. These resulted in lower evapotranspiration losses and more stormflow on the severely burned watershed compared with the moderately burned watershed. In the spring snowmelt period, the lower tree density of the severely burned watershed allowed more of the snowpack to be lost to evaporation. As a result, less stormflow occurred than on the more shaded, moderately burned watershed.

In the first year after a 150 ha watershed was burned over by a wildfire in southern France, streamflow discharge increased by 30% to nearly 60 mm (Lavabre *et al.*, 1993). The pre-fire vegetation on the watershed was primarily a mixture of maquis, cork oak (*Quercus suber*) and chestnut oak (*Quercus prinus*) trees. The increase was attributed to the reduction in evapotranspiration due to the loss of vegetation by the fire.

While increase in streamflow is most common after wildfires, mountain ash (*Eucalyptus regnans*) catchments in south-east Australia experienced a significant decrease in streamflow starting 3–5 years after severe wildfire in 1939 (Langford, 1976; Kuczera, 1987). The decrease was attributed to the increase in transpiration which coincides with rapid, vigorous regeneration of the ash-type eucalypt forests. The decrease in streamflow discharge is a long-term consequence which peaks 15–20 years after the fire and streamflow may not return to pre-disturbance conditions for 100–150 years. These particular catchments are the water supply for highly populated Melbourne. This case highlights an additional vulnerability to an urban water supply due to wildfire (Kuczera, 1987).

13.4.4 Peak flow

The effects of wildfire on storm peak flows are highly variable and complex. Some of the most profound impacts besides the wildfire itself can be the post-fire peak flow response. (Neary *et al.*, 2005) (Fig. 13.2). One to three orders of magnitude increase in peak flows is related to the occurrence of short-duration but intense rainfall events, steep watersheds and high-severity burn areas. These peak flows are instrumental in channel formation, sediment transport and sediment redistribution within stream corridors. The timing of these peak flow events is often very short, producing 'flash floods'. Such peak flow events increase in frequency after the fire.

One aspect of the peak flow is the size of the area (watershed) being affected by the rainfall event and the burn severity within the watershed. Cannon *et al.* (2001a,b) suggest that areas of about 1 km^2 or less will produce the greatest specific discharge as that size often will have the combination of high soil burn severity, steep slopes and the intense rainfall from a single storm cell. Peak flows are also an important consideration in management and design of structures (bridges, dams, levees, buildings, cultural sites, etc.).

13.4.5 Baseflow

The removal of forest canopy cover decreases interception and transpiration and this generally increases annual water yields including baseflow (MacDonald and Stednick, 2003). The increases in annual water yield following forest harvest are usually assumed to be proportional to the amount of forest cover removed, but at least 15 to 20% of the trees must be removed to produce a statistically detectable effect; this would be analogous to tree loss from a moderate or more severe burn. In areas where the annual precipitation is less than 450 to 500 mm, removal of the forest canopy is unlikely to increase annual water yields significantly. In drier areas, the decrease in interception and transpiration is generally offset by the increase in soil

Fig. 13.2. Channel scour after a high-intensity rainfall event on the 2011 Wallow Fire in Arizona, USA.

evaporation, and there is no net change in runoff as long as there is no change in the underlying runoff processes (MacDonald and Stednick, 2003).

Baseflows are often increased after wildfires as the evapotranspiration and interception decrease. Local soils and geology play an important role in determining if the excess water goes to springs, baseflow or groundwater recharge. These may be driven by the seasonal patterns of the amount and timing of precipitation (Neary *et al.*, 2005).

13.4.6 Water quality

Fire-affected watersheds often increase their flows which, in turn, will affect the water quality. Suspended fines (ash and sediment) and bed-load material are the most visible effects, often increasing by several orders of magnitude (Fig. 13.3).

Smith's *et al.* (2011) review reported first year post-fire suspended sediment exports varied from 0.017 to 50 t/ha/year across a large range of catchment sizes (0.021–1655 km^2). This represented an estimated increase of 1–1460 times unburned exports. Maximum reported concentrations of total suspended solids in streams for the first year after fire ranged from 11 to 500,000 mg/l. Similarly, there was a large range in first year post-fire stream exports of total N (1.1–27 kg/ha/year) and total P (0.03–3.2 kg/ha/year), representing a multiple change of 0.3–430 times unburned, while NO_3^- exports of 0.04–13.0 kg/ha/year (3–250 times unburned) have been reported. NO_3-N, NH_4-N, PO_4^{3-}, K^+ and alkalinity increased in stream water following ash input, yet concentrations of each returned to pre-fire conditions within 4 months (Earl and Blinn, 2003). Mineral nutrients such as Ca^{2+}, Mg^{2+} and K^+ are typically converted to oxides (often a major component of the light-coloured ash remaining after fire) that are relatively soluble (Ice *et al.*, 2004). The amount of Ca^{2+} typically found in ash-contaminated runoff can be used as a marker to define water contaminated with ash runoff. Elevated Na^+, Cl^- and SO_4^{2-} solute yields

Fig. 13.3. Ash and sediment deposits after a high-intensity rainfall event on the 2012 High Park Fire in Colorado, USA.

have been observed soon after fire in coniferous forests (Smith *et al.*, 2011). Crouch *et al.* (2006) found that NH_4-N, P and total CN^- concentrations were significantly correlated with Ca^{2+} concentrations, indicating an association of chemicals with ash-related inputs.

13.5 Fire Effects on Sediment Yield

13.5.1 Soil erosion

Variability in post-wildfire erosion responses is caused by differences in the runoff and transport processes which are affected by topography, temporal and spatial variability of fire-affected soils (water storage, infiltration, fine root breakdown, etc.), precipitation and runoff (Moody *et al.*, 2013). Complexity in the erosion response is due to non-uniformity in the spatial distribution of sediment sources and from sediment that is being transported on hillslopes leading to changes in the surface roughness. This in turn leads to deviations in runoff patterns and sediment transport (Kirkby, 2011). Soil erodibility in post-fire environments can be particularly variable in response to changes caused by heating during wildfires and changes in soil moisture conditions after wildfires. Burning creates an additional erodible layer (ash or char) and also destroys soil structure and cohesiveness, increasing the soil available for overland flow transport (Nyman *et al.*, 2013). Thus, the sediment availability, which is a function of the sediment supply and its associated erodibility and mobility, determines vulnerability of the hillslope to erosion processes (Moody *et al.*, 2013).

Characteristics of climate, topography, soils, vegetation, degree and extent of soil burn severity, and channel proximity create high variability in post-fire responses and recovery rates (DeBano *et al.*, 1998; Robichaud, 2000). More specifically, post-fire runoff, peak flow rates and erosion rates are highly dependent on rainfall intensity and amount, as well as on the magnitude and spatial distribution of fire-induced soil disturbances. Fire effects on soil include decreases in soil organic matter and surface litter, reduction in soil aggregates resulting in less soil structure, loss of interceptive and transpiring vegetation, changes in hydraulic roughness, and alteration or formation of water-repellent soil conditions (DeBano *et al.*, 1998; Certini, 2005; Shakesby and Doerr, 2006). High-severity fires tend to be larger and have more homogeneous patches of soil disturbance than low- or moderate-severity fires. Increased spatial extent or patch size of disturbed soil may result in greater overland flow, increased potential for rilling and larger amounts of sediment transport (Moody *et al.*, 2008).

In severely burned areas, high-intensity, short-duration rain events have increased peak flows from two to 2000 times (DeBano *et al.*, 1998; Neary *et al.*, 2005). Published sediment yields after high-severity wildfires range from 0.01 to over 110 t/ha/year in the first year after burning (Benavides-Solorio and MacDonald, 2005; Moody and Martin, 2009; Robichaud *et al.*, 2000) (Fig. 13.4). In most cases, vegetative regrowth and the decline in soil water repellency means that these large increases in runoff and erosion diminish quite rapidly. Most long-term studies show no detectable increase in erosion by about the fourth year after burning (Benavides-Solorio and MacDonald, 2005).

13.5.2 Debris flows

Debris flows sometimes occur after wildfires and can be described as a sediment-laden flow with unconsolidated sediment concentration of 50–77% by volume capable of supporting gravel and boulders while flowing; this is often referred to as 'sediment bulking' (Cannon *et al.*, 2008). Post-wildfire triggering mechanisms for debris flows include: (i) progressive entrainment of soil eroded from hillslopes and channels by overland flow (Cannon *et al.*, 2001b; Santi *et al.*, 2008) coupled with ash to provide sufficient fine-grained material (Gabet and Sternberg, 2008) to support the sediment; (ii) saturation of soil above the fire-induced water-repellent 'layer', which initiates 'thin debris flows'; and (iii) shallow soil slumps induced by low infiltration into soils. This infiltration increases pore pressures, resulting in liquefaction and mobilization.

Cannon *et al.* (2008) suggest that short-duration, high-intensity events in the intermountain west and longer-duration, frontal precipitation events in southern California, USA cause the majority of the debris flows. Nyman *et al.* (2011, 2015) also observed short-duration, high-intensity

Fig. 13.4. Sediment and debris after a high-intensity rainfall event on the 2011 Wallow Fire in Arizona, USA.

rainfall events combined with converging hillslopes with ample available sediment as the cause of observed debris flows in south-eastern Australia.

Santi *et al.* (2008) reviewed 46 post-fire debris flows and suggested significant bulking by scour and erosion in the channels, with debris flow rates ranging from 0.3 to 9.9 m^3 of debris produced for every metre of channel length. Debris flow inputs for short reaches of channels (up to several hundred metres) were as high as 22.3 m^3/m. They also found increased debris yields downstream in 87% of the channels studied. Debris was contributed from side channels into the main channels for 54% of the flows, with an average of 23% of the total debris coming from those side channels. These results show that channel erosion and scour are the dominant sources of debris in burned areas, with yield rates increasing significantly partway down the channel. In contrast, Staley *et al.* (2014) emphasized the importance of hillslope erosional processes in contributing material to post-fire debris flows where there is no discrete material source or initiation point. In Australia, Smith *et al.* (2012) found that hillslopes contributed 22–74% the deposition material after bushfires.

13.6 Stabilization and Rehabilitation

The USDA Forest Service and US Department of the Interior land management agencies develop watershed rehabilitation plans with Burned Area Emergency Response (BAER) teams after severe wildfires. The aim of these is to reduce the effects of soil erosion and flooding (Robichaud and Ashmun, 2013). Many factors are taken into consideration including soil burn severity, climate and values at risk from the fires. Many post-fire assessment procedures and tools that have been developed for the USA are being used by managers in other countries. Concerted efforts have been made to adapt portions of the US post-fire assessment and treatment recommendation protocol for use within countries throughout the world (Australia, Canada, Portugal, Greece, Spain, Argentina, etc.).

The general procedure is to map the soil burn severity, determine if there are values at risk, model potential increase in watershed response, select treatments and implement treatments before the first damaging storms. Since 2002, the USDA Forest Service, Remote Sensing Applications Center and the US Geological Survey, Center for Earth Resources Observation and Science have used pre- and post-fire Landsat (US remote sensing satellite programme) satellite images of the burned area to derive a preliminary classification of landscape change. The differences between the pre- and post-fire image data form a continuous raster GIS layer that is classified into four burn severity classes: unburned, low, moderate and high, referred to as the Burned Area Reflectance Classification (BARC) map and is usually the starting point for the soil burn severity map (Parsons *et al.*, 2010). The BARC is validated using a field guide directing the user to make five observations (ground cover, ash colour and depth, soil structure, roots, soil water repellency) at various data-collection locations for each site within a fire (Parsons *et al.*, 2010).

Estimation of potential post-fire erosion is often accomplished using the Forest Service Water Erosion Prediction Project (FSWEPP) interfaces (Elliot, 2004; Elliot and Robichaud, 2011), adaptations of WEPP for forest and rangeland environments. WEPP and the full suite of FSWEPP interfaces can be found online (http://forest.moscowfsl.wsu.edu/fswepp, accessed 10 September 2015). There are several FSWEPP interfaces that calculate potential post-fire erosion rates. The Erosion Risk Management Tool (ERMiT; Robichaud *et al.*, 2006) is the most common. ERMiT predicts the probability associated with a given hillslope sediment yield (untreated, treated with seeding, dry agricultural straw mulching, or erosion barriers) from a single storm in each of 5 years following wildfire.

Alternatively, the Kinematic Runoff and Erosion Model (KINEROS2), a spatially distributed, event-based watershed rainfall–runoff and erosion model, and the companion ArcGIS-based Automated Geospatial Watershed Assessment (AGWA) tool are also used in the post-fire environment (Goodrich *et al.*, 2012). AGWA automates the time-consuming tasks of watershed delineation into distributed model elements and initial parameterization of these elements using commonly available, national GIS data layers. After wildfires, model parameters are changed in the model as a function of burn severity and pre-fire land cover type. AGWA has a differencing function with which the stored results from pre- and post-fire simulations can be subtracted over all the spatially distributed model elements. These differences, in absolute or percentage change terms, can then be mapped back into the GIS display to provide a quick visual indication of watershed 'hot spots' where large changes between the two simulations have taken place.

Post-fire debris flow probability and volume estimates are provided by an empirical (logistic regression) model (Cannon *et al.*, 2011; Kean *et al.*, 2011) based on historical debris flow occurrence and magnitude data, rainfall, terrain and soils information, and soil burn severity conditions. The US Geological Survey provides post-fire debris flow estimates for many large fires in the western USA each fire season (http://landslides.usgs.gov/hazards/postfire_debrisflow/, accessed 16 December 2015). The output interactive map displays estimates of the probability of debris flow (%), potential volume of debris flow (m^3), and combined relative debris flow hazard at the scale of the drainage basin and individual stream segment based upon a designed 25-year recurrence interval rainfall event.

A large body of empirical data and related physical understanding now exists concerning post-wildfire runoff and erosion processes for many different post-wildfire locations throughout the world (Moody *et al.*, 2013). Recent research syntheses have included measured post-fire erosion rates (Moody and Martin, 2009), fire effects on soils (Cerdà and Robichaud, 2009) and effects of fire on soil and water (Neary *et al.*, 2005). The range of post-wildfire response is the result of the combination of the spatially distributed rainfall and fire-affected soil properties which change and interact on different temporal and spatial scales. Post-fire treatment syntheses and treatment catalogues provide current information on various treatments in formats that are easily used by post-fire assessment teams. Recently published treatment syntheses (Napper, 2006; Foltz *et al.*, 2009; Robichaud *et al.*, 2010; Peppin *et al.*, 2011) are available at the BAERTOOLS website (http://forest.moscowfsl.wsu.edu/BAERTOOLS, accessed 10 September 2015).

References

Baker, M.B. Jr (1990) Hydrologic and water quality effects of fire. In: Krammes, J.S. (Tech. Coord.) *Proceedings: Effects of Fire Management of Southwestern Natural Resources.* General Technical Report RM-GTR-191. USDA Forest Service, Rocky Mountain Forest and Range Experiment Station, Fort Collins, Colorado, pp. 31–42.

Benavides-Solorio, J.D. and MacDonald, L.H. (2005) Measurement and prediction of post-fire erosion at the hillslope scale, Colorado Front Range. *International Journal of Wildland Fire* 14, 457–474.

Bren, L. (2014) *Forest Hydrology and Catchment Management*, An Australian Perspective. Springer, Dordrecht, The Netherlands.

Brown, J.K., Mardsen, M.A., Ryan, K.C. and Reinhardt, E.D. (1985) *Predicting Duff And Woody Fuel Consumed by Prescribed Fire in Northern Rocky Mountains.* Research Paper INT-337. USDA Forest Service, Intermountain Forest and Range Research Station, Ogden, Utah.

Burles, K. and Boon, S. (2011) Snowmelt energy balance in a burned forest plot, Crowsnest Pass, Alberta, Canada. *Hydrological Processes* 25, 3012–3029.

Butzen, V., Seeger, M., Marruedo, A., de Jonge, L., Wengel, R., Ries, J.B. and Casper, M.C. (2015) Water repellency under coniferous and deciduous forest – experimental assessment and impact on overland flow. *Catena* 133, 255–265.

Campbell, R.E., Baker, M.B. Jr, Ffolliott, P.F., Larson, F.R. and Avery, C.C. (1977) *Wildfire Effects on Ponderosa Pine Ecosystems: An Arizona Case Study.* Research Paper RM-191. USDA Forest Service, Rocky Mountain Forest and Range Experiment Station, Fort Collins, Colorado.

Cannon, S.H., Bigio, E.R. and Mine, E. (2001a) A process for fire-related debris flow initiation, Cerro Grande fire, New Mexico. *Hydrological Processes* 15, 3011–3023.

Cannon, S.H., Kirkham, R.M. and Parise, M. (2001b) Wildfire-related debris-flow initiation processes, Storm King Mountain, Colorado. *Geomorphology* 39, 171–188.

Cannon, S.H., Gartner, J.E., Wilson, R.C. and Laber, J.L. (2008) Storm rainfall conditions for floods and debris flows from recently burned areas in southwestern Colorado and southern California. *Geomorphology* 96, 250–269.

Cannon, S.H., Boldt, E.M., Laber, J.L., Kean, J.W. and Staley, D.M. (2011) Rainfall intensity–duration thresholds for postfire debris-flow emergency-response planning. *Natural Hazards* 59, 209–236.

Cerdà A. and Robichaud, P.R. (2009) *Fire Effects on Soils and Restoration Strategies*. Science Publishers, Enfield, New Hampshire.

Certini, G. (2005) Effects of fire on properties of forest soils: a review. *Oecologia* 143, 1–10.

Crouch, R.L., Timmenga, H.J., Barber, T.R. and Fuchsman, P. (2006) Post-fire surface water quality: comparison of fire retardant versus wildfire-related effects. *Chemosphere* 62, 874–889.

DeBano, L.F. (1981) *Water Repellent Soils: A State-Of-The Art*. General Technical Report PSW-46. USDA Forest Service, Pacific Southwest Forest and Range Experiment Station, Berkeley, California.

DeBano, L.F. (2000) The role of fire and soil heating on water repellency in wildland environments: a review. *Journal of Hydrology* 231–232, 195–206.

DeBano, L.F., Savage, S.M. and Hamilton, D.A. (1976) The transfer of heat and hydrophobic substances during burning. *Soil Science Society of America Journal* 40, 779–782.

DeBano, L.F., Neary, D.G. and Ffolliott, P.F. (1998) *Fire's Effects on Ecosystems*. Wiley, New York.

Dekker, L.W. and Ritsema, C.J. (2000) Wetting patterns and moisture variability in water repellent Dutch soils. *Journal of Hydrology* 231–232, 148–164.

Dekker, L.W., Doerr, S.H., Oostindie, K., Ziogas, A.K. and Ritsema, C.J. (2001) Water repellency and critical soil water content in dune sand. *Soil Science Society of America Journal* 65, 1667–1674.

Doerr, S.H. and Thomas, A.D. (2000) The role of soil moisture in control patterns and moisture variability in water repellent Dutch soils. *Journal of Hydrology* 231–232, 134–147.

Doerr, S.H., Shakesby, R.A. and Walsh, R.P.D. (2000) Soil water repellency: its causes, characteristics and hydro-geomorphological significance. *Earth-Science Reviews* 51, 33–65.

Doerr, S.H., Shakesby, R.A., Dekker, L.W. and Ritsema, C.J. (2006) Occurrence, prediction and hydrological effects of water repellency amongst major soil and land-use types in a humid temperate climate. *European Journal of Soil Science* 57, 741–754.

Dore, S., Montes-Helo, M., Hart, S.C., Hungate, B.A., Koch, G.W., Moon, J.B., Finkral, A.J. and Kolb, T.E. (2012) Recovery of ponderosa pine ecosystem carbon and water fluxes from thinning and stand-replacing fire. *Global Change Biology* 18, 3171–3185.

Earl, S.R. and Blinn, D.W. (2003) Effects of wildfire ash on water chemistry and biota in South-Western USA streams. *Freshwater Biology* 48, 1015–1030.

Ebel, B.A. (2013) Wildfire and aspect effects on hydrologic states after the 2010 Four Mile Canyon Fire. *Vadose Zone Journal* 12, 19.

Ebel, B.A., Moody, J.A. and Martin, D.A. (2012) Hydrologic conditions controlling runoff generation immediately after wildfire. *Water Resources Research* 48, WO3529, doi: 10.1029/2011WR011470 (accessed 9 April 2016).

Elliot, W.J. (2004) WEPP internet interfaces for forest erosion prediction. *Journal of American Water Resources Association* 40, 299–309.

Elliot, W.J. and Robichaud, P.R. (2011) Risk-based erosion assessment: application to forest watershed management and planning. In: Morgan, R.P.C. and Nearing, M.A. (eds) *Handbook of Erosion Modelling*. Wiley, Chichester, UK, pp. 313–323.

Elliot, W.J., Liebenow, A.M., Laflen, J.M. and Kohl, K.D. (1989) *A Compendium of Soil Erodibility Data from WEPP Cropland Soil Field Erodibility Experiments, 1987 and 1988*. NSERL Report No. 3. Ohio State University and USDA Agricultural Research Service National Soil Erosion Research Laboratory, Lafayette, Indiana.

Foltz, R.B., Robichaud, P.R. and Rhee, H. (2009) *A Synthesis of Postfire Road Treatments for BAER Teams: Methods, Treatment Effectiveness, and Decision Making Tools for Rehabilitation*. General Technical Report RMRS-GTR-228. USDA Forest Service, Rocky Mountain Research Station, Fort Collins, Colorado.

Fosberg, M.A. (1977) *Heat and Water Transport Properties in Conifer Duff and Humus*. Research Paper RM-195. USDA Forest Service, Rocky Mountain Forest and Range Experiment Station, Fort Collins, Colorado.

Gabet, E.J. and Sternberg, P. (2008) The effects of vegetative ash on infiltration capacity, sediment transport, and the generation of progressively bulked flows. *Geomorphology* 101, 666–673.

Gleason, K.E., Nolin, A.W. and Roth, T.R. (2013) Charred forests increase snowmelt: effects of burned woody debris and incoming solar radiation on snow ablation. *Geophysical Research Letters* 40, 4654–4661.

Goodrich, D.C., Burns, I.S., Unkrich, C.L., Semmens, D.J., Guertin, D.P., Hernandez, M., Yatheendradas, S., Kennedy, J.R. and Levick, L.R. (2012) KINEROS2/AGWA: model use, calibration, and validation. *Transactions of the ASABE* 55, 1561–1574.

Huffman, E.L., MacDonald, L.H. and Stednick, J.D. (2001) Strength and persistence of fire induced soil hydrophobicity under ponderosa and lodgepole pine, Colorado Front Range. *Hydrological Processes* 15, 2877–2892.

Ice, G.G., Neary, D.G. and Adams, P.W. (2004) Effects of wildfire on soils and watershed processes. *Journal of Forestry* 102, 16–20.

Kean, J.W., Staley, D.M. and Cannon, S.H. (2011) *In situ* measurements of post-fire debris flows in southern California: comparisons of the timing and magnitude of 24 debris-flow events with rainfall and soil moisture conditions. *Journal of Geophysical Research* 116, F04019.

Kirkby, M.J. (2011) A personal synthesis and commentary, hydro-geomorphology, erosion and sedimentation. In: Kirkby, M.J. (ed.) *Hydro-Geomorphology, Erosion and Sedimentation*. Benchmark Papers in Hydrology Vol. 6. International Association of Hydrological Sciences, Wallingford, UK, pp. 1–36.

Klock, G.O. and Helvey, J.D. (1976) Soil–water trends following a wildfire on the Entiat experimental forest. In: *Proceedings of the Tall Timbers Fire Ecology Conference, October 16–17, 1994, Portland, Oregon*. Proceedings No. 15. Tall Timbers Research Station, Tallahassee, Florida, pp. 193–200.

Kuczera, G.A. (1987) Prediction of water yield reductions following a bushfire in ash–mixed species eucalypt forest. *Journal of Hydrology* 94, 215–236.

Langford, K.J. (1976) Change in yield of water following a bushfire in a forest of *Eucalyptus regnans*. *Journal of Hydrology* 29, 87–114.

Larsen, I.J., MacDonald, L.H., Brown, E., Rough, D., Welsch, M., Pietraszek, J.H., Libohova, Z., Benavides-Solorio, J.D. and Schaffrath, K. (2009) Causes of post-fire runoff and erosion; water repellency, cover or soil sealing? *Soil Science Society of America Journal* 73, 1393–1407.

Lavabre, J., Torres, D.S. and Cernesson, F. (1993) Changes in the hydrological response of a small Mediterranean basin a year after a wildfire. *Journal of Hydrology* 142, 273–299.

MacDonald, L.H. and Stednick, J.D. (2003) *Forests and Water: A State-of-the-Art Review for Colorado*. Completion Report No. 196. Colorado Water Resources Research Institute, Colorado State University, Fort Collins, Colorado.

Moffet, C.A., Pierson, F.B., Robichaud, P.R. and Spaeth, K.E. (2007) Modeling soil erosion on steep sagebrush rangeland before and after prescribed fire. *Catena* 71, 218–228.

Moody, J.A. and Martin, D.A. (2009) Synthesis of sediment yields after wildland fire in different rainfall regimes in the western United States. *International Journal of Wildland Fire* 18, 96–115.
Moody, J.A., Martin, D.A., Haire, S.L. and Kinner, D.A. (2008) Linking runoff response to burn severity after wildfire. *Hydrological Processes* 22, 2063–2074.
Moody, J.W., Shakesby, R.A., Robichaud, P.R., Cannon, S.H. and Martin, D.A. (2013) Current research issues related to post-wildfire runoff and erosion processes. *Earth-Science Review* 122, 10–37.
Napper, C. (2006) *Burned Area Emergency Response Treatments Catalog*. Watershed, Soil, Air Management 0625 1801 – STTDC. USDA Forest Service, National Technology and Development Program, San Dimas, California.
Neary, D.G., Ryan, K.C. and DeBano, L.F. (eds) (2005) *Wildland Fire in Ecosystems: Effects of Fire on Soil and Water*. General Technical Report RMRS-GTR-42, Vol. 4. USDA Forest Service, Rocky Mountain Research Station, Ogden, Utah.
Nolan, R.H., Lane, P.N.J., Benyon, R.G., Bradstock, R.A. and Mitchell, P.J. (2014) Changes in evapotranspiration following wildfire in resprouting eucalypt forests. *Ecohydrology* 7, 1363–1377.
Nyman, P., Sheridan, G., Smith, H.G. and Lane, P.N.J. (2011) Evidence of debris flow occurrence after wildfire in upland catchments of south-east Australia. *Geomorphology* 125, 383–401.
Nyman, P., Sheridan, G.J., Moody, J.A., Smith, H.G., Noske, P.J. and Lane, P.N.J. (2013) Sediment availability on burned hillslopes. *Journal of Geophysical Research: Earth Surface* 118, 2451–2467.
Nyman, P., Smith, H.G., Sherwin, C.B., Langhans, C., Lane, P.N.J. and Sheridan, G.J. (2015) Predicting sediment delivery from debris flows after wildfire. *Geomorphology* 250, 173–186.
Parsons, A., Robichaud, P.R., Lewis, S.A., Napper, C. and Clark, J.T. (2010) *Field Guide for Mapping Post-Fire Soil Burn Severity*. General Technical Report RMRS-GTR-243. USDA Forest Service, Rocky Mountain Research Station, Fort Collins, Colorado.
Peppin, D.L., Fulé, P.Z., Sieg, C.H., Beyers, J.L., Hunter, M.E. and Robichaud, P.R. (2011) Recent trends in post-wildfire seeding in western US forests: costs and seed mixes. *International Journal of Wildland Fire* 20, 702–708.
Pierson, F.B., Robichaud, P.R., Moffet, C.A., Spaeth, K.E., Williams, C.J., Hardegree, S.P. and Clark, P.E. (2008) Soil water repellency and infiltration in coarse-textured soils of burned and unburned sagebrush ecosystems. *Catena* 74, 98–108.
Robichaud, P.R. (1996) Spatially-varied erosion potential from harvested hillslopes after prescribed fire in the interior Northwest. PhD dissertation, University of Idaho, Moscow, Idaho.
Robichaud, P.R. (2000) Fire effects on infiltration rates after prescribed fire in northern Rocky Mountain forests, USA. *Journal of Hydrology* 231–232, 220–229.
Robichaud, P.R. and Ashmun, L.E. (2013) Tools to aid post-wildfire assessment and erosion-mitigation treatment decisions. *International Journal of Wildland Fire* 22, 95–105.
Robichaud, P.R. and Miller, S.M. (1999) Spatial interpolation and simulation of post-burn duff thickness after prescribed fire. *International Journal of Wildland Fire* 9, 137–143.
Robichaud, P.R., Beyers, J.L. and Neary, D.G. (2000) *Evaluating the Effectiveness of Postfire Rehabilitation Treatments*. General Technical Report RMRS GTR-63. USDA Forest Service, Rocky Mountain Research Station, Fort Collins, Colorado.
Robichaud, P.R., Elliot, W.J., Pierson, F.B., Hall, D.E. and Moffet, C.A. (2006) Erosion Risk Management Tool (ERMiT), Version 2006.01.18. USDA Forest Service, Rocky Mountain Research Station, Moscow, Idaho. Available at: http://forest.moscowfsl.wsu.edu/fswepp/ (accessed 9 April 2016).
Robichaud, P.R., Elliot, W.J., Pierson, F.B., Hall, D.E. and Moffet, C.A. (2007) Predicting postfire erosion and mitigation effectiveness with a web-based probabilistic erosion model. *Catena* 71, 229–241.
Robichaud, P.R., Ashmun, L.E. and Sims, B. (2010) *Post-Fire Treatment Effectiveness for Hillslope Stabilization*. General Technical Report RMRS GTR-240. USDA Forest Service, Rocky Mountain Research Station, Fort Collins, Colorado.
Ryan, K.C. (2002) Dynamic interactions between forest structure and fire behavior in boreal ecosystems. *Silva Fennica* 36, 13–39.
Santi, P.M., de Wolf, V.G., Higgins, J.E., Cannon, S.H. and Gartner, J.E. (2008) Sources of debris flow material in burned areas. *Geomorphology* 96, 310–321.
Savage, S.M., Osborn, J., Letey, J. and Heaton, C. (1972) Substances contributing to fire-induced water repellency in soils. *Soil Science Society of America Proceedings* 36, 674–678.
Shakesby, R.A. and Doerr, S.H. (2006) Wildfire as a hydrological and geomorphological agent. *Earth-Science Reviews* 74, 269–307.

Sheridan, G.J., Lane, P.N.J. and Noske, P.J. (2007) Quantification of hillslope runoff and erosion processes before and after wildfire in a wet Eucalyptus forest. *Journal of Hydrology* 343, 12–28.

Smith, H.G., Sheridan G.J., Lane, P.N.J., Nyman, P. and Haydon, S. (2011) Wildfire effects on water quality in forest catchments: a review with implications for water supply. *Journal of Hydrology* 396, 170–192.

Smith, H.G., Sheridan, G.J., Nyman, P., Child, D.P., Lane, P.N.J., Hotchkis, M.A.C. and Jacobsen, G.E. (2012) Quantifying sources of fine sediment supplied to post-fire debris flows using fallout radionuclide tracers. *Geomorphology* 139, 403–415.

Soto, B. and Diaz-Fierros, F. (1997) Soil water balance as affected by throughfall in gorse (*Ulex europaeus* L.) shrubland after burning. *Journal of Hydrology* 195, 218–231.

Soto, B., Basanta, R., Benito, E., Perez, R. and Diaz-Fierros, F. (1994) Runoff and erosion from burnt soils in northwest Spain. In: Sala, M. and Rubio, J.L. (eds) *Proceedings: Soil Erosion and Degradation as a Consequence of Forest Fires, Barcelona, Spain*. Geoforma Ediciones, Logroño, Spain, pp. 91–98.

Staley, D.M., Wasklewicz, T.A. and Kean, J.W. (2014) Characterizing the primary material sources and dominant erosional processes for post-fire debris-flow initiation in a headwater basin using multi-temporal terrestrial laser scanning data. *Geomorphology* 214, 324–338.

Stoof, C.R., Moore, D., Ritsema, C.J. and Dekker, L.W. (2011) Natural and fire-induced soil water repellency in a Portuguese shrubland. *Soil Science Society of America Journal* 75, 2283–2295.

Wells, C.G., Campbell, R.E., DeBano, L.F., Lewis, C.E., Fredrickson, R.L., Franklin, E.C., Froelich, R.C. and Dunn, P.H. (1979) *Effects of Fire on Soil: A State-of-the-Knowledge Review*. General Technical Report W0-7. USDA Forest Service, Washington, DC.

Westerling, A.L., Hidalgo, H.G., Cayan, D.R. and Swetnam, T.W. (2006) Warming and earlier spring increases western US forest wildfire activity. *Science* 313, 940–943.

Winkler, R., Boon, S., Zimonick, B. and Baleshta, K. (2010) Assessing the effects of post-pine beetle forest litter on snow albedo. *Hydrological Processes* 24, 803–812.

14 Hydrological Processes of Reference Watersheds in Experimental Forests, USA

D.M. Amatya[1]*, J. Campbell[2], P. Wohlgemuth[3], K. Elder[4], S. Sebestyen[5], S. Johnson[6], E. Keppeler[7], M.B. Adams[8], P. Caldwell[9] and D. Misra[10]

[1]*USDA Forest Service, Cordesville, South Carolina, USA;* [2]*USDA Forest Service, Durham, New Hampshire, USA;* [3]*USDA Forest Service, Riverside, California, USA;* [4]*USDA Forest Service, Fort Collins, Colorado, USA;* [5]*USDA Forest Service, Grand Rapids, Minnesota, USA;* [6]*USDA Forest Service, Corvallis, Oregon, USA;* [7]*USDA Forest Service, Fort Bragg, California, USA;* [8]*USDA Forest Service, Morgantown, West Virginia, USA;* [9]*USDA Forest Service, Otto, North Carolina, USA;* [10]*University of Alaska–Fairbanks, Fairbanks, Alaska, USA*

14.1 Introduction

Long-term research at small, gauged, forested watersheds within the USDA Forest Service, Experimental Forest and Range network (USDA-EFR) has contributed substantially to our current understanding of relationships between forests and streamflow (Vose *et al.*, 2014). Many of these watershed studies were established in the early to mid-20th century and have been used to evaluate the effects of forest disturbances such as harvesting, road construction, wild and prescribed fire, invasive species and changes in tree species composition on hydrological responses including stormflows, peak flows, water yield, ground water table and evapotranspiration. Forest hydrologists and natural resources managers are still working to fully understand the effects of watershed disturbances on hydrology, water quality and other ecosystem services (Zegre, 2008). Much of our knowledge on this topic is derived from steep, mountainous watersheds where these studies were initially conducted. An assessment by Sun *et al.* (2002) has shown that low-gradient watersheds with forested wetlands generally have lower water yields, lower runoff ratios and higher evapotranspiration than upland-dominated watersheds, adding to our knowledge of forest hydrology, particularly on the effects of topography on streamflow patterns and stormflow peaks and volumes.

While paired watershed studies (Bosch and Hewlett, 1982; Brown *et al.*, 2005) have been invaluable in understanding the hydrological response to disturbances, reference watersheds can provide valuable insight into hydrological processes in relatively undisturbed forest ecosystems. The term 'reference' watershed is favoured over the term 'control' because reference watersheds also change over time in response to natural (e.g. windthrow, insects, fire, hurricanes, climatic extremes) and human-induced disturbances (e.g. atmospheric pollution, invasive species, climate change). However, reference watersheds experience disturbances that are typically minor compared with most experimental treatments. Several recent studies have synthesized data from small reference watersheds, including

*Corresponding author; e-mail: damatya@fs.fed.us

This chapter is in the public domain. *Forest Hydrology: Processes, Management and Assessment* (eds D.M. Amatya, T.M. Williams, L. Bren and C. de Jong)

those in the USDA-EFR network, highlighting important insights that can be gained from long-term data (Jones *et al.*, 2012; Argerich *et al.*, 2013; Creed *et al.*, 2014).

This chapter provides an overview and comparison of factors influencing hydrological processes, especially streamflow dynamics and evapotranspiration, at ten relatively undisturbed reference watersheds in the USDA-EFR network (Fig. 14.1, Table 14.1). We demonstrate the breadth of the hydrogeological, topographic, climatic and ecological characteristics of reference watersheds by discussing how factors such as climate, topography, geology, soils and vegetation influence runoff generation (Fig. 9.1, Chapter 9, this volume) of these reference watersheds. We also briefly consider how site factors influence evapotranspiration, which determines water balance and regulates streamflow. This enhances our current understanding of the hydrological behaviour of these watersheds enabling us to better predict responses to, and prepare for, future management and environmental changes (Jones *et al.*, 2009; Vose *et al.*, 2014).

Located in vastly different ecohydrological regions, these watersheds have multiple factors influencing the streamflow (Q) regimes. Therefore we chose to assess differences in streamflow magnitudes and frequencies using flow duration curves (FDCs) and their flow percentiles (Searcy, 1959; Vogel and Fennessey, 1995). FDCs have been used to study integrated streamflow responses to different types and distributions of storm runoff events (i.e. rainstorms, snowmelt) and landscape characteristics, and have been applied extensively to evaluate streamflow responses to changing climate and other disturbances (e.g. Arora and Boer, 2001). An FDC with a steep slope throughout indicates a stream with more variable flow, whereas a flat slope is indicative of more stable flow with less variability. A steep slope at the upper end indicates more flashy streams with direct runoff characterizing a flood regime, while a flatter slope indicates

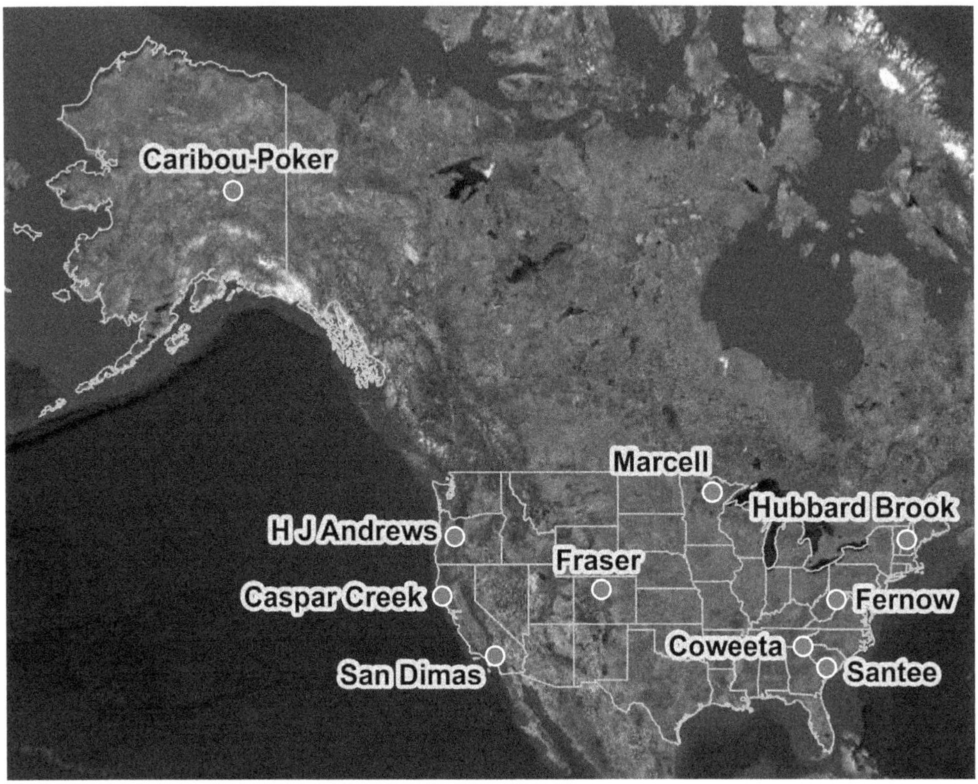

Fig. 14.1. Map of the ten USDA Forest Service Experimental Forests included in this chapter.

Table 14.1. Comparative characteristics of reference watersheds at ten long-term paired experimental forest watersheds in the USA.

Watershed characteristics	Caribou-Poker (CPCRW), Alaska	Caspar Creek (CCEW), California	Coweeta (CHL), North Carolina	Fernow (FnEF), West Virginia	Fraser (FrEF), Colorado	H.J. Andrews (HJAEF), Oregon	Hubbard Brook (HBEF), New Hampshire	Marcell (MEF), Minnesota	San Dimas (SDEF), California	Santee (SEF), South Carolina
Physiographical region as per classification by Fenneman	Yukon–Tanana Northern Plateaus Province (12)	Pacific Mountain System, 23f, Pacific Border Province	Appalachian Highlands, 5b, Blue Ridge Province	Appalachian Highlands, 8c, Appalachian Plateau	Rocky Mountain Systems, 15, Southern Rocky Mountain	Pacific Mountain System, 22b, Sierra-Cascade Mountain	Appalachian Highlands, 9b, New England Province	12b, Interior Plain, Central Lowland	Pacific Mountain System, 23g, Pacific Border Province	Atlantic Plain Coastal Plain
Climatic region as per classification by Köppen (Peel *et al.*, 2007)	Dfc, continental subarctic or boreal taiga	Csb, temperate/ mesothermal, Mediterranean	Cfb, marine temperate	Dfb, continental warm summer	Dsc, continental cold winter & cool, short, dry summer	Csb, temperate/ mesothermal dry summer	Dfb, continental warm summer	Dfb, continental warm summer	Csa, Mediterra-nean hot summer	Cfa, temperate humid subtropical
Watershed #/ year gauging started	(CPCRW – C2) 1969	(North Fork, NF) 1962	(WS18) 1936	(WS4) 1951	(East St Louis, ESL) 1943	(WS02) 1952	(WS 3) 1957	(S2) 1961	(Bell 3) 1938	(WS80) 1968
Latitude/ longitude	65.17°N, 147.50°W	39.35°N, 123.73°W	35.05°N, 83.43° W	39.03°N, 79.67°W	39.89°N, 105.88°W	44.21°N, 122.23°W,	44.0°N, 71.7°W	47.514°N, 93.473°W	34.20°N, 117.78°W	33.17°N, 79.77°W
Elevation (m amsl)[a]	210–826	30–322	726–993	670–866	2907–3719	572–1079	527–732	420–430	755–1080	3.7–10
Average slope (%)	31	49	52	20	16	41	21	3	34	<3
Drainage area (ha)	520	473	12.5	38.7	803	61	42.4	9.7	25	160
Vegetation type/ average leaf area index (LAI) (m^2/m^2)	Boreal forest/ LAI = 4.1	Second-growth coast redwood/ Douglas fir forest/ LAI = 11.7	Mixed deciduous forest/ LAI = 6.2	Mixed deciduous hardwoods/ LAI = 4.5	Mixed spruce/fir and pine/ LAI = 3.44	Conifer primarily Douglas fir and western hemlock/ LAI = 12	Northern hardwood/ LAI = 6.3	Deciduous uplands; black spruce–*spaghnum* bog/ LAI = N/A	Mixed chaparral/ LAI = 2.2	Pine mixed hardwood/ LAI = 2.8
Dominant geology/ aquifer	Yukon–Tanana metamorphic complex/ discontinuous permafrost	Marine shales & sandstones, Coastal Belt of the Franciscan Complex	Quartz dioritic gneiss predominant, Coweeta Group	Sedimentary; Hampshire formation sandstone and shales	Gneiss and schist, glacial till	Volcanic tuffs and breccias covered with andesite colluvium	Meta-sedimentary/ mica schist, calc-silicate granulite, Silurian Rangeley formation	Glacial till overlying deep outwash sands above Precambrian bedrock	Precambrian metamor-phics and Mesozoic granitics	Sedimentary/ Santee limestone

Continued

Table 14.1. Continued.

Watershed characteristics	Caribou-Poker (CPCRW), Alaska	Caspar Creek (CCEW), California	Coweeta (CHL), North Carolina	Fernow (FnEF), West Virginia	Fraser (FrEF), Colorado	H.J. Andrews (HJAEF), Oregon	Hubbard Brook (HBEF), New Hampshire	Marcell (MEF), Minnesota	San Dimas (SDEF), California	Santee (SEF), South Carolina
Dominant soil type/depth	Olnes Silt Loam – *Typic Cryorthents*; Fairplay Silt Loam – *Fluvaquentic Endoaquolls*; Ester Silt Loam – *Typic Histoturbels*/ 0.2–0.5 m	Vandamme Series Ultisols (*Typic Haplohumults*)/ 1–1.5 m	Coweeta–Evard–Saunook complex (fine-loamy, mixed, *Mesic* and *Humic Hapludults*)/ >1.5 m	Loamy-skeletal, mixed mesic *Typic Dystrudepts*/ 1 m	Leighcan Series, loamy-skeletal, *Typic Dystrocyepts*/ <1.5 m	50% andesite colluvium, (unnamed soil series), 20% Limberlost series, loam to green breccia/ up to 1.2 m	Lyman–Tunbridge–Becket Series, Typic *Haplorthods*/ C horizon depth up to 9 m	Warba Series fine loamy, mixed, superactive, frigid haplic Glossudalfs (0.5 m); Loxley peat *Dysic*, frigid *Typic Haplosaprists* (≤7 m)	Trigo–Exchequer Series loamy, mixed, thermic, shallow, *Typic Xerorthents*/ 0.1–0.5 m	Wahee Series clayey, mixed, thermic *Aeric Ochraquults*/ 1.5 m
Long-term mean precipitation (mm)	412	1316	2010[b]	1458	750	2300	1350	780	715	1370
Long-term mean temperature (°C)	–3.0	10.7	12.9[b]	9.3	1.0	8.4	5.9	3.4	14.4	18.3
Long-term mean potential evapotranspiration (PET) (mm)	466 (Thornthwaite)[c]	660 (Thornthwaite)[c]	1,013 (Hamon)[d]	560 (pan)[e]	383 (Thornthwaite)[c]	546 (Thornthwaite)[c]	550 (Thornthwaite)[c]	552 (Hamon)[d]	753 (Thornthwaite)[c]	967 (Thornthwaite)[c]
Dryness index (DI)	1.13	0.50	0.50	0.38	0.51	0.24	0.41	0.71	1.05	0.71
Long-term mean streamflow (mm)	80 (1978–2003)	659	997	640 (1951–1990)[b]	337	1321	860	170	84	280
Period of streamflow record	1969–present	1962–present	1936–present	1951–present	1943–present	Nov 1952–present	1957–present	1961–present	1938–1960; 1964–2001; 2013–present	1969–1981; 1989–1999; 2003–present

Surface runoff/ flow generation	Saturation excess flow	Infiltration-excess overland flow limited to compacted surfaces	Rare surface runoff, direct channel and fast shallow subsurface flow from VSA'	Minimal surface runoff	Rare, only during snowmelt	Minimal surface runoff – high porosity	Minimal surface runoff	Infiltration excess over frozen & saturation excess flow over unfrozen soils	Rare hillslope flow except after fire when infiltration excess flow	Saturation excess flow
Subsurface flow/ drainage	Shallow subsurface flow	Transient subsurface stormflow and soil pipe preferential flow	Shallow lateral flow via soils with high conductivity	Lateral subsurface flow to channel	Shallow subsurface (macropores, coarse soils) and ground-water	Shallow subsurface lateral flow	Lateral subsurface flow	Shallow subsurface with some seepage to an underlying groundwater aquifer	Groundwater flow unknown but presumably high rate of shallow lateral flow	Shallow lateral subsurface flow with negligible deep seepage
Average water table dynamics/ depth (m)	Unknown	1–8 m	>1.5 m except near stream	Unknown	Unknown	Unknown	Variable water table depth	~0.3 m in the bog; 0.5 m in uplands	Unknown; potentially very deep	Shallow, 0.9 m
Riparian areas for hydrology	None	1%	Limited due to steep topography	Limited due to steep topography	Limited to valleys, fens, bogs	Limited	Limited due to steep topography	33% of area is a peatland	~2 %	15% area
Major or extreme natural disturbance	1967 Fairbanks Flood	Windstorm (1995); landslide (2006)	Chestnut blight (1920s–1930s); drought; hurricanes; hemlock wooly adelgid (2003–present)	Chestnut blight; hurricanes; windstorms; SuperStorm Sandy (2012)	Pine bark beetle epidemic	None in reference	Hurricane (1938); ice storm (1998)	Peatland wildfire (1864); potential for derecho, tornados, wildfires	Wildfire	Hurricane Hugo (1989)
Other specific hydrological features	3% permafrost underlain	Fog input, soil pipes			Snowmelt-dominated hydrological regime		Discontinuous dense pan C horizon	Drainage from bog dome and uplands; some deep seepage to the aquifer	Extremely high levels of nitrate from chronic air pollution	Compared with pre-Hugo, flow reversal in paired watersheds after Hugo
Key publica-tion(s) on forest hydrological processes	Haugen *et al.* (1982); Hinzman *et al.* (2002)	Ziemer (1998); Reid and Lewis (2009)	Hewlett and Hibbert (1967); Swift *et al.* (1988)	Reinhart *et al.* (1963); Adams *et al.* (1994)	Alexander *et al.* (1985); Troendle and King (1985)	Rothacher *et al.* (1967); Post and Jones (2001)	Detty and McGuire (2010b); Gannon *et al.* (2014)	Sebestyen *et al.* (2011); Verry *et al.* (2011)	Riggan *et al.* (1985); Meixner and Wohlgemuth (2003)	Harder *et al.* (2007); Jayakaran *et al.* (2014)

Continued

Table 14.1. Continued.

Watershed characteristics	Caribou-Poker (CPCRW), Alaska	Caspar Creek (CCEW), California	Coweeta (CHL), North Carolina	Fernow (FnEF), West Virginia	Fraser (FrEF), Colorado	H.J. Andrews (HJAEF), Oregon	Hubbard Brook (HBEF), New Hampshire	Marcell (MEF), Minnesota	San Dimas (SDEF), California	Santee (SEF), South Carolina
Forest experimental watershed contact	Jamie Hollingsworth jhollingsworth@alaska.edu	Elizabeth Keppeler ekeppeler@fs.fed.us	Peter Caldwell pcaldwell02@fs.fed.us	Mary Beth Adams mbadams@fs.fed.us	Kelly Elder kelder@fs.fed.us	Sherri Johnson sherrijohnson@fs.fed.us	John Campbell jlcampbell@fs.fed.us	Stephen Sebestyen ssebestyen@fs.fed.us	Pete Wohlgemuth pwohlgemuth@fs.fed.us	Devendra Amatya damatya@fs.fed.us
Experimental forest website	http://www.lter.uaf.edu/bnz_cpcrw.cfm	http://www.fs.fed.us/psw/topics/water/caspar	http://www.srs.fs.usda.gov/coweeta/	http://www.nrs.fs.fed.us/ef/locations/wv/fernow	http://www.fs.usda.gov/efr/fraser	http://andrewsforest.oregonstate.edu	http://www.hubbardbrook.org	http://www.nrs.fs.fed.us/ef/marcell/	http://www.fs.fed.us/psw/ef/san_dimas	http://www.srs.fs.usda.gov/charleston/santee

[a]amsl = above mean sea level.
[b]Water years taken as starting in May and ending in April.
[c]PET estimated from Thornthwaite (1948) method
[d]PET estimated from Hamon (1963) method with corrections.
[e]PET estimated from evaporation pan (Patric and Goswami, 1968).
[f]VSA = Variable Source Area

flood modulation due to surface storage and/or highly permeable soils. If the lower end of a curve is flat, the watershed sustains baseflow during dry periods, through release from a stored water source (e.g. groundwater), whereas a steep slope indicates a tendency for streams to dry up due to seasonality in precipitation and/or evapotranspiration and relative lack of storage. Because FDCs depict these streamflow attributes, they are important for water resources planning, especially for water uses that are influenced by extreme high and low flows. We also use the ratio of the 90th and 50th percentile daily flow (Q_{90}/Q_{50}) as an index of baseflow to assess its pattern among the watersheds, with higher values representing relatively higher baseflow or more stable flow.

Long-term (>25 years) mean daily flows are averaged for each month to characterize seasonal variability within and among sites, which assists in identifying controlling factors that cannot otherwise be captured by FDCs. The dryness index (DI; ratio of mean annual potential evapotranspiration to precipitation) is used as an indicator of energy-limited (DI < 1) versus moisture-limited (DI > 1) watersheds (e.g. Creed *et al.*, 2014). In the next section, we describe the setting and environmental features of each of the ten USDA-EFR reference watersheds evaluated. Key characteristics are compared in Table 14.1.

14.2 Site Description

14.2.1 Caribou-Poker Creek Research Watershed (CPCRW), reference sub-watershed C2, Alaska

The CPCRW is located near Chatanika in interior Alaska (Fig. 14.1) and is representative of the northern boreal forest. The 520 ha C2 reference watershed is isolated and free of any human intervention. The vegetation in CPCRW is dominated by birch and aspen on the south-facing slopes and black spruce forests on the north-facing slopes. The climate is typically continental with warm summers and cold winters.

The CPCRW is unique among the watersheds in this cross-site comparison because it is underlain by discontinuous permafrost. The permafrost distribution within the watershed exerts a strong influence over hydrological patterns (Jones and Rinehart, 2010). Studies show that as the areal extent of permafrost increases, peak discharge increases, baseflow decreases and response to precipitation events increases (Bolton *et al.*, 2004). The C2 watershed was chosen as a reference watershed because it is underlain by only about 3% permafrost compared with the adjacent C3 and C4 watersheds which are underlain by 53% and 19%, respectively.

Total mean precipitation in the C2 watershed is 412 mm, with mean snowfall and rainfall being 130 mm and 280 mm, respectively (Bolton *et al.*, 2004). Annual maximum snow depth averages 750 mm with a snow water equivalent of 110 mm. Of the total precipitation, nearly 20% becomes streamflow while evapotranspiration makes up over 75% (Bolton *et al.*, 2004). About 35% of the total precipitation falls as snow between October and April. Snowfall peaks around January while rainfall peaks around July. The spatial distribution of rainfall amount is influenced by elevation.

The relatively flat FDC for the C2 watershed (Plates 11 and 12, Table 14.2) may be attributed to the relatively well-drained soils that allow infiltration to deeper subsurface reservoirs. Runoff is generated only when the infiltration capacity is met. Streamflow in the watershed is generated by shallow subsurface storm runoff from permafrost-dominated areas, but steady groundwater baseflows with the highest Q_{90}/Q_{50} of all the sites (Table 14.2) are produced from permafrost-free areas such as C2. Spring snowmelt is usually the major hydrological event of the year, although the annual peak flow usually occurs during summer rainstorm events, as the highest rainfall intensities are greater than the maximum snowmelt rate on a daily timescale (Kane and Hinzman, 2004). It may be noted from Fig. 14.2 that the mean monthly streamflow of C2 is relatively even over the months of April through October. During winter the gauges are mostly frozen and any flow is hardly recorded, except for relatively warm temperatures. Although rainfall peaks around July, there is an increase in mean flow from July to September due to an increase in baseflow.

Table 14.2. Daily flow values for various percentage time exceedance of the flow at the ten study sites.

Watershed #/name/location	No. of daily records	Daily flow, *Q* (mm), for percentiles									
		0.1	1	5	10	25	50	75	90	95	Q_{90}/Q_{50}
C2/CPCRW/Alaska	4,058	3.5	2.3	1.6	1.2	0.78	0.51	0.32	0.22	0.17	0.43
NF/CCEW/California	7,671	68.0	25.3	8.9	4.5	1.13	0.27	0.08	0.04	0.03	0.15
WS18/CHL/North Carolina	27,482	22.6	11.8	7.0	5.5	3.70	2.04	1.06	0.62	0.47	0.30
WS4/FnEF/West Virginia	21,430	34.6	15.4	6.8	4.4	2.00	0.78	0.14	0.02	0.00	0.026
ESL/FrEF/Colorado	11,687	14.5	9.6	7.1	5.4	2.79	1.16	0.63	0.41	0.26	0.35
WS02/HJAEF/Oregon	22,280	66.6	29.1	15.1	9.3	4.01	1.43	0.38	0.18	0.13	0.126
WS3/HBEF/New Hampshire	20,181	51.4	24.2	9.8	5.5	2.33	0.92	0.31	0.06	0.03	0.067
S2/MEF/Minnesota	19,723	14.1	5.7	2.4	1.3	0.30	0.02	0.00	0.00	0.00	0.00
Bell 3/SDEF/Californa	18,518	30.8	4.7	1.0	0.4	0.12	0.01	0.00	0.00	0.00	0.00
WS80/SEF/South Carolina	11,256	41.7	16.8	4.2	2.1	0.42	0.03	0.00	0.00	0.00	0.00

CPCRW = Caribou-Poker Creek Research Watershed; CCEW = Caspar Creek Experimental Watershed; CHL = Coweeta Hydrologic Laboratory; FnEF = Fernow Experimental Forest; FrEF = Fraser Experimental Forest; HJAEF = H.J. Andrews Experimental Forest; HBEF = Hubbard Brook Experimental Forest; MEF = Marcell Experimental Forest; SDEF = San Dimas Experimental Forest; SEF = Santee Experimental Forest.

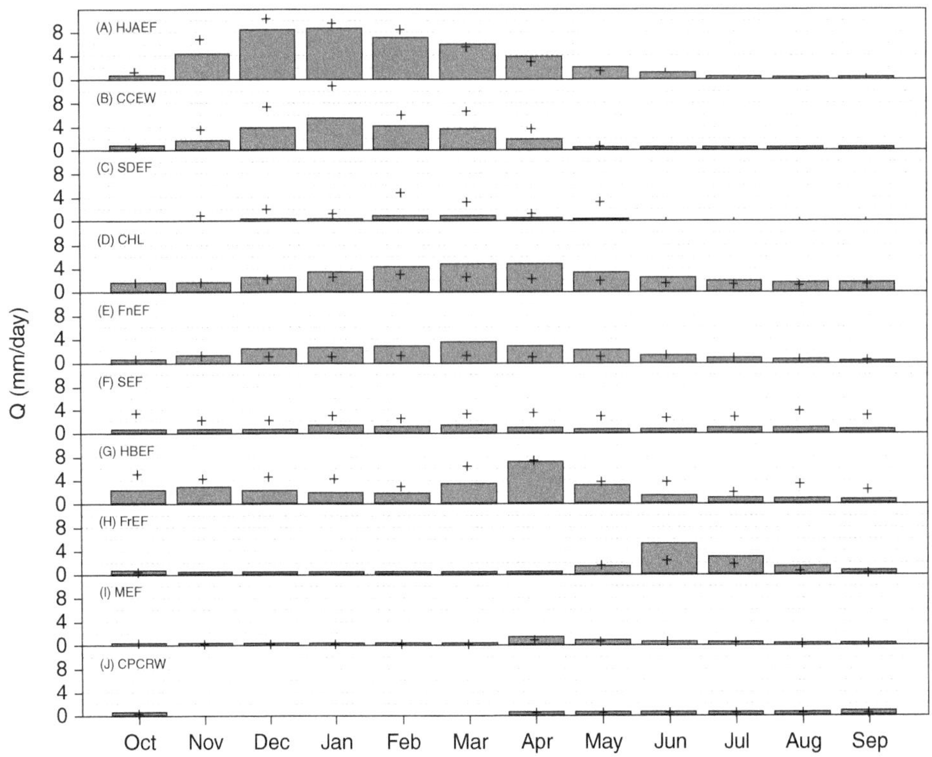

Fig. 14.2. Monthly mean daily streamflow, *Q*, averaged over the record period for each month, arranged by climate and region. '+' sign indicates standard deviation (SD) of daily flow by month. FrEF mean flow was estimated by regression of baseflow for November to May and SDs are not presented. Sample size was insufficient for flow at CPCRW for the months of November to May (HJAEF = H.J. Andrews Experimental Forest; CCEW = Caspar Creek Experimental Watershed; SDEF = San Dimas Experimental Forest; CHL = Coweeta Hydrologic Laboratory; FnEF = Fernow Experimental Forest; SEF = Santee Experimental Forest; HBEF = Hubbard Brook Experimental Forest; FrEF = Fraser Experimental Forest; MEF = Marcell Experimental Forest; CPCRW = Caribou-Poker Creek Research Watershed).

14.2.2 Caspar Creek Experimental Watershed (CCEW), reference watershed North Fork (NF), California

Located in a coast redwood and Douglas fir forest on the Jackson Demonstration State Forest in north-western California (Fig. 14.1), the CCEW hosts research designed to evaluate the effects of timber management on watershed processes. Initially, the entire 473 ha NF watershed served as the reference watershed, but after portions were logged in 1985, three NF sub-watersheds (16 to 39 ha) were designated as long-term reference watersheds. Bedrock is marine sandstone and shale of the Franciscan Complex. Most soils are 1–2 m deep loams and clay-loams and underlain by saprolite at depths of 3–8 m near ridgetops. Only about one-fifth of the 4.6 km/km^2 drainage density supports perennial streamflow. Timber production has been the major land use, and evidence of 19th century logging and the impacts of this legacy persist.

Snow is hydrologically insignificant and 95% of rainfall occurs in October–April. Fog occurs on about one-third of days in June–September, reducing summer transpiration (Keppeler, 2007). The marine influence ensures that summer air temperatures rarely exceed 20°C and winter minimums seldom drop below 0°C.

Stream runoff is about half of the average annual rainfall (Reid and Lewis, 2009). Transpiration and canopy evaporation account for nearly equal portions of the remainder (Fig. 9.1, Chapter 9, this volume). Actual evapotranspiration is limited by soil moisture deficits in May–September. Analysis of climate-related trends suggests that autumn rainfall and streamflow have declined, but with no change in annual totals.

The FDC for CCEW spans a wide range of streamflow compared with most of the other USDA-EFR sites (Plates 11 and 12) due to the strong seasonal pattern of large, episodic winter rain events that typically produce multiple, short-duration peak flows while extended summer droughts result in a long, slow recession for about half the year (Fig. 14.2). Summer streamflow is generated primarily from groundwater, and by autumn about 300 mm of precipitation is needed to mitigate moisture deficits sufficiently to generate stormflow. Stormflow (total flow based on difference between initial discharge at start of runoff and the discharge at 3 days following the cessation of the rainfall event) comprises about two-thirds of annual runoff (Reid and Lewis, 2009). Infiltration is rapid on uncompacted soils and vertical throughflow dominates near the surface. A deeper clay-rich argillic horizon can impede downward flow and generate lateral subsurface flow, although preferential flow through interconnected soil macropores limits pore-pressure increases and the extent of this perched flow. Perennial and intermittent soil pipes occur in the upper 2 m of the regolith and are frequently encountered near channel heads. When transient groundwater tables rise to the elevation of these pipes, they rapidly transmit subsurface flow to channels, mitigating pore-pressure increases upslope (Keppeler and Brown, 1998). Saturation-excess overland (return) flow is limited, but can occur on valley bottoms during large storms.

14.2.3 Coweeta Hydrologic Laboratory (CHL), reference watershed WS18, North Carolina

The CHL is located in western North Carolina (Fig. 14.1) and is representative of southern Appalachian mixed deciduous hardwoods. The 13 ha WS18 watershed was last selectively harvested in the early 1920s prior to the establishment of the CHL (Douglass and Hoover, 1988). Although the watershed has not been actively managed for more than 80 years, there have been several natural disturbances that have altered forest structure and species composition, including Chestnut blight fungus (*Endothia parasitica*) in the 1920s–1930s, drought in the 1980s and 2000s, Hurricane Opal in 1995, and hemlock woolly adelgid (*Adelges tsugae*) defoliation from 2002 to the present (Boring *et al.*, 2014).

Precipitation in WS18 averages 2010 mm/year; it is highest in the late winter months and lowest in the autumn, although a disproportionate amount of large events associated with tropical storms occurs during this season. Less than 10% of precipitation occurs as snow. The variability in precipitation has been increasing over time resulting in more frequent extremely wet years and extremely dry years, while annual average air temperature has been increasing by 0.5°C/decade since 1981 (Laseter *et al.*, 2012).

Annual precipitation in WS18 is approximately equally partitioned into streamflow (49.6%) and evapotranspiration (50.4%). During the growing season, transpiration accounts for 55% of total evapotranspiration with evaporation from canopy interception making up the balance, approximately 15% of precipitation (Ford *et al.*, 2011). Streamflow is typically highest in March–April and lowest in September–October but never ceases, even during extreme drought. Seasonal patterns in streamflow reflect the combined effects of the seasonality in precipitation and evapotranspiration (Fig. 14.2).

Baseflows are relatively high, producing the third largest Q_{90}/Q_{50} ratio among sites (Table 14.2). Baseflows are sustained by lateral movement of water through deep unsaturated soil (Fig. 9.1, Chapter 9, this volume), driven by large hydraulic gradients induced by steep slopes (Hewlett and Hibbert, 1963). On average, approximately 5% of annual precipitation (9% of annual streamflow) is discharged as stormflow (Swift *et al.*, 1988). Stormflow originates primarily from small portions of the watershed located adjacent to the stream in coves and in riparian zones where the water table may be near the surface (Hewlett and Hibbert, 1967). Shallow lateral subsurface discharge from upslope landscape positions to streams can also contribute to stormflow where large soil macropores exist. Overland flow is extremely rare or non-existent because of the presence of well-developed forest floors and subsurface macropores.

14.2.4 Fernow Experimental Forest (FnEF), reference watershed WS4, West Virginia

The FnEF is located in eastern West Virginia (Fig. 14.1) and is representative of the 'unmanaged' forests of the central Appalachian region. The 39 ha WS4 watershed is forested with an approximately 100-year-old stand of mixed deciduous hardwoods. The bedrock is acidic sandstone and shale. Depth to bedrock is generally less than 1 m and the topography is steep.

Precipitation is distributed evenly throughout the year and averages 1458 mm. Although snow is common in winter, snowpack generally lasts no more than a few weeks; snow contributes approximately 14% on average of precipitation (Adams *et al.*, 1994). Large rainfall events can occur during extra-tropical hurricanes in the summer and autumn, but about half of the largest storms have occurred during the dormant season (1 November–30 April), when streams are most responsive to rainfall because evapotranspiration losses are low (Fig. 14.2).

The stream channel is intermittent near the top of the watershed. Streamflow may cease during the late summer and early autumn (about 10% of daily flows), in response to high evapotranspirative demand and low precipitation. Although baseflow contributes relatively little to Q_{90}/Q_{50} (Table 14.2), it dominates stream discharge in WS4. Most discharge occurs during the dormant season (Fig. 14.2) due to greater precipitation and decreased evapotranspirative demand from deciduous forests. Baseflow is sustained by lateral subsurface flow to channels; DeWalle *et al.* (1997) characterized the mean transit time for baseflow on WS4 as 1.4–1.6 years, which suggests a dominance of slow movement through the soil matrix.

The water balance on WS4 was well quantified by Patric (1973) with runoff accounting for about 40% of precipitation, 27% of the balance being lost through transpiration and about 16% to canopy evaporation. Seasonal differences in losses from canopy interception due to leaf development and leaf drop were detected.

Stormflow discharge is fairly flashy (Plates 11 and 12), with the storm hydrograph responding rapidly to storm precipitation inputs and then returning quickly to baseflow conditions, and streamflow generation occurs via saturation excess flow. Stormflow discharge typically occurs less than 15% of the time. There is little to no infiltration-excess overland flow even during the largest storms because of the high infiltration capacity of an intact forest floor.

14.2.5 Fraser Experimental Forest (FrEF), reference watershed East St Louis (ESL), Colorado

The FrEF is located in the Rocky Mountain cordillera of Colorado (Fig. 14.1) and is representative of subalpine watersheds over a large portion of the central Rockies. It spans the subalpine to alpine zone; a zone that is characterized by relatively low temperatures and moderate precipitation (Love, 1960). The area is dominated by Engelmann spruce and subalpine fir on higher-elevation and shaded slopes, lodgepole pine on lower-elevation sunny slopes and alpine tundra above the treeline. The 803 ha ESL watershed has received no significant treatment in over 90 years (Retzer, 1962).

Precipitation is dominated heavily by snowfall (about 75%) from October through May (Alexander *et al.*, 1985) and runoff is dominated by snowmelt (about 90%) from May through August (Fig. 14.2). Significant summertime convective rainfall events may also temporarily increase flow. The main stem is perennial but baseflow is low, stable and unmeasured during the winter months due to logistical difficulties of stream measurements in winter.

The runoff coefficient for annual flow is about 45% with significant wintertime sublimation losses from the canopy and summertime evapotranspiration. Summertime rainfall is primarily used on site by vegetation, with high evaporative losses due to dry air masses and wind.

High-elevation stream reaches are intermittent with spring and summertime flows fed by snowmelt (Fig. 14.2). The hydrological regime is dominated by a typical seasonal snowmelt hydrograph with a rapid rising limb in May and June, followed by a long recession, returning to baseflow (second largest Q_{90}/Q_{50}, Table 14.2) in August (Alexander *et al.*, 1985; Troendle and King, 1985). Extensive spring networks feed the drainage systems as the annual snowmelt pulse moves through the basin (Retzer, 1962). Rainfall events punctuate the snowmelt hydrograph, but contribute insignificant amounts to the annual runoff. Infiltration-excess overland flow is rare, but may occur under the snowpack during the melt season when frozen ground impedes infiltration. Saturation-excess overland flow is extremely rare as infiltration rates for the porous soils and glacial till typically exceed maximum rainfall and snowmelt rates (Retzer, 1962).

The ESL represents the highest elevation range, largest snowpack and largest watershed of this cross-site comparison. Maximum snowmelt rates are limited by incoming energy and can never reach extreme rainfall rates. Rain-on-snow flood events can alter flow statistics, but are rare in this portion of the Rockies. The relatively large size of the basin also reduces flashy response or high runoff per unit area observed in smaller basins.

14.2.6 H.J. Andrews Experimental Forest (HJAEF), reference watershed WS02, Oregon

The HJAEF is located in the western Cascade Mountains of central Oregon (Fig. 14.1) and is representative of Pacific Northwest moist conifer forests. Watershed 2 (WS02) is 60 ha and the geology is dominated by bedrock of volcanic origin. Stream channels are steep and confined with unsorted sediment dominated by cobbles and boulders, with patches of silt and exposed bedrock. Shallow hillslope soils (generally less than 1 m deep) are loam and clay loam. Stone content ranges from 35 to 80%, increasing on south-facing slopes. The steep hillslopes in WS02 are dominated by 500- to 550-year-old Douglas fir (*Pseudotsuga mensiesii*) forests with western hemlock (*Tsuga heterophylla*) and western red cedar (*Thuja plicata*) (Rothacher *et al.*, 1967). The canopy is greater than 60 m tall. The climate is continental with cold winters and cool, short, dry summers.

Annual precipitation averages 2300 mm, falling primarily as rain between November and April and with occasional snow at higher elevations. Soil temperatures remain above freezing. The annual hydrograph in WS02 has a strong seasonal pattern with a high winter baseflow and frequent autumn, winter and spring stormflows in contrast to very low flows in summer (Fig. 14.3).

Approximately 57% of the precipitation is streamflow (Post and Jones, 2001). Baseflow accounts for only 43% of the discharge (Q_{90}/Q_{50} = 0.126) (Table 14.2) whereas quickflow comprises the remainder (Fig. 9.1, Chapter 9, this volume). McGuire *et al.* (2005) estimated that mean baseflow residence time for WS02, based on δO^{18} of water, was approximately 2.2 years. They suggested that topography and steepness may be exerting greater control on residence times than watershed area. Although there are no detectable trends in streamflow from 1987 to 2007, in more recent time periods (1996–2007) slight decreasing trends have been observed (Argerich *et al.*, 2013).

The relatively steep FDC for WS02 (Plates 11 and 12) has been attributed to highly permeable soils and strong seasonal precipitation patterns. Fast percolation rates, typically greater than 0.12 m/h, are influenced by high stone content and large pore spaces (Rothacher *et al.*, 1967). These characteristics also lead to substantial hyporheic flows lateral to and beneath the streams (Kasahara and Wondzell, 2003).

14.2.7 Hubbard Brook Experimental Forest (HBEF), reference watershed W3, New Hampshire

The HBEF is located in New Hampshire (Fig. 14.1) and is representative of mature northern hardwood stands. Vegetation at W3 is composed mainly of sugar maple (*Acer saccharum*), American beech (*Fagus grandifolia*) and yellow birch (*Betula alleghaniensis*). The 42 ha watershed is mostly second growth and much of the HBEF was harvested in the 1910s (Table 14.1). Additional salvage harvesting occurred at the HBEF following the Great New England Hurricane of 1938. More recently, trees incurred some damage during the North American Ice Storm of 1998, with no apparent impact on annual runoff.

The climate at the HBEF is cool and humid. On average, W3 receives 1350 mm of precipitation annually, which is distributed evenly throughout the year. Precipitation has increased by 25% during the record period, which is consistent with broader regional trends (Brown *et al.*, 2010). Approximately one-third of precipitation falls as snow (Fig. 9.1, Chapter 9, this volume) and a snowpack generally persists from late December until mid-April. Soil frost forms during winter two out of every three years with an average annual maximum depth of 6 cm.

The annual hydrograph shows a strong seasonal pattern with a peak during snowmelt runoff. Despite the higher flow during spring, floods can occur at any time of year when soil water deficits are reduced (Fig. 14.2). An increasing trend in precipitation has resulted in increasing trends in the magnitude of both low and high streamflows (Campbell *et al.*, 2011).

Approximately 64% of the precipitation that falls on the watershed becomes streamflow, with evapotranspiration comprising the remainder. Slight, but statistically significant declines in evapotranspiration have occurred in W3 (14% over 56 years) for reasons that are unknown. This decline appears to be due to local influences since similar trends are not consistently found at a larger regional scale.

The relatively steep FCD for W3 (Plates 11 and 12) has traditionally been attributed to coarse, well-drained soils and mountainous topography that produce a flashy runoff response. Overland flow is also minimal because of the high infiltration capacity of the forest floor. In recent years, a more complete understanding of complex flow generation processes at the site has emerged. Data from a network of wells in W3 have revealed an intermittent, discontinuous water table (Detty and McGuire, 2010a; Gannon *et al.*, 2014; Gillin *et al.*, 2015). Stormflow generation is the result of lateral subsurface flow in the solum. Under some soil moisture conditions, small changes in groundwater can produce large changes in runoff, suggesting a threshold response that is related to flowpaths and soil transmissivity (Detty and McGuire, 2010b; Gannon *et al.*, 2014). During low flows, only the near-stream zone is consistently hydrologically connected to the stream network. As the watershed wets up, more distal, previously isolated portions of the water table become hydrologically connected.

14.2.8 Marcell Experimental Forest (MEF), reference watershed S2, Minnesota

The MEF is located along the southern fringe of the boreal biome, in northern Minnesota (Fig. 14.1). The landscape includes uplands, peatlands, lakes and streams. Unlike mountainous research watersheds, streamflow typically is not bedrock controlled in the western lakes section where outwash sands, some >50 m deep, form large aquifers (Verry *et al.*, 2011). Aquifer–peatland connectivity varies between two peatland types: bogs and fens (Bay, 1967). In watersheds with either type, streamflow may originate from precipitation and flow along near-surface and shallow surface flowpaths in upland mineral soils (Verry *et al.*, 2011). Bog watersheds may be perched due to loamy clay tills that retard the vertical flow of water from soils to the outwash aquifer (Verry *et al.*, 2011). In fen watersheds, most streamflow, which may exceed streamflow from bogs by orders of magnitude during low flow, originates as discharge from aquifers and is perennial (Bay, 1967).

The 10 ha S2 study watershed, with a bog (33% of the area), has low topographic relief (Table 14.1) with upland mineral soils that drain through peatland margins to an intermittent stream. Eleven to 33% of annual precipitation (456–981 mm) occurs as streamflow and 5–17% recharges the underlying aquifer (Nichols and Verry, 2001) (Fig. 9.1, Chapter 9, this volume). Calculated evapotranspiration (precipitation − streamflow − recharge) has been 372–605 mm/year. Nine of the ten highest daily streamflows have occurred during rainfall–runoff events, not snowmelt or rain-on-snow events. Periods of no streamflow occur during any month and there has been no flow during 38% of the record (Plates 11 and 12), consistent with the zero value of Q_{90}/Q_{50} (Table 14.2).

Although most of the S2 area is uplands, most of the annual water budget (58%) comes from direct precipitation on the peatland (Verry *et al.*, 2011). If the water table is >5–10 cm below the peatland surface, streamflow ceases and that storage must be replenished before resumption. Rainfall amount during summer exceeds snow water equivalents during winter and stormflows recess rapidly to no flow due to evapotranspiration. Melt from snow accumulation (November/December to March/April) results in several weeks of high flows (Sebestyen *et al.*, 2011) (Fig. 14.2). Winter and spring frost in upland soils, exceeding 50 cm, prevents infiltration (Verry *et al.*, 2011). Snowmelt waters flow overland until soils thaw in the spring, after which flow mostly occurs in the shallow subsurface through sandy loams above loamy clay horizons (Verry *et al.*, 2011). Subsurface flow may persist for weeks until the upland deciduous forest begins transpiring. During large summer rainfall events, subsurface flow may last for several hours, but rarely longer.

14.2.9 San Dimas Experimental Forest (SDEF), reference watershed Bell 3, California

The 25 ha watershed at SDEF is located in southern California (Fig. 14.1) and is representative of the chaparral forests of the US Southwest. Chaparral forest is a dense, drought-tolerant shrubland with a closed canopy some 3–5 m in height. Chaparral is a fire-prone ecosystem and wildfires have burned the SDEF about every 40 years.

Regional hydrology is controlled by climate and geology: cool, wet winters followed by long summer droughts; and ongoing tectonic uplift that has produced steep topography and exposed fractured crystalline basement rocks that weather to thin, coarse-textured, azonal soils (Dunn *et al.*, 1988) (Table 14.1). Precipitation falls almost exclusively as rain from winter frontal storms and rare summer thunderstorms. Nearly 90% of the annual rainfall occurs between December and April with the most runoff in February (Fig. 14.2).

Streamflow accounts for only roughly 11% of the rainfall, with the remainder apportioned to evapotranspiration and groundwater recharge. Groundwater dynamics on the SDEF are virtually unknown, rendering the closure of any water balance exercise moot. However, groundwater recharge is potentially large through the fractured substrate, reducing any calculated value of actual evapotranspiration. Soil moisture is at or below the wilting point by the end of the summer and the drought-adapted plants likely get their water from fractures in the bedrock.

Stream runoff is generated by saturation excess flow in riparian zones, presumably as shallow throughflow moves laterally through the coarse soil mantle (Fig. 9.1, Chapter 9, this volume). Infiltration-excess overland flow on hillside slopes is rare and occurs only during the most intense rainstorms, reflecting the high infiltration rates of the soil and percolation into bedrock. However, after wildfire, with the combustion of the canopy and surface litter layer as well as changes in soil properties (bulk density and water repellency), hillslope hydrology shifts to pervasive overland flow after saturation of the very thin surface wettable layer (Rice, 1974; DeBano, 1981). Water that formerly slowly flowed by subsurface pathways now moves quickly into the stream channels, increasing runoff for comparable storms by up to four orders of magnitude over pre-fire levels (Wohlgemuth, 2016). The effects of fire on the forest hydrology can persist for several years.

14.2.10 Santee Experimental Forest (SEF), reference watershed WS80, South Carolina

The SEF is located in eastern South Carolina (Fig. 14.1) and is representative of the subtropical coastal watersheds throughout much of the US Southeast, with hot and humid summers and moderate winter seasons. The 155 ha WS80 watershed is covered with a pine/mixed hardwood forest (Table 14.1), which has been undisturbed by management activities since 1936, but was heavily affected by Hurricane Hugo in 1989 that damaged >80% of the forest canopy (Hook *et al.*, 1991).

Seasonally, the winter is generally wet with low-intensity, long-duration rain events and rare snowfall. Summer is characterized by short-duration, high-intensity storm events and tropical depression storms are common. The seasonal runoff response to rain events is shown in Fig. 14.2.

Approximately 22%, on average, of annual precipitation becomes runoff (Amatya *et al.*, 2006), resulting in about 78% evapotranspiration, assuming negligible seepage (Fig. 9.1, Chapter 9, this volume). Approximately 60% of the runoff is contributed by shallow surface or runoff/rainwater, the rest by subsurface flow (Epps *et al.*, 2013).

Based on the FDC analysis this watershed produces flow only 56.3% of the time and hence has a zero value of Q_{90}/Q_{50} (Plates 11 and 12, Table 14.2). The principal flow generation mechanism is driven by the shallow water table (Fig. 9.1, Chapter 9, this volume) (Harder *et al.*, 2007; Epps *et al.*, 2013), controlled primarily by rainfall and evapotranspiration, and minimally by deeper groundwater underlain by Santee Limestone approximately 20 m below the ground surface. The formation of an argillic horizon with poorly drained clayey subsoil provides a dynamic shallow groundwater table that has a complex non-linear relationship with streamflow (Harder *et al.*, 2007). Saturation-excess surface and shallow subsurface runoff with rapid lateral transfers within the highly permeable upper soil layer may occur along reaches with flat topography. Surface depressional storage was shown to affect the surface runoff rate (Amoah *et al.*, 2012). Runoff and peak flow at this watershed are dependent on both rainfall amount and intensity, as well as antecedent conditions reflected by initial water table positions (Epps *et al.*, 2013).

A key observation from WS80 is the reversal of the flow relationship between this and the treatment watershed, compared with the earlier calibration period, for a decade beginning three years after Hurricane Hugo severely damaged vegetation on both watersheds. As a result reduced evapotranspiration in selected hurricane-affected vegetation on the reference watershed enhanced its streamflow (Jayakaran *et al.*, 2014). Long-term data also indicate rising air temperature and increasing frequency of large storms (Dai *et al.*, 2013).

14.3 Discussion of Hydrological Processes

14.3.1 Flow duration curves

FrEF and CPCRW host the largest reference watersheds among our study sites (Table 14.1). FrEF has the highest elevation range, deepest snowpack and the largest drainage area. These factors, combined with the snowmelt-driven hydrological regime, explain the somewhat different behaviour in flow duration with higher flow values for FrEF than for CPCRW (Plates 11 and 12, Table 14.2). The muted high flows, with their greater influence at CPCRW potentially due to its relatively well-drained soil conditions (see Section 14.2.1 above), are most likely attributed to the large size of these watersheds. However, this does not hold true for CCEW which, although comparable in size to CPCRW (Table 14.1), has a steep FDC for low exceedance, perhaps due to its much larger seasonal precipitation, deep clay horizon and soil pipes that contribute to a rapid runoff response (see Section 14.2.2 above).

In comparison, SDEF has the second smallest reference watershed and forth steepest watershed examined (Table 14.1). As a result, its FDC shows very flashy storm responses followed by long, declining flows that eventually are zero for 47.5% of the record. Similarly, MEF, characterized by deep peat and possibly high storage capacity, and SEF with shallow sandy clay loam soils generate no surface flow for 44% of their periods of record, with $Q_{90}/Q_{50} = 0$ for all three sites (Table 14.2). Although SEF is the lowest gradient watershed, the high flow range that occurs for less than 1% of the time is greater than at most of the other sites, except for HBEF, HJAEF and CCEW. The highest flows at this site result from storm runoff from saturated clay-rich soils (Epps *et al.*, 2013; Griffin *et al.*, 2014).

Along the Coast Range of the western USA, HJAEF and CCEW have FDCs that are similar in shape, likely related to seasonal climatic patterns. The HJAEF has the third steepest basin slope after CHL and CCEW (Table 14.1) but the highest FDC slope for low flows occurring more than 0.2 to 30% of the time, above which the CHL has the highest low flow (Plates 11 and 12). Although WS02 at HJAEF is smaller than the watershed at CCEW (Table 14.1), it generally sustains higher flows, except at the lowest exceedance frequencies, likely because it receives 1.75 times more precipitation than the CCEW. Both of these western watersheds have similar forest species and leaf area index (LAI) (Table 14.1) as well as frequent large storms in winter and dry summers. Weiler and McDonnell (2004) suggest additional factors including lateral soil conductivity and drainable porosity may explain variability in streamflow response, specifically at HJAEF.

CHL has the steepest basin slope (52%) of all the watersheds in this analysis and a 95th percentile flow (Q_{95}) of 0.47 mm/day, which is the largest of all the sites (Table 14.2). Of the three sites in the Appalachian Mountains (i.e. CHL, FnEF and HBEF), CHL also has the smallest drainage area and is more southerly than FnEF and HBEF (Table 14.1). Interestingly, this reference watershed also has the highest Q_{90}/Q_{50} values (indicative of sustained baseflow) and lowest flow values for the higher flow ranges ($Q_{0.1}$ or lower exceedance) but has equal or higher flows at and above Q_{25} compared with FnEF or HBEF (Table 14.2). The higher flow in the lower exceedance range in the more northern HBEF site could be partially attributed to snowmelt and the higher flow in lower exceedance range at the CHL site is likely due to sustained baseflows caused by high storage of deep soils (Table 14.1).

Although on opposite coasts, the 61 ha HJAEF site yields consistently higher percentile flows (Table 14.2) compared with the 42.4 ha HBEF site at almost the same latitude, similar elevations, potential evapotranspiration, and surface and subsurface flow generation mechanisms (Table 14.1). The exception is the extreme high end of discharges at or below 0.01% exceedance when both exhibit a similar pattern (Plates 11 and 12), which is attributed to the HJAEF having higher slope and 41% higher precipitation than the HBEF. In their analysis of threshold hydrological response across northern catchments, Ali *et al.* (2015) found some similarities in rainfall- and snowmelt-driven events between these two watersheds.

14.3.2 Long-term mean daily flow

Figure 14.2 (plots A–C) shows long-term mean daily flow by month for west-coast watersheds which all have strongly seasonal rainfall. Oregon's HJAEF (plot A) has the greatest monthly flows, with a longer winter rainy season than the more southerly sites. In California, coastal CCEW (plot B) reflects the transition from the wetter north-west to the arid Mediterranean climate of SDEF (plot C). These three western sites show highly variable winter flow patterns due to the episodic nature of the Pacific frontal systems with increased coefficient of variation further south where large winter storms are less frequent. These patterns are also consistent with the relative variability defined by the upper and lower exceedance percentiles of the FDCs (Plates 11 and 12, Table 14.2).

Similarly, the east-coast watersheds in Fig. 14.2 (plots D–G) range from high mean flow in the winter to low flow in the summer and early autumn, with the exception of HBEF (plot G). CHL (plot D) shows a smooth annual hydrograph that peaks in late spring following the seasonal rainfall pattern. FnEF (plot E) and SEF (plot F) have similar mean annual precipitation, but the SEF produces less than half of the runoff generated at FnEF, primarily due to higher potential evapotranspiration (Table 14.1). The seasonal signal for the FnEF and CHL reflects their inland locations and a more pronounced dormant season relative to SEF. Both CHL and FnEF show relatively little streamflow variability due to relatively consistent precipitation with little variance. The relatively high streamflow variability at the SEF results from a dynamic water table regulated by coastal climate and shallow clayey argillic horizon. HBEF (plot G) is well north of the other east-coast basins, putting it in a location where snow plays a greater role in the hydrological regime. It is the only watershed in the study that shows a significant double peak in annual flow: a rainfall peak in November and a snowmelt or rain-on-snow peak in April.

Snowmelt and continentality have a dominant influence on annual water budgets in the last three study areas: FrEF (plot H), MEF (plot I) and CPCRW (plot J) (Fig. 14.2). FrEF receives most of its precipitation in the form of wintertime snow. The CPCRW (plot J) represents an extreme in almost every metric used (Table 14.1) including the annual precipitation and runoff. All of the snowmelt-dominated watersheds show lower relative variance in flow because the peak flows are regulated by both the amount of snow and the maximum amount of energy available to melt snow, with the occasional exception at the MEF where some peak flows occur during rain-on-snow events. In general, higher mean monthly flows are observed in basins close to coastal moisture sources or at lower latitudes, although there are exceptions (SDEF, SEF). Higher variances are also observed near coasts, where large, episodic rainfall events are more influential. Snowmelt processes reduce variance (FrEF, MEF and CPCRW), while inland watersheds also exhibit less variability in daily mean flows (FnEF and CHL).

14.3.3 Other watershed characteristics affecting hydrology

Data from these ten sites show that none of the parameters in Table 14.1 (temperature, potential evapotranspiration, drainage area, altitude, latitude) has a significant influence on annual streamflow, except for annual precipitation, which is found to be a strong driver ($R^2 = 0.85$), as expected. However, annual evapotranspiration, calculated as the difference between precipitation and streamflow (i.e. not considering groundwater recharge), correlates well ($R^2 = 0.72$) with an independent estimate of potential evapotranspiration, and also with temperature ($R^2 = 0.76$) and latitude (inversely, $R^2 = 0.53$), as expected. Another interesting finding is that sites (CPCRW, MEF, SDEF and SEF) with DI values higher than 0.71 closer to soil moisture limited have a much lower (0.12–0.22) average runoff coefficient (streamflow/precipitation) than the remaining energy-limited sites (0.44–0.64) which have a DI < 0.50 (Table 14.1). Although most of the site characteristics for the HBEF and HJAEF are similar, except for precipitation which is higher at the HJAEF, the streamflow as a percentage of precipitation for the HJAEF is actually lower than that of the HBEF. This is possibly due to the higher evapotranspiration of its conifer forest, with its LAI almost twice that of the northern hardwood forest at the HBEF site. However, other factors such as geology and lithology besides the evapotranspiration might also be influencing losses. FrEF receives similar precipitation to SDEF and MEF, but has two to four times the annual streamflow because of much lower potential evapotranspiration as well as runoff occurring in a relatively steep basin, over a concentrated period, when a significant portion of the vegetation is dormant. However, some seepage to a regional aquifer at the MEF and possible groundwater recharge at the SDEF are also factors in their lower flow.

14.3.4 Implications of hydrological processes

Improved understanding of runoff generation and flowpaths helps land managers identify hydrologically connected areas that contribute to streamflow and pollutant discharge. The synthesis of runoff patterns across sites (Plates 11 and 12, Fig. 14.2) is important for identifying relationships between streamflow and nutrients that aid in developing load duration curves used to establish water quality standards (Argerich *et al.*, 2013). This important information is being used to assess the impacts of forest disturbance and restoration projects, and will help to better predict hydrological and chemical responses and transport. For example, monitoring procedures developed at the CCEW site are widely used to assess sediment and pollutant loads. This information is helpful in evaluating potential timber harvest impacts and in the development of forest management regulations and best management practices (Cafferata and Reid, 2013).

Knowledge of processes derived using long-term records from these diverse watersheds (Table 14.1) enables scientists to better understand their interrelationships with climate, forest vegetation and water use, and ecosystem dynamics (Vose *et al.*, 2012). For example, intensively monitored plots at CHL are providing new insights into relationships between soil moisture, carbon and nitrogen cycling, and vegetation allocation processes along topographic gradients. Furthermore, these records are being used to study hydrological recovery from disturbances such as the catastrophic mountain pine bark beetle infestation at FrEF, extreme hurricanes at SEF and historic land use at CCEW.

14.4 Summary

This cross-site comparison has used long-term hydrometeorological patterns, basin hydromorphological parameters and other attributes (Table 14.1) to compare and contrast forest hydrological processes (Fig. 9.1, Chapter 9, this volume) at ten reference watersheds in the USDA-EFR network. The response of streamflow to variation in annual precipitation magnitude, form and seasonality, and evapotranspiration at each watershed was evaluated by using daily FDCs (Plates 11 and 12), as well as the long-term mean daily flow for each month (Fig. 14.2).

Statistical results (Plates 11 and 12, Fig. 14.2 and Table 14.2) in the context of key watershed variables (Table 14.1) show that these watersheds have distinct hydrological processes and, therefore, can help frame our conceptual understanding of forest runoff processes (Fig. 9.1, Chapter 9, this volume), with precipitation as a driving variable for both high and low flows. While some seasonal flow patterns were observed among sites along the eastern and western near-coastal areas, flowpaths of rain and snowmelt water were shown to vary greatly across and within reference watersheds, potentially affecting the timing and peak of storm runoff, as illustrated by the FDCs (Plates 11 and 12, Table 14.2) and long-term monthly mean daily flows (Fig. 14.3). A DI value of about 0.50–0.70 was found to be an approximate break range for identifying sites with high runoff or low runoff, relative to the precipitation received. The analysis also revealed that larger watersheds do not necessarily yield higher baseflows and damped high flows. In addition, the presence of an argillic horizon, large topographic depressions and riparian area, preferential flowpaths, pipeflow, steep slopes and certain soil physical properties also significantly affect flowpaths, the magnitude and variation of runoff generation, and possibly the water balance (Weiler and McDonnell, 2004; Griffin *et al.*, 2014; Gillin *et al.*, 2015; Klaus *et al.*, 2015). Furthermore, the results also demonstrate that a better hydrological understanding of low topographic relief sites such as MEF and SEF is needed because these areas are common but not well represented by EFR sites, which are mostly in mountainous terrain.

Although this comparative study helps advance our understanding of runoff generation mechanisms across these diverse watersheds, increased evidence in recent years supports a non-linear rainfall–runoff response both on hillslopes and low-gradient coastal landscapes, highlighting the need to better quantify hydrological thresholds and understand physical controls (Spence, 2010; la Torre Torres *et al.*, 2011; Epps *et al.*, 2013; Ali *et al.*, 2015; Klaus *et al.*, 2015). Research on linkages between hydrology and soil development (e.g. Gillin *et al.*, 2015), peatland watershed responses to environmental and climatic change (Kolka *et al.*, 2011), rainfall–runoff relationships in chaparral vegetation, interactive effects of vegetation and stand type on streamflow (Jayakaran *et al.*, 2014), hydrological processes on tidally affected riparian forested wetlands (Czwartacki, 2013), etc. is advancing in some of these watersheds. Incorporation of this new information into ecohydrological models (Dai *et al.*, 2010; Amatya and Jha, 2011) will improve predictions of runoff generation and our ability to assess responses to future disturbances.

Long-term data from this spectrum of watersheds demonstrate the value of the USDA-EFR network for studies of a variety of hydrological processes and their interactions in different environments, which is not possible at individual sites or using short-term studies. This variability across sites will also be critical in future studies for process-level, statistical and modelling research relating to impacts, vulnerability and risk assessments of climate and land-use change, and forest disturbance on hydrology, biogeochemistry and water supply. These reference watersheds also continue to be important for use in paired watershed studies to evaluate effects of disturbances such as forest harvesting, prescribed burning, devegetation, changes in forest structure and species composition, fertilization and other land management practices on water yield, evapotranspiration, flowpath routing, nutrient cycling and sediment transport. Indeed, the research is being used to chart long-term effects and the data collected have been essential for cross-site syntheses (e.g. Kolka *et al.*, 2011; Jones *et al.*, 2012; Creed *et al.*, 2014; Gottfried *et al.*, 2014; Vose *et al.*, 2014). However, additional studies are also warranted to examine consistency of these long-term data and results from the reference watersheds used in various hydrological analyses herein and elsewhere for their potential deviation, if any, due to unforeseen external factors including climate change (Alila *et al.*, 2009; Ali *et al.*, 2015).

Therefore, there is a critical need for continued monitoring of these long-term watersheds, as they are well suited for documenting and detailing baseline hydrological conditions and also serve as valuable benchmarks for addressing emerging forest and water issues of the 21st century.

References

Adams, M.B., Kochenderfer, J.N., Wood, F., Angradi, T.R. and Edwards, P.J. (1994) *Forty Years of Hydrometeorological Data from the Fernow Experimental Forest*. General Technical Report NE-184. USDA Forest Service, Northeastern Forest Experiment Station, Radnor, Pennsylvania.

Alexander, R., Troendle, C., Kaufmann, M., Shepperd, W., Crouch, G. and Watkins, R. (1985) *The Fraser Experimental Forest, Colorado: Research Program and Published Research 1937–1985*. General Technical Report GTR-RM-118. USDA Forest Service, Rocky Mountain Research Station, Fort Collins, Colorado.

Ali, G., Tetzlaff, D., McDonnell, J.J., Soulsby, C., Carey, S., Laudon, H., McGuire, K., Buttle, J., Seibert, J. and Shanley, J. (2015) Comparison of threshold hydrologic response across northern catchments. *Hydrological Processes* 29, 3575–3591.

Alila, Y.P., Kuras, K., Schnorbus, M. and Hudson, R. (2009) Forests and floods: a new paradigm sheds light on age-old controversies. *Water Resources Research* 45, W08416, doi: 10.1029/2008WR007207 (accessed 9 April 2016).

Amatya, D.M. and Jha, M.K. (2011) Evaluating SWAT model for a low gradient forested watershed in coastal South Carolina. *Transactions of the ASABE* 54, 2151–2163.

Amatya, D.M., Miwa, M., Harrison, C.A., Trettin, C.C. and Sun, G. (2006) *Hydrology and Water Quality of Two First Order Forested Watersheds in Coastal South Carolina*. ASABE Paper #06-2182. American Society of Agricultural and Biological Engineers, St Joseph, Michigan.

Amoah, J., Amatya, D.M. and Nnaji, S. (2012) Quantifying watershed depression storage: determination and application in a hydrologic model. *Hydrological Processes* 27, 2401–2413.

Argerich, A., Johnson, S.L., Sebestyen, S.D., Rhoades, C.C., Greathouse, E., Knoepp, J.D., Adams, M.B., Likens, G.E., Campbell, J.L. and McDowell, W.H. (2013) Trends in stream nitrogen concentrations for forested reference catchments across the USA. *Environmental Research Letters* 8, 014039.

Arora, V.K. and Boer, G.J. (2001) Effects of simulated climate change on the hydrology of major river basins. *Journal of Geophysical Research* 106, 3335–3348.

Bay, R.R. (1967) Ground water and vegetation in two peat bogs in northern Minnesota. *Ecology* 48, 308–310.

Bolton, W.R., Hinzman, L.D. and Yoshikawa, K. (2004) Water balance dynamics of three small catchments in a sub-Arctic boreal forest. In: Kane, D.L. and Yang, D. (eds) *Northern Research Basins Water Balance (Proceedings of a workshop held Victoria, BC, 15–19 March 2004)*. IAHS Publication No. 290. International Association of Hydrological Sciences, Wallingford, UK, pp. 213–223.

Boring, L.R., Elliott, K.J. and Swank, W.T. (2014) Successional forest dynamics 30 years following clearcutting. In: Swank, W.T. and Webster, J.R. (eds) *Long-Term Response of a Forest Watershed Ecosystem. Clearcutting in the Southern Appalachians.* Oxford University Press, New York, pp. 11–35.

Bosch, J.M. and Hewlett, J.D. (1982) A review of catchment experiments to determine the effect of vegetation changes on water yield and evapotranspiration. *Journal of Hydrology* 55, 3–23.

Brown, A.E., Zhang, L., McMahon, T.A., Western, A.W. and Vertessy, R.A. (2005) A review of paired catchment studies for determining changes in water yield resulting from alterations in vegetation. *Journal of Hydrology* 310, 28–61.

Brown, P.J., Bradley, R.S. and Keimig, F.T. (2010) Changes in extreme climate indices for the northeastern US, 1870–2005. *Journal of Climate* 23, 6555–6572.

Cafferata, P.H. and Reid, L.M. (2013) *Application of Long-term Watershed Research to Forest Management in California: 50 Years of learning from Caspar Creek Experimental Watersheds*. California Forestry Report No. 5. State of California Natural Resources Agency, Department of Forestry and Fire Protection, Sacramento, California.

Campbell, J.L., Driscoll, C.T., Pourmokhtarian, A. and Hayhoe, K. (2011) Streamflow responses to past and projected future changes in climate at the Hubbard Brook Experimental Forest, New Hampshire, USA. *Water Resources Research* 47, W02514, doi:10.1029/2010WR009438 (accessed 9 April 2016).

Creed, I.F., Spargo, A.T., Jones, J.A., Buttle, J.M., Adams, M.B., Beall, F.D., Booth, E.G., Campbell, J.L., Clow, D., Elder, K., Green, M.B., *et al.* (2014) Changing forest water yields in response to climate warming: results from long-term experimental watershed sites across North America. *Global Change Biology* 20, 3191–3208.

Czwartacki, B.J. (2013) Time and tide: understanding the water dynamics in a tidal freshwater forested wetland. MSc. thesis, College of Charleston, Charleston, South Carolina.

Dai, Z., Li, C., Trettin, C.C., Sun, G., Amatya, D.M. and Li, H (2010) Bi-criteria evaluation of the MIKE SHE model for a forested watershed on the South Carolina coastal plain. *Hydrology and Earth System Science* 14, 1033–1046.

Dai, Z., Trettin, C.C. and Amatya, D.M. (2013) *Effects of Climate Variability on Forest Hydrology and Carbon Sequestration on the Santee Experimental Forest in Coastal South Carolina*. General Technical Report SRS-172. USDA Forest Service, Southern Research Station, Asheville, North Carolina.

DeBano, L.F. (1981) *Water Repellent Soils: A State-of-the-Art*. General Technical Report PSW-46. USDA Forest Service, Pacific Southwest Forest and Range Experiment Station, Berkeley, California.

Detty, J.M. and McGuire, K.J. (2010a) Topographic controls on shallow groundwater dynamics: implications of hydrologic connectivity between hillslopes and riparian zones in a till mantled catchment. *Hydrological Processes* 24, 2222–2236.

Detty, J.M. and McGuire, K.J. (2010b) Threshold changes in storm runoff generation at a till mantled headwater catchment. *Water Resources Research* 46, W07525, doi: 10.1029/2009WR008102 (accessed 9 April 2016).

DeWalle, D.R., Edwards, P.J., Swistock, B.R., Aravena, R. and Drimmie, R.J. (1997) Seasonal isotope hydrology of three of three Appalachian forest catchments. *Hydrological Processes* 11, 1895–1906.

Douglass, J.E. and Hoover, M.D. (1988) History of Coweeta. In: Swank, W.T. and Crossley, D.A. (eds) *Forest Hydrology and Ecology at Coweeta*. Ecological Studies Vol. 66. Springer, New York, pp. 17–31.

Dunn, P.H., Barro, S.C., Wells, W.G. II, Poth, M.A., Wohlgemuth, P.M. and Colver, C.G. (1988) *The San Dimas Experimental Forest: 50 Years of Research*. General Technical Report PSW-105. USDA Forest Service, Pacific Southwest Forest and Range Experiment Station, Berkeley, California.

Epps, T., Hitchcock, D., Jayakaran, A.D., Loflin, D., Williams, T.M. and Amatya, D.M. (2013) Characterization of storm flow dynamics of headwater streams in the South Carolina lower coastal plain. *Journal of the American Water Resources Association* 49, 76–89.

Ford, C.R., Hubbard, R.M. and Vose, J.M. (2011) Quantifying structural and physiological controls on variation in canopy transpiration among planted pine and hardwood species in the southern Appalachians. *Ecohydrology* 4, 183–195.

Gannon, J.P., Bailey, S.W. and McGuire, K.J. (2014) Organizing groundwater regimes and response thresholds by soils: a framework for understanding runoff generation in a headwater catchment. *Water Resources Research* 50, 8403–8419.

Gillin, C.P., Bailey, S.W., McGuire, K.J. and Gannon, J.P. (2015) Mapping of hydropedologic spatial patterns in a steep headwater catchment. *Soil Science Society of America Journal* 79, 440–453.

Gottfried, G.J., Ffolliott, P.F., Brooks, K.N., Kolka, R.K., Raish, C.B. and Neary, D.G. (2014) Contributions of studies on experimental forests to hydrology and watershed management. In: Hayes, D. (ed.) *Long-Term Research on Experimental Forests and Ranges*. USDA Forest Service, Washington, DC, pp. 387–403.

Griffin, M.P., Callahan, T.J., Vulava, V.M. and Williams, T.M. (2014) Storm-event flow pathways in lower coastal plain forested watersheds of the southeastern United States. *Water Resources Research* 50, 8365–8280.

Hamon, W.R. (1963) Computation of direct runoff amounts from storm rainfall. *International Association of Scientific Hydrology Publication* 63, 52–62.

Harder, S.V., Amatya, D.M., Callahan, T.J., Trettin, C.C. and Hakkila, J. (2007) A hydrologic budget of a first-order forested watershed, coastal South Carolina. *Journal of the American Water Resources Association* 43(3), 1–13.

Haugen, R.K., Slaughter, C.W., Howe, K.E. and Dingman, S.L. (1982) *Hydrology and Climatology of the Caribou-Poker Creeks Research Watershed, Alaska*. CRREL Report 82-26. US Army Corps of Engineers, Washington, DC.

Hinzman, L.D., Nobuyoshi, I., Yoshikawa, K., Bolton, W.R. and Petrone, K.C. (2002) Hydrologic studies in Caribou-Poker Creeks Research Watershed in support of long term ecological research. *Eurasian Journal Forest Research* 5(2), 67–71.

Hewlett, J.D. and Hibbert, A.R. (1963) Moisture and energy conditions within a sloping soil mass during drainage. *Journal of Geophysical Research* 68, 1081–1087.

Hewlett, J.D. and Hibbert, A.R. (1967) Factors affecting the response of small watersheds to precipitation in humid areas. In: Sopper, W.E. and Lull, H.W. (eds) *Forest Hydrology*. Pergamon Press, New York, pp. 275–291.

Hook, D.D., Buford, M.A. and Williams, T.M. (1991) Impact of Hurricane Hugo on the South Carolina coastal pine forest. *Journal of Coastal Research* (SI-8), 291–300.

Jayakaran, A., Williams, T.M., Ssegane, H.S., Amatya, D.M., Song, B. and Trettin, C. (2014) Hurricane impacts on a pair of coastal forested watersheds: implications of selective hurricane damage to forest structure and streamflow dynamics. *Hydrology and Earth System Science* 18, 1151–1164.

Jones, J.A., Achterman, G.L., Augustine, LA., Creed, I.F., Ffolliott, P.F., MacDonald, L. and Wemple, B.C. (2009) Hydrologic effects of a changing forested landscape – challenges for the hydrological sciences. *Hydrological Processes* 23, 2699–2704.

Jones, J.A., Creed, I.F., Hatcher, K.L., Warren, R.J., Adams, M.B., Benson, M.H., Boose, E., Brown, W.A., Campbell, J.L., Covich, A., *et al.* (2012) Ecosystem processes and human influences regulate streamflow response to climate change at long-term ecological research sites. *BioScience* 62, 390–404.

Jones, J.B. and Rinehart, A.J. (2010) The long term response of stream flow to climatic warming in headwater streams of interior Alaska. *Canadian Journal of Forest Research* 40, 1210–1218.

Kane, D.L. and Hinzman, L.D. (2004) Monitoring extreme environments: arctic hydrology in transition. *Water Resources Impact* 6, 24–27.

Kasahara, T. and Wondzell, S.M. (2003) Geomorphic controls on hyporheic exchange flow in mountain streams. *Water Resources Research* 39, 1005, doi: 10.1029/2002WR001386 (accessed 10 April 2016).

Keppeler, E. (2007) *Effects of Timber Harvest on Fog Drip and Streamflow, Caspar Creek Experimental Watersheds, Mendocino County, California*. General Technical Report PSW GTR-194. USDA Forest Service, Pacific Southwest Research Station, Albany, California, pp. 85–93.

Keppeler, E.T. and Brown, D. (1998) *Subsurface Drainage Processes and Management Impacts*. General Technical Report PSW GTR-168. USDA Forest Service, Pacific Southwest Research Station, Albany, California, pp. 25–34.

Klaus, J., McDonnell, J.J., Jackson, C.R., Du, E. and Griffiths, N.A. (2015) Where does streamwater come from in low-relief forested watersheds? A dual-isotope approach. *Hydrology and Earth System Science* 19, 125–135.

Kolka, R.K., Sebestyen, S.D. and Bradford, J.H. (2011) An evolving research agenda of the Marcell Experimental Forest. In: Kolka, R.K., Sebestyen, S.D., Verry, E.S. and Brooks, K.N. (eds) *Peatland Biogeochemistry and Watershed Hydrology at the Marcell Experimental Forest*. CRC Press, Boca Raton, Florida, pp. 73–91.

Laseter, S.H., Ford, C.R., Vose, J.M. and Swift, L.W. (2012) Long-term temperature and precipitation trends at the Coweeta Hydrologic Laboratory, Otto, North Carolina, USA. *Hydrology Research* 43, 890–901.

La Torre Torres, I., Amatya, D.M., Callahan, T.J. and Sun, G. (2011) Seasonal rainfall–runoff relationships in a lowland forested watershed in the southeastern USA. *Hydrological Processes* 25, 2032–2045.

Love, L. (1960) *The Fraser Experimental Forest … Its Work and Aims*. Station Paper No. 8. USDA Forest Service, Rocky Forest and Range Experiment Station, Fort Collins, Colorado.

McGuire, K.J., McDonnell, J.J., Weiler, M., Kendall, C., McGlynn, B.L., Welker, J.M. and Seibert, J. (2005) The role of topography on catchment-scale water residence time. *Water Resources Research* 41, W05002, doi: 10.1029/2004WR003657 (accessed 10 April 2016).

Meixner, T. and Wohlgemuth, P.M. (2003) Climate variability, fire, vegetation recovery, and watershed hydrology. In: Renard, K.G., McElroy, S.A., Gburek, W.J., Canfield, H.E. and Scott, R.L. (eds) *The First Interagency Conference on Research in the Watersheds, October 27–30, 2003, Benson, Arizona.* US Department of Agriculture, Agricultural Research Service, Washington, DC, pp. 651–656.

Nichols, D.S. and Verry, E.S. (2001) Stream flow and ground water recharge from small forested watersheds in north central Minnesota. *Journal of Hydrology* 245, 89–103.

Patric, J.H. (1973) *Deforestation Effects on Soil Moisture, Streamflow, and Water Balance in the Central Appalachians*. Research Paper No. NE-259. USDA Forest Service, Northeastern Forest Experiment Station, Upper Darby, Pennsylvania.

Patric, J.H. and Goswami, N. (1968) Evaporation pan studies – forest research at Parsons. *West Virginia Agriculture and Forestry* 1(4), 6–10.

Peel, M.C., Finlayson, B.A. and McMahon, T.A. (2007) Updated world map of the Köppen–Geiger climate classification. *Hydrology and Earth System Science* 11, 1633–1644.

Post, D.A. and Jones, J.A. (2001) Hydrologic regimes of forested, mountainous, headwater basins in New Hampshire, North Carolina, Oregon, and Puerto Rico. *Advances in Water Resources* 24, 1195–1210.

Reid, L.M. and Lewis, J. (2009) Rates, timing, and mechanisms of rainfall interception loss in a coastal redwood forest. *Journal of Hydrology* 375, 459–470.

Reinhart, K.G., Eschner, A.R. and Trimble, G.R. (1963) *Effect on Streamflow of Four Forest Practices in the Mountains of West Virginia*. Research Paper No. NE-1. USDA Forest Service, Northeastern Forest Experiment Station, Upper Darby, Pennsylvania.

Retzer, J. (1962) *Soil Survey: Fraser Alpine Area, Colorado.* Soil Survey Series 1956, No. 20. USDA Forest Service, Soil Conservation Service, Colorado Agricultural Experiment Station, Washington, DC.

Rice, R.M. (1974) The hydrology of chaparral watersheds. In: Rosenthal, M. (ed.) *Proceedings of the Symposium on Living with the Chaparral, March 30–31, 1973, Riverside, California.* Sierra Club, San Francisco, California, pp. 27–34.

Riggan, P.J., Lockwood, R.N. and Lopez, E.N. (1985) Deposition and processing of airborne nitrogen pollutants in Mediterranean-type ecosystems in southern California. *Environmental Science and Technology* 19, 781–789.

Rothacher, J., Dyrness, C.T. and Fredriksen, R.L. (1967) Hydrologic and Related Characteristics of Three Small Watersheds in the Oregon Cascades. USDA Forest Service, Pacific Northwest Forest and Range Experiment Station, Portland, Oregon. Available at: http://andrewsforest.oregonstate.edu/pubs/pdf/pub344.pdf (accessed 14 April 2016).

Searcy, J.K. (1959) Flow-duration curves. In: *Manual of Hydrology: Part 2: Low-Flow Techniques*. US Geological Survey Water-Supply Paper 1542-A. US Government Printing Office, Washington, DC.

Sebestyen, S.D., Dorrance, C., Olson, D.M., Verry, E.S., Kolka, R.K., Elling, A.E. and Kyllander, R. (2011) Long-term monitoring sites and trends at the Marcell Experimental Forest. In: Kolka, R.K., Sebestyen, S.D., Verry, E.S. and Brooks, K.N. (eds) *Peatland Biogeochemistry and Watershed Hydrology at the Marcell Experimental Forest*. CRC Press, Boca Raton, Florida, pp. 15–71.

Spence, C. (2010) A paradigm shift in hydrology: storage thresholds across scales influence catchment runoff generation. *Geography Compass* 47, 819–833.

Sun, G., McNulty, S.G., Amatya, D.M., Skaggs, R.W., Swift, L.W. Jr, Shepard, J.P. and Riekerk, H. (2002) A comparison of the watershed hydrology of coastal forested wetlands and the mountainous uplands in the southern US. *Journal of Hydrology* 263, 92–104.

Swift, L.W., Cunningham, G.B. and Douglass, J.E. (1988) Climate and hydrology. In: Swank, W.T. and Crossley, D.A. (eds) *Forest Hydrology and Ecology at Coweeta*. Ecological Studies Vol. 66. Springer, New York, pp. 35–55.

Thornthwaite, C.W. (1948) An approach toward a rational classification of climate. *The Geographical Review* 38, 55–94.

Troendle, C. and King, R. (1985) The effect of timber harvest on the Fool Creek watershed, 30 years later. *Water Resources Research* 21, 1915–1922.

Verry, E.S., Brooks, K.N., Nichols, D.S., Ferris, D.R. and Sebestyen, S.D. (2011) Watershed hydrology. In: Kolka, R.K., Sebestyen, S.D., Verry, E.S. and Brooks, K.N. (eds) *Peatland Biogeochemistry and Watershed Hydrology at the Marcell Experimental Forest*. CRC Press, Boca Raton, Florida, pp. 193–212.

Vogel, R.M. and Fennessey, N.M. (1995) Flow duration curves II: a review of applications in water resources planning. *Journal of the American Water Resources Association* 31, 1029–1039.

Vose, J.M., Ford, C.R., Laseter, S.H., Dymond, S.F., Sun, G., Adams, M.B., Sebestyen, S.D., Campbell, J.L., Luce, C., Amatya, D.M., *et al.* (2012) Can forest watershed management mitigate climate change effects on water resources? In: Webb, A.A., Bonell, M., Bren, L., Lane, P.J.N., McGuire, D., Neary, D.J., Nettles, J., Scott, D.F., Stednick, J. and Wang, Y. (eds) *Revisiting Experimental Catchment Studies in Forest Hydrology (Proceedings of a Workshop held during the XXV IUGG General Assembly in Melbourne, June–July 2011)*. IAHS Publication No. 353. International Association of Hydrological Sciences, Wallingford, UK, pp. 12–25.

Vose, J.M., Swank, W.T., Adams, M.B., Amatya, D., Campbell, J., Johnson, S., Swanson, F.J., Kolka, R., Lugo, A., Musselman, R. and Rhoades, C. (2014) The role of experimental forests and ranges in the development of ecosystem science and biogeochemical cycling research. In: Hayes, D. (ed.) *Long-Term Research on Experimental Forests and Ranges*. USDA Forest Service, Washington, DC, pp. 387–403.

Weiler, M. and McDonnell, J. (2004) Virtual experiments: a new approach for improving process conceptualization in hillslope hydrology. *Journal of Hydrology* 285, 3–18.

Wohlgemuth, P.M. (2016) Long-term hydrologic research on the San Dimas Experimental Forest: lessons learned and future directions. In: Stringer, C.E., Krauss, K.W. and Latimer, J.S. (eds) *Headwaters to Estuaries: Advances in Watershed Science and Management – Proceedings of the Fifth Interagency Conference on Research in the Watersheds, March 2–5, 2015, North Charleston, South Carolina*. e-General Technical Report SRS-211. USDA Forest Service, Southern Research Station, Asheville, North Carolina, pp. 227–232.

Zegre, N. (2008) The history and advancement of change detection methods in forest hydrology. In: *Society of American Foresters Water Resources Working Group Newsletter.* Society of American Foresters, Bethesda, Maryland.

Ziemer, R.R. (1998) Stormflow response to roadbuilding and partial cutting in small streams of northern California. *Water Resources Research* 17, 907–917.

15 Applications of Forest Hydrological Science to Watershed Management in the 21st Century

J.M. Vose[1]*, K.L. Martin[1] and P.K. Barten[2]

[1]*USDA Forest Service, Raleigh, North Carolina, USA and North Carolina State University, Raleigh, North Carolina, USA;* [2]*University of Massachusetts Amherst, Amherst, Massachusetts, USA*

15.1 Introduction

The remainder of the 21st century will present significant challenges for forest watershed management, as rapid and compounded environmental, economic and social change contribute to an increasingly uncertain future. Many of these changes portend a growing risk of water scarcity for a growing human population and greater vulnerability to extreme droughts and more intense storms. Forest hydrological science has a strong tradition of providing the information required for restoring and managing disturbed and stressed landscapes, positioning the field well to guide management to provide essential ecosystem services in the future. Indeed, the origins of the establishment of public forest lands for the protection of watersheds in the USA under the Weeks Act of 1911 (http://www.foresthistory.org/ASPNET/Policy/WeeksAct/index.aspx, accessed 10 April 2016) reflected a strong recognition of the role of forests in regulating water supply and providing high-quality surface water for aquatic ecosystems and human consumption. Knowledge gained from watershed research and management experience over the past century provides a solid foundation to prepare for future management decisions; however, it is not clear if the past alone will serve as an adequate model. Rates of contemporary landscape modification, climate change and altered disturbance regimes are unprecedented and few watershed ecosystems remain beyond the influence of human activity (e.g. Likens, 2001; Seastedt *et al.*, 2008; Hobbs *et al.*, 2009). Hydrological cycles have already been altered and changes will continue as climate change, population growth, water diversion and numerous other environmental changes continue (Huntington, 2006; Naiman, 2013). At the same time, there are societal expectations that watershed ecosystems can be managed to maintain functional states (Naiman, 2013). An assessment of how forest hydrology can be applied, adapted and expanded to address these challenges is critical for ensuring that water-based ecosystem services can be sustained in the future.

In this chapter we examine the role of forest hydrological science in the development and application of watershed management in the 21st century. We provide a brief synthesis of anticipated biophysical and socio-economic changes expected to occur over the coming decades and discuss critical watershed science needs and management responses to maintain watershed

*Corresponding author; e-mail: jvose@fs.fed.us

 Forest Hydrology: Processes, Management and Assessment (eds D.M. Amatya, T.M. Williams, L. Bren and C. de Jong)

ecosystem services in the coming decades. We build on several recent discussions (e.g. National Research Council, 2008; Riveros-Iregui *et al.*, 2011; Vose *et al.*, 2012; Wang *et al.*, 2012; Egginton *et al.*, 2014) on the role of ecohydrology in addressing water resource challenges now and in the future. We focus our examples on forest watersheds in the southern US forests, as the complex mixture of public and private forest land ownership creates substantial challenges for watershed management at larger spatial scales. Despite the focus on the southern USA, the general principles are applicable to forest watersheds across the globe.

The last century of forest watershed research has provided a fundamental understanding of watershed hydrological processes and best management practices (BMPs) that either protect or restore these processes when watersheds are managed. The state-of-the-knowledge on forest watershed science has been summarized by Ice and Stednick (2004) for the continental USA, Lockaby *et al.* (2013) for the southern USA and de la Crétaz and Barten (2007) for the north-eastern USA. These summaries provide the following five key lessons for watershed management:

1. Forests provide the cleanest and most stable flows of surface water and groundwater recharge among all land uses.
2. Flow amount (water yield) and timing can be altered by forest management; flows can increase or decrease depending upon post-disturbance successional patterns.
3. Nutrient levels in forested watersheds are generally low; however, sediment loading can increase when disturbance results in erosion and sediment delivery.
4. Riparian areas and forested wetlands are especially important for regulating flows and protecting water quality.
5. The implementation of BMPs is critical for ensuring that forests can be managed to avoid or minimize adverse effects on water resources.

A recent National Research Council (2008) review assessed the applicability of these cornerstones and concluded that detailed understanding of hydrological processes and land-use effects at the experimental watershed scale is strong for comparatively short time periods (i.e. 5 to 15 years), but our understanding diminishes rapidly as spatial and temporal scales increase (e.g. to eco-regions, multiple decades). Hence, in this chapter we ask: (i) how will large-scale changes in land use and long-term changes in climatic conditions affect our ability to formulate and implement watershed management policies, plans and practices; and (ii) what new research questions and approaches will be needed to address critical information gaps and uncertainty? We address these questions by focusing on how our current understanding of forest watershed responses to management practices can be applied to sustain water resources and what new management approaches might be required. We consider the shift from a forest management philosophy in the eastern USA of avoiding water quality impacts to a more comprehensive view of forests as a vitally important land cover and land use required to sustain aquatic ecosystems, water supplies and public health (*sensu* Postel and Richter, 2003).

15.2 Biophysical and Socio-Economic Changes Expected to Occur Over the Coming Decades

Changes in earth systems including atmospheric chemistry, nutrient and hydrological cycling resulting from human activities are significant enough to define a new geological epoch, the Anthropocene. Marking the end of the Holocene, the most recent 10,000- to 12,000-year interglacial period, there is some debate as to whether the Anthropocene began circa 1800 with the Industrial Revolution, in the post-war era of the 1950s, or about using those dates as the beginning of two stages (Steffen *et al.*, 2007). This is because human effects on atmospheric chemistry can be traced back to initial industrialization and the associated use of fossil fuels, beginning with coal-powered steam engines. By 1950, the concentration of atmospheric CO_2 had increased to 310 ppm from pre-industrial levels of 270–275 ppm (Steffen *et al.*, 2007). During the second half of the 20th century, the world population doubled and became more urbanized, global economic activity increased 15-fold and anthropogenic sources of reactive nitrogen (fertilizers, fossil fuel combustion) surpassed the sum of all natural production (Steffen *et al.*, 2007). Atmospheric CO_2 has surpassed 400 ppm and is increasing at an accelerating rate, accumulating an additional 2.25 ppm/year today – compared with 0.75 ppm/year in 1959 (Field *et al.*, 2014).

The Anthropocene will continue to be an era of significant and rapid change as the primary driving factors, human population growth and development, continue to accelerate. By 1950, temperate broadleaf and mixed forests covered less than half the earth surface capable of supporting this biome, and by 2050 it is estimated that another 10% will be lost (Millennium Ecosystem Assessment, 2005). This is due in part to an expected world population of 9.5 billion by 2050, a 36% increase from 2010 (United Nations, 2013). As the population grows, urban growth and development will continue as, by 2050, 66% of the world population is expected to live in urban areas, a 12% increase from 2014 (United Nations, 2014). In the Southeast USA, 12–17 million ha of additional development are expected by 2060, which represents at least twice the present area of urban land cover (Wear, 2013). Urban growth will be particularly concentrated in the Southern Appalachian Piedmont, creating a connected urban corridor from Richmond, Virginia through Raleigh–Durham, North Carolina to Atlanta, Georgia (Wear and Greis, 2013; Terrando *et al.*, 2014). The increasing population will result in increased water demand. If current patterns of development continue across the region, the effects of urbanization will be exacerbated by low-density development ('sprawl') that increases the connectivity of developed areas while fragmenting and isolating natural areas (Terrando *et al.*, 2014). This translates to a loss of between 7 and 13% of regional forestland across the Southeast, with losses up to 21% in the Southern Appalachian Piedmont subsection (Wear, 2013). As forest is replaced by urban uses, concentrations of sediment, nutrients, pollutants and pathogens all increase and degrade water quality (Lockaby *et al.*, 2013). Population growth and development also affect water availability; by 2050, water stress (defined as human demand divided by water supply) is expected to increase by 10% across the Southeast. As forest is lost, forest types are also expected to shift, with planted pine replacing much of the remaining natural pine (Huggett *et al.*, 2013).

In addition to development and rapid land-use change, the Anthropocene is an era of rapid climate change. Global average temperatures are estimated to have risen by 0.65–1.06°C between 1880 and 2012 (Field *et al.*, 2014). This trend is expected to continue, with an additional increase of up to 4.8°C in global average temperature by the end of the century (Field *et al.*, 2014). In the Southeast, average temperatures have increased by just over 1°C since 1970, with greater increases during the summer (Carter *et al.*, 2014). In the near future, the Southeast is expected to have a more variable climate with temperatures increasing by approximately 2 to 4°C and more days exceeding 35°C by the end of the century (McNulty *et al.*, 2013). Precipitation forecasts are more variable and while some models suggest minimal change, this could be an artefact of the regional position between the Southwest, where precipitation is expected to decrease, and the Northeast, where precipitation is expected to increase (Carter *et al.*, 2014). Even with the uncertainty of precipitation models, greater evaporative loss from increased temperatures may increase water stress. Most general circulation models predict that the frequency of extreme precipitation events will increase worldwide as the climate warms (O'Gorman and Schneider, 2009), likely increasing the magnitude and frequency of both flood events (overbank flow) and drought (both meteorological and hydrological). Many regions of the USA have recorded an increased frequency of precipitation extremes (i.e. more droughts and larger, high-intensity rainfall events) during the last 50 years (Easterling *et al.*, 2000; Huntington, 2006; Field *et al.*, 2014). The timing and spatial distribution of extreme or low-probability events are among the most uncertain aspects of future climate scenarios. Forecasts are complicated by natural variability of inter- and intra-annual precipitation across the continental USA related to large-scale global climate teleconnections (e.g. El Niño Southern Oscillation, Pacific Decadal Oscillation, North Atlantic Oscillation) (Karl *et al.*, 1995; Allen and Ingram, 2002).

Extreme precipitation events are not the only sources of future uncertainty and variation; novel and compounded disturbances are expected to accelerate in the future. For example, climate change is expected to increase fire activity (Marlon *et al.*, 2008). Increasing temperatures and the resulting drier conditions will also increase wildfire risk, in part because of extended fire seasons. By 2060 in the Southeast USA, climate change is expected to increase the frequency and intensity of wildfires and extend fire seasons by up to 3 months (Liu *et al.*, 2013). Large, high-intensity wildfires are uncommon in the region because of the effectiveness of long-term prescribed fire, fuel load reduction and wildfire suppression programmes. Hotter, drier conditions could limit the number of days that meet the criteria for controlled burning, thereby limiting the opportunity for fuel load reduction and proactive fire management (Melvin, 2012; Liu *et al.*, 2013; Mitchell *et al.*, 2014). Further, under extreme future conditions, some fires will likely burn at high intensity regardless of prescribed fire management. This was the case in 2007, when the Georgia Bay Complex fires burned as crown fires through the Osceola National Forest, even in stands that had been treated with prescribed fire in the previous five years (Fites *et al.*, 2007). Severe wildfires can cause increased runoff and erosion by removing litter and duff layers, altering soil permeability and reducing evapotranspiration because of high tree mortality (Ice *et al.*, 2004; Certini, 2005; Doerr *et al.*, 2006). Future fire risk is also dependent upon future fuel dynamics, of which pests, diseases and invasive species are an increasingly important component. Pests and diseases amplify fire risk by causing mortality and, thus, increasing fuel loads. This is also true of some invasive species. For example, cogongrass (*Imperata cylindrica*) is a highly flammable invasive spreading rapidly throughout the Southeast (Bradley *et al.*, 2010).

In the Southeast, 9% of the forest contains at least one invasive species, with an annual spread rate exceeding 58,000 ha (Miller *et al.*, 2013). Many successful plant invaders are rapidly growing species with high evapotranspiration rates that increase water fluxes. In the western USA, *Tamarix* invasions have increased transpirational water fluxes and, in consequence, decreased streamflow (Ehrenfeld, 2010). Although invasive plants with such a substantial effect are not yet present in the Southeast, the invasion of hemlock woolly adelgid (*Adelges tsugae*) alters hydrological cycling by removing eastern hemlock (*Tsuga canadensis*), an evergreen riparian tree (Ford and Vose, 2007; Brantley *et al.*, 2015). The effects of invasive species on ecosystem structure and function are increasingly well documented (Lovett *et al.*, 2006; Ehrenfeld, 2010). However, less is understood about how novel or hybrid communities of multiple invasive species affect ecosystem functions (Hobbs *et al.*, 2006, 2009).

Novel ecosystems are increasing across landscapes as consequences of the Anthropocene, where nearly all ecosystems are affected by human activity. As defined by Hobbs *et al.* (2006), novel ecosystems are characterized by species combinations that have not previously occurred in a biome, are a result of either direct or indirect human actions, and have become self-perpetuating. Many of these ecosystems are so different from earlier successional patterns and species assemblages that restoration efforts are unlikely or very difficult. It is, therefore, more likely that urban and suburban areas will, in some cases, need to be managed as novel or hybrid ecosystems to maintain ecological services, including water resources (Hobbs *et al.*, 2014).

15.3 Management Responses to Maintain Watershed Ecosystem Services in the 21st Century

While we expect that many of the principles of forest hydrological science derived from the previous century will continue to be highly relevant and applicable for the remainder of the 21st century, we propose that the rapid pace and scale of biophysical and socio-economic changes expected over the coming decades will require a combination of modified and new management approaches to maintain ecosystem services. For example, modifications of current BMPs to address greater precipitation variability might include wider riparian buffers, larger culverts at road crossings, and more efficient and stable road design. The need for new management approaches is driven in large part by the growing demand for freshwater. Water derived from forests has always been considered a valuable ecosystem service and watershed protection to maintain water quality was a primary focus. In the future, increasing demand for freshwater will likely place a greater emphasis on managing forests for water yield. Large-scale management may be necessary to meet the needs of an increasingly urbanized landscape. In the following section, we discuss two critical areas where forest hydrological science will need to advance to inform and support management responses (Vose *et al.*, 2012).

15.3.1 Managing species composition and stand structure to optimize water yield

The potential impacts of increased climate variability such as droughts and heavy rainfall events will be determined by the balance between precipitation (*P*) inputs versus tree water demand (potential evapotranspiration or *PET*) in the future. For example, forested areas in the arid Southwest are characterized by low *P*/*PET* ratios (<1), forested areas in the Northeast and Northwest are characterized by high *P*/*PET* ratios (>1), and large areas in the South USA have *P*/*PET* ratios near unity (Plate 13). Current ecological, socio-economic and watershed management systems have evolved in response to this balance between precipitation and potential evapotranspiration. In areas where precipitation greatly exceeds potential evapotranspiration, water is generally abundant, mesic species are favoured, and the management focus is on flooding and water quality protection. In contrast, in areas where potential evapotranspiration greatly exceeds precipitation, water is limiting, xeric species are favoured, and the management focus is on managing dry periods and associated disturbances such as wildfire and on developing reliable water supply sources for agriculture and human needs. Future scenarios suggest it is likely that at mid-latitudes, wet areas will generally get wetter and the dry areas will generally get drier (Field *et al.*, 2014). However, even if overall precipitation does not change, higher air temperatures will amplify the effects of droughts when they occur (Breshears *et al.*, 2005). In the southern USA, the high diversity of tree species and the ability to actively manage forests over much of the landscape provide a unique opportunity to develop or refine optimal watershed management strategies to protect water quality and, potentially, to increase or sustain water yields.

Forests in the eastern USA are changing and these changes can affect water yield, quality and timing. Many areas of the Southeast have gained additional forested area over the 20th century as agricultural land use declined, and evapotranspiration has likely increased as a result (Kim *et al.*, 2014). In addition, forest species transitions have occurred due to purposeful management activities such as the establishment of plantation forests, but also due to successional processes and altered disturbance regimes. For example, from the early part of the 20th century, species composition in oak and oak–pine forests has transitioned throughout the eastern USA, a process that has been characterized as mesophication (Nowacki and Abrams, 2015). This term is used to describe a shift in dominance away from species that tended to thrive in more xeric conditions with shorter fire rotations (e.g. thick-barked oak species). There are many potential factors that have contributed to this change, including wetter conditions, fire suppression and the maturation of much of the forest following widespread harvests during the 20th century (McEwan *et al.*, 2011; Nowacki and Abrams, 2008, 2015).

This change in species composition alters vulnerability to drought and the relative magnitude of water balance components through changes in evapotranspiration, both in terms of interception and transpiration (Zhang *et al.*, 2001). Physical canopy architecture, tree height and duration (evergreen versus deciduous) all affect interception rates (Calder *et al.*, 2003). Shorter trees have higher interception rates than taller trees of the same species, and evergreen species tend to have higher interception rates compared with deciduous species (Rutter *et al.*, 1975; Calder *et al.*, 2003; Ford *et al.*, 2011). In the Southeast, Ford *et al.* (2011) found that interception was almost twice as high in plantation pine stands (*Pinus strobus*) compared with mixed hardwood stands. Ford *et al.* (2011) also found that transpiration has a greater effect on evapotranspiration than interception, and transpiration is particularly important in dry years. Xylem anatomy and the resulting sapwood area are important determinants of stand transpiration (Wullschleger *et al.*, 2001; Ford *et al.*, 2007). Mesophytic species are typically diffuse porous and have greater sapwood area than ring- or semi-ring porous species. As sapwood area increases, potential water transport increases (Enquist *et al.*, 1998; Meinzer *et al.*,

2005). For example, transpiration rates for a given diameter yellow poplar (*Liriodendron tulipifera*) (diffuse porous xylem) are nearly twofold greater than for hickory (*Carya* spp.) (semi-ring porous) and fourfold greater than for oaks (*Quercus* spp.) (ring porous xylem) (Fig. 15.1). In addition, transpiration and stomatal conductance rates of diffuse porous species are also much more responsive to climatic variation compared with ring-porous species such as oaks and hickories (Ford *et al.*, 2007). When droughts are severe, diffuse porous, mesophytic species have higher mortality rates than ring porous species (Klos *et al.*, 2009). Watershed data also suggest that pine forests in general, and southern pine plantations specifically, have greater evapotranspiration, due to higher interception and transpiration, than corresponding hardwood forests (Ford *et al.*, 2011) and are more vulnerable to drought (Domec *et al.*, 2015).

Taken together, these findings suggest that forests in the southern USA are using more water now than in the past and that they could be more vulnerable to drought in the future. As such, it might be expected that streamflow gauges would detect decreasing trends in long-term streamflow; however, numerous factors influence streamflow, so establishing a simple cause-and-effect relationship is challenging, especially at large spatial scales. For example, studies have detected both decreasing and increasing flows in the southern USA, which could be due in part to precipitation variability (Patterson *et al.*, 2013; Kim *et al.*, 2014; Yang *et al.*, 2014). Despite this variability in observations, management that shifts southern forests back towards more ring porous, drought-tolerant species might increase water yield and provide resilience to future drought. This change in species could be encouraged through increased use of prescribed fire, which should favour *Quercus* spp. and reduce mesophytic species. However, treatments must be repeated regularly and in some cases combined with manipulation such as thinning to achieve changes in relative species abundances (Green *et al.*, 2010; Martin *et al.*, 2011; Arthur *et al.*, 2015). In addition to prescribed fire, particularly in areas that cannot be burned, forest thinning could remove mesophytic species and favour water-efficient and drought-tolerant species.

Pine plantations are an important forest type in the southern USA, providing fibre and solid timber products for the region, nation and globe (Wear and Greis, 2013). Decades of research have resulted in genetic improvements and silvicultural practices (e.g. site preparation, fertilization, thinning, weed control) that have

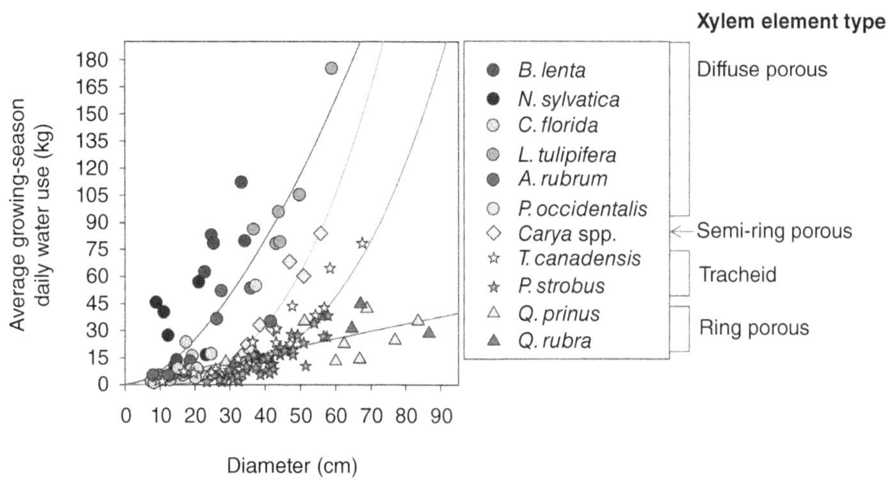

Fig. 15.1. Mean growing-season daily water use across forest species with different xylem anatomy in the southern USA. Full names for the species listed are: *Betula lenta, Nyssa sylvatica, Cornus florida, Liriodendron tulipifera, Acer rubrum, Platanus occidentalis, Tsuga canadensis, Pinus strobus, Quercus prinus* and *Quercus rubra*. (From Ford *et al.*, 2011.)

substantially increased productivity (Fox *et al.*, 2007) of southern pine plantations; however, these highly productive forests also tend to use more water (Jackson *et al.*, 2005; Ford *et al.*, 2011; King *et al.*, 2013) and are more vulnerable to drought (Domec *et al.*, 2015; Ward *et al.*, 2015). Some suggest that the acres planted in pine will increase in the future (e.g. Huggett *et al.*, 2013), with values ranging from an increase of 20 to 25 million ha by 2060, depending upon assumptions related to climatic and economic conditions (e.g. global forest products markets, biomass energy). Where and when water is plentiful, it is unlikely that this expansion will have adverse effects on water resources. However, under more variable rainfall or in areas where water is (or will be) increasingly scarce, expansion of pine plantations or other fast-growing trees could have negative effects on water resources (Calder *et al.*, 2009; King *et al.*, 2013; Vose *et al.*, 2015). Alternatives include managing plantations with lower stocking (Sun *et al.*, 2015) or managing for species that use less water (Calder *et al.*, 2009), such as restoring longleaf pine (Lockaby *et al.*, 2013).

Management actions can also be implemented to minimize the impacts of drought on water quality. In more developed areas, an obvious measure is to limit stream water withdrawals (Webb and Nobilis, 1995; Meier *et al.*, 2003) and, if possible, wastewater discharge during periods of low flow, and encourage re-use of treated wastewater to help reduce higher-temperature effluent volume entering streams (Kinouchi, 2007). In forested areas, efforts should focus on minimizing inputs of sediments and nutrients into the stream. It may be beneficial to plan the timing of management activities so they do not disturb streams during low flow. Since removal and alteration of riparian vegetation increases stream temperatures (Beschta *et al.*, 1987; Groom *et al.*, 2011) following timber harvest (Swift and Messer, 1971; Swift and Baker, 1973; Wooldridge and Stern, 1979; Sun *et al.*, 2004) and wildfires (Dunham *et al.*, 2007; Isaak *et al.*, 2010), maintaining or increasing shading effects of riparian forest canopy reduces fluctuations in water temperature, dissolved oxygen concentrations and stress (both acute and chronic) on aquatic organisms (Burton and Likens, 1973; Swift and Baker, 1973; Peterson and Kwak, 1999; Kaushal *et al.*, 2010).

15.3.2 Managing at larger spatial scales

Forest management for water resources should attempt to address the landscape scale of major river systems. In the Southeast USA, growing metropolitan areas of the Piedmont are dependent upon watersheds that originate in the largely forested Mountain region. Downstream of the rapidly urbanizing Piedmont, the Coastal Plain includes large areas of agriculture and plantation forestry. Water supply and management systems are embedded in this matrix of forest, urban and agricultural landscapes. This complex, but interconnected landscape provides a broader context for forest management. As the growing human population becomes increasingly urbanized and demand for freshwater increases, we expect a greater need for forest watershed management options to provide a stable supply of freshwater. The concept of managing forests at large spatial scales to augment annual streamflow is not new (Douglass, 1983); however, recent severe drought in many areas of the USA has increased awareness of the relationship among forest disturbance and management, drought and streamflow (Ford *et al.*, 2011; Jones *et al.*, 2012). Since harvesting often increases annual water yield, it has been suggested that the effects of drought could be mitigated by maintaining lower-density forests (McLaughlin *et al.*, 2013). Less-dense forests might provide increased water yield while reducing water stress on trees during drought.

While we have a good understanding of the effects of disturbance and forest management on water yield from studies on small watersheds, it is not clear if effects can simply be scaled up and results extrapolated over larger spatial scales (National Research Council, 2008). Tools such as remote sensing, GIS and networks of sensors can facilitate studies across larger spatial scales. In addition, hydrological models are an important tool for scaling across space and time, and they can also be used for retrospective analyses of complex systems and to generate future scenarios, identify critical knowledge gaps and generate new hypotheses. As an example, we used RHESSys, a regional hydro-ecological simulation system (Tague and Band, 2004) to further examine the potential for using forest management to increase water yield at larger spatial scales. The RHESSys model has been used to assess the effects of climate, fire and urbanization on water resources across multiple ecosystems (e.g. Tague *et al.*, 2009; Mittman *et al.*, 2012; Godsey *et al.*, 2014; Hwang *et al.*, 2014; Vicente-Serrano *et al.*, 2015). As a case study, we used the Beetree Creek watershed, which is a 1414 ha watershed in the Appalachian Mountains of western North Carolina where streamflow has been recorded daily by a US Geological Survey gauge since 1926. Runoff from Beetree Creek collects in a reservoir that serves as a secondary drinking-water source for the city of Asheville, North Carolina. Over a 6-year simulation period, we found that a 50% reduction of forest density, with a 30 m riparian buffer, mitigated the effects of a 20% reduction in precipitation, although the effects seemed to decline in the last year (Fig. 15.2a). When we removed all precipitation in June–August to simulate an extreme summer drought, the same 50% reduction in forest density with a riparian buffer still exhibited a mitigating effect, particularly during the dormant season, likely due to soil water storage (Fig. 15.2b). This might be a significant contribution during dry periods, particularly in watersheds such as this one that are part of a municipal water supply.

Although it is clear that streamflow can be altered with forest management, major challenges remain in managing forests to enhance water supply. First, a large proportion of the watershed has to be cut in order to increase

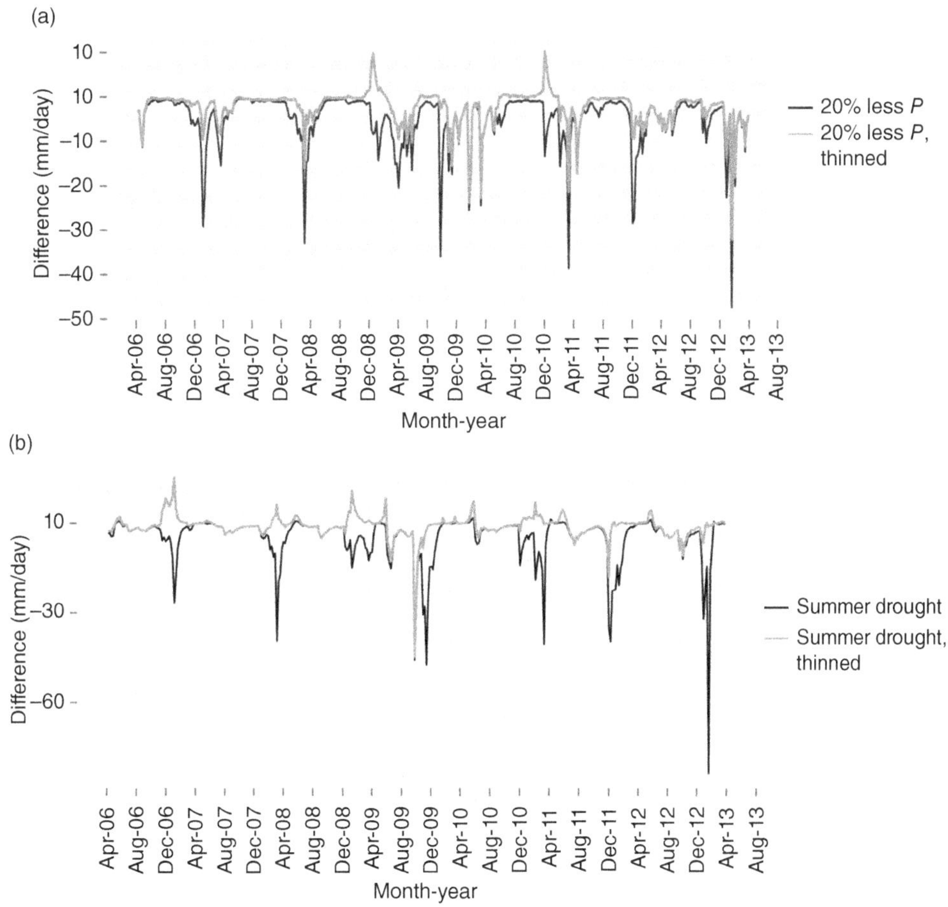

Fig. 15.2. RHESSys simulations comparing baseline water yield (*Q*) to harvest treatments (50% forest reduction, leaving 30 m riparian buffer) under (a) a 20% reduction in precipitation (*P*) and (b) an extreme summer drought, with no precipitation in June–August.

annual water yield at large spatial scales (Bosch and Hewlett, 1982; Ice and Stednick, 2004). Consequently, the potential increases in streamflow through forest cutting are minimal due to limitations on the amount of land that can be harvested at any given time (Kattelmann *et al.*, 1983). Our simulation experiment was conducted on a small (1414 ha) watershed; this harvest likely had little or no detectable effect downstream or on the overall 481,220 ha Upper French Broad River watershed. In addition, streamflow responses are often short-lived due to rapid forest regrowth (especially in the eastern USA; Swank *et al.*, 2014) and the aggrading post-cut forest may actually have lower streamflow than the uncut forest (Ford *et al.*, 2011). And, because of the unpredictable nature of droughts, it is impractical to time harvesting operations as a drought response strategy to maintain streamflow. In contrast to management actions that are intended to augment streamflow, increasing drought stress in some forest ecosystems may warrant management strategies that retain water (and hence reduce streamflow) on the landscape in order to minimize tree mortality (Grant *et al.*, 2013). Even in cases where thinning might not increase streamflow, lower-density forests are likely to be more resistant and resilient to drought conditions, allowing the majority of trees to survive and resume ecosystem service production in the future. Further, replanting or regenerating harvested forests with species that consume less water is a longer-term solution that may be more effective in some cases, so long as it is economically feasible and does not adversely affect other forest management objectives, such as forest productivity, carbon sequestration, wildlife habitat and water quality (King *et al.*, 2013).

Overall, our experiments simulating reduced precipitation and an extreme drought (Fig. 15.2a and b) support suggestions that future conditions might at times exceed the capacity of forests to provide ecosystem services, including water resources. Therefore, management of coupled social–ecological systems must include water use and storage strategies to bridge the gaps created by extreme conditions during severe or extended droughts. For example, municipalities will need strategies to maintain water supplies, such as reducing consumption, increasing conservation, adding additional emergency sources or creating additional storage.

15.4 Conclusions and Recommendations

The remainder of the 21st century will present significant challenges for forest watershed management, as rapid and compounded biophysical and socio-economic changes contribute to an increasingly uncertain future. Many of these changes portend a growing risk of water scarcity for a growing human population and greater vulnerability to extreme droughts and more intense storms. A century of forest watershed science has been critical for ensuring the sustainability of water resources derived from forest watersheds. We know with certainty that forest vegetation has a strong influence over the water balance and hydrological and biogechemical cycling processes and that BMPs must be implemented to protect water resources in managed forests. A key question is whether our current understanding, tools and management practices will be applicable in the remainder of the 21st century.

We propose that much of our understanding of forest hydrological processes and how to manage forest watersheds accordingly will continue to be applicable; however, the rapid pace and magnitude of change will constrain management outcomes. We expect that forests will continue to remain a better land-use choice compared with non-forest alternatives for clean, stable water resources, but new adaptive management regimes may be needed to reduce water demand and maintain forests on the landscape. Although it is understood that processes like evapotranspiration, water yield and timing are affected by forest management, the duration and spatial scale of these effects merit further investigation (National Research Council, 2008). Projections indicate a future of increasing pine plantations and expansion of fast-growing species for carbon sequestration and bioenergy, but landscape-scale effects on water yield and quality, and the magnitude of potential trade-offs between managing for carbon and water, have not been systematically explored across time and space (Jackson *et al.*, 2005; King *et al.*, 2013). The likelihood of increasing water scarcity will require a better understanding of how to manage forest structure and species composition for both maximum water yield and minimized tree mortality. Forests changes in the eastern USA (i.e. via succession and intensive forest management) have created forests that require more water and are more drought-prone. The challenge of managing forests at large spatial scales suggests a need to identify if and where management would be particularly effective in increasing water yield and, thus, water supplies. Further research could also identify the most drought-vulnerable areas so that management could be prioritized to increase forest resilience. Modelling studies provide a valuable tool for examining potential short- and long-term consequences of forest management on water resources, forest resilience and other ecosystem services, including carbon sequestration and wood and fibre production, at landscape scales. However, modelling must be accompanied by continued or additional monitoring not only of small watersheds, but large ones as well. When and where possible, experiments nested across larger watersheds using an adaptive management approach would provide the most realistic and useful information.

References

Allen, M.R. and Ingram, W.J. (2002) Constraints on future changes in climate and the hydrologic cycle. *Nature* 419, 224–232.

Arthur, M.A., Blankenship, B.A., Schorgendorfer, A., Loftis, D.L. and Alexander, H.D. (2015) Changes in stand structure and tree vigor with repeated prescribed fire in an Appalachian hardwood forest. *Forest Ecology and Management* 340, 46–61.

Beschta, R.L., Bilby, R.E., Brown, G.W., Holtby, L.B. and Hofstra, T.D. (1987) Stream temperature and aquatic habitat: fisheries and forestry interactions. In: Salo, E.O. and Cundy, T.W. (eds) *Streamside Management: Forestry and Fishery Interactions*. Contribution No. 57. University of Washington Institute of Forest Resources, Seattle, Washington, pp. 191–232.

Bosch, J.M. and Hewlett, J.D. (1982) A review of catchment experiments to determine the effect of vegetation changes on water yield and evapotranspiration. *Journal of Hydrology* 55, 3–23.

Bradley, B.A., Wilcove, D.S. and Oppenheimer, M. (2010) Climate change increases risk of plant invasion in the Eastern United States. *Biological Invasion* 12, 1855–1872.

Brantley, S.T., Miniat, C.F., Elliott, K.J., Laseter, S.H. and Vose, J.M. (2015) Changes to southern Appalachian water yield and stormflow after loss of a foundation species. *Ecohydrology* 8, 518–528.

Breshears, D.D., Cobb, N.S., Rich, P.M., Price, K.P., Allen, C.D., Ballice, R.G., Romme, W.H., Kastens, J.H., Floyd, M.L., Belnap, J., *et al.* (2005) Regional vegetation die-off in response to global-change-type drought. *Proceedings of the National Academy of Science USA* 102, 15144–15148.

Burton, T.M. and Likens, G.E. (1973) Effect of strip-cutting on stream temperatures in Hubbard Brook Experimental Forest, New Hampshire. *BioScience* 23, 433–435.

Calder, I.R., Reid, I., Nisbet, T.R. and Green, J.C. (2003) Impact of lowland forests in England on water resources: application of the Hydrological Land Use Change (HYLUC) model. *Water Resources Research* 39, 1319, doi: 10.1029/2003WR002042 (accessed 10 April 2016).

Calder, I.R., Nisbet, T. and Harrison, J.A. (2009) An evaluation of the impacts of energy tree plantations on water resources in the United Kingdom under present and future UKCIP02 climate scenarios. *Water Resources Research* 45, W00A17, doi: 10.1029/2007WR006657 (accessed 10 April 2016).

Carter, L.M., Jones, J.W., Berry, L., Burkett, V., Murley, J.F., Obeysekera, J., Schramm, P.J. and Wear, D. (2014) Southeast and the Caribbean. In: Melillo, J.M., Richmond, T.C. and Yohe, G.W. (eds) *Climate Change Impacts in the United States: The Third National Climate Assessment*. US Global Change Research Program, Washington, DC, pp. 396–417.

Certini, G.G. (2005) Effects of fire on properties of forest soils: a review. *Oecologia* 143, 1–10.

De la Crétaz, A.L. and Barten, P.K. (2007) *Land Use Effects on Streamflow and Water Quality in the Northeastern United States*. CRC Press/Taylor & Francis, Boca Raton, Florida.

Doerr, S.H., Shakesby, R.A., Blake, W.H., Chafer, C.M., Humphreys, G.S. and Wallbrink, P.J. (2006) Effects of differing wildfire severities on soil wettability and implications for hydrological response. *Journal of Hydrology* 319, 295–311.

Domec, J.J., King, J.S., Ward, E., Oishi, A.C., Palmroth, S., Radecki, A., Bell, D.M., Miao, G., Gavazzi, M., Johnson, D.M., *et al.* (2015) Conversion of natural forests to managed forest plantations decreases tree resistance to prolonged droughts. *Forest Ecology and Management* 355, 58–71.

Douglass, J.E. (1983) The potential for water yield augmentation from forest management in the eastern United States. *Water Resources Bulletin* 9, 351–358.
Dunham, J.B., Rosenberger, A.E. and Luce, C.H. (2007) Influences of wildfire and channel reorganization on spatial and temporal variation in stream temperature and the distribution of fish and amphibians. *Ecosystems* 10, 335–346.
Easterling, D.R., Evans, J.L., Groisman, P.Y., Karl, T.R., Kunkel, K.E. and Ambenje, P. (2000) Observed variability and trends in extreme climate events: a brief review. *Bulletin of the American Meteorological Society* 81, 417–425.
Egginton, P., Beall, F. and Buttle, J. (2014) Reforestation – climate change and water resource implications. *The Forestry Chronicle* 90, 516–524.
Ehrenfeld, J.G. (2010) Ecosystem consequences of biological invasions. *Annual Review of Ecology, Evolution and Systematics* 41, 59–80.
Enquist, B.J., Brown, J.H. and West, G.B. (1998) Allometric scaling of plant energetics and population density. *Nature* 395, 163–165.
Field, C.B., Barros, V.R., Dokken, D.J., Mach, K.K. Masterandrea, M.D., Bilir, T.E., Chatterjee, M., Ebi, K.L., Estrada, Y.O., Genova, R.C., *et al.* (eds) (2014) *Climate Change 2014: Impacts, Adaptation, and Vulnerability. Part A: Global and Sectorial Aspects. Contribution of Working Group II to the Fifth Assessment Report of the Intergovenmental Panel on Climate Change*. Cambridge University Press, Cambridge/New York.
Fites, J.A., Carol, E., Vaillant, N., Campbell, M., Courson, M., Decker, T., Lew, C., Marouk, S., Taylor, Z. and Anderson, R. (2007) Big Turnaround and Georgia Bay Complexes Fire Behavior Assessment Report. USDA Forest Service Fire Behavior Assessment Team. Available at: http://www.fs.fed.us/adaptivemanagement/reports/fbat/GA_Wildfires_2007_detail_report_AMSET.pdf (accessed 10 April 2016).
Ford, C.R. and Vose, J.M. (2007) *Tsuga canadensis* (L.) Carr. mortality will impact hydrologic processes in southern Appalachian forest ecosystems. *Ecological Applications* 17, 1156–1167.
Ford, C.R., Hubbard, R.M., Kloeppel, B.D. and Vose, J.M. (2007) A comparison of sap flux-based evapotranspiration estimates with catchment-scale water balance. *Agricultural and Forest Meteorology* 145, 176–185.
Ford, C.R., Hubbard, R.M. and Vose, J.M. (2011) Quantifying structural and physiological controls on canopy transpiration of planted pine and hardwood stand species in the Southern Appalachians. *Ecohydrology* 4, 183–195.
Fox, T.R., Jokela, E.J. and Allen, H.L. (2007) The development of pine plantation silviculture in the Southern United States. *Journal of Forestry* 105, 337–347.
Godsey, S.E., Kirchner, J.W. and Tague, C.L. (2014) Effects of changes in winter snowpacks on summer low flows: case studies in the Sierra Nevada, California, USA. *Hydrological Processes* 28, 5048–5064.
Grant, G.E., Tague, C.L. and Allen, C.D. (2013) Watering the forest for the trees: an emerging priority for managing water in forest landscapes. *Frontiers in Ecology and the Environment* 11, 314–321.
Green, S.R., Arthur, M.A. and Blankenship, B.A. (2010) Oak and red maple seedling survival and growth following periodic prescribed fire on xeric ridgetops on the Cumberland Plateau. *Forest Ecology and Management* 259, 2256–2266.
Groom, J.D., Dent, L., Madsen, L.J. and Fleuret, J. (2011) Response of western Oregon (USA) stream temperatures to contemporary forest management. *Forest Ecology and Management* 262, 1618–1629.
Hobbs, R.J., Arico, S., Aronson, J., Baron, J.S., Bridgewater, P., Cramer, V.A., Epstein, P.R., Ewel, J.J., Klink, C.A., Lugo, A.E., *et al.* (2006) Novel ecosystems: theoretical and management aspects of the new ecological world order. *Global Ecology and Biogeography* 15, 1–7.
Hobbs, R.J., Higgs, E. and Harris, J.A. (2009) Novel ecosystems: implications for conservation and restoration. *Trends in Ecology and Evolution* 24, 599–605.
Hobbs, R.J., Higgs, E., Hall, C.M., Bridgewater, P.M., Chapin, F.S. III, Ellis, E.C., Ewel, J.J., Hallett, L.M., Harris, J., Hulvey, K.B., *et al.* (2014) Managing the whole landscape: historical, hybrid, and novel ecosystems. *Frontiers of Ecology and the Environment* 12, 557–564.
Huggett, R., Wear, D.N., Li, R., Coulston, J. and Liu, S. (2013) Forecasts of forest conditions. In: Wear, D.N. and Greis, J.G. (eds) *The Southern Forest Futures Project*. General Technical Report 178. USDA Forest Service, Southern Research Station, Asheville, North Carolina, pp. 73–101.
Huntington, T.G. (2006) Evidence for intensification of the global water cycle: review and synthesis. *Journal of Hydrology* 319, 83–95.
Hwang, T., Band, L.E., Miniat, C.F., Song, C., Bolstad, P.V., Vose, J.M. and Love, J.P. (2014) Divergent phonological response to hydroclimate variability in forested mountain watersheds. *Global Change Biology* 20, 2580–2595.

Ice, G.G. and Stednick, J.D. (2004) *A Century of Forest and Wildland Watershed Lessons*. Society of American Foresters, Bethesda, Maryland.

Ice, G.G., Neary, D.G. and Adams, P.W. (2004) Effects of wildfire on soils and watershed processes. *Journal of Forestry* 102, 16–20.

Isaak, D.J., Luce, C.H., Rieman, B.E., Nagel, D.E., Peterson, E.E., Horan, D.L., Parkes, S. and Chandler, G.L. (2010) Effects of climate change and wildfire on stream temperatures and salmonid thermal habitat in a mountain river network. *Ecological Applications* 20, 1350– 1371.

Jackson, R.B., Jabbogy, E.G., Avissar, R., Roy, S.B., Barrett, D.J., Cook, C.W., Farley, K.A., le Maitre, D.C., Carl, B.A. and Murray, B.C. (2005) Trading water for carbon with biological carbon sequestration. *Science* 310, 1944–1947.

Jones, J.A., Creed, I.F., Hatcher, K.L., Warren, R.J., Adams, M.B., Benson, M.H., Boose, E., Brown, W.A., Campbell, J.L., Covich, A., *et al.* (2012) Ecosystem processes and human influences regulate streamflow response to climate change at long-term ecological research sites. *BioScience* 62, 390–404.

Karl, T.R., Knight, R.W. and Plummer, N. (1995) Trends in high-frequency climate variability in the twentieth century. *Nature* 377, 217–220.

Kattelmann, R.C., Berg, N.H. and Rector, J. (1983) The potential for increasing streamflow from Sierra-Nevada watersheds. *Water Resources Bulletin* 19, 395–402.

Kaushal, S.S., Likens, G.E., Jaworski, N.A., Pace, M.L., Sides, A.M., Seekell, D., Belt, K.T., Secor, D.H. and Wingate, R.L. (2010) Rising stream and river temperatures in the United States. *Frontiers in Ecology and the Environment* 8, 461–466.

Kim, Y., Band, L.E. and Song, C. (2014) The influence of forest regrowth on the stream discharge in the North Carolina Piedmont watersheds. *Journal of the American Water Resources Association* 50, 57–73.

King, J.S., Ceulemans, R., Albaugh, J.M., Dillen, S.Y., Docmec, J.C., Fichot, R., Fischer, M., Leggett, Z., Sucre, E., Trnka, M. and Zenone, T. (2013) The challenge of lignocellulosic bioenergy in a water-limited world. *BioScience* 63, 102–117.

Kinouchi, T. (2007) Impact of long-term water and energy consumption in Tokyo on wastewater effluent: implications for the thermal degradation of urban streams. *Hydrological Processes* 21, 1207–1216.

Klos, R.J., Wang, G.G., Bauerle, W.L. and Rieck, J.R. (2009) Drought impact on forest growth and mortality in the southeast USA: an analysis using Forest Health and Monitoring data. *Ecological Applications* 19, 699–708.

Likens, G. (2001) Biogeochemistry, the watershed approach: some uses and limitations. *Marine and Freshwater Research* 52, 5–12.

Liu, Y.Q., Goodrick, S.L. and Stanturf, J.A. (2013) Future US wildfire potential trends projected using a dynamically downscaled climate change scenario. *Forest Ecology and Management* 294, 120–135.

Lockaby, G., Nagy, C., Vose, J.M., Ford, C.R., Sun, G., McNulty, S., Caldwell, P., Cohen, E. and Myers, J.M. (2013) Forests and water. In: Wear, D.N. and Greis, J.G. (eds) *The Southern Forest Futures Project.* General Technical Report 178. USDA Forest Service, Southern Research Station, Asheville, North Carolina, pp. 309–339.

Lovett, G.M., Canham, C.D., Arthur, M.A., Weathers, K.C. and Fitzhugh, R.D. (2006) Forest ecosystem responses to exotic pests and pathogens in eastern North America. *BioScience* 56, 395–405.

Marlon, J.R., Bartlein, P.J., Caracillet, C., Gavin, D.G., Harrison, S.P., Higurera, P.E., Joos, F., Power, M.J. and Prentice, I.C. (2008) Climate and human influences on global biomass burning over the past two millennia. *Nature Geosciences* 1, 697–702.

Martin, K.L., Hix, D.M. and Goebel, P.C. (2011) Coupling of vegetation layers and environmental influences in a mature, second-growth Central Hardwood forest landscape. *Forest Ecology and Management* 261, 720–729.

McEwan, R.W., Dyer, J.M. and Pederson, N. (2011) Multiple interacting ecosystem drivers: toward an encompassing hypothesis of oak forest dynamics across eastern North America. *Ecography* 34, 244–256.

McLaughlin, D.L., Kaplan, D.A. and Cohen, M.J. (2013) Managing forests for increased regional water yield in the Southeastern Coastal Plain. *Journal of the American Water Resources Association* 49, 953–965.

McNulty, S., Myers, J.M., Caldwell, P. and Sun, G. (2013) Climate change summary. In: Wear, D.N. and Greis, J.G. (eds) *The Southern Forest Futures Project*. General Technical Report 178. USDA Forest Service, Southern Research Station, Asheville, North Carolina, pp. 27–44.

Meier, W., Bonjour, C., Wuest, A. and Reichert, P. (2003) Modeling the effect of water diversion on the temperature of mountain streams. *Journal of Environmental Engineering – ASCE* 129, 755–764.

Meinzer, F.C., Bond, B.J., Warren, J.M. and Woodruff, D.R. (2005) Does water transport scale universally with tree size? *Functional Ecology* 19, 558–565.

Melvin, M. (2012) National Prescribed Fire Use Survey Report. Technical Report 01-12. National Association of State Foresters and Coalition of Prescribed Fire Councils, Inc. Available at: http://www.stateforesters.org/sites/default/files/publication-documents/2012_National_Prescribed_Fire_Survey.pdf (accessed 14 April 2016).

Millennium Ecosystem Assessment (2005) *Ecosystems and Human Well-Being: Current State and Trends*. Island Press, Washington, DC.

Miller, J.H., Lemke, D. and Coulston, J. (2013) The invasion of southern forests by nonnative plants: current and future occupation, with impacts, management strategies and mitigation approaches. In: Wear, D.N. and Greis, J.G. (eds) *The Southern Forest Futures Project*. General Technical Report 178. USDA Forest Service, Southern Research Station, Asheville, North Carolina, pp. 397–456.

Mitchell, R.J., Youngqiang, L., O'Brien, J.J., Elliott, K.J., Starr, G., Miniat, C.F. and Hiers, J.K. (2014) Future climate and fire interactions in the southeastern region of the United States. *Forest Ecology and Management* 327, 316–326.

Mittman, T., Band, L.E., Hwang, T. and Smith, M.L. (2012) Distributed hydrologic modeling in the suburban landscape: assessing parameter transferability from gauged reference catchments. *Journal of the American Water Resources Association* 48, 546–557.

Naiman, R.J. (2013) Socio-ecological complexity and the restoration of river ecosystems. *Inland Waters* 3, 391–410.

National Research Council, Water Science and Technology Board (2008) *Hydrologic Effects of a Changing Forest Landscape*. National Academies Press, Washington, DC.

Nowacki, G.J. and Abrams, M.D. (2008) The demise of fire and 'mesophication' of forests in the eastern United States. *BioScience* 58, 123–138.

Nowacki, G.J. and Abrams, M.D. (2015) Is climate an important driver of post-European vegetation change in the Eastern United States? *Global Change Biology* 21, 314–334.

O'Gorman, P.A. and Schneider, T. (2009) The physical basis for increases in precipitation extremes in simulations of 21st-century climate change. *Proceedings of the National Academy of Sciences USA* 106, 14773–14777.

Patterson, L.A., Lutz, B. and Doyle, M.W. (2013) Climate and direct human contributions to changes in mean annual streamflow in the South Atlantic, USA. *Water Resources Research* 49, 7278–7291.

Peterson, J.T. and Kwak, T.J. (1999) Modeling the effects of land use and climate change on riverine smallmouth bass. *Ecological Applications* 9, 1391–1404.

Postel, S. and Richter, B. (2003) *Rivers for Life: Managing Water for People and Nature*. Island Press, Washington, DC.

Riveros-Iregui, D.A., Wang, L. and Wilcox, B. (2011) Ecohydrology and the challenges of the coming decades. *American Geophysical Union, Hydrology Section Newsletter* (July), 32–34.

Rutter, A.J., Morton, A.J. and Robins, P.C. (1975) Predictive model of rainfall interception in forests 2. Generalizations of model and comparison with observations in some coniferous and hardwood stands. *Journal of Applied Ecology* 12, 367–380.

Seastedt, T.R., Hobbs, R.J. and Suding, K.N. (2008) Management of novel ecosystems: are novel approaches required? *Frontiers in Ecology and the Environment* 6, 547–553.

Steffen, W., Crutzen, P.J. and McNeill, J.R. (2007) The Anthropocene: are humans now overwhelming the great forces of nature? *Ambio* 36, 614–621.

Sun, G., Riedel, M. and Jackson, R. (2004) Influences of management of Southern forests on water quantity and quality. In: Rauscher, H.M. and Johnsen, K. (eds) *Southern Forest Sciences: Past, Current, and Future*. General Technical Report SRS-75. USDA Forest Service, Southern Research Station, Asheville, North Carolina.

Sun, G., Caldwell, P.V. and McNulty, S.G. (2015) Modelling the potential role of forest thinning in maintaining water supplied under a changing climate across the conterminous United States. *Hydrological Processes* 29, 5016–5030.

Swank, W.T., Knoepp, J.D., Vose, J.M., Laseter, S.H. and Webster, J.R. (2014) Response and recovery of water yield and timing, stream sediment, abiotic parameters, and stream chemistry following logging. In: Swank, W.T. and Webster, J.R. (eds) *Long-Term Response of a Forest Watershed Ecosystem. Clearcutting in the Southern Appalachians*. Oxford University Press, New York, pp. 36–56.

Swift, L.W. and Baker, S.E. (1973) *Lower Water Temperatures Within a Streamside Buffer Strip*. Research Note SE-193. USDA Forest Service, Southeastern Forest Experiment Station, Asheville, North Carolina.

Swift, L.W. and Messer, J.B. (1971) Forest cuttings raise temperatures of small streams in Southern Appalachians. *Journal of Soil and Water Conservation* 26, 111–116.

Tague, C.L. and Band, L.E. (2004) RHESSys: Regional Hydro-Ecologic Simulation System – an object-oriented approach to spatially distributed modeling of carbon, water, and nutrient cycling. *Earth Interactions* 8, 1–42.

Tague, C.L., Seaby, L. and Hope, A. (2009) Modeling the eco-hydrologic response of a Mediterranean type ecosystem to the combined impacts of projected climate change and altered fire frequencies. *Climatic Change* 93, 137–155.

Terrando, A.J., Costanza, J., Belyea, C., Dunn, R.R., McKerrow, A. and Collazo, J.A. (2014) The southern megalopolis: using the past to predict the future of urban sprawl in the southeast US. *PLoS One* 9, e102261.

United Nations, Department of Economic and Social Affairs, Population Division (2013) *World Population Prospects: The 2012 Revision, Key Findings and Advance Tables*. Working Paper No. ESA/P/WP.288. United Nations, New York.

United Nations, Department of Economic and Social Affairs, Population Division (2014) *World Urbanization Prospects, The 2014 Revision*. Paper ST/ESA/SER.A/366. United Nations, New York.

Vicente-Serrano, S.M., Camarero, J.J., Zebalza, J., Sanguesa-Barreda, G., Lopez-Moreno, J.I. and Tague, C.L. (2015) Evapotranspiration deficit controls net primary production and growth of silver fir: implications for circum-Mediterranean forests under forecasted warmer and drier conditions. *Agricultural and Forest Meteorology* 206, 45–54.

Vose, J.M., Ford, C.R., Laseter, S.H., Dymond, S.F., Sun, G., Adams, M.B., Sebestyen, S.D., Campbell, J.L., Luce, C., Amatya, D.M., *et al.* (2012) Can forest watershed management mitigate climate change effects on water resources? In: Webb, A.A., Bonell, M., Bren, L., Lane, P.J.N., McGuire, D., Neary, D.J., Nettles, J., Scott, D.F., Stednick, J. and Wang, Y. (eds) *Revisiting Experimental Catchment Studies in Forest Hydrology (Proceedings of a Workshop held during the XXV IUGG General Assembly in Melbourne, June–July 2011)*. IAHS Publication No. 353. International Association of Hydrological Sciences, Wallingford, UK, pp. 12–25.

Vose, J.M., Miniat, C.F., Sun, G. and Caldwell, P.V. (2015) Potential implications for expansion of freeze-tolerant Eucalyptus plantations on water resources in the Southern United States. *Forest Science* 61, 509–521.

Wang, L., Liu, J., Sun, G., Wei, X., Liu, S. and Dong, Q. (2012) Water, climate and vegetation: ecohydrology in a changing world. *Hydrology and Earth System Science* 16, 4633–4636.

Ward, E.J., Domec, J.C., Laviner, M.A., Fox, T.R., Sun, G., McNulty, S., King, J. and Noormets, A. (2015) Fertilization intensifies drought stress: water use and stomatal conductance of *Pinus taeda* in a mid-rotation fertilization and throughfall reduction experiment. *Forest Ecology and Management* 355, 72–82.

Wear, D.N. (2013) Forecasts of land uses. In: Wear, D.N. and Greis, J.G. (eds) *The Southern Forest Futures Project*. General Technical Report 178. USDA Forest Service, Southern Research Station, Asheville, North Carolina, pp. 45–72.

Wear, D.N. and Gries, J.G. (eds) (2013) *The Southern Forest Futures Project*. General Technical Report 178. USDA Forest Service, Southern Research Station, Asheville, North Carolina.

Webb, B.W. and Nobilis, F. (1995) Long-term water temperature trends in Austrian rivers. *Hydrological Sciences Journal* 40, 83–96.

Wooldridge, D.D. and Stern, D. (1979) *Relationships of Silvicultural Activities and Thermally Sensitive Forest Streams*. Report DOE 79-5a-5. College of Forest Resources, University of Washington, Seattle, Washington.

Wullschleger, S.D., Hanson P.J. and Todd, D.E. (2001) Transpiration from a multi-species deciduous forest as estimated by xylem sap flow techniques. *Forest Ecology and Management* 143, 205–213.

Yang, Q., Hanqin, T., Friedrichs, M.A.M., Liu, M., Li, X. and Yang, J. (2014) Hydrological responses to climate and land-use changes along the North American east coast: a 110-year historical reconstruction. *Journal of the American Water Resources Association* 51, 47–67.

Zhang, L., Dawes, W.R. and Walker, G.R. (2001) Response of mean annual evapotranspiration to vegetation changes at catchment scale. *Water Resources Research* 37, 701–708.

16 Hydrology of Taiga Forests in High Northern Latitudes

A. Onuchin[1]*, T. Burenina[1], A. Shvidenko[1,2], G. Guggenberger[3] and A. Musokhranova[1]

[1]Siberian Branch of the Russian Academy of Sciences, Krasnoyarsk, Russia; [2]International Institute for Applied Systems Analysis, Laxenburg, Austria; [3]Leibniz University, Hannover, Germany

16.1 Introduction

Today's estimates of the hydrological role of the taiga forest (also known as boreal forest) and its water cycle characteristics are, from our perspective, especially contradictory regarding the influence of boreal forests on annual runoff. Unlike tropical and deciduous temperate forests, which are known to evaporate more moisture than other land types and, hence, to reduce river runoff, the situation with taiga forest is much more ambiguous. Some researchers generalize conclusions made for tropical and temperate forests and believe that these conclusions could be applied to all types of forests, and that the assumption about increasing runoff from forested watersheds is based on a wrong interpretation of forest ecosystem hydrological cycles (Hamilton, 2008). This concept is based on viewing forest ecosystems as water consumers that reduce groundwater level and reduce river runoff through evaporation of intercepted precipitation and transpiration. This chapter attempts to present a different view of this problem in forests of the taiga zone.

Hydrological cycles in the forest are determined by many factors and conditions, including environmental factors, size of forest area, its share in catchments and other forest characteristics. This brings up a critically important question in terms of estimating the hydrological role of forests: why, under some conditions, does the forest contribute to increasing river runoff, whereas it reduces the runoff in other conditions through increasing evapotranspiration?

Clearly, researchers of ecosystems and complex natural processes should base their studies on holistic approaches. However, seeking to understand a specific phenomenon by reductionism simplifies the reality and separates the phenomenon from more complex contextual systems and processes; scaling up often results in a loss of experiment precision and robustness. Today, improving our understanding of hydrological processes in the forest requires development of scientifically based approaches to the determination of their scales (Cohen and Bredemeier, 2011). Comparison of local versus large-area study results brings to light the current gaps in the knowledge of the water–forest system. We hope that the results of our local and regional studies of forest hydrology will contribute to the general understanding of water–forest system functioning and will help to eliminate contradictions of views on the hydrological role of taiga forests.

*Corresponding author; e-mail: onuchin@ksc.krasn.ru

16.2 Boreal Forest Characteristics and Growth Conditions

The taiga zone is the broad circumpolar vegetation zone of the high northern latitudes. The definition of 'boreal forest' varies in different countries and often depends on different jurisdictions. The boundary of the boreal forest zone in the northern hemisphere is deemed to coincide with July isotherms: the boundary goes along +13°C and +18°C isotherms in the north and south, respectively. A commonly used Canadian definition of the 'boreal biome' (a major life zone) is vegetation that is composed primarily of cone-bearing, needle-leaved or scale-leaved evergreen trees found in regions that have long winters and moderate to high annual precipitation (Burton *et al.*, 2003). Boreal forests make up over 30% of the global forest area (FAO, 2010), covering vast areas of North America and Eurasia with cold climates and mostly podzolic soils that provide absolute predominance of conifers over all other tree species. Existing estimates of the area of boreal forests vary substantially: from 1161.6 million ha for ten countries (Shvidenko and Apps, 2006) to 1214 million ha (including 920 million ha of stocked forests) (FAO, 2010) to 1444.4 million ha for seven countries that include woodlands (Burton *et al.*, 2003). Basically concentrated in the taiga zone, boreal forests occupy relatively small areas (mostly along rivers) in tundra and forest-tundra. The most northern boreal forests are described at 72°30'N in Central Siberia.

Climatic conditions vary considerably across the boreal forest area. However, taiga forests have a number of common characteristics. The frost period, sometimes with very low air temperatures, is well expressed, the climate varies in continentality among localities, annual precipitation ranges from 300 to 900 mm, and summer is fairly cool. As a rule, annual precipitation exceeds evapotranspiration. Substantial areas are covered by permafrost, particularly in Northern Asia. The taiga zone thermal conditions grow colder, and evaporation decreases, from south to north. For these reasons, peat formation becomes more widespread and numerous lakes occur, especially on permafrost.

Coniferous species generally dominate circumpolar taiga forests, mainly represented by trees of the four most widespread genera – pine (*Pinus*), larch (*Larix*), spruce (*Picea*) and fir (*Abies*) – although these forests vary in area and species composition between the continents. In Eurasia, particularly in Northern Asia, larch dominates (36% of all forest land in Russia), with the largest areas being occupied by *Larix sibirica*, *Larix sukaczewii*, *Larix gmelini* and *Larix cajanderi*, followed by pine (16%), mainly *Pinus sylvestris*, which occurs practically across the entire forest zone. Dark conifer taiga is composed largely of mixed spruce–fir stands (12.5%). Here *Picea sibirica* in Siberia and *Picea ajanensis* in the Russian Far East usually dominate, while fir species occur in various proportions and are geographically separated (Strakhov *et al.*, 2001; Ermakov, 2003).

The geographic location and specifics of individual forest habitats control the composition of green forest floor vegetation, including different genera of grasses, herbs, mosses and lichens. The proportion of any tree species in a habitat depends on climate continentality, landforms, site topography and human-caused levels of disturbance. In non-mountain landscapes, larch decreases and pine increases in proportion from north to south; whereas in the mountains, dark conifer tree species dominate and forest vegetation clearly reflects an altitudinal zonality (Ermakov, 2003).

In North America, mainly in Alaska, USA and Canada, boreal forests are formed by white spruce (mainly *Picea glauca*), black spruce (*Picea mariana*), Sitka spruce (*Picea sitchensis*), balsam fir (*Abies balsamifera*), white fir (*Abies alba*), grand fir (*Abies grandis*), ponderosa pine (*Pinus ponderosa*), western white pine (*Pinus monticola*), jack pine (*Pinus banksiana*) and many other pine species. Common juniper (*Juniperus communis*), red cedar (*Juniperus virginiana*), Alaska cypress (*Chamaecyparis nootkatensis*), thuya, hemlock and sequoia are also widespread (Spurr and Barnes, 1980).

In Europe, zonal taiga forests are found mostly in Scandinavian countries. The major forest-forming tree species are much the same as in north-eastern Russia, namely spruce (*Picea abies*), European larch (*Larix decidua*), yew, noble fir (*A. alba*), juniper, Scots pine (*Pinus sylvestris*) and black pine (*Pinus nigra*).

Regional differences in boreal forests are largely climate driven. An oceanic or moderately continental humid climate with a warm summer and a short-term and mild frost period in North America and north-western Eurasia promotes mixed conifer–broadleaved forests that favour the formation of a sub-taiga ecotone, which gradually gives place to temperate forests. The physiognomy of mixed forests depends on broad-leaved tree species distribution; these species are intolerant of continental climate, but because they are able, more or less, to tolerate low air temperatures, they may have spread far to the north (Strakhov *et al.*, 2001).

Over two-thirds of the taiga zone (northern areas of Alaska, Canada and Russia) is covered by permafrost. East of the Yenisei River in Russia, permafrost stretches from the Arctic Ocean to the southern border of the country. The cryolite-zone forests are largely formed by larch (Plate 14), which are replaced by stands of other tree species where environmental conditions are less severe (Osawa *et al.*, 2010).

16.3 Literature Review

Different aspects of the hydrological role of forests are described in detail in the literature. Hamilton (2008) reviewed many publications concerned with the role of tropical and temperate forests. The hydrological role of taiga forests is among the highest researched subjects because these forests make up a considerable proportion of the global temperate forests, accounting in Russia, for example, for three-quarters of the total forest area. Ground data that help estimate the hydrological importance of forests found across the forest zone of the northern hemisphere are obtained mainly using a network of paired experimental runoff stations. The methodology and results are detailed in a number of papers (Fedorov, 1977; Bosch and Hewlett, 1982; Gu *et al.*, 2013). Many research projects carried out in forest and non-forest areas ranging from elementary plots to large-branched basins show that woody vegetation causes water cycle changes (Molchanov, 1961; Voronkov, 1988; Johnson, 1998; Bond *et al.*, 2008) and is ideally suited to allow rainfall infiltration to groundwater flow (Lebedev, 1982; Waldenmeyer, 2003; Hegg *et al.*, 2004).

Because combinations of factors, including features of the environment, determine hydrological cycles in the forest, many forest hydrologists (Keller, 1988; Whitehead and Robinson, 1993; Johnson, 1998; Onuchin *et al.*, 2006; Onuchin and Burenina, 2008; Burenina *et al.*, 2014) use a landscape–hydrological approach taking into account scales of forest vegetation-caused changes of hydrological regimes. Forest hydrology studies conducted in Northern Eurasia and North America (Krestovskiy, 1984; Hornbeck *et al.*, 1997, 2014; Buttle *et al.*, 2005; Campbell *et al.*, 2013; Burenina *et al.*, 2014; Winkler *et al.*, 2015) show that severe disturbances, such as large phytophagous insect outbreaks, large forest fires and large-scale clearcutting, impact water source formation and hydrological cycles of river basins of any size and level of complexity.

Hydrological cycles in the taiga zone are determined to a large extent by interception of snow by tree crowns, by how long snow remains in crowns (i.e. its residence time) and by how much snow drops to the ground or is evaporated from crowns. The contributions of the above factors to snow cover formation depend on the regional climate, weather and biometric parameters of forest stands (Hedstrom and Pomeroy, 1998; Jones *et al.*, 2001).

Different views exist of the importance of canopy-intercepted snow in water budget computations. Schmidt *et al.* (1988), Lundberg and Halldin (1994) and many other researchers have reported that the interception and the subsequent evaporation of snow are critically important controls of the amount of snow accumulated under the forest canopy. In addition to sublimation of tree-crown-intercepted snow, Lundberg and Halldin (1994) emphasize the importance of wind-caused redistribution of the snow that has dropped from the forest canopy to the ground.

Therefore, interception by the canopy of taiga forests in a cold climate differs from that in the southern part of the temperate climate zone. In a cold climate, snow may remain in tree crowns from several days to several months (Pomeroy and Schmidt, 1993), whereas in a warmer climate the canopy-intercepted snow usually disappears completely by the next snowfall.

Environmental conditions (air temperature, relative humidity and wind speed) have a marked influence on the snow water budget. The results of our studies show that snow water flows depend largely on the combination of the above factors and on water cycle non-linearity. According to some authors, this is where a contradiction occurs. For example, interception of snow by tree crowns increases with increasing air temperature, because the warm and moist snow becomes more adhesive to the crowns (Miller, 1964; Onuchin, 2001). At the same time, metamorphism of the intercepted snow may increase and the snow may become less solid (Kobayashi, 1987; Gubler and Rychetnik, 1991). Some researchers (Bunnell *et al.*, 1985; Wheeler, 1987; Schmidt and Gluns, 1991) note that low air temperature-induced wind speed and low snow density contribute to snow interception by the forest canopy, and a warm spell following a snowfall enhances the amount of intercepted snow falling to the ground.

The relationship between air temperature and snowfall from crowns may vary because of the effect of temperature on branch rigidity and snow adhesion (Pomeroy and Gray, 1995). At air temperatures close to the snowmelt point, tree branches become elastic and are unable to hold the snow accumulated during a cold period. As a result, the snow falls from tree crowns to the ground (Schmidt and Pomeroy, 1990).

Most taiga forests are confined to the cryolite zone. Estimating the hydrological role of these forests, especially those on continuous permafrost, presents certain difficulties caused by unstable water budget associated with the seasonal thawing of frost soil (Georgiyevsky and Shiklomanov, 2003). Our studies show that the hydrological regime of the rivers of the cryolite zone differs markedly from that of the taiga forest immediately south of this zone. We analysed the runoff for northern rivers to find that it varies both spatially and temporally. There is a clear geographical influence on the seasonal hydrological behaviours of the rivers. The further north the river lies, the more pronounced the snowmelt flood and rain-caused stream rise peaks. The flow of the river from small basins may increase many times after even small rains (Burenina *et al.*, 2015).

16.4 Local and Regional Variations in Hydrology

16.4.1 The snow moisture balance in the taiga zone

Season and regional peculiarities

In the taiga zone, where snow cover is an important hydrological cycle component, the estimates of the hydrological role of forests should accurately consider the specific character of snow accumulation in the forest versus open sites. Forest hydrologists realize that, to do this, the terms 'forest' and 'open site' should reflect landscape characteristics (Kolomyts, 1975; Schleppi, 2011) and should be interpreted with regard to forest and open site vegetation parameters, as well as the microclimatic conditions.

LaMalfa and Ryle (2008) note that in Utah, USA more snow is accumulated in meadows than in conifer forests, with the accumulation in deciduous forests being intermediate. The snowmelt runoff is formed accordingly. Snow accumulation in the forest depends considerably on the forest species composition and canopy closure. In conifer stands, snow cover water decreases with increasing canopy closure (Berris and Harr, 1987; Onuchin, 2001). Snow accumulation on open sites depends on site area, shape and location (Golding and Swanson, 1978; Onuchin, 1984).

Some researchers believe that glades (clearings or open areas) in the forest do not add to the total precipitation received by a watershed; that is, glades promote runoff (Kattelmann *et al.*, 1983; Folliott *et al.*, 1989). This effect is due to reduction of transpiration-caused water losses, accumulation of meltwater in glades and the presence of large amounts of the previous year's water in the soil (Gary and Troendle, 1982). River runoff increases are very prominent in wet years, whereas the increase becomes practically unidentifiable in dry years.

Based on their analyses of snow accumulation and melt, some researchers (López-Moreno and Stahli, 2008; Schleppi, 2011) state that the differences found between forest and open sites with respect to these processes are induced by differences in radiation regime, relative air humidity, convection and wind speed. These forest/open site differences vary with the geographic environment, including latitude, elevation above sea level and other climate controls. It has also been established that, depending on their geographical location, forests may have higher or lower snow amounts as compared with open sites and snow may melt in the forest earlier or later than in open sites.

The time that snow remains on the ground and the snow cover water content depend on both climate and forest cover parameters. According to Alewell and Bebi (2011), forest cover disturbance may have opposite effects in different climatic conditions. The hydrological effects of the disturbance are determined by snow accumulation and melt, which processes are sensitive to forest cover changes (Hibbert, 1969; Kattelmann *et al.*, 1983; Stednick, 1996). Hydrological differences of forests are, thus, largely controlled by the budget of snow water, the importance of which increases with increasing snowfall in annual precipitation.

Increasing forest area and forest cover density generally increase snow cover duration in cold areas, where the forest canopy acts as a filter to incoming shortwave radiation (Hardy *et al.*, 1997; Link and Marks, 1999). In warmer climates, the forest canopy accumulates more heat and thereby favours snowmelt (Davis *et al.*, 1997; López-Moreno and Latron, 2008).

When estimating the hydrological role of taiga forests, it is necessary to realize that, in winter, water cycling processes are determined largely by ice and snow properties. These may vary considerably with environmental conditions and are particularly influenced by air temperature and relative humidity. In many studies of taiga forest hydrological cycles, the discussion of how environment-caused changes of snow properties influence the intensity and directions of moisture flows in terrestrial ecosystems and in the surface layer is insufficient without further consideration of these variables.

In summer, when water is mainly in liquid and gaseous states, vertical moisture flows dominate in the surface atmospheric layer and all components of the ecosystems, including soil, participate in the active water cycle (physical evaporation, transpiration and runoff). Moisture flow intensity and direction are controlled by soil and vegetation characteristics, as well as by plant biomass, including the amounts of transpiring needles and leaves.

In winter, when precipitation occurs as snow and water is preserved in snow cover for a long time without transpiration, active water cycling moves largely to the surface atmospheric layer. The major controls of snow water flows and, hence, of the wintertime water budget are precipitation interception by the forest canopy, surface snow evaporation, wind-caused horizontal snow cover redistribution and surface snow evaporation during blizzards. In this season, the intensity and direction of moisture flows do not depend on vegetation productivity. They are determined largely by the vegetation cover type (a forest or an open site) and by environmental conditions.

Forest

Estimating snow accumulation in the forest commonly uses relative values such as snow storage or snow accumulation coefficients. These coefficients are the ratios of the snowpack in the forest to snowpack in relatively small forest glades or in deciduous forest sites, which show the capability of forest stands accumulating snow. These enable one to draw conclusions about canopy-intercepted snow amounts.

Many studies have attempted to model snow accumulation in the forest (Miller, 1964; Rosman, 1974; Harested and Bunnell, 1981; Rubtsov *et al.*, 1986; Hedstrom and Pomeroy, 1998). Today's models are mostly case models that consider canopy closure influences on snow accumulating and do not reflect regional differences properly. Our previous studies (e.g. Onuchin, 2001) revealed that snow interception by the forest canopy is to a large extent determined by winter air temperatures, which vary widely among regions. We attempted to build a generalized model of snow accumulation in the forest based on data obtained in different geographical conditions, including Canada (British Columbia, Alberta and Saskatchewan) and the USA (Alaska, Oregon, Montana, Idaho, Michigan and Minnesota). We also covered the following regions of North Eurasia: the mountains of Central Asia; the Republic of Belorussia; Mongolia; and different regions of Russia's forests from European Russia to the Russian Far East and from the mountain forests of southern Siberia to northern tundra open woodlands (Murashev and Kuznetsova, 1939; Molchanov, 1961; Harested and Bunnell, 1981; Rubtsov *et al.*, 1986; Voronkov, 1988; Hedstrom and Pomeroy, 1998; Onuchin, 2001; Buttle *et al.*, 2005; Konstantinos *et al.*, 2009). We processed this substantial amount of data on snow storage in the forest and developed a highly generalized model. This, along with considering forest stand parameters, indicated the importance of wintertime air temperatures.

Research

At the preliminary stage of modelling, we had considered as model parameters canopy closure, age and composition of stands, site class, average stand height and growing stock volume. However, based on the results of estimating the significance of the regression coefficients of the model, we limited the parameters used to stand canopy closure, age and tree species composition.

Overall, the developed model is based on the snow storage data for 243 forest stands that differ in species composition and age, with the period of data collection ranging from 1 to 12 years and with snowfall ranges from 30 to 830 mm and January air temperature ranging from −4 to −40°C. The model is:

$$K = 118.1 + 0.016S_0 - \frac{15.8\ln(A)}{\ln(T)} - 1.3LC - 2.4XC \quad (16.1)$$

$$\left(R^2 = 0.71, \sigma = 9.6, F = 209.7\right)$$

where K is the snow storage coefficient (%); S_0 is the amount of snow precipitation (mm); A is the stand age (years); T is the absolute average January air temperature (°C); L and X are coefficients in species composition formulae expressed in tens of per cent of the total growing stock for larch and other conifer trees in a stand, respectively; and C is the stand canopy closure (in units from 0 to 1).

The model (Eqn 16.1) enables us to quantify changes of snow storage in the forest depending on tree species composition and canopy closure for a range of weather and climatic conditions. The model analysis showed that snow interception by tree crowns increases with increasing stand age, wintertime air temperature, canopy closure and proportion of conifer tree species.

Dry snow, frost weather and rare but heavy snowfalls enhance snow penetration under the forest canopy. Increasing air temperature, in combination with frequent and light snowfalls, promotes snow interception by tree crowns (Plate 15). The role of the wind in this process varies. In dry and frost weather, even low wind results in the intercepted snow dropping from tree crowns and, hence, contributes to the snowpack. Moist snow holds well in tree crowns and wind enhances its evaporation, thereby providing the conditions for the next snowfall interception and, hence, resulting in decreasing snowpack. However, under the high relative humidity of a marine climate, or in the mountains, winds associated with snowfalls result in snow adhesion to tree crowns (Miller, 1964).

Our model shows that an increase in falling snow interception by tree crowns occurring from increasing air temperature is more prominent in mature stands than in young stands. This is because in any geographical conditions moist snow intercepted by tree crowns holds better to old branches, as they are stronger and more resistant to bending than young tree branches. According to the model, snowpack clearly tends to decrease at tree ages from 10 to 80 years, whereas tree ageing over 150 years has little influence on snowpack coefficient values.

We did numerical experiments using the model. Figure 16.1 presents the response of snowpack coefficients to changing air temperature and canopy closure of 100-year-old conifer stands, with a snowfall amount of 200 mm. Snow interception by tree crowns increases drastically with January temperature increasing over −15°C, and in this case air temperature influence on the interception is even higher than that of canopy closure.

Inconsistencies with the general trend of decreasing snowpack with increasing stand density of stocking and canopy closure observed for certain years are attributable to winter thaw spells, when more intensive snowmelt in stands of low stocking as compared with stands of high stocking offsets the difference in snow interception by tree crowns (Rutkovsky and Kuznetsova, 1940). As mentioned above, a similar effect occurs in a warmer climate, where the forest canopy accumulates heat, which enhances snowmelt (Davis *et al.*, 1997; López-Moreno and Latron, 2008).

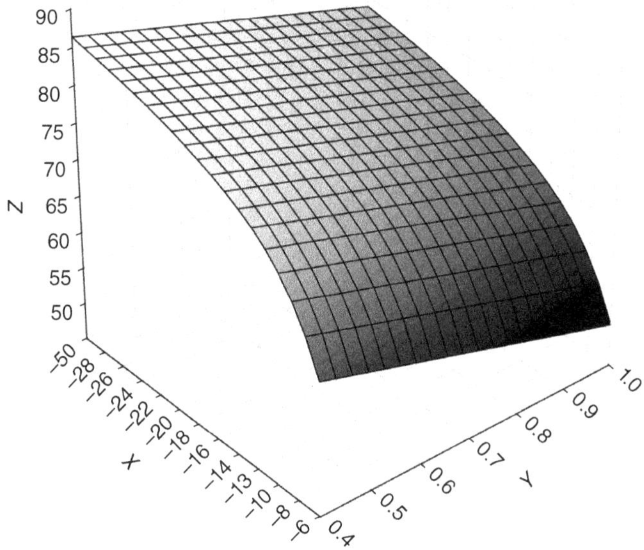

Fig. 16.1. Dependence of snowpack coefficient (*Z*, %) on conifer stand canopy closure (*Y*, shares of unit) and average January air temperature (*X*, °C).

When modelling snow interception, it is necessary to consider that, along with average air temperature, relative humidity and wind speed values, the changes of these values are important. For example, when a decrease in air temperature follows a snowfall that occurred at about 0°C background temperature and high relative humidity, even strong wind is unlikely to result in the intercepted snow falling to the ground. On the contrary, when a thaw follows the snowfall, most intercepted snow will slide to the ground.

Even models with a high level of generalization have application limits depending on specifics of individual regions and the range of variability of each predictor used. From this, we could conclude that our model assures reliable estimates of snow interception by the forest canopy for a cold continental climate without frequent and sudden thaw spells in winter.

Open sites

The estimation of the taiga zone hydrological cycles would be incomplete without analysing the water budget of open (treeless) sites. Earlier in the text, we discussed the characteristics of the water cycles of different types of landscape of the taiga zone in summer. It should be noted that the ratio between evaporation and runoff in the summer season is largely determined by vegetation biomass. In the cold season, snow water flows on open sites are controlled by many factors, of which background climatic conditions, site size, shape, aspect and location relative to prevailing winds are very important. Estimating the open site capability of accumulating snow traditionally uses a precipitation preservation coefficient (Sosedov, 1967; Onuchin, 1984). This coefficient is the ratio between snowpack and precipitated snow amount:

$$K = \frac{S_n}{X} \tag{16.2}$$

where *K* is precipitation preservation coefficient; S_n is water amount in the open site snow cover (mm); and *X* is the total amount of solid precipitation at the time of measuring snowpack (mm). This coefficient is, in essence, analogous to the coefficient used in estimating the snow accumulation functions of the forest.

Our analysis of the experimental data collected for the south-eastern areas adjacent to Lake Baikal, for central districts of Krasnoyarsk Region, and for Taimyr and other parts of Siberia revealed that snow cover formation in open sites is much controlled by snowstorms. Along with making snow more compacted, snowstorms promote snow evaporation and induce its redistribution. However, the contribution of snowstorms to snow evaporation remains a little-studied issue. Although there exists indirect evidence obtained by a balanced method of fairly intensive snow evaporation during snowstorms (Dyunin, 1961; Osokin, 1962; Berkin and Filippov, 1972), the published experimental data confirming this viewpoint are scarce.

On open sites, snow falls right on to the ground. However, wind blows the fallen snow out more intensively than in the forest, and this is often the case with snow evaporation and melting. Our studies conducted in various regions of Siberia show that, by the onset of the snowmelt period, less snow is present on large open sites or on sites located on windward slopes as compared with small-sized open sites protected from the prevailing wind, with the background snow precipitation being equal. Snowpack on frequent-wind open sites is 30 to 60% less than on wind-protected sites (Fig. 16.2). This decrease in snowpack is a result of wind-caused snow removal and more intensive snow evaporation during snowstorms. It is important to know what causes a decrease in snowpack in individual open sites, because the snow removed by wind from open sites still contributes to the regional river runoff formation, whereas evaporated snow water does not participate in this process.

Our studies of snow cover formation on open sites and in adjacent forest stands at the basin of Lake Baikal (Onuchin, 1984) showed that on north-facing slopes snowstorms inducing snow deflation occur on open sites of any size, provided that the sites are located in the upper parts of north-west-facing slopes, and also on open sites exceeding 15 to 20 ha in area, whatever their aspects. On other open sites of

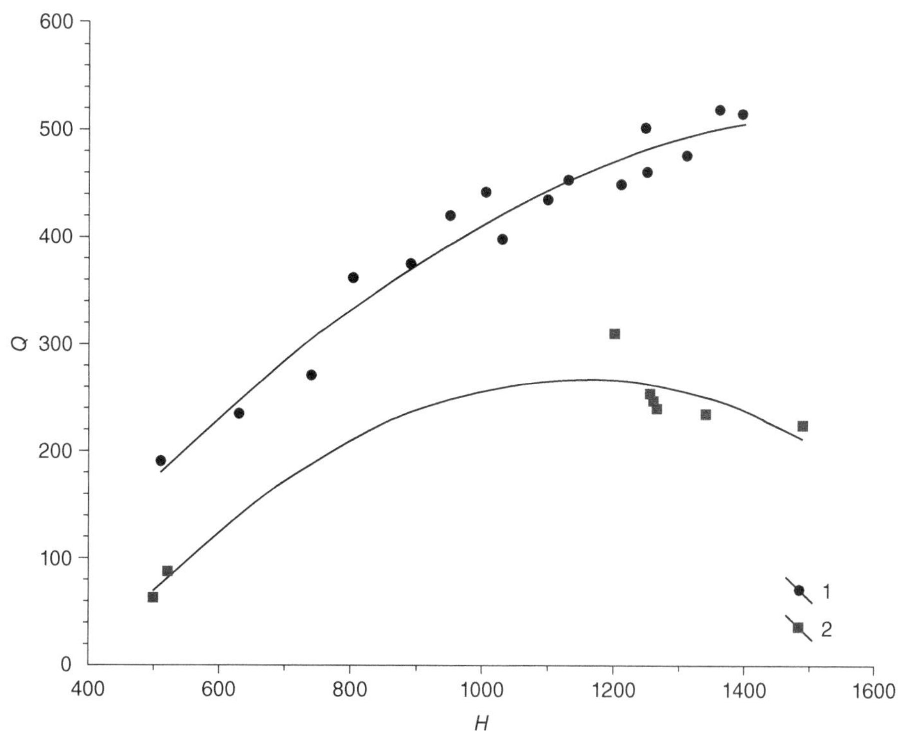

Fig. 16.2. Dependence of snow water equivalent (*Q*, mm), average for 1981–1983, on altitude (*H*, m): 1 = open areas protected from wind; 2 = open areas exposed to wind.

the northern slope and on sites located on the south-facing slope, deflation-inducing snowstorms are much less active, and on some sites they do not occur at all.

On large open sites and on open sites located on windward slopes, the snowpack variability coefficients usually range from 22 to 35%, compared with only 6 to 10% for relatively small and wind-protected forest glades. On frequent-wind sites, the variability of snow density and the amount of water stored in snow cover are higher than on wind-protected sites.

Our snowpack measurements showed that the contribution of the snow accumulated near the forest outskirt to snowpack, when converted to the total site area, varies from 8 to 21 mm (2 to 16%) depending on site size. Evaporated snow water derived from the difference between the background amount of solid precipitation and snowpack on open sites, with regard for the contribution by the snow accumulated near forest outskirts, amounted to 160 mm. This agrees with the results of other snow evaporation studies, which found that the evaporation might be as high as 140 to 200 mm in mountainous areas of Siberia (Osokin, 1962; Berkin and Filippov, 1972).

The experimental data (Onuchin, 1987) showed that under the same conditions evaporation of dry, fine snow was 1.5 to 3 times the evaporation of a dense snow monolith. When airflow velocity increased from 2 to 12 m/s, evaporation intensity grew from 0.3 to 2.0 mm/day and from 0.2 to 0.65 mm/day, respectively. The results were analysed to reveal that the significance of the same factors varies depending on locality conditions. For windward slopes, the distance between forest outskirts in the direction of the prevailing wind appeared to be the major factor, whereas the most important factor for wind-protected slopes was open site size.

We used our own and other published data (Pruitt, 1958; Miller, 1966; Sosedov, 1967; Berkin and Filippov, 1972; Golding and Swanson, 1978; Onuchin, 1984; Onuchin *et al.*, 2008) to identify general snow accumulation trends for open sites (Fig. 16.3). As is clear from Fig. 16.3, snowpack decreases with increasing open site area, because the wind activity increases. This relationship is most pronounced for cold winters. In the case of warm winters, open site area has little influence on snow accumulation. The influence of open

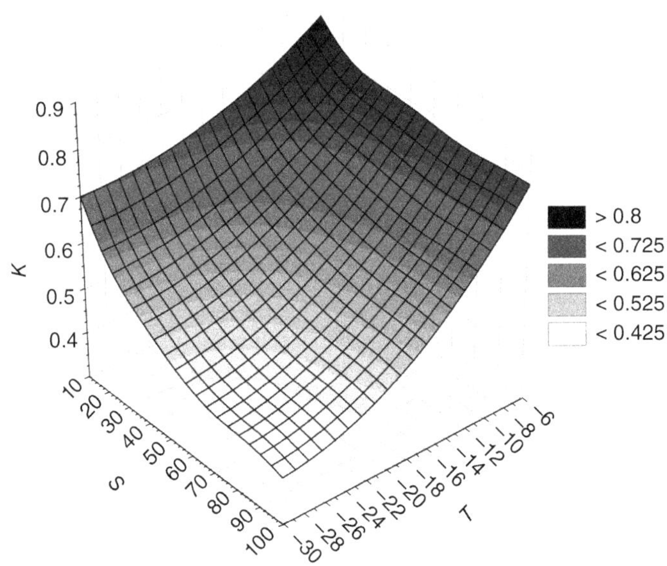

Fig. 16.3. Dependence of snow accumulation coefficient (*K*, shares of unit) on open site size (*S*, ha) and January subzero air temperatures (*T*, °C).

site area on snow accumulation also decreases under a stable, windless anticyclone.

16.4.2 Impact of forest cover on the water yield

After forest harvesting, while a new generation of forest grows, the forest ecosystem experiences continuous structural changes. Therefore, future hydrological scenarios for river basins are determined by climatic parameters and post-logging forest succession. There exists a wide range of probable responses of a geosystem water budget to forest cover disturbances, even under relatively homogeneous geographical conditions.

In some parts of Siberia, where post-logging forest regeneration may take a very long time, hydrological regime transformations in river basins are specific due to the extremely continental climate. In the first several years after clearcutting, increased wind activity on vast cut sites promotes snowstorms and snow evaporation and reduces snowpack accumulation. Under the same background climatic conditions, this results in decreasing annual runoff from the river basins subjected to clearcutting. As woody vegetation gradually recovers after logging, especially where the recovery occurs through the vegetation conversion, the capability of the recovering stands to accumulate snow recovers and even increases compared with pre-logging. The runoff from logged basins, which deciduous species usually gradually occupy, also increases (Krestovskiy, 1984; Onuchin *et al.*, 2006, 2009).

As climate becomes less continental, the response of the forest hydrological regime to forest logging changes considerably. In the first years after logging, runoff increases due to increased snow accumulation. The runoff is then reduced for a period of up to 50 years due to increased evapotranspiration of dense, highly productive young conifer stands that predominate during that time (Krestovskiy, 1984). Similar data were obtained for logged dark conifer sites of the north-facing macroslope of West Sayan in Siberia (Burenina, 1982; Lebedev, 1982). These studies showed that forest logging conducted in excessively moist conditions results in a sudden increase in runoff, which becomes stable as the forest regenerates. The time taken to reach stability is usually greater than 100 years.

We studied the hydrological effects of changes in the size of forest areas over the river basins of the mountains around Lake Issyk-Kul, Kyrgyzstan (formerly Kirgizia) (Onuchin *et al.*, 2008). In this region characterized by mountain climate and well-pronounced cycling of wet and dry periods, the effect appeared to be highly specific. For wet cycles, the river runoff was found to decrease with increasing forested area percentage as opposed to dry cycles, when evapotranspiration decreased and the total runoff increased with increasing forest area percentage.

We may thus state that the hydrological role of forests changes with changes in their structure and background climatic conditions. In a cold climate, reducing forest area results in increasing snowstorm activity and snow evaporation and, hence, decreases the total runoff (Fig. 16.4a). In a warmer climate, forest evapotranspiration becomes a factor, reducing river runoff (Fig. 16.4b). Therefore, forest logging results in a drastic runoff increase, because the water budget components, such as snow interception by tree crowns and stand transpiration, are reduced on logged sites, but snowpack evaporation does not differ much between the open sites and under the canopy.

The differentiation of boreal climate into 'cold' and 'warm' when estimating the hydrological role of forests is conditional. For instance, Fig. 16.4a and b shows the results of the numerical experiments with the runoff formation models for two areas: (i) the Kureyka River basin, Middle Siberia (an example of 'cold climate'), with January air temperature of −28.6°C (Fig. 16.4a); and (ii) the area around Lake Issyk-Kul, Kyrgyzstan (an example of 'warm climate'), where January air temperature averages −7°C (Fig. 16.4b) (Onuchin *et al.*, 2006, 2008). The studies to date do not cover the whole diversity of the boreal climatic conditions. Moreover, it should be remembered that water budget transformations of a river basin also depend on its size, geology and structure as well as on weather conditions. Numerical experiments with

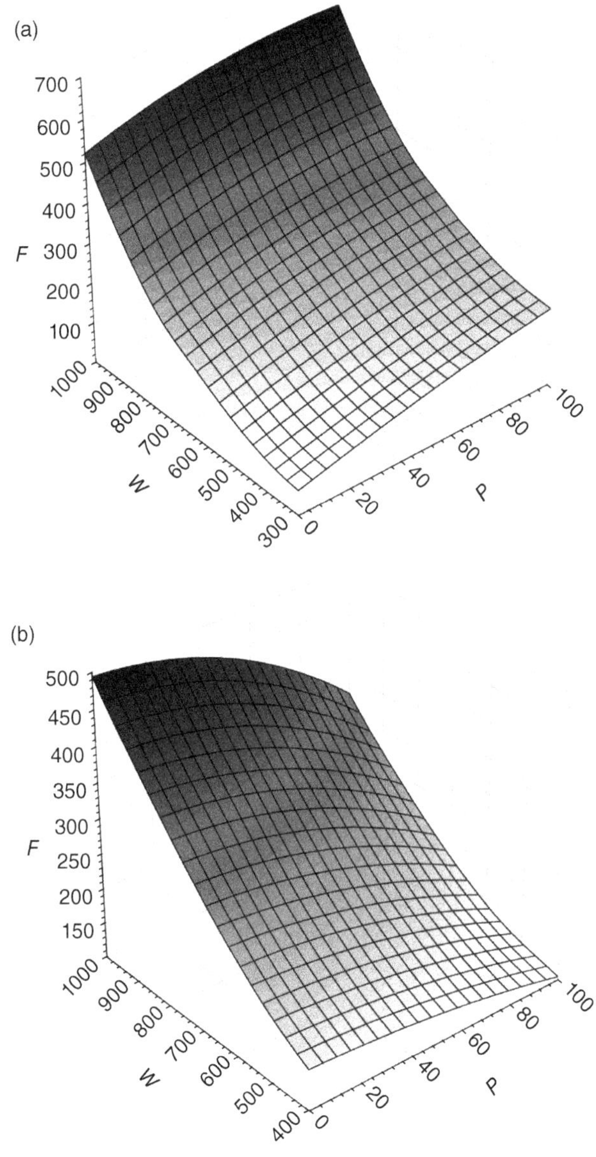

Fig. 16.4. Dependence of river runoff (*F*, mm) on basin forested area percentage (*P*, %) and total moisture (*W*, mm) in (a) cold and (b) warm climates.

the models developed by the authors of this chapter (Onuchin *et al.*, 2006, 2009; Onuchin, 2015) enabled us to identify the climatic thresholds beyond which the forest changes from being a factor that reduces river runoff to a factor that promotes river runoff and reduces water evaporation (Fig. 16.5).

16.5 Conclusion

The studies discussed above show that the hydrological role of taiga forests varies among river basins depending on snow water budget, which is controlled by forest vegetation parameters and background climatic conditions, including air

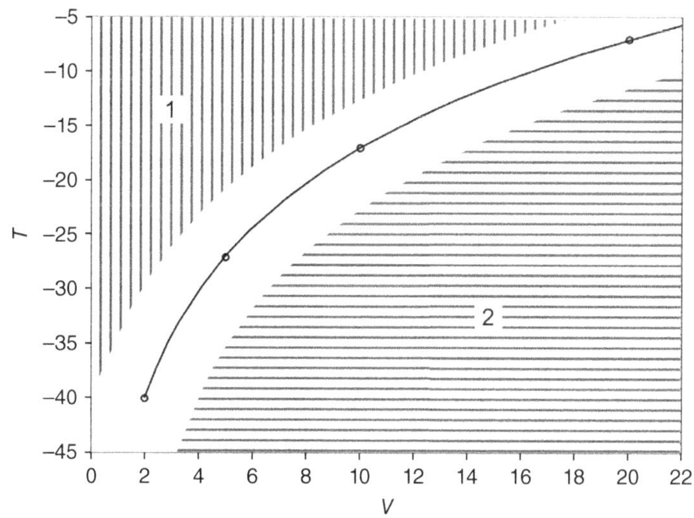

Fig. 16.5. Weather and climatic threshold controlling the water production function of the forest. *V* is wind speed (m/s); *T* is air temperature (°C); 1 = area in which the forest is a factor in evapotranspiration increasing and river runoff declining; 2 = area in which the forest is a factor in river runoff increasing.

temperature, relative humidity, wind speed and amount of snow in precipitation.

We may state with certainty that background climatic conditions induce the most prominent changes of the hydrological role of forests of the taiga zone. In the extremely continental climate characterized by low relative humidity and high wind activity, snow evaporation is always higher for open sites than for forested sites. High snow evaporation from open sites is due to snowstorms, whereas the snow that has dropped from the forest canopy to the ground is protected from deflation and evaporation. The difference between open and forest sites in snow evaporation may amount to hundreds of millimetres, which increases with increasing wind speed and decreasing relative humidity.

In warm winters of continental climate with higher relative humidity, snow evaporation is always less on open sites than in the forest, where tree crowns intercept much moist snow. In these conditions, wind promotes more snow evaporation from the forest canopy than from open sites. Dense and moist snow covering open sites is neither lifted from the ground nor redistributed by snowstorms, and its evaporating surface area is, therefore, fairly small as compared with the surface area of the snow intercepted by rough forest canopy. Intercepted snow remains in tree crowns for a long time, during which much of it evaporates.

The hydrological role of taiga forests is determined to a large extent by wintertime water cycling. Snow water flows are controlled by a combination of wind speed, air temperature and relative humidity. Disregarding these factors is the major reason for apparent contradictions of the estimates of the hydrological role of the forest.

Our studies have identified the major climatic thresholds beyond which the forest changes its water production function; that is, ceases to be a factor in reducing river runoff and becomes the cause of decreasing evapotranspiration and, hence, increasing river runoff. However, not all of the factors and their combinations influencing the hydrological role of forests have been thoroughly studied.

The system approach to analysing hydrological processes in the forest enables us to discover the roots of the contradictions of the estimates and to develop models that would predict water budget changes based on forest formation trends and background climatic conditions. This approach helps to develop a geographically specific concept of the hydrological role of forests. Such a concept will consider the water cycling mechanisms that determine hydrological effects of changing forest cover with regard to the geophysical background (Onuchin, 2015). The application of this concept in hydrological cycle models requires interpretation of the terms 'forest' and 'open area' in the context of landscape and quantitative evaluation. Including local water cycle models and the data on the vegetation cover dynamics into global hydrological models allows researchers to obtain consistent and spatially distributed water budget estimates for vast areas. This system approach will enable us to predict future changes of the hydrological regimes of different areas as related to global climate change and land-use regimes and may become an effective tool of sustainable natural resource use, including forest use.

16.6 Acknowledgement

The Government of the Russian Federation (grant number 14.B25.31.0031) supported this work.

References

Alewell, C. and Bebi, P. (2011) Forest development in the European Alps and potential consequences on hydrological regime. In: Bredemeier, M., Cohen, S., Godbold, D.L., Lode, E., Pichler, V. and Schleppi, P. (eds) *Forest Management and the Water Cycle: An Ecosystem-Based Approach*. Ecological Studies Vol. 212. Springer, Dordrecht, the Netherlands, pp. 111–126.

Berkin, N.S. and Filippov, A.K. (1972) Value of winter evaporation in Predbaikalie region. In: *Meteorology and Climatology of Pribaikalie*. Publishing House of ISU, Irkutsk, USSR, pp. 22–31 (in Russian).

Berris, S.N. and Harr, R.D. (1987) Comparative snow accumulation and melt during rainfall in forested and clear-cut plots in the western Cascades of Oregon. *Water Resources Research* 23, 135–142.

Bond, B.J., Meinzer, F.C. and Brooks, J.R. (2008) How trees influence the hydrological cycle in forest ecosystems. In: Wood, P.J., Hannah, D.M. and Sadler, J.P. (eds) *Hydroecology and Ecohydrology: Past, Present and Future*. Wiley, Chichester, UK, pp. 7–35.

Bosch, J.M. and Hewlett, J.D. (1982) A review of catchment experiments to determine the effect of vegetation changes on water yield and evapotranspiration. *Journal of Hydrology* 55, 3–23.

Bunnell, F.L., McNay, R.S. and Shank, C.C. (1985) *Trees and Snow: The Deposition of Snow on the Ground – A Review and Quantitative Synthesis*. IWIFR-17. Research Branch, Ministries of Environment and Forests, Victoria, British Columbia, Canada.

Burenina, T.A. (1982) Dynamics of hydrological regime restoration at the experimental felled area in West Sayan. In: Protopopov, V.V. (ed.) *Environmental Role of Siberian Forest Ecosystems*. Publishing House of Institute of Forest and Wood, USSR Academy of Sciences, Krasnoyarsk, USSR, pp. 106–114 (in Russian).

Burenina, T.A., Onuchin, A.A. and Fedotova, E.V. (2014) Hydrological effect of forest logging in boreal zone of Siberia. In: Roberts, N. (ed.) *Forest Ecosystems: Biodiversity, Management and Conservation*. Nova Science Publishers, Inc., New York, pp. 117–147.

Burenina, T., Onuchin, A., Guggenberger, G., Musokhranova, A. and Prisov, D. (2015) Dynamics of hydrological regime in permafrost zone of Central Siberia. In: Liu Juan (ed.) *Environment, Chemistry and Biology IV*. International Proceedings of Chemical Biological & Environmental Engineering, Vol. 90, IACSIT Press, Singapore, pp. 125–132.

Burton, P.J., Messier, C., Smith, D.W. and Adamowicz, W.L. (2003) *Towards Sustainable Management of the Boreal Forests*. NRC Research Press, Ottawa.

Buttle, J.M., Creed, I.F. and Moore, R.D. (2005) Advances in Canadian forest hydrology, 1999–2003. *Hydrological Processes* 19, 169–200.

Campbell, J.L., Bailey, A.S., Eagar, C., Green, M.B. and Battles, J.J. (2013) Vegetation treatments and hydrologic responses at the Hubbard Brook Experimental Forest, New Hampshire. In: Camp, A.E., Irland, L.C. and Carroll, C.J.W. (eds) *Long-Term Silvicultural & Ecological Studies: Results for Science and Management*, Vol. 2. GISF Research Paper 013. Global Institute of Sustainable Forestry, Yale University, New Haven, Connecticut, pp. 1–9.

Cohen, S. and Bredemeier, M. (2011) Synthesis and outlook. In: Bredemeier, M., Cohen, S., Godbold, D.L., Lode, E., Pichler, E. and Schleppi, E. (eds) *Forest Management and the Water Cycle: An Ecosystem-Based Approach*. Ecological Studies Vol. 212. Springer, Dordrecht, the Netherlands, pp. 507–512.

Davis, R.E., Hardy, J.P., Ni, W., Woodcock, C., McKenzie, J.C., Jordan, R. and Li, X. (1997) Variation of snow cover ablation in the boreal forest: a sensitivity study on the effects of conifer canopy. *Journal of Geophysical Research: Atmospheres* 102, 29389–29395.

Dyunin, A.K. (1961) *Evaporation of Snow.* Publishing House of the Siberian Branch of the USSR Academy of Sciences, Novosibirsk, USSR (in Russian).

Ermakov, N.B. (2003) *Diversity of Boreal Vegetation of Northern Asia. Hemiboreal Forests. Classification and Ordination.* Publishing House of the Siberian Branch of the Russian Academy of Sciences, Novosibirsk, Russia (in Russian).

FAO (2010) *Global Forest Resources Assessment 2010: Main Report.* FAO Forestry Paper No. 163. Food and Agriculture Organization of the United Nations, Rome.

Fedorov, S.F. (1977) *Water Balance Research in the Forest Zone of European Territory of the USSR.* Hydrometeoizdat, Leningrad, USSR (in Russian).

Folliott, P.F., Gottfried, G.J. and Baker, M.B. (1989) Water yield from forest snowpack management: research findings in Arizona and New Mexico. *Water Resources Research* 25, 1999–2007.

Gary, H.L. and Troendle, C.A. (1982) *Snow Accumulation and Melt under Various Stand Densities in Lodgepole Pine in Wyoming and Colorado.* Research Note RM-417. USDA Forest Service, Rocky Mountain Forest and Range Experimental Station, Fort Collins, Colorado.

Georgiyevsky, V.Y. and Shiklomanov, I.A. (2003) Climate change and water resources. In: Shiklomanov, I.A. and Rodda, J.C. (eds) *World Water Resources at the Beginning of the Twenty-First Century.* Cambridge University Press, Cambridge, pp. 390–413.

Golding, D.L. and Swanson, D.H. (1978) Snow accumulation and melt in small forest openings in Alberta. *Canadian Journal of Forest Research* 8, 380–388.

Gu, W.Z., Liu, J.-F., Lu, J.-J. and Frentress, J. (2013) Current challenges in experimental watershed hydrology. In: Bradley P.M. (ed.) *Current Perspectives in Contaminant Hydrology and Water Resources Sustainability.* InTech, Rijeka, Croatia, pp. 299–333.

Gubler, H. and Rychetnik, J. (1991) Effects of forests near the timberline on avalanche formation. In: Bergmann, H., Lang, H., Frey, W., Issler, D. and Salm, B. (eds) *Snow, Hydrology and Forests in High Alpine Areas.* IAHS Publication No. 205. IAHS Press, Wallingford, UK, pp. 19–38.

Hamilton, L.S. (2008) *Forests and Water.* FAO Forestry Paper No. 155. Food and Agricultural Organization of the United Nations, Rome.

Hardy, J.P., Davis, R.E., Jordan, R., Li, X., Woodcock, C., Ni, W. and McKenzie, J.C. (1997) Snow ablation modeling at the stand scale in a boreal jack pine forest. *Journal of Geophysical Research: Atmospheres* 102, 29397–29405.

Harested, A.S. and Bunnell, F.L. (1981) Prediction of snow-water equivalents in coniferous forests. *Canadian Journal of Forest Research* 11, 854–857.

Hedstrom, N.R. and Pomeroy, J.W. (1998) Measurements and modelling of snow interception in the boreal forest. *Hydrological Processes* 12, 1611–1625.

Hegg, Ch., Badaux, A., Luscher, P. and Witzig, R. (2004) Zur Schutzwirkang des waldes gegen Honwassser. *Forum für Wissen* 2004, 15–20.

Hibbert, A.R. (1969) Water yield changes after converting a forested catchment to grass. *Water Resources Research* 5, 634–640.

Hornbeck, J.W., Martin, C.W. and Eagar, C. (1997) Summary of water yield experiments at Hubbard Brook Experimental Forest, New Hampshire. *Canadian Journal of Forest Research* 27, 2043–2052.

Hornbeck, J.W., Bailey, A.S., Eagar, C. and Campbell, J.L. (2014) Comparisons with results from the Hubbard Brook Experimental Forest in the northern Appalachians. In: Swank, W.T. and Webster, J.R. (eds) *Long-Term Response of a Forest Watershed Ecosystem: Clearcutting in the Southern Appalachians.* Oxford University Press, New York, pp. 213–228.

Johnson, R. (1998) The forest cycle and low river flows: a review of UK and international studies. *Forest Ecology and Management* 109, 1–7.

Jones, H.G., Pomeroy, J.W., Walker, D.A. and Hoham, R.W. (2001) *Snow Ecology: An Interdisciplinary Examination of Snow-Covered Ecosystems.* Cambridge University Press, Cambridge.

Kattelmann, R.C., Berg, N.H. and Rector, J. (1983) The potential for increasing streamflow from Sierra Nevada watersheds. *Journal of the American Water Resources Association* 19, 395–402.

Keller, H.M. (1988) European experiences in long-term forest hydrology research. In: Swank, W.T. and Crossley, D.A. (eds) *Forest Hydrology and Ecology at Coweeta.* Ecological Studies Vol. 66. Springer, New York, pp. 407–414.

Kobayashi, D. (1987) Snow accumulation on a narrow board. *Cold Regions Science and Technology* 13, 239–245.
Kolomyts, E.G. (1975) Formation and distribution of snow cover of Sos'vinsk Priobie. In: *The Sos'vinsk Priobie*. Publishing House of ISU, Irkutsk, USSR, pp. 158–214 (in Russian).
Konstantinos, M.A., Pascal, S. and Lettenmaier, D.P. (2009) Modeling snow accumulation and ablation processes in forested environments. *Water Resources Research* 45(5), W05429, doi: 10.1029/2008WR007042 (accessed 10 April 2016).
Krestovskiy, O.I. (1984) Logging and forestation impact on water-yield in southern and middle taiga of ETU. *Journal of Water Resources* 5, 125–135 (in Russian).
LaMalfa, E.M. and Ryle, R. (2008) Differential snowpack accumulation and water dynamics in aspen and conifer communities: implications for water yield and ecosystem function. *Ecosystems* 11, 569–581.
Lebedev, A.V. (1982) *Hydrological Role of Siberian Mountain Forests*. Nauka, Siberian Branch, Novosibirsk, USSR (in Russian).
Link, T.E. and Marks, D. (1999) Distributed simulation of snow cover mass- and energy-balance in the boreal forest. *Hydrological Processes* 13, 2439–2452.
López-Moreno, J.I. and Latron, J. (2008) Influence of canopy density on snow distribution in a temperate mountain range. *Hydrological Processes* 22, 117–126.
López-Moreno, J. and Stahli, M. (2008) Statistical analysis of the snow cover variability in a subalpine watershed: assessing the role of topography and forest interactions. *Journal of Hydrology* 348, 379–394.
Lundberg, A. and Halldin, S. (1994) Evaporation of intercepted snow: analysis of governing factors. *Water Resources Research* 30, 2587–2598.
Miller, D.H. (1964) *Interception Processes during Snowstorms*. Research Paper PSW-18. USDA Forest Service, Pacific Southwest Forest and Range Experiment Station, Berkeley, California.
Miller, D.H. (1966) *Transport of Intercepted Snow from Trees during Snow Storms*. Research Paper PSW-RP-33. USDA Forest Service, Pacific Southwest Forest and Range Experiment Station, Berkeley, California.
Molchanov, A.A. (1961) *Forest and Climate*. Academy of Sciences of the USSR, Moscow, USSR (in Russian).
Murashev, S.I. and Kuznetsova, Z.I. (1939) Influence of the composition, age and completeness plantations on snow regime. In: *Water Regime in the Woods*. VNIILH, Pushkino, USSR, pp. 27–44 (in Russian).
Onuchin, A.A. (1984) Snow accumulating role of forest openings in forest ecosystems of Khamar-Daban. In: Protopopov V.V. (ed.) *Transformation of Environmental Factors by Forest Ecosystems*. Institute of Forest and Wood, Siberian Branch of the USSR Academy of Sciences, Krasnoyarsk, USSR, pp. 75–86 (in Russian).
Onuchin, A.A. (1987) Transformation of solid precipitation by mountain forests of Khamar-Daban. Abstract of dissertation for Candidacy in Biological Sciences, Krasnoyarsk, USSR (in Russian).
Onuchin, A.A. (2001) General snow accumulation trends in boreal forests. *Proceedings of the Academy of Sciences: Geographic Series* 2, 80–86 (in Russian).
Onuchin, A. (2015) The reasons of conceptual contradictions in evaluating hydrological role of boreal forests. *Siberian Journal of Forest Science* 2, 41–54 (in Russian).
Onuchin, A. and Burenina, T. (2008) The hydrological role of forests in Siberia. In: Prescott, A. and Barkely, T. (eds) *Trends in Water Resources Research*. Nova Science Publishers, New York, pp. 67–92.
Onuchin, A., Balster, H., Borisova, H. and Blitz, E. (2006) Climatic and geographic patterns of river runoff formation in Northern Eurasia. *Advances in Water Resources* 29, 1314–1327.
Onuchin, A., Gaparov, K. and Mikheeva, N. (2008) Forest and climate factors impacts on annual runoff of Issik-Kul basin rivers. *Journal of Forestry* 6, 45–52 (in Russian).
Onuchin, A., Burenina, T., Gaparov, K. and Ziryukina, N. (2009) Land use impacts on river hydrological regimes in Northern Asia. In: Cluckie, I.D., Chen, Y., Babovic, V., Konikow, L., Mynett, A., DeMuth, S. and Savic, D. (eds) *Hydroinformatics in Hydrology, Hydrogeology and Water Resources. Proceedings of Symposium JS4 at the Joint Convention of the International Association of Hydrological Sciences (IAHS) and the International Association of Hydrogeologists (IAH) held in Hyderabad, India, 6–12 September 2009*. IAHS Publication No. 331. International Association of Hydrological Sciences, Wallingford, UK, pp. 163–170.
Osawa, A., Zyryanova, O.A., Matsuura, Y., Kajimoto, T. and Wein, W. (eds) (2010) *Permafrost Ecosystems: Siberian Larch Forests*. Ecological Studies Vol. 209. Springer, Dordrecht, the Netherlands.
Osokin, I.M. (1962) On varying snow cover density with height in mountain regions of the USSR. *Proceedings of Zabaikalie Branch, the All-Union Geographical Society, Chita* 18, 26–31 (in Russian).
Pomeroy, J.W. and Gray, D.M. (1995) *Snow Accumulation, Relocation and Management*. NHRI Science Report No. 7. National Hydrology Research Institute, Environment Canada, Saskatoon, Saskatchewan, Canada.

Pomeroy, J.W. and Schmidt, R.A. (1993) The use of fractal geometry in modelling intercepted snow accumulation and sublimation. In: Proceedings of the 50th Eastern Snow Conference, Quebec City. Available at: http://www.usask.ca/hydrology/papers/Pomeroy_Schmidt_1993.pdf (accessed 10 April 2016). pp. 1–10.
Pruitt, W.O. Jr (1958) Qali, a taiga snow formation of ecological importance. *Ecology* 39, 169–172.
Rosman, A.P. (1974) Influence of crown density on distribution of snow cover in pine phytocenoses of middle Sikhote-Alin. In: *Water-Protective Importance of Forests*. Far East Publishing House, Vladivostok, USSR, pp. 47–48 (in Russian).
Rubtsov, M.V., Deryugin, A.A. and Gurtsev, V.I. (1986) Forest impact on snow accumulation and snow melt in middle taiga of the European north. *Journal of Forestry* 2, 11–16 (in Russian).
Rutkovsky, V.I. and Kuznetsova, Z.I. (1940) Impact of stands on snow regime. In: *Water Protective Role of Forest*. VNIILKH, Pushkino, USSR, pp. 149–179 (in Russian).
Schleppi, P. (2011) Forested water catchments in a changing environment. In: Bredemeier, M., Cohen, S., Godbold, D.L., Lode, E., Pichler, V. and Schleppi, P. (eds) *Forest Management and the Water Cycle: An Ecosystem-Based Approach*. Ecological Studies Vol. 212. Springer, Dordrecht, the Netherlands. pp. 89–110.
Schmidt, R.A. and Gluns, D.R. (1991) Snowfall interception on branches of three conifer species. *Canadian Journal of Forest Research* 21, 1262–1269.
Schmidt, R.A. and Pomeroy, J.W. (1990) Bending of a conifer branch at subfreezing temperatures: implications for snow interception. *Canadian Journal of Forest Research* 20, 1250–1253.
Schmidt, R.A., Jairell, R.L. and Pomeroy, J.W. (1988) Measuring snow interception and loss from an artificial conifer. In: *Proceedings of the 56th Western Snow Conference, Kalispell, Montana, April 19–21, 1988*. Colorado State University, Fort Collins, Colorado, pp. 166–169.
Shvidenko, A. and Apps, M. (2006) The International Boreal Forest Research Association: understanding boreal forests and forestry in a changing world. *Mitigation and Adaptation Strategies for Global Change* 11, 5–32.
Sosedov, I.S. (1967) *Studying of the Snow Water Balance at Mountain Slopes of Zailiyskiy Alatau*. Publishing House 'Nauka', Alma-Ata, USSR (in Russian).
Spurr, S.H. and Barnes, B. V. (1980) *Forest Ecology*, 3rd edn. Wiley, New York.
Strakhov, V.V., Pisarenko, A.I. and Borisov, V.A. (2001) The forests of the world and Russia. *Bulletin of the Ministry of Natural Resources RF 'Using and protection of natural resources of Russia'* 9, 49–63 (in Russian).
Stednick, J.D. (1996) Monitoring the effects of timber harvest on annual water yield. *Journal of Hydrology* 176, 79–95.
Voronkov, N.A. (1988) *Significance of Forests for Water Protection*. Gydrometeoizdat, Leningrad, USSR (in Russian).
Waldenmeyer, G. (2003) *Abflussbildubg und Regionalisierung in einem fortlicch genutzten Einzagsgebit (Durreychtal, Nordschwarrwald)*. Universität Karlsruhe/Karlsruher Schriften Band 20. Institut für Geographie und Geoökologie, Universität Karlsruhe, Karlsruhe, Germany.
Wheeler, K. (1987) Interception and redistribution of snow in a subalpine forest on a storm by storm basis. In: *Proceedings of the 55th Annual Western Snow Conference, Vancouver, British Columbia, Canada*. Western Snow Conference, Vancouver, British Columbia, Canada, pp. 78–87.
Whitehead, P.G. and Robinson, M. (1993) Experimental basin studies – an international and historical perspective of forest impacts. *Journal of Hydrology* 145, 217–230.
Winkler, R., Spittlehouse, D., Boon, S. and Zimonick, B. (2015) Forest disturbance effects on snow and water yield in interior British Columbia. *Hydrology Research* 46, 521–532.

17 Future Directions in Forest Hydrology

T.M. Williams[1]*, D.M. Amatya[2], L. Bren[3], C. de Jong[4] and J.E. Nettles[5]

[1]Clemson University, Georgetown, South Carolina, USA; [2]USDA Forest Service, Cordesville, South Carolina, USA; [3]The University of Melbourne, Creswick, Victoria, Australia; [4]University of Strasbourg, Strasbourg, France; [5]Weyerhaeuser Company, Columbus, Mississippi, USA

17.1 Forest Hydrology: What Have We Learned?

17.1.1 Hydrological cycle in forests

Forest hydrology is a separate and unique branch of hydrology due to the special conditions caused by trees, and the understorey beneath them, comprising a forest. Understanding the forest, with trees that can grow over 100 m tall, may have crowns up to 20–30 m in diameter with roots 5–10 m deep and spread as widely as the crowns, and have lifespans from 50 to 5000 years, presents unique challenges to science. Forests cover approximately 26.2% of the world, with 45.7% of Latin America and the Caribbean being covered, 35% of East Asia and the Pacific, and 35% of the European Union. Canada and the USA combined account for only 6.8% of the world's forests, while Africa has even less at 5.7% (About.com, 2013). The wide distribution of forests makes it difficult to generalize about the role of trees and forest ecosystems in the global hydrological cycle.

The 16 chapters organized in this book deal with major hydrological processes such as runoff, drainage and evapotranspiration on various forest types from northern boreal forests to tropical forests, from snow-dominated temperate mountain forest watersheds to low-gradient humid coastal plain forests, small- to large-scale watersheds, and most other forest types including flooded and wetland forests. Most forests lose water through evaporation of precipitation intercepted by crowns (Chapter 3), with losses greatest for conifers in regions of frequent low-intensity rainfall separated by dry periods (Chapter 14). Yet in some tropical montane forests, water condenses from the atmosphere on to tree leaves and the resulting drip may increase precipitation by up to 20% (Chapters 2 and 6). In very cold continental climates (Chapter 16) open areas are more likely to lose water (snow) by sublimation and wind than forests; but in warmer regions openings have greater melt and produce more water than forests. Transpiration is a dominant process in the forest hydrological cycle, but is generally difficult to measure directly (Chapter 3), except on a single tree basis. Estimates of transpiration on a stand, hillslope or watershed basis cannot be separated from evaporation leading to the coined word 'evapotranspiration', which Savenije (2004) suggested hampers our understanding of the process. Although these are only a few of the problems associated with trying to explain forest hydrology that varies with temperature, rainfall, species, tree age, slope, drainage and soil type, this

*Corresponding author; e-mail: tmwllms@clemson.edu

summary strives to outline some of the major findings across the forests of the world.

Figure 10.1 shows the distribution of forests spread over much of the earth. Tropical forests are concentrated in Africa, South America and South-East Asia; subtropical forests in south-eastern North America and south-eastern Asia. Temperate forests are concentrated in Europe, coastal Australia, eastern North America and far eastern Asia; and the vast northern boreal forests circle the earth in a band across Europe, Asia and North America. Temperature limits alpine forests with treeline elevations varying from 700 m in Sweden (68°N) to 4000 m in Mexico (19°N) and Ecuador (0°) but all with mean annual surface temperature between 6 and 7°C (Körner and Paulsen, 2004). Likewise, forests are absent in regions of permafrost that occurs with mean annual air temperature below about –6°C (Smith and Riseborough, 2002). The other major limit to forest distribution is the balance of annual rainfall and potential evaporation. In the tropics forests are absent when rainfall is below 1000 mm (Staver *et al.*, 2011) yet in the arctic (61°N) forests are present with 260 mm of annual precipitation (Carey and Woo, 2001). In both the subtropical and temperate zones forests are also constrained by rainfall and evaporation, but patterns of forest are highly fragmented due to human land uses. As we see in the discussions in Chapters 1 and 5, the interaction of man with forest land and hydrology provides great incentive to resolve issues in forest hydrology. The long recorded history of European settlement may provide insight into variations due to long-term human–forest interaction. Similarly, Chapter 8 provides an overview of runoff dynamics of drained forests managed for silvicultural production.

While Chapters 9 and 10 deal with reviews of modelling tools and applications of geospatial technologies for understanding hydrological processes, impact assessments and decision support systems on forested landscapes, Chapter 15 addresses challenges in forest hydrological science for watershed management in the remainder of 21st century. Below we provide some critical highlights of what we learned from each chapter and where we go from here regarding various aspects of forest hydrological science, its applications, limitations, challenges, and opportunities for advancing it to address the issues of changing land use and climate change.

Boreal forests

Although absent in the southern hemisphere, boreal forests are the most widespread forests in the world. The type includes both maritime and continental climatic regions and tends to occur somewhat further south in eastern Eurasia and North America. Chapters 4 and 16, and related parts of Chapter 14, discuss the ways that forests cause snow hydrology to vary with temperature and winds during and shortly after the snow falls. In colder regions snow does not adhere to tree crowns as well and is easily dislodged by wind. On the ground it is subject to further wind distribution and sublimation (Chapter 16). In warmer and more maritime regions interception of snow is greater due to the tendency of snow near 0°C to adhere to foliage and there is greater likelihood of partial melting and refreezing. These effects reverse the normal impact of forest removal in far northern regions of Siberia, causing a postharvest reduction of water available for streamflow (Chapter 16). While snowmelt is the most important factor causing streamflow, in the southern areas rain-on-snow events are often associated with largest flows (Chapter 14). Streamflow may cease in small watersheds due to freezing in winter and/or increased evapotranspiration rates in the summer (Chapter 14), while flow beneath the ice in larger rivers is difficult to measure. Wetlands of this region are discussed quite extensively in Chapter 7 and data from Caribou-Poker Creek watershed in Chapter 14 illustrate one example of hydrology of this forest region.

Temperate forests

The vast majority of temperate forests lie in the northern hemisphere, with other areas found in the South Island of New Zealand, southern and eastern Australia (including Tasmania), and Chile. These forest types have been studied most extensively and forest hydrology as a science originated in the temperate forests of central Europe in the 18th century. Most of the long-term US forest hydrology data (Chapter 14) originate in temperate forest types. Nearly all of the runoff process studies cited in Chapter 2 occurred in temperate forests of North America, Europe, New Zealand and Australia. Likewise, the bulk of paired watershed research outlined in Chapter 12 also took place in those areas.

The most distinct characteristics of this forest region are a long dormant season due to low temperatures and extensive deciduous forests in the northern hemisphere. High evapotranspiration rates during late spring and summer generally result in a considerable deficit of soil moisture as precipitation minus evapotranspiration in late summer and early autumn. Precipitation during the winter may be stored as snow or as recharge of soil moisture, when forest vegetation is dormant and deciduous species are leafless. Streamflow is generally seasonal with highest flows during the spring, due to snowmelt, rain on snow, or high soil moisture.

The data in Chapter 14 demonstrate the wide variety of temperate forest hydrology associated with geographic and geological conditions. Marcell, Minnesota has hydrology similar to the boreal region with long cold winters due to its low relief and northern mid-continental location. Hubbard Brook, New Hampshire and Fraser, Colorado both have hydrology dominated by snowmelt despite Fraser being 5° further south. Snowmelt is a large component of runoff in Colorado due to elevation and mid-continental location. Strong maritime influences lessen the impact of snow accumulation and melt at the H.J. Andrews, Oregon and Casper Creek, California watersheds, as does the southern locations of Fernow, West Virginia and Coweeta, North Carolina. Pacific heavy winter rains create high runoff during the winter in the western watersheds while spring rains on moist ground are more likely to produce high runoff in the eastern ones. While high summer evapotranspiration is important in all four of these watersheds, the eastern watersheds are more likely to encounter runoff-producing summer thunderstorms and the occasional impact of tropical cyclones.

The above discussion of variations in forest hydrology of temperate North America is likely to be equally important in Europe (Chapter 5, this volume). However, experimental catchments have been more concentrated on temperate and alpine forests such as the Swiss Sperbel and Rappengraben, the German Eberswalde (temperate), the Welsh Plynlimon and German Harz (boreal). The review of European studies in Chapter 5 suggests that lack of large nationally owned forests and complex European policies precluded the type of coordinated collection of long-term basin-scale forest hydrological data as was presented in Chapter 14. Europe also has high mountains, maritime climatic regions, and more continental climatic regions in Eastern Europe and Russia. The exception to the generalization has been the work done in Scandinavia, much of which involves drainage of wetland forests as outlined in Chapter 8.

Subtropical forests

Forests in the subtropical climatic zones have two distinct climatic patterns: (i) rainfall well distributed throughout the year; and (ii) winter rainfall with hot dry summers. Forests in the well-distributed rainfall zone occur in the south-eastern USA, south-eastern China, southern Japan, north-eastern Australia and the North Island of New Zealand. The second type occurs around the Mediterranean Sea, which gives this climatic type its common name. Mediterranean climate is also common in parts of south-western North America and southern Australia.

The northern and southern (in each hemisphere) edges of the subtropical zone have similarity to the temperate zone, with a winter season of plant dormancy and lower evaporative demands. With a Mediterranean climate lower winter evaporative demand combines with higher rainfall rates for a highly seasonal runoff pattern. The data from San Dimas, California in Chapter 14 exemplify forest hydrology of this climatic type. Although forest fires are common in all climate types, the dry summers of the Mediterranean type make the hydrological impacts of forest fire (Chapter 13) an important aspect in this region.

In eastern North America and Asia the subtropical climatic zone is also influenced by maritime climate from the Northern Tropical Convergence Zone during summer and autumn. Tropical cyclones (hurricanes and typhoons) are the most spectacular aspect of this influence, but higher summer rainfall is common (see Fig. 7.4 for example). Summer runoff is generally small or non-existent, as demonstrated by data from Santee, South Carolina (Chapter 14). Yet, rainfalls of 100–600 mm associated with tropical storms produce large areas (approaching 100%) of saturation-excess overland flow throughout coastal low-gradient watersheds.

Tropical forests

One important aspect of tropical forest hydrology is the lack of dormancy due to cold temperature. Trees are evergreen and transpire the entire year, as long as there is adequate rainfall. Tropical forests are divided into rainforests, with high year-round rainfall, and seasonal monsoonal forests with a pronounced dry season (see Plate 5). High energy associated with direct solar angles causes high rates of evaporation and intense thunderstorms where high atmospheric moisture is available. High energy results in extreme rates of hydrological processes which may bring into question the validity of principles that have been tested primarily in the temperate zone.

The tropical rainforests are located primarily in the Amazon and Congo Basins and South-East Asia as well as insular and montane forests where prevailing winds cause advection of coastal moisture. These forests are generally close to the equator and have relatively constant daily temperature fluctuation throughout the year. Despite multi-layered evergreen forests interception losses can be low as 9% of rainfall in the Amazonian rainforest (Lloyd and Marques, 1988).

Seasonal monsoons are most typically associated with India and South-East Asia, but seasonality is fairly high between 15° and 20° north and south latitude, associated with seasonal movement of the Inter-Tropical Convergence Zone (Plate 5). During the rainy season these regions may have high-intensity rainfall over sustained periods. Bonell and Gilmour (1978) found surface runoff and rainfall intensity were factors in forest hydrology of northern Australia, in contradiction to Hewlett *et al.*'s (1977) contention that rainfall intensity did not explain streamflow volume or peak discharge in humid temperate forests (see Chapter 1). Elsenbeer (2001) suggested that occurrence of surface runoff on tropical watersheds was determined by rainfall intensity and vertical conductivity of subsurface soil layers (see Chapter 6).

17.2 Where Do We Go from Here?

17.2.1 What will forest hydrology become?

Forest hydrology emerged as an effort to understand how human changes to the forests altered the amount of water in our rivers during floods and droughts. Now humans are changing both landscape and (likely) climate. At the same time forests are an integral component of the landscape and maintaining their functional integrity is necessary for the sustainability of both ecosystems and societies (Amatya *et al.*, 2011). There is an urgent need for better understanding of the linkages between trees, forests and water, for awareness raising and capacity building in forest hydrology, and for embedding this knowledge and research findings in policies (Hamilton *et al.*, 2008; Chapter 5, this volume). Many of the challenges of the coming decades discussed in the context of Europe and the south-eastern USA (Chapters 5 and 15) are equally applicable for many parts of the world. Forest hydrology over the last century has been concerned primarily with the effects of various forms of forest management on water quantity and quality. Over the next century the role of forests in mitigating climate change may become the greatest challenge. As we see in Chapter 3, forest carbon assimilation and transpiration are controlled by the same physiological mechanism, stomatal opening. Rapidly growing forests can provide sustainable carbon-neutral energy. Trees also assimilate carbon and can sequester that carbon for centuries to millennia. However, intake of CO_2 requires exposing internal leaf tissue to the atmosphere, with transpiration occurring when vapour pressure is lower. Only by understanding the variation in water use per unit of assimilated carbon can we understand and manage forests to balance growth for wood products, energy production and carbon assimilation with water use.

In addition to carbon assimilation, streamflow from forested watersheds produces high-quality water requiring minimal treatment for drinking-water. Forests play a role in aquifer recharge by affecting the processes by which rain is partitioned into recharge and runoff. Although those processes have been well defined (Chapter 2), understanding how climate, forest characteristics and geology determine the pathways and quantities of water movement, from crowns to stream or aquifer, is still far from our grasp. Because forests make up a relatively large portion of many of our watersheds, it is important to understand the hydrology, processes and their pathways on both natural and managed forests, while considering the contribution of other land uses (Amatya *et al.*, 2015).

Much of our present understanding of forest hydrology is limited mainly to research on temperate forests, so even the most well-established tenets do not always apply universally. We have given examples of contrasting situations; for example, cutting forests of parts of Siberia may decrease streamflow rather than increase it. In another example, forest floor infiltration may not exceed rainfall intensity during intense tropical thunderstorms. To extend forest hydrology, the underlying principles must be found by extending research into all forested regions.

17.2.2 Evaporation and transpiration

Evapotranspiration (ET), the word that is dear to the hearts of many forest hydrologists and land and water managers, reveals how very little we really know about the principles that drive movement of water from forests into the atmosphere. ET accounts for the greatest flow in most forested ecosystems (Chapter 3), but is measured well only on particular forest stands and/or watersheds where there is no loss of water from the watershed, other than that measured at the weir. ET has been estimated for nearly every paired watershed experiment, but always as the residual in water balance so that it includes all the errors and unknowns. ET measurement (or lack of direct measurement) may well be the reason for the '$R^2 = 0.8$' dilemma posed in Chapter 1. Does rainfall and evaporative potential (PET) explain about 80% streamflow in all forests? Until we can quantify how actual evaporation (E) and transpiration (T) change with forest characteristics, climatic drivers and weather conditions, we may have no hope of doing better than $R^2 = 0.8$, regardless of the model form we use.

E and T measurements may be the most rapidly expanding part of forest hydrology. Sapflow measurements have great potential for understanding the relationship of forest ecology to hydrology. Wide differences in sapflow are evident between different tree species and sizes as can be seen in Plate 2. Understanding how these species differences relate to autecology of those species will become a productive avenue for future research. Also, new remote sensing techniques of airborne, or ground-based, LiDAR (Vauhkonen *et al.*, 2016), addressed in Chapter 10, may produce better estimates of crown dynamics than diameter at breast height and leaf area index. Such advances will potentially allow understanding landscape-scale ET. Novel approaches like the one studied by Good *et al.* (2015), who combined two distinct stable-isotope flux partitioning techniques to quantify ET subcomponents (interception, transpiration, soil evaporation and surface water evaporation) and the hydrological connectivity of bound, plant-available soil waters with more mobile surface waters, can also be explored for forest systems.

Scaling E and T measurements over plot, hillslope, watershed and regional space presents another challenge. Sapflow produces an accurate estimate of transpiration for a single tree. Eddy-covariance towers sample integrated areas depending on fetch. Water balance works only for gauged basins with minimal deep seepage. Remote sensing from satellites can measure worldwide data for estimates of evapotranspiration but their current resolution limits application on plot or small watershed scale unless high-resolution images with ground-truthing are used (Chapter 10). A method is needed to integrate and compare results from these methods such that E and T can be measured at any scale appropriate to a societal need.

17.2.3 Condensation

Condensation is the process that is least pursued in forest hydrology despite the fact that it may represent an important part of water exchange. Makarieva *et al.* (2014) have put forward the thoery that air passage over forests yields more rainfall since forest areas with the highest evaporation drive both upwelling and condensation. However, rather than merely influencing the moisture content in the air that is passing over a forest, the process of evapotranspiration can impact regional atmospheric dynamics by enhancing rainfall and thus modifying large-scale pressure gradients. They argue that this, in turn, enhances and stabilizes precipitation in a positive feedback loop.

Scientists at the WSL Birmensdorf (Herzog *et al.*, 1995) carried out long-term experiments on water exchange in Norway spruce in an alpine environment based on measurements of diurnal variation in stem radii. A daily temporary decline in sapflow at mid-crown before midday was observed but not explained. This phenomenon could be linked to effects of condensation before the onset of transpiration as measured in the shrub zone above the treeline (de Jong, 2005). In future, measurement techniques shedding more light on condensation and evapotranspiration such as radial stem variations should be more fully expoited (Zweifel, 2015).

Coordinated simultaneous measurements of evaporation, transpiration and atmospheric dynamics are needed to determine the linkage of local and regional air mass transfer and movements in relation to local precipitation.

17.2.4 Runoff processes

We have a good qualitative understanding of processes that produce runoff from rainfall on forested systems. Basic processes, depicted in Figs 2.2, 6.5, 9.1 and 9.3, reveal a common understanding of alternative paths of rainfall to streamflow. However, quantitative estimates of flow pathways are dependent on the location of the research site. Where paths have been altered by human intervention, providing artificial drainage to provide trafficability and increased tree growth on poorly drained soils (Figs 8.1 and 8.2), we find quantitative analysis requires alternative hydraulic conductivity estimates for differing stages of the forest regeneration cycle. Most of our quantitative understanding has come from isotope or geochemical tracer analysis to streamflow. An outstanding discussion of the use and limitations of isotopes can be found in Klaus and McDonnell (2013). While end-member mixing has been a common technique using geochemical tracers, the recent ability to differentiate dissolved organic carbon fractions of stream natural organic matter may provide alternatives to examine flow through the forest floor (Yang *et al.*, 2015).

A path to developing a unified explanation of forest subsurface processes is beginning to emerge. McDonnell (2013) began to explore an idea that subsurface processes may behave in a manner similar to infiltration-excess overland flow. As discussed in Chapter 2, incoming rain may travel in several alternative pathways to become streamflow. Vertical flow to groundwater represents the highest-gradient pathway. Jackson *et al.* (2014) present an elegant mathematical depiction of partitioning between vertical and slope-parallel flow above an impeding layer based on ratios of lateral and vertical gradients and hydraulic conductivities with the thickness of saturated material above the impeding layer. This analysis is similar to the arguments made by Elsenbeer (2001) for classifying tropical soils that would produce overland flow. The analysis is exact only for planar slopes with slope-parallel impeding layers, but does express an idea that could be more inclusive of conditions normally found in forested systems. Uchida *et al.* (2005) developed a decision tree to evaluate the prevalence and flow rate of hillslopes, dominated by pipeflow, based on both rainfall amount and intensity. Their decision tree depends on quantity of rain to initiate pipeflow and intensity of rain, in relation to maximum pipeflow rate, to determine the rate of hillslope pipeflow. In the case of pipeflow both vertical and slope-parallel hydraulic conductivities are functions of active soil macropores and pipes.

17.2.5 Merging forest hydrology and ecohydrology

As defined by Smettem (2008):

> Ecohydrology seeks to understand the interaction between the hydrological cycle and ecosystems. The influence of hydrology on ecosystem patterns, diversity, structure and function coupled with ecological feedbacks on elements of the hydrological cycle and processes are central themes of ecohydrology. The scope covers both terrestrial and freshwater ecosystems and the management of our relationship with the environment.

That definition also fits forest hydrology as a subset of that wider discipline. Jackson *et al.* (2009) cited the Swiss watershed experiments discussed in Chapters 1 and 5 as early ecohydrological experiments. One could argue that afforestation and some deforestation experiments are examples of ecohydrology since they deal with transformation of grassland to/from forest.

Bonnell (2002) also pointed out that ecohydrology was not a completely new science but does incorporate new connections between hydrological processes and stream hydrobiology. Coupling with the stream hyporheic process is new and has not been addressed in earlier studies of hillslope subsurface flow phenomena. The wider science of ecohydrology can couple forest hydrology with wider studies of the interaction of forests with more arid grasslands as well as wetlands and streams.

Ecohydrology may provide the tools needed to answer the century old question of 'do forests bring rain or merely respond to increased rain?' This question may become more important in tropical forests. Over much of the tropics the balance between forest, savannah or grassland is not determined climatically but may be in ecological alternative states that may be easily altered by fire or fire exclusion as well as many other human activities. If forest cover increases rainfall then a change in biome may become difficult to reverse (Staver *et al.*, 2011).

17.3 Broader Dimensions of Forest Hydrology

Advancing forest hydrology is critical to forest ecosystem management, as it drives contaminant cycling and loading dynamics in the soil, through plants, animals, precipitation inputs, and surface and subsurface flow networks that support downstream water quality, besides serving as a reference for assessing developmental impacts. Although it is understood that water yield and timing are affected by forest management, the duration and spatial scale of these effects merit further investigation (NRC, 2008).

Vose *et al.* (Chapter 15) state that:

> Projections indicate a future of increasing pine plantations and expansion of fast-growing species for carbon sequestration and bioenergy, but landscape-scale effects on water yield and quality, and the magnitude of potential trade-offs between managing for carbon and water, have not been systematically explored across time and space (Jackson *et al.*, 2005; King *et al.*, 2013).

The challenge of addressing forest hydrology and managing forests at large spatial scales requires also an understanding of large-scale processes and interactions with landscapes within and outside, usually accomplished by modelling approaches. However, the uncertainties in the variability of field circumstances, measurements and the modelling approaches must also be considered (Harmel *et al.*, 2010).

Intelligent, field-based, real-time monitoring of forest hydrological processes will improve data collection at a much finer spatial and temporal scale than traditional research methods (Sun *et al.*, 2016). Recent advancements in monitoring and mapping technology using LiDAR, satellite imageries, stable isotopes for partitioning water flux sources (Good *et al.*, 2015; Klaus *et al.*, 2015) and other sensor technologies, together with increased computing speed, should also be used as opportunities to address these complex processes. This will allow further investigation of the relationships between forest ecohydrological processes and remote sensing products which are currently poorly understood.

Jones *et al.* (2009) emphasized a need to address forest hydrology as a landscape hydrology that embraces the interactive effects of various land-based activities on water supplies. In order to improve the efficiency and effectiveness of designing landscapes to ensure sustainability, models commensurate with those available for agricultural lands are needed to characterize the biological, chemical and physical processes of forested lands. The fact that the hydrology and water quality of undisturbed forested lands are generally used as a baseline reference (Chapter 14) for determining anthropogenic impacts adds further emphasis to the importance of testing and, where necessary, further developing models for application to forested catchments.

The scope of forest hydrological science has to be expanded from understanding the meteorological and hydrological influences of forests based on small watershed research of the 20th century (Hewlett, 1982), to quantifying the ecohydrological impacts of global changes today (Amatya *et al.*, 2011; Vose *et al.*, 2011). It must also advance to address current complex issues, including urbanization, climate change, wildfires, invasive species, instream flow, floods, droughts, beneficial water uses, changing patterns of development and ownership, and changing societal values. In that context, there is a critical need for continued monitoring of existing long-term forest watersheds worldwide, as they are well suited for documenting and detailing baseline hydrological conditions and also serve as valuable benchmarks for advancing forest hydrological science and addressing emerging forest and water issues of the 21st century.

17.3.1 Meeting forest management needs

Global water demand is expected to increase 55% by 2050, primarily in developing countries (WWAP, 2014), where rising standards of living are likely to also increase demand for wood products and energy. Climate change and natural variability may reduce water availability, even in areas unaccustomed to drought. These conditions may put strong pressure on forest managers to sustain and somehow increase water yields of forested watersheds for municipal and other downstream uses, while water stress leaves the forest itself more vulnerable to dieback, pests and fire (Grant *et al.*, 2013). As cities grow, large forested municipal watersheds will have to be managed to meet as yet undefined benchmarks of both water yield and water quality, experiences described by Barten *et al.* (2012). Climate change mitigation and energy security initiatives will rely on increased forest biomass growth and utilization. As increasing forest growth requires higher water uptake on a plant basis, forest managers will need reliable planning tools to manage these requirements from a tree to a landscape basis. To develop these tools, we not only should advance forest hydrological science for understanding complex interactions but also must learn to scale currently available research and model results to define reasonably achievable benchmarks of water quantity and quality. We must also understand forest management practices and estimation techniques that allow such benchmarks to be achieved within the constraints of the forest owner. Challenges include changes in forests and water yield associated with climate change, land-use change, resistance of the public to forest modification, and the ever-present effects associated with disturbances such as fires, the age distribution of forests, insects and diseases, and forest regeneration impacts besides the natural ones. As the world demands more clean water supply, wood, energy and carbon storage from forests, forest hydrology becomes equally critical to sustainably providing services while protecting water resources.

References

About.com (2013) Forest Types of the World – Maps of the World's Forest. Available at: http://www.forestry.about.com (accessed 27 February 2013).

Amatya, D.M., Douglas-Mankin, K.R., Williams, T.M., Skaggs, R.W. and Nettles, J.E. (2011) Advances in forest hydrology: challenges and opportunities. *Transactions of the ASABE* 54, 2049–2056.

Amatya, D.M., Sun, G., Rossi, C.G., Ssegane, H., Nettles, J.E. and Panda, S. (2015) Forests, land use, and water. In: Zolin, C.A. and de A.R. Rodrigues, R. (eds) *Impact of Climate Change on Water Resources in Agriculture*. CRC Press, Boca Raton, Florida, pp. 116–153.

Barten, P.K., Ashton, M.S., Boyce, J.K., Brooks, R.T., Buonaccorsi, J.P., Campbell, J.S., Jackson, S., Kelly, M.J., King, M.J., Sham, C.H., *et al.* (2012) *Review of the Massachusetts DWSP Watershed Forestry Program.* DWSP Science and Technical Advisory Committee, Massachusetts Division of Water Supply Protection, Boston, Massachusetts.

Bonell, M. (2002) Ecohydrology – a completely new idea? *Hydrological Sciences Journal* 47, 809–810.

Bonell, M. and Gilmour, D.A. (1978) The development of overland flow in a tropical rainforest catchment. *Journal of Hydrology* 39, 365–382.

Carey, S.K. and Woo, M. (2001) Slope runoff processes and flow generation in a subarctic, subalpine catchment. *Journal of Hydrology* 253, 110–129.

De Jong, C. (2005) The contribution of condensation to the water cycle under high-mountain conditions. *Hydrological Processes* 19, 2419–2435.

Elsenbeer, H. (2001) Hydrologic flowpaths in tropical soilscapes – a review. *Hydrological Processes* 15, 1751–1759.

Good, S.P., Noone, D. and Bowen, G. (2015) Hydrologic connectivity constrains partitioning of global terrestrial water fluxes. *Science* 349, 175–178.

Grant, G.E., Tague, C.L. and Allen, C.D. (2013) Watering the forest for the trees: an emerging priority for managing water in forest landscapes. *Frontiers in Ecology and the Environment* 11, 314–321.
Hamilton, L.S., Dudley, N., Greminger, G., Hassan, N., Lamb, D., Stolton, S. and Tognetti, S. (2008) *Forests and Water*. FAO Forestry Paper No. 155. Food and Agriculture Organization of the United Nations, Rome.
Harmel, R.D., Smith, P.K. and Migliaccio, K.W. (2010) Modifying goodness-of-fit indicators to incorporate both measurement and model uncertainty in model calibration and validation. *Transactions of the ASABE* 53, 55–63.
Herzog, K.M., Häsler, R. and Thum, R. (1995) Diurnal changes in the radius of a subalpine Norway spruce stem: their relation to the sap flow and their use to estimate transpiration. *Trees* 10, 94–101.
Hewlett, J.D. (1982) *Principles of Forest Hydrology*. University of Georgia Press, Athens, Georgia.
Hewlett, J.D., Forston, J.C. and Cunningham, G.B. (1977) The effect of rainfall intensity on storm flow and peak discharge from forest land. *Water Resources Research* 13, 259–266.
Jackson, C.R., Bitew, M. and Du, E. (2014) When interflow also percolates: downslope travel distances and hillslope process zones. *Hydrological Processes* 28, 3195–3200.
Jackson, R.B., Jabbogy, E.G., Avissar, R., Roy, S.B., Barrett, D.J., Cook, C.W., Farley, K.A., le Maitre, D.C., Carl, B.A. and Murray, B.C. (2005) Trading water for carbon with biological carbon sequestration. *Science* 310, 1944–1947.
Jones, J.A., Achterman, G.L., Augustine, LA., Creed, I.F., Ffolliott, P.F., MacDonald, L. and Wemple, B.C. (2009) Hydrologic effects of a changing forested landscape – challenges for the hydrological sciences. *Hydrological Processes* 23, 2699–2704.
King, J.S., Ceulemans, R., Albaugh, J.M., Dillen, S.Y., Docmec, J.C., Fichot, R., Fischer, M., Leggett, Z., Sucre, E., Trnka, M. and Zenone, T. (2013) The challenge of lignocellulosic bioenergy in a water-limited world. *BioScience* 63, 102–117.
Klaus, J. and McDonnell, J.J. (2013) Hydrograph separations using stable isotopes: review and evaluation. *Journal of Hydrology* 505, 47–64.
Klaus, J., McDonnell, J.J., Jackson, C.R., Du, E. and Griffiths, N.A. (2015). Where does streamwater come from in low-relief forested watersheds? A dual-isotope approach. *Hydrolology and Earth System Science* 19, 125–135.
Körner, C. and Paulsen, J. (2004) A world-wide study of high altitude treeline temperatures. *Journal of Biogeography* 31, 713–732.
Lloyd, C.R. and Marques, A.d.O. (1988) Spatial variability of throughfall and stemflow measurements in Amazonian rainforest. *Agricultural and Forest Meteorology* 42, 63–73.
Makarieva, A.M., Gorshkov, V.G., Sheil, D., Nobre, A.D., Bunyard, P. and Li, B.L. (2014) Why does air passage over forest yield more rain? Examining the coupling between rainfall, pressure, and atmospheric moisture content. *Journal of Hydrometeorology* 15, 411–426.
McDonnell, J.J. (2013) Are all runoff processes the same? *Hydrological Processes* 27, 4103–4111.
NRC (National Research Council) (2008) *Hydrologic Effects of a Changing Forest Landscape*. The National Academies Press, Washington, DC.
Savenije, H.H.G. (2004) The importance of interception and why we should delete the term evapotranspiration from our vocabulary. *Hydrological Processes* 18, 1507–1511.
Smettem, K.R.J. (2008) Welcome address for the new 'Ecohydrology' journal. *Ecohydrology* 1, 1–2.
Smith, M.W. and Riseborough, D.W. (2002) Climate and the limits of permafrost: a zonal analysis. *Permafrost and Periglacial Processes* 13, 1–15.
Staver, A.C., Archbald, S. and Levin, S.A. (2011) The global extent and determinants of savanna and forest as alternative biome states. *Science* 344, 230–232.
Sun, G., Amatya, D.M. and McNulty, S.G. (2016) Forest hydrology. In: Singh, V.P. (ed.) *Chow's Handbook of Applied Hydrology*, 2nd edn. McGraw-Hill, New York, Part 7: Systems Hydrology, Chapter 85 (in press).
Uchida, T., Tromp-van Meerveld, I. and McDonnell, J.J. (2005) The role of lateral pipe flow in hillslope runoff response: an intercomparison of nonlinear hillslope response. *Journal of Hydrology* 311, 117–133.
Vauhkonen, J., Holopainen, M., Kankare, V., Vastaranta, M. and Vitale, R. (2016) Geometrically explicit descriptions of forest canopy based on 3D triangulations of airborne laser scanning data. *Remote Sensing of Environment* 173, 248–257.
Vose, J.M., Sun, G., Ford, C.R., Bredemeier, M., Otsuki, K., Wei, X.H., Zhang, Z.Q. and Zhang, L. (2011) Forest ecohydrological research in the 21st century: what are the critical needs? *Ecohydrology* 4, 146–158.

WWAP (United Nations World Water Assessment Programme) (2014) *The United Nations World Water Development Report 2014: Water and Energy*. United Nations Educational, Scientific and Cultural Organization, Paris.

Yang, L., Chang, S.W., Shin, H.S. and Hur, J. (2015) Tracking the evolution of stream DOM source during storm events using end member mixing analysis based on DOM quality. *Journal of Hydrology* 523, 333–341.

Zweifel, R. (2015) Radial stem variations – a source of tree physiological information not fully exploited yet. *Plant, Cell & Environment* 39, 231–232.

Index

Note: bold page numbers indicate figures and tables.